Lecture Notes in Computer Science   9103

Commenced Publication in 1973
Founding and Former Series Editors:
Gerhard Goos, Juris Hartmanis, and Jan van Leeuwen

More information about this series at http://www.springer.com/series/7409

Chris Biemann · Siegfried Handschuh
André Freitas · Farid Meziane
Elisabeth Métais (Eds.)

Natural Language Processing and Information Systems

20th International Conference on Applications
of Natural Language to Information Systems,
NLDB 2015
Passau, Germany, June 17–19, 2015
Proceedings

 Springer

Editors
Chris Biemann
Technische Universität Darmstadt
Darmstadt
Germany

Siegfried Handschuh
Universität Passau
Passau
Germany

André Freitas
Universität Passau
Passau
Germany

Farid Meziane
University of Salford
Salford
UK

Elisabeth Métais
Conservatoire National des Arts et Métiers
Paris
France

ISSN 0302-9743   ISSN 1611-3349 (electronic)
Lecture Notes in Computer Science
ISBN 978-3-319-19580-3   ISBN 978-3-319-19581-0 (eBook)
DOI 10.1007/978-3-319-19581-0

Library of Congress Control Number: 2015939997

LNCS Sublibrary: SL3 – Information Systems and Applications, incl. Internet/Web, and HCI

Springer Cham Heidelberg New York Dordrecht London
© Springer International Publishing Switzerland 2015

Printed on acid-free paper

Springer International Publishing AG Switzerland is part of Springer Science+Business Media
(www.springer.com)

Preface

We are living in a fascinating period, when the applications of natural language processing (NLP) are going mainstream. The maturity of the field has reached a stage in which its solidity can be witnessed in the adoption and uptake of its technologies outside the core NLP academic community. The success and visibility of software systems such as IBM Watson, Siri, and Google Knowledge graph are incentivizing the increase of interest in the field and the adoption of NLP in diverse areas.

The past two decades in which NLDB has been active were fundamental for the establishment of the foundations and the further maturation of this field and NLDB can claim the share of its contributions. During this time, NLDB has been one of the main conferences for applications in NLP, in which researchers could find a sweet spot between rigorous scientific contributions and openness to new ideas and perspectives. Another notable characteristic that has been consistently present at NLDB is multidisciplinarity: NLDB has welcomed contributions from applications of NLP into different areas, and has helped in bridging NLP to different communities and application fields.

This year's NLDB featured a special track on natural language and its connection to semantic and cognitive computing. Semantic computing aims at connecting the user's information need with the meaning of content in a multidisciplinary fashion. Cognitive computing systems naturally interact with people and learn over time.

The NLDB 2015 program spanned a wide range of topics that all revolve around the use of natural language to access information. Three invited speakers covered topics as diverse as distributional semantics, computational humanities, and linked open data. Sessions comprised the following topics: unsupervised and semi-supervised machine learning, information extraction, event extraction and named entity recognition, multilingual alignment and translation, sentiment detection and user-generated content processing, indexing and the lexicon, query processing, question answering, speech processing, and dialog systems. Last, but not least, the special session on semantic and cognitive computing attracted position papers and featured a sponsored talk by IBM.

NLDB 2015 had a higher number of submissions compared with previous years. Out of 100 submissions, 18 papers were accepted as full papers (18 % acceptance rate), 15 as short papers (33 % acceptance rate) and 14 as posters and demos (47 % acceptance rate). A lot of work was involved in the careful selection of the papers and we would like to thank the Program Committee and reviewers for their hard work and dedication. We would also like to thank the invited speakers for their inspiring contributions to the program.

NLDB 2015 was held in the picturesque town of Passau. Dating back to Roman times, Passau is located at the German-Austrian border and it is notable for being at the convergence of three rivers (the Danube, the Inn, and the Ilz). Its location at the heart of Europe and its proximity to different borders made Passau a natural confluence for different cultures and influences, a fact that is embodied in its architecture that harmoniously blends German, Austrian, and Italian styles. Despite its openness, Passau is deeply connected to its German and Bavarian traditions and symbols and offers a great entry point for the German culture.

The conference was generously supported financially by the Insight Centre for Data Analytics, the SSIX EU Project, IBM (Platinum Sponsors), by the University of Passau, which hosted the conference, and by 3rdParty (Silver Sponsor). Adamantios Koumpis, Stephanie Pauli, Elfried Kronawitter and Ulrike Holzapfel were fundamental for the organization of the conference, supporting the sponsorship, and in the local coordination of the event.

In its 20th anniversary we would like to celebrate the effort behind the construction of the NLDB community, expressing our gratitude to authors, Program Committee members, and organizers for the past editions of NLDB.

April 2015

Chris Biemann
Siegfried Handschuh
André Freitas
Farid Meziane
Elisabeth Métais

Organization

Organizing Committee

Program Chair

Chris Biemann   TU Darmstadt, Germany

Conference Chairs

Siegfried Handschuh   University of Passau, Germany
Elisabeth Métais   CNAM, France
Farid Meziane   University of Salford, UK

Organization Chair

André Freitas   University of Passau, Germany

Tutorials and Workshop Chairs

Christin Seifert   University of Passau, Germany
André Freitas   University of Passau, Germany

Sponsorship Chair

Adamantions Koumpis

Senior Program Committee

Gerard de Melo   Els Lefever   Christina Unger
Valia Kordoni   Simone Paolo Ponzetto   Torsten Zesch
Mathieu Lafourcade   Mathieu Roche   Michael Zock
Johannes Leveling   Maguelonne Teisseire

Program Committee

Hidir Aras   Sandra Bringay   Kees van Deemter
Imran Sarwar Bajwa   Cornelia Caragea   Bart Desmet
Pierpaolo Baslie   Diego Ceccarelli   Olivier Ferret
Nicolas Béchet   Christian Kop   Antske Fokkens
Johan Bos   Philipp Cimiano   Vladimir Fomichov
Gosse Bouma   Christian Chiarcos   Thierry Fontenelle
Mihaela Bornea   Kostadin Cholakov   Debasis Ganguly
Yacine Rezgui   ErnestoWilliam De Luca   Yaakov Hacohen-Kerner

Sebastian Hellmann
Michael Herweg
Helmut Horacek
Dino Ienco
Ashwin Ittoo
Paul Johannesson
Richard Johansson
Epaminondas Kapetanios
Saurabh Kataria
Sophia Katrenko
Zoubida Kedad
Eric Kergosien
Valia Kordoni
Leila Kosseim
Jochen Leidner
Deryle W. Lonsdale
Cdric Lopez
John McCrae
Marie-Jean Meurs
Luisa Mich
Claudiu Mihaila

Shamima Mithun
Andres Montoyo
Andrea Moro
Rafael Muoz
Guenter Neumann
Jan Odijk
Alexander Panchenko
Heiko Paulheim
Davide Picca
Pascal Poncelet
Violaine Prince
Gábor Prószéky
Behrang Qasemizadeh
Shaolin Qu
Reinhard Rapp
Martin Riedl
André Freitas
Eric Ringger
Mathieu Roche
Mike Rosner
Paolo Rosso

Patrick Saint Dizier
Bahar Sateli
Roman Schneider
Khaled Shaalan
Max Silberztein
Vijayan Sugumaran
Krishnaprasad
Thirunarayan
Juan Trujillo
Dan Tufis
L. Alfonso Ureña-López
Sunil Vadera
Panos Vassiliadis
Andreas Vlachos
Joachim Wagner
Tonio Wandmacher
Feiyu Xu
Wlodek Zadrozny
Fabio Massimo Zanzotto
Erqiang Zhou

Platinum sponsors

Gold sponsor

Silver sponsor

Fig. 1. NLDB 2015 sponsors.

Sponsors

The NLDB organizers are grateful for the NLDB sponsors and their generous contributions:

Insight Centre for Data Analytics, Ireland (Platinum), SSIX EU Project (Platinum), IBM (Platinum), Lionbridge (Gold), and 3rdPlace (Silver).

Contents

Context-Aware NLP

Cognitive and Semantic Computing

Sentiment and Opinion Analysis

Information Extraction and Social Media

NLP and Usability

Text Classification and Extraction

Posters and Demonstrations

Information Extraction

Improving Supervised Classification
Using Information Extraction

Mian Du, Matthew Pierce, Lidia Pivovarova[(✉)], and Roman Yangarber

Department of Computer Science, University of Helsinki, Helsinki, Finland
lidia.pivovarova@cs.helsinki.fi

Abstract. We explore supervised learning for multi-class, multi-label text classification, focusing on real-world settings, where the distribution of labels changes dynamically over time. We use the PULS Information Extraction system to collect information about the distribution of class labels over named entities found in text. We then combine a knowledge-based rote classifier with statistical classifiers to obtain better performance than either classification method alone. The resulting classifier yields a significant improvement in macro-averaged F-measure compared to the state of the art, while maintaining comparable micro-average.

1 Introduction

We present PULS, a framework for Information Extraction (IE) from text, designed for decision support in various domains and scenarios, including business intelligence. In the PULS project, we work with large corpora collected continuously from multiple online sources, and consisting of millions of news articles, collected over several years. The Information Extraction (IE) system is used to extract structured events related to the Business domain from the corpus. In the Business domain, events of interest typically focus on activities that involve companies or persons—e.g., corporate acquisitions, product launches, investments, contracts, leadership changes, etc. The IE system extracts thousands of such events daily. We then try to categorize the events according to their industry sector, e.g., *Telecommunications*, *Dairy Foods*, or *Energy*. We consider a document's labels to be the industry sectors that apply to any events extracted from it; thus, we treat the problem as a document classification task.

Our main goal in this paper is to investigate how knowledge automatically extracted from text can help in text categorization. We use company names and company *descriptors* to classify documents according to their industry sectors.

The PULS IE system processes the documents using a pipeline of modules. One of these modules—the named entity recognition (NER) module—finds companies mentioned in the text and their associated descriptors; a descriptor is a noun phrase linked to a company name—e.g., "the smartphone giant Apple." Information about names and descriptors is stored in a knowledge base, together with the ID of the document where the company was found. The documents

C. Biemann et al. (Eds.): NLDB 2015, LNCS 9103, pp. 3–18, 2015.
DOI: 10.1007/978-3-319-19581-0_1

have been hand-labeled with their true industry sectors, providing a link from company names to sector labels in the knowledge base. We assume that each company has its own label "preferences," that is, the set of industries in which it usually operates. Using this assumption, we collect the co-occurrence counts of company names with industry sectors in the corpus, and use these counts to predict the sector labels for new documents. It is similarly possible to use company descriptors to predict the sector labels; for example, we can assume that "mobile phone manufacturer" is an indicator of the *Telecommunication* sector and "dairy company" is most likely to co-occur with *Dairy Foods*.

The paper is organized as follows: in Sect. 2 we give a brief overview of PULS. Section 3 introduces related work. In Sect. 4 we describe the data we use for training and testing the classifiers. In Sect. 5, we present an array of statistical classifiers and describe the training and classification processes. We then present the knowledge-based rote classifier (Sect. 6) and how it can be combined with the statistical classifiers (Sect. 7), followed by experiments and evaluation of the results, in Sect. 8. We conclude with a discussion of the results and plans for future work, in Sect. 9.

2   PULS Overview

PULS (the Pattern Understanding and Learning System[1]) is designed to discover, aggregate, verify, and visualize information obtained from the Web, and deliver it to the user in a concise and easy-to-access form. PULS's news analysis methodology has been applied to several knowledge-intensive domains, including business intelligence, tracking information about outbreaks of infectious diseases, and security and cross-border crime [1,13,19,42].

In the business-intelligence domain, PULS tracks *entities* (such as companies and persons) and *events*, such as investments, acquisitions, contracts, layoffs, etc., which it automatically extracts from large amounts of business news using information retrieval (IR), information extraction, machine learning, and data mining techniques.

Building upon the extracted information, PULS acts as a decision-support system, which provides deeper semantic analysis than general-purpose search engines, and automatically maintains up-to-date profiles for companies and industry sectors. Another aspect of the system is its ability to track complex networks of relationships in the business domain through time and across multiple news sources.

A high-level architecture of the system is given in Fig. 1: it contains (a) an IR module; (b) a natural language processing (NLP) engine, which performs information extraction, inference, and aggregation; (c) a machine learning module, including classifiers and pattern discovery modules; and (d) a component to collect information from social media sources.

---

[1] http://puls.cs.helsinki.fi/home.

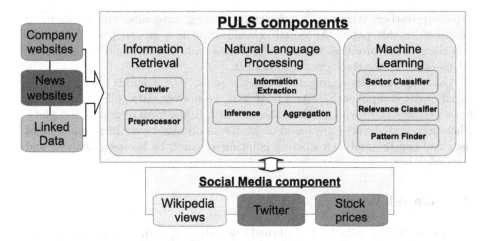

Fig. 1. PULS Information analysis platform

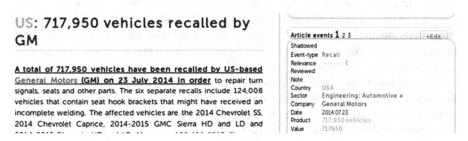

Fig. 2. Components of the user interface: input document, and a RECALL event extracted by PULS

First, the IR module obtains unstructured raw text data from various sources on the Web. Currently, PULS collects RSS feeds from news websites and company websites, and extracts the text from the Web links provided in the RSS. PULS uses over a thousand news websites which provide an RSS feed related to the business domain (e.g., BBC Business News, New York Times Business Day, etc.). Every 10 min the crawler extracts links of news from these RSS feeds, downloads the HTML files, extracts the text, identifies the language, and stores the news into a database.

The NLP engine is a key component of the PULS platform. Information Extraction transforms facts found in plain text into a structured form. An example event is shown in Fig. 2. The text mentions a product recall event, conducted by General Motors in July 2014. For each event, the system extracts a set of related *entities*: companies, industry sector(s), products, location, date, and other attributes of the event. This is structured information; it is stored in the database for subsequent querying and downstream analysis.

The particular industry sector involved in the event—e.g., *"Engineering: Automotive"* in the GM example—is typically not mentioned in the text explicitly; rather, it has to be determined using automatic classification, as described in this paper. Automatic classification is a crucial part of the system since PULS produces thousands of events daily and it would be impossible for users to browse these events without it.

Using the entities aggregated from the texts, PULS builds queries for the social media component [7]. As a final step, we present data collected from the news websites and social media to the end user, in the form of graphs and plots. These aggregated views are based on statistics obtained over large amounts of data and can be used as a starting point for research by business analysts and Web scientists.

3   Related Work

Multi-label text classification is a broad research area, with surveys in, e.g., [36–38]. Here we focus on work most related to ours.

A commonly used *data representation* for text categorization is the "bag of words" (BOW) model, which ignores the document structure and assumes that words occur independently, [22]. This model can be extended by using n-grams [2,9,43]. We use the bag-of-words model with a combination of unigrams and bigrams.

Information Extraction (IE) can be used to obtain additional features for classification [18–20,30]. We use company names extracted from the text by a named-entity recognition system, to build a baseline "rote" classifier (see Sect. 6). The difference between the cited papers and our work is that we use information extracted from the corpus and stored in the knowledge base in addition to the data extracted from a single document. Thus, we follow the recent line of study in the area of cross-document IE, which is focused on the validation and summarization of data obtained from multiple sources [24,26,28,29,41]. Cross-document IE is also similar to the knowledge-base population and entity linking tasks, [6,16,21,33–35]. In this paper we focus on knowledge base *utilization* for text classification, rather than on knowledge base population as a separate task.

Text datasets are typically "naturally skewed" [25], since topics differ both in frequency and importance, depending on where the data originates; additional skew may be introduced by annotator bias. Such imbalance poses a challenge for categorization, especially when the classes have a high degree of overlap [31]. One possible solution for this problem is balancing of the training-set or re-sampling, [5,10,39]. In a previous paper, we demonstrate that classifiers trained on balanced data perform better, on average, than classifiers trained using the original distribution of labels in the corpus [8]. In this paper we use the same balancing techniques.

4   Data

We focus on supervised-learning techniques to classify news articles into industry sectors. Although we are primarily interested in the PULS document collection, as mentioned in Sect. 1, all experiments we present here are conducted on the publicly available Reuters corpus (RCV1),[2] to allow meaningful comparison and to assure replicability. RCV1 contains 800,000 news stories published by Reuters between 1996–1997. Documents are labeled using 103 *Topic* labels, 350 *Industry* labels and 296 *Region* codes; the labels are organised hierarchically. In this paper we use a subset of 200 industry sectors.[3]

Although RCV1 is a popular dataset, relatively few papers use its sector classification, and not all of them are directly comparable with our study. For example, [14] simultaneously classify documents by topics, sectors, and locations. Crammer et al. [4] build classifiers to distinguish confusable industry pairs (e.g., *Life* and *Non-Life Insurance*), and use only 6 sector labels in their paper. Gabrilovich and Markovitch [12] use only 16 of the 350 industry labels; Hatami et al. [17] do not report standard evaluation measures, such as F-measure.

To our knowledge, five papers are directly comparable to our work, in that they use a large number of sector labels and report micro- and/or macro-averaged F-measures: [3,23,27,32,44]. In the Results section (Table 4) we present a detailed comparison between the results on RCV1 industry labels from these papers and our results.

We use the raw text data from RCV1. We only use documents that have sector labels, of which there are 351,810 in total. These documents were manually classified by Reuters editors into 350 industry sectors. There are seven- and five-digit industry codes; seven-digit codes are children of the corresponding five-digit codes: e.g., *Fruit Growing* (I0100206), *Vegetable Growing* (I0100216) and *Soya Growing* (I0100223) are all children of *Horticulture* (I01002).

This sector classification has some inconsistencies, as observed by others, e.g., [23]. We map all seven-digit codes to their corresponding parent codes, and merge labels that have the same name but different code.[4] After this pre-processing, 245 distinct sector labels remain.

5 Array of Binary Classifiers

We split the multi-label classification task into many binary classification subtasks, carried out by an array of statistical classifiers, one trained for each individual sector. All classifiers in the array use exactly the same training set, where all documents labeled with a given sector are used as positive instances for that sector's classifier, while all remaining training documents are used as negative instances. We try two supervised-learning algorithms: Naive Bayes and Support Vector Machines (SVM). We use implementations from the open-source WEKA toolkit [15].

---

[2] http://about.reuters.com/researchandstandards/corpus/.

[3] Henceforth we use the terms *label*, *class* and *(industry) sector* interchangeably.

[4] For example, we merge I64000 and I65000, both called *Retail Distribution*.

5.1   Text Representation

Each training and test document is represented using bag-of-words features from the text. We use only nouns, adjectives, and verbs in our feature set, and apply simple filters to remove all stop-words, proper names, locations, dates, and common verbs such as "have" and "do."[5] We also generate bigrams that consist of these three parts of speech. When indexing documents after feature selection, we use a unigram as a feature only if it appears *outside of any bigram features* extracted from that document. For example if the phrase "power plant" appears in a document we will consider "power" or "plant" as independent features, only if they also appear elsewhere in the document (and not in another extracted bigram). This allows us to resolve ambiguity to some extent; for example, we can more easily distinguish documents containing the feature "SIM card," which may be relevant for *Telecommunications*, from "credit card," which is relevant for *Commercial Banking*.

In total, 77,636 training instances (documents) yield 49,262 unique features, used by the binary classifiers. We use two feature-selection methods—, and Bi-Normal Separation (BNS), [11]. We then try several learning algorithms and feature selection methods to find the combination which yields the best performance.

5.2   Training and Test Data Pools

If a particular sector is dominant in the training set, the negative features for other classifiers could become dominated by features drawn from this sector, which may hurt performance on some other sector since it won't learn negative features from other, "minor" sectors (those having fewer documents in the corpus). If some sector is also over-represented in the test set, we run the risk of over-fitting. For these reasons we try to keep the training data as balanced as possible across sectors, and ensure that the test set will contain a sufficient number of instances for every binary classifier in the array. To construct the training set we use an algorithm previously described in [8]; the process starts document collection from the sector that has the smallest number of instances in the corpus and thus guarantee that each sector will have a sufficient number of instances in the training and test pools. However, it is impossible to construct a dataset with an equal number of instances for each label due to the massive overlap between sectors.

Table 1 shows the most frequent sectors in the balanced training pool. We can see, e.g., that although we only collected 450 positive training instances for *Diversified Holding Companies*, it still receives 3644 positive instances in the pool, most of which were picked up when collecting data for other sectors.

For comparison, in [8], we used an *unbalanced* training pool, which is simply half of the corpus.

All data *outside* the balanced and unbalanced training pools—called the "test pool"—are available for the construction of test sets. From the test pool, we

---

[5] Some proper names may be used by IE-based classifiers, Sect. 6.

Table 1. Number of *positive* instances in the training pool, for the ten most frequent sectors

Code	Sector	Instances	Code	Sector	Instances
I83960	Diversified Holding Companies	3644	I16101	Electricity Production	1986
I81402	Commercial Banking	3153	I01001	Agriculture	1980
I13000	Petroleum and Natural Gas	2628	I33020	Computer Systems and Software	1805
I79020	Telecommunications	2145	I75000	Air Transport	1754
I21000	Metal Ore Extraction	2099	I35101	Passenger Cars	1713

generate 11 samples of 10,000 documents each, using the original distribution in the corpus. We use one of these samples as a held-out *development* set for parameter tuning (Sect. 5.3), and nine as test sets. Using the averaged scores from these nine test sets we find the best classifier (Sect. 8). The eleventh test set is used to obtain a final result, using the best classifier, for comparison with previous works (Sect. 4).

5.3   Classification

The SVM classifiers output a binary decision for every document. For Naive Bayes, the output for each sector is a confidence score between 0.01 and 1; thus a decision threshold is required to make a classification. We learn the best threshold over a range of thresholds (in increments of 0.01), using a held-out *development* set (one of the test sets, described in Sect. 5.2). We then evaluate on the remaining test sets using the learned threshold.

6   IE-based Classifiers

We use PULS IE system to build a knowledge base that contains sector distribution information for each company mentioned in the corpus. In this paper we investigate ways to use this information for text categorization.

The IE system finds mentions of companies in the corpus, using a named-entity recognition (NER) module. It distinguishes company names from other proper names in the text, e.g., persons and locations. The NER module also merges variants of the same name, for example, "Apple," "Apple Inc.," "Apple Computer, Inc.," etc.

The NER module is based on a cascade of low-level patterns that find noun groups within a text. This means that the module finds not only named entities but also their *descriptors*, i.e. noun and adjective modifiers of a given name. For example, Apple can be described in the text as "computer maker" or "software giant". As can be seen in this example, a descriptor always consists of two main components: *domain*, an area in which the company works (i.e. "computer", "software") and *type*, a word that is synonymous with "company" (i.e. "maker", "giant"). A descriptor may also contain other components, such as a

Table 2. Sector distribution for company "Apple"

Sector	Freq	Prob
Computer Systems and Software	549	0.61
Electronic Active Components	61	0.07
Data-communications and Networking	36	0.04
Telecommunications	19	0.02
Electrical and Electronic Engineering	13	0.01

geographic marker (i.e. "English company", "Swedish company") or some additional information, (i.e. "big company", "local company", etc.). A descriptor may contain all of these components, or only some of them. We use a short list of approximately 20 company words—such as "corporation", "firm", and "manufacturer"—to determine the company type. We also filter out generic words, when finding the company domain.

The knowledge base contains the following many-to-many relations:

– document-sector
– document-company
– company-descriptor

We try using various combinations of these relationships to build a rote classifier. We use the IE system to process documents from the training set and build a knowledge base, then use this knowledge to classify documents from the test set.

We assume that each company has its sector preferences, i.e. the set of industries in which it usually operates. As a consequence, company names in the corpus co-occur with particular sectors. For example, Table 2 shows the top sectors that co-occur with "Apple."; it shows the frequency (the co-occurrence count of the company with the sector), and the proportion, which is the normalized count. It can be seen from the table that in 60 % of cases Apple is mentioned in documents labeled with *Computer Systems and Software* sector, thus it is natural to suggest that documents that mention Apple belong to this sector.

However, each document may belong to more than one sector, therefore, instead of choosing only the top-most frequent sector the classifier should return the entire sector distribution, which can be calculated using the evidence from all companies mentioned in the text. Thus the probability that document D belongs to sector S, in the simplest case, can be defined by the formula:

$$P(S|D) = \frac{1}{|C_D|} \times \sum_{c \in C_D} P(S|c) \tag{1}$$

where C_D is the set of companies mentioned in the document, and $P(S|c)$ is the proportion of times c co-occurs with S in the knowledge base; e.g.,

$$P(Computer\ Systems\ and\ Software|Apple) = 0.61 \tag{2}$$

(from Table 2). Note that although the company may be mentioned in the document several times, we currently ignore the frequency of mentions of a company within a document.

This method would be reliable if the knowledge base contains sufficient evidence to associate the company with particular sector(s). Therefore, we only use companies that appear in the corpus three or more times. This means that if a document discusses a new (or little-known) company, the name-based classifier will be unable to find a sector for the document. In this case we can use descriptors to label the document, as descriptors allow us to use evidence gained from *other* companies in the corpus. For example, if company X is described in the text as "software company" we can assume that the sector distribution for this company would be similar to the sector distribution for "Apple". In this case the probability that document D belongs to sector S can be described by the formula:

$$P(S|D) = \frac{\sum\limits_{c \in C_D} P(S|c) + \sum\limits_{d \in d_D} P(S|d)}{|C_D| + |d_D|} \tag{3}$$

where d_D is the set of all descriptors mentioned in the document. Note that $|C_D| \neq |d_D|$ because in this case we can use a company descriptor even when the company does not appear in any other document in the corpus.

This estimate of $P(S|c)$ based on co-occurrence may be inaccurate: for rare companies, some sectors may dominate the distribution by mere chance. Moreover, sector overlap may lead to a situation where the company belonging to one sector frequently co-occurs with another. Descriptors, therefore, may sometimes be more reliable for predicting the sector. To check this assumption, we define the probability that a company belongs to a particular sector as follows:

$$P(S|c) = \sum_{d \in d_C} P(d|C) \times P(S|d) \tag{4}$$

where d_C is the set of all descriptors associated with company c in the knowledge base. We then use (4) in (1) to obtain the final sector distribution for the document:

$$P(S|D) = \frac{1}{|C_D|} \times \sum_{c \in C_D} \sum_{d \in d_C} P(d|C) \times P(S|d) \tag{5}$$

Note that in this case the company name is substituted by a set of descriptors; however it is possible to use the company name in combination with company descriptors:

$$P(S|D) = \frac{\sum\limits_{c \in C_D} \sum\limits_{d \in d_C} P(d|C) \times P(S|d) + \sum\limits_{c \in C_D} P(S|c)}{2 \times |C_D|} \tag{6}$$

7   Combined Classifiers

We experiment with several methods of combining the rote classifier, described in Sect. 6, with the balanced probabilistic classifiers, described in Sect. 5, to see if the combination can produce better *overall* predictions. One method of combining is a simple two-stage process: for each document, we first try to identify sectors using the rote classifier; if that does not return any sectors, we then attempt to classify using the statistical classifiers. We also experiment with the reverse order of these classification stages. The motivation for this method is to give the overall system a "second chance" at classification, in the hope that together the two methods may overcome their respective shortcomings. Another method of combining classifiers is to return the *union* of the results of the two classifiers—rote and probabilistic. Again, we learn the optimal threshold for each classifier in the combination using the development set.

8   Experiments and Results

8.1   Evaluation Measures

Common measures in text classification are precision, recall, and F-measure. For a given class c, these are calculated as:

$$Rec_c = \frac{TP_c}{TP_c + FN_c} \qquad\qquad Prec_c = \frac{TP_c}{TP_c + FP_c}$$

$$F1_c = \frac{2 \times Rec \times Prec}{Rec + Prec}$$

where TP_c, TN_c, FP_c and FN_c are the number of true positive, true negative, false positive, and false negative classified instances for the class, respectively.

In evaluating multi-label classification, *macro-averaging* and *micro-averaging* are commonly reported [5,40]. In micro-average evaluation, first the numbers of true- and false-positives, and true- and false-negatives are counted for all instances in the test set, and then the standard measures, e.g., recall or precision, are calculated using these numbers:

$$Rec_\mu = \frac{\sum\limits_{i \in S} TP_i}{\sum\limits_{i \in S} (TP_i + FN_i)} \qquad\qquad Prec_\mu = \frac{\sum\limits_{i \in S} TP_i}{\sum\limits_{i \in S} (TP_i + FP_i)}$$

$$\mu\text{-}F1 = \frac{2 \times Rec_\mu \times Prec_\mu}{Rec_\mu + Prec_\mu}$$

where S is the set of all classes. In the macro-average evaluation scheme, the measures are calculated for each class *separately* first, and then these are averaged across all classes:

$$Rec_M = \frac{\sum\limits_{i \in S} Rec_i}{|S|} \qquad\qquad Prec_M = \frac{\sum\limits_{i \in S} Prec_i}{|S|} \qquad M\text{-}F1 = \frac{\sum\limits_{i \in S} F1_c}{|S|}$$

We report both evaluation schemes, although we focus more on the macro-average scores, as explained below, since they are less dependent on the particular distribution of labels in the corpus. Henceforth we denote the macro-averaged F-measure by M-F1, and micro-averaged F-measure by μ-F1.

Table 3. Results from all classifiers and feature selection methods, averaged across 9 test sets randomly sampled from original distribution. For each classifier, the best threshold is trained on one random, originally-distributed development set. Rote classifier names correspond to the following formulae from Sect. 6: **name** – (1), **name+desc** – (3), **name⤳desc** – (5), **name+name⤳desc** – (6). For combined classifiers → and ∪ denote the two-stage and union combining methods, respectively (Sect. 7).

Classifier	M-average			μ-average		
	Rec	Pre	F1	Rec	Pre	F1
Statistical classifiers						
NB+IG	31.3±0.9	21.9±0.6	19.7±0.6	31.5±0.5	22.4±0.6	26.2±0.5
NB+BNS	34.2±1.1	16.6±0.6	15.8±0.5	**33.1±0.7**	13.4±0.4	19.0±0.5
SVM+IG	31.9±1.3	**59.2±1.1**	**37.1±1.2**	30.5±0.4	**72.7±0.6**	**42.9±0.4**
SVM+BNS	**32.7±0.9**	55.2±1.0	36.2±0.7	30.1±0.5	70.8±0.6	42.2±0.5
Rote classifiers						
name	36.8±0.8	**65.2±1.0**	44.5±0.7	45.9±0.5	**60.5±0.4**	52.2±0.5
descriptor	8.8±0.3	38.4±1.2	11.6±0.3	16.4±0.2	29.0±0.3	20.9±0.4
name+desc	**39.4±0.8**	63.3±0.7	**46.2±0.7**	**48.5±0.5**	57.8±0.5	**52.8±0.4**
name⤳desc	11.9±0.2	48.0±0.9	16.0±0.3	20.6±0.4	39.0±0.4	27.0±0.4
name+name⤳desc	39.2±0.8	60.0±0.8	44.8±0.6	**48.5±0.5**	54.5±0.4	51.3±0.4
Combined classifiers						
name→SVM+IG	46.2±1.0	**73.7±0.8**	55.1±0.8	52.5±0.5	**75.9±0.4**	62.0±0.4
SVM+IG→name	47.0±1.2	67.7±0.9	53.7±1.1	49.9±0.3	73.9±0.3	59.6±0.3
name ∪ SVM+IG	52.2±1.1	66.3±0.8	56.9±0.9	57.7±0.4	71.1±0.3	**63.7±0.4**
name+desc→SVM+IG	48.4±1.1	69.2±0.7	55.5±0.9	56.2±0.5	70.0±0.3	62.4±0.4
SVM+IG→name+desc	46.7±1.0	70.2±0.8	54.6±0.8	53.8±0.5	71.2±0.4	61.3±0.4
name+desc ∪SVM+IG	**53.7±1.0**	64.5±0.8	**57.2±0.8**	**59.7±0.4**	68.1±0.3	63.6±0.3

8.2   Comparison of Classifiers and Feature Selection Methods

Results obtained by all classifiers are shown in Table 3. As seen from the table, the SVM classifier yields higher performance than NB, independently of the feature selection method used. IG performs better than BNS with both Naive Bayes and SVM.

The basic rote classifier that uses only company names (denoted by **name** in Table 3) performs better than any statistical classifier alone. This classifier has high precision, which supports the intuition that each company has particular sector preferences (Sect. 6). This classifier also has relatively high recall— higher than the best single statistical classifier, SVM+IG, which suggests that the majority of documents in the Reuters corpus contain a company name.

By contrast, the rote classifier that uses only descriptors (**descriptor**), performs poorly. Recall is particularly low, suggesting that descriptors are more sparse than company names, in RCV1. A company has only one name but may be described in a variety of ways; therefore, a descriptor-based classifier requires significantly more data to be accurate than a company-name-based classifier.

Table 4. Classification results on RCV1 industry sectors, compared with state of the art.

Reference	Algorithm	M-F1	μ-F1
[23]	SVM	29.7	51.3
[44]	SVM	30.1	52.0
[27]	SVM + re-ranking	34.1	62.8
[32]	Naive Bayes	-	70.5
[3]	Bloom Filters	47.8	**72.4**
Our work:	**name+desc ∪ SVM+IG**	**57.7**	63.8

Despite poor performance on their own, however, descriptors used in conjunction with company names (**name+desc**) result in better performance than either method alone. In particular, adding descriptors gives a slight boost to recall.

Although the rote classifier that uses descriptors from the knowledge base (**name⤳desc**) has higher precision relative to the classifier using descriptors from the document, it does not perform well in general. The explanation for this may again relate the size of the corpus and sparsity of descriptors in the data.

In summary, the rote classifier that uses company names and descriptors from the document (**name+desc**) yields the highest F-measure among single classifiers. Combining it with SVM+IG yields the best overall performance. To save space we show only selected classifier combinations in Table 3; it can be seen from the table that the classifiers that have higher scores alone work better in combination, and that, for combined classification, taking the union of classified sectors gives better results than the two-stage method. A possible explanation is that recall is a weak point for all reported classifiers; it can be seen from the table that two-stage combination improves precision performance, while union combination boosts recall.

Finally, while the combination of SVM+IG with the **name+desc** rote classifier yields the highest M-F1, the combination with the **name** rote classifier yields the highest μ-F1. As mentioned previously, we consider macro-averaging to be more meaningful as an indicator of performance in a dynamic, real-world environment; therefore we consider the former classifier best. We then apply this classifier to the eleventh dataset, which has not been used in other experiments. M-F1 obtained by this classifier is higher than the best previously reported results, as shown in Table 4. It also can be seen from the table that the difference between M-F1 and μ-F1 for our classifiers is smaller than that reported in prior work. This supports the claim that classifiers trained on balanced data are less sensitive to changes in label distribution—which is one of our main objectives.

9   Conclusion

We have presented experiments with supervised learning for labeling business-news documents with multiple industry sectors. We treat the multi-class, multi-label problem as a set of binary sub-tasks, with one binary classifier for each sector. We explore several combinations of learning algorithms and feature selection methods, and evaluate them using a large amount of manually-labeled data. Further, we focus on building robust classifiers, suitable for real-world classifications—rather than on improving performance on a single, static corpus—by balancing the data given to each classifier during training.

The main contribution of this paper is that combining a named-entity-based rote classifiers with the balanced classifiers yields better results than either classifier alone. This method improves on the best M-F1 previously reported, while using the same amount of training data for the rote classifier, and considerably less for the statistical classifiers.

Using company descriptors inferred from the knowledge base does not improve performance in comparison with using descriptors and company names extracted from the document. One possible reason for that is the relatively small size of the corpus and high sparsity of descriptors. We plan to explore this issue further by using larger datasets and leveraging a richer set of semantic features, which can be provided by higher-level event attributes, obtained via IE.

The μ-F1 in our experiments is lower than the best μ-F1 reported in the literature on RCV1. This is likely due to the fact that both Puurula (2012) [32] and Cisse et al. (2013) [3] try to model inter-dependencies among the labels in the corpus. This is not done in [23] or [44]. We plan to investigate this further in future work.

References

1. Atkinson, M., Piskorski, J., van der Goot, E., Yangarber, R.: Multilingual real-time event extraction for border security intelligence gathering. In: Wiil, U.K. (ed.) Counterterrorism and Open Source Intelligence. Lecture Notes in Social Networks, vol. 2, pp. 355–390. Springer, Vienna (2011)
2. Bekkerman, R., Allan, J.: Using bigrams in text categorization. Technical Report IR-408, Department of Computer Science, University of Massachusetts, Amherst (December 2004)
3. Cisse, M.M., Usunier, N., Arti, T., Gallinari, P.: Robust bloom filters for large multilabel classification tasks. In: Advances in Neural Information Processing Systems, pp. 1851–1859 (2013)
4. Crammer, K., Dredze, M., Pereira, F.: Confidence-weighted linear classification for text categorization. J. Mach. Learn. Res. **13**, 1891–1926 (2012)
5. Dendamrongvit, S., Vateekul, P., Kubat, M.: Irrelevant attributes and imbalanced classes in multi-label text-categorization domains. Intell. Data Anal. **15**(6), 843–859 (2011)
6. Dredze, M., McNamee, P., Rao, D., Gerber, A., Finin, T.: Entity disambiguation for knowledge base population. In: Proceedings of the 23rd International Conference on Computational Linguistics, pp. 277–285. Association for Computational Linguistics (2010)

7. Du, M., Kangasharju, J., Karkulahti, O., Pivovarova, L., Yangarber, R.: Combined analysis of news and Twitter messages. In: Joint Workshop on NLP&LOD and SWAIE: Semantic Web, Linked Open Data and Information Extraction, pp. 41–48 (2013)

8. Du, M., Pierce, M., Pivovarova, L., Yangarber, R.: Supervised classification using balanced training. In: Besacier, L., Dediu, A.-H., Martín-Vide, C. (eds.) SLSP 2014. LNCS, vol. 8791, pp. 147–158. Springer, Heidelberg (2014)

9. Dhondt, E., Verberne, S., Weber, N., Koster, C., Boves, L.: Using skipgrams and pos-based feature selection for patent classification. In: Computational Linguistics in the Netherlands (2012)

10. Erenel, Z., Altınçay, H.: Improving the precision-recall trade-off in undersampling-based binary text categorization using unanimity rule. Neural Comput. Appl. **22**(1), 83–100 (2013)

11. Forman, G.: An extensive empirical study of feature selection metrics for text classification. J. Mach. Learn. Res. **3**, 1289–1305 (2003)

12. Gabrilovich, E., Markovitch, S.: Feature generation for text categorization using world knowledge. IJCAI **5**, 1048–1053 (2005)

13. Grishman, R., Huttunen, S., Yangarber, R.: Information extraction for enhanced access to disease outbreak reports. J. Biomed. Inform. **35**(4), 236–246 (2003)

14. Gullo, F., Domeniconi, C., Tagarelli, A.: Projective clustering ensembles. Data Min. Knowl. Disc. **26**(3), 452–511 (2013)

15. Hall, M., Frank, E., Holmes, G., Pfahringer, B., Reutemann, P., Witten, I.H.: The WEKA data mining software: an update. ACM SIGKDD Explor. Newsl. **11**(1), 10–18 (2009)

16. Han, X., Sun, L.: An entity-topic model for entity linking. In: Proceedings of the 2012 Joint Conference on Empirical Methods in Natural Language Processing and Computational Natural Language Learning, pp. 105–115. Association for Computational Linguistics (2012)

17. Hatami, N., Chira, C., Armano, G.: A route confidence evaluation method for reliable hierarchical text categorization. arXiv preprint (2012). arXiv:1206.0335

18. Huang, R., Riloff, E.: Classifying message board posts with an extracted lexicon of patient attributes. In: Proceedings of the 2013 Conference on Empirical Methods in Natural Language Processing, pp. 1557–1562 (2013)

19. Huttunen, S., Vihavainen, A., Du, M., Yangarber, R.: Predicting relevance of event extraction for the end user. In: Poibeau, T., Saggion, H., Piskorski, J., Yangarber, R. (eds.) Multi-source, Multilingual Information Extraction and Summarization. Theory and applications of natural language processing, pp. 163–176. Springer, Berlin (2012)

20. Huttunen, S., Vihavainen, A., von Etter, P., Yangarber, R.: Relevance prediction in information extraction using discourse and lexical features. In: Proceedings of NoDaLiDa: the 18th Nordic Conference on Computational Linguistics. Riga, Latvia (2011)

21. Ji, H., Grishman, R., Dang, H.T., Griffitt, K., Ellis, J.: Overview of the tac 2010 knowledge base population track. In: Third Text Analysis Conference (TAC 2010) (2010)

22. Koller, D., Sahami, M.: Hierarchically classifying documents using very few words. Technical report 1997–75, Stanford InfoLab (February 1997)

23. Lewis, D.D., Yang, Y., Rose, T.G., Li, F.: RCV1: a new benchmark collection for text categorization research. J. Mach. Learn. Res. **5**, 361–397 (2004)

24. Liao, S., Grishman, R.: Using document level cross-event inference to improve event extraction. In: Proceedings of the 48th Annual Meeting of the Association for Computational Linguistics, pp. 789–797. Association for Computational Linguistics (2010)

25. Liu, Y., Loh, H.T., Sun, A.: Imbalanced text classification: a term weighting approach. Expert Syst. Appl. **36**(1), 690–701 (2009)

26. Mann, G.S., Yarowsky, D.: Multi-field information extraction and cross-document fusion. In: Proceedings of the 43rd annual meeting on association for computational linguistics, pp. 483–490. Association for Computational Linguistics (2005)

27. Moschitti, A., Ju, Q., Johansson, R.: Modeling topic dependencies in hierarchical text categorization. In: Proceedings of the 50th Annual Meeting of the Association for Computational Linguistics: Long Papers, vol. 1, pp. 759–767. Association for Computational Linguistics (2012)

28. Patwardhan, S., Riloff, E.: Effective information extraction with semantic affinity patterns and relevant regions. EMNLP-CoNLL **7**, 717–727 (2007)

29. Piskorski, J., Tanev, H., Atkinson, M., van der Goot, E., Zavarella, V.: Online news event extraction for global crisis surveillance. In: Nguyen, N.T. (ed.) Transactions on Computational Collective Intelligence V. LNCS, vol. 6910, pp. 182–212. Springer, Heidelberg (2011)

30. Pokkunuri, S., Ramakrishnan, C., Riloff, E., Hovy, E., Burns, G.A.: The role of information extraction in the design of a document triage application for biocuration. In: Proceedings of BioNLP 2011 Workshop, pp. 46–55. Association for Computational Linguistics (2011)

31. Prati, R.C., Batista, G.E.A.P.A., Monard, M.C.: Class imbalances *versus* class overlapping: an analysis of a learning system behavior. In: Monroy, R., Arroyo-Figueroa, G., Sucar, L.E., Sossa, H. (eds.) MICAI 2004. LNCS (LNAI), vol. 2972, pp. 312–321. Springer, Heidelberg (2004)

32. Puurula, A.: Scalable text classification with sparse generative modeling. In: Anthony, P., Ishizuka, M., Lukose, D. (eds.) PRICAI 2012. LNCS, vol. 7458, pp. 458–469. Springer, Heidelberg (2012)

33. Rao, D., McNamee, P., Dredze, M.: Entity linking: finding extracted entities in a knowledge base. In: Poibeau, T., Saggion, H., Piskorski, J., Yangarber, R. (eds.) Multi-source, pp. 93–115. Multilingual Information Extraction and Summarization. Springer, Heidelberg (2013)

34. Roth, D., Yih, W.t.: Probabilistic reasoning for entity & relation recognition. In: Proceedings of the 19th international conference on Computational linguistics, vol. 1, pp. 1–7. Association for Computational Linguistics (2002)

35. Sil, A., Cronin, E., Nie, P., Yang, Y., Popescu, A.M., Yates, A.: Linking named entities to any database. In: Proceedings of the 2012 Joint Conference on Empirical Methods in Natural Language Processing and Computational Natural Language Learning, pp. 116–127. Association for Computational Linguistics (2012)

36. Sorower, M.S.: A literature survey on algorithms for multi-label learning. Technical report, Oregon State University, Corvallis, OR, USA, December 2010

37. Tsoumakas, G., Katakis, I.: Multi-label classification: an overview. Int. J. Data Warehouse. Min. (IJDWM) **3**(3), 1–13 (2007)

38. Tsoumakas, G., Katakis, I., Vlahavas, I.: Mining multi-label data. In: Maimon, O., Rokach, L. (eds.) Data Mining and Knowledge Discovery Handbook, pp. 667–685. Springer, Heidelberg (2010)

39. Wang, S., Li, D., Zhao, L., Zhang, J.: Sample cutting method for imbalanced text sentiment classification based on BRC. Knowl.-Based Syst. **37**, 451–461 (2013)

40. Yang, Y.: An evaluation of statistical approaches to text categorization. Inf. Retrieval **1**(1–2), 69–90 (1999)
41. Yangarber, R., Jokipii, L.: Redundancy-based correction of automatically extracted facts. In: Proceedings of HLT-EMNLP: Conference on Empirical Methods in Natural Language Processing, Vancouver, Canada, pp. 57–64 (2005)
42. Yangarber, R., Steinberger, R.: Automatic epidemiological surveillance from online news in MedISys and PULS. In: Proceedings of IMED-2009: International Meeting on Emerging Diseases and Surveillance, Vienna, Austria (2009)
43. Zhang, W., Yoshida, T., Tang, X.: A comparative study of TF*IDF, LSI and multi-words for text classification. Expert Syst. Appl. **38**(3), 2758–2765 (2011)
44. Zhuang, D., Zhang, B., Yang, Q., Yan, J., Chen, Z., Chen, Y.: Efficient text classification by weighted proximal SVM. In: Fifth IEEE International Conference on Data Mining (2005)

Supervised Machine Learning Techniques to Detect *TimeML* Events in French and English

Béatrice Arnulphy[1], Vincent Claveau[2](✉), Xavier Tannier[3], and Anne Vilnat[3]

[1] Inria - Rennes-Bretagne Atlantique, Rennes, France
`beatrice.arnulphy@inria.fr`
[2] IRISA-CNRS, Rennes, France
`vincent.claveau@irisa.fr`
[3] LIMSI-CNRS, University of Paris Sud, Orsay, France
`{xavier.tannier,anne.vilnat}@limsi.fr`

Abstract. Identifying events from texts is an information extraction task necessary for many NLP applications. Through the *TimeML* specifications and TempEval challenges, it has received some attention in recent years. However, no reference result is available for French. In this paper, we try to fill this gap by proposing several event extraction systems, combining for instance Conditional Random Fields, language modeling and k-nearest-neighbors. These systems are evaluated on French corpora and compared with state-of-the-art methods on English. The very good results obtained on both languages validate our approach.

Keywords: Event identification · Information extraction · TimeML · TempEval · CRF · Language modeling · English · French

1 Introduction

Extracting events from texts is a keystone for many applications concerned with information access (question-answering systems, dialog systems, text mining...). During the last decade, this task received some attention through the *TempEval*[1] conference series (2007, 2010, 2013). In these challenges, participants were provided with corpora annotated with *TimeML* features (cf. Sec. 2.1) in several languages, as well as an evaluation framework. It allowed to obtain reference results and relevant comparison between event-detection systems.

Yet, despite the success of the multilingual *TempEval-2* challenge, no participant proposed systems for French, for any task. Up to now, the situation is such that:

- the few studies dealing with detecting events in French cannot be compared since they use different evaluation materials;
- the performance of the systems cannot be compared to state-of-the-art systems, mainly developed for English.

---

[1] http://www.timeml.org/tempeval2/.

© Springer International Publishing Switzerland 2015
C. Biemann et al. (Eds.): NLDB 2015, LNCS 9103, pp. 19–32, 2015.
DOI: 10.1007/978-3-319-19581-0_2

The work presented in this paper aims at addressing these two shortcomings by proposing several systems for detecting events in French. These systems are evaluated within different frameworks/languages so that they can be compared with state-of-the-art systems, in particular those developed for English. More precisely, the tasks that we are tackling are the identification of events and of nominal markers of events. The systems we propose are versatile enough to be easily adapted to different languages or data types. They are based on usual machine learning techniques – decision trees, conditional random fields (CRFs), k-nearest neighbors (kNNs) – but make use of lexical resources, either existing, or semi-automatically built. These systems are tested on different evaluation corpora, including those of *TempEval-2* challenge. They are applied to both English and French data sets; the English data allow us to assess their performance relative to other published approaches. Whereas the French data provide reference results for this language.

The paper is structured as follows: in Sect. 2, the context of this work is presented, including the *TempEval* extraction tasks and the TimeML standard. In Sect. 3, we propose a review of the state-of-the-art systems developed for these tasks. Our own extraction systems are then detailed (Sect. 4) and their results on English and French are respectively reported in Sects. 5 and 6.

2   Extracting Events: The TempEval Framework

The *TempEval* challenges offered a unique framework dedicated to event detection tasks. The tasks rely on the *TimeML* specification language. In the remaining of this section, we give insights into this standard and we detail the *TempEval* challenges.

2.1   TimeML

Event definition used in *TempEval* follows the *ISO-TimeML* language specification [21]. It was developed to annotate and standardize events and temporal expressions in natural language texts. According to this standard, an event is described in a generic way as *"a cover term for situations that happen or occur"* [20]. For instance, this annotation scheme considers[2]:

- event expressions (<EVENT>), with their class and attributes (time, aspect, polarity, modality). There are 7 classes of events: ASPECTUAL, I_ACTION, I_STATE, OCCURRENCE, PERCEPTION, REPORTING and STATE;
- temporal expressions and their normalized values (<TIMEX3>);
- temporal relations between events and temporal expressions (<TLINK>);
- aspectual (<ALINK>) and modal (<SLINK>) relations between events;
- linguistic markers introducing these relations (<SIGNAL>).

This annotation scheme was first applied to English, and then to other languages (with small changes in the scheme and adaptations to the annotation

---

[2] For details and examples, see [23].

guide for each considered language). The *TimeML* annotated corpora are called TimeBank: *TimeBank 1.2* [19] for English, *FR-TimeBank* [7] for French, and so on. In practice, it is noteworthy that events in these corpora are mostly verbs and dates. Nominal events, though important for many applications, are less frequent, which may cause specific problems when trying to identify them (cf. Sects. 5 and 6).

In this article, we focus on identifying events as defined by the *TimeML* tag <EVENT> [29], which is the purpose of task B in *TempEval-2*. An example of such an event, from the TimeBank-1.2 annotated corpora[3], is given below: line 1 is the sentence with 2 events annotated, lines 2 and 3 describe the attributes of these events.

(1) The financial <EVENT eid="e3" class="OCCURRENCE">assistance</EVENT> from the World Bank and the International Monetary Fund are not <EVENTeid="e4" class="OCCURRENCE">helping</EVENT>.

(2) <MAKEINSTANCE eventID="e3" eid="ei377" tense="NONE" aspect="NONE" po-. larity="POS" pos="NOUN"/>

(3) <MAKEINSTANCE eventID="e4" eid="ei378" tense="PRESENT". aspect="PROGRESSIVE" polarity="NEG" pos="VERB"/>

2.2 TempEval Challenges

Up to now, there have been three editions of *TempEval* evaluation campaign (organized during *SemEval*[4]).

TempEval-1[5] [28] focused on detecting relations between provided entities. In this first edition, only English texts were proposed. *TempEval-2*[6] [29] focused on detecting events, temporal expressions and temporal relations. This campaign was multilingual (including English, French and Spanish) and the tasks were more precisely defined than for *TempEval-1*.

TempEval-3[7] [27] consisted again in the evaluation of event and temporal relation extraction, but only English and Spanish tracks were proposed. Moreover, a new focus of this third edition was to evaluate the impact of adding automatically annotated data to the training set.

As previously mentioned, in this paper, we mainly focus on extracting events (marked by verbs or nouns) as initially defined in *TempEval-2* challenge. Besides, as our goal is to produce and evaluate systems for French, we use the dataset developed for *TempEval-2* (as well as other French datasets that will be described below).

3 Related Work

Several studies have been dedicated to the annotation and the automatic extraction of events in texts. Yet, most of them were carried out in a specific framework,

---

[3] http://www.TimeBank-1.2/data/timeml/ABC19980108.1830.0711.html.

[4] http://semeval2.fbk.eu/semeval2.php.

[5] http://www.timeml.org/tempeval/.

[6] http://semeval2.fbk.eu/semeval2.php?location=tasks#T5.

[7] http://www.cs.york.ac.uk/semeval-2013/task1/.

with a personal definition of what could be an event. This is the case for example in monitoring tasks (for example on seismic events [11]), popular event detection from tweets [5] or in sports [14]. These task-based definitions of events are not discussed in this paper, as they often lead to dedicated systems and can hardly be evaluated in other contexts. In this section, we focus on the closest studies, either done within the *TempEval-2* framework or not, but relying on the generic and linguistically motivated definition of events as proposed in *TimeML*.

3.1   Extracting TimeML Events

EVITA system [23] aims to extract *TimeML* events in *TimeBank1.2*, combining linguistic and statistical approaches, using *WordNet* as external resource. STEP [6] aims at classifying every *TimeML* items with a machine learning approach based on linguistic features, without any external resources. They also develop two baseline systems (MEMORIZE and a simulation of EVITA). Although every *TimeML* elements were searched for, the authors focus specifically on nominal events. They reached the conclusion that the automatic detection of these events (*i.e.* nouns or noun phrases tagged <event>) is far from being trivial, because of the high variability of expressions, and consequently because of the lack of training data covering all the possible cases.

Parent et al. [18] worked on the extraction of *TimeML* structures in French. Their corpus of biographies and novels was manually annotated before *FR-TimeBank*'s publication. These studies primarily concern the adverbial phrases expressing temporal localization. Their model is mainly based on parsing and pattern matching of syntactic segments. Concerning nouns, they used their own reviewed version of the *VerbAction* lexicon [25] and few syntactic rules. To the best of our knowledge, this work is the only one concerning *TimeML* events on French.

3.2   Work Within Scope of TempEval-2

Several systems participated in *TempEval-2* campaign, most of them on the English dataset. The best ranked, TIPSEM [16], learns CRF models from training data and the approach is focused on semantic information. The evaluation exercise is divided into four groups of problems to be solved. In the recognition problem group, the features are morphological (lemma, part-of-speech (PoS) context from *TreeTagger* [24]), syntactical (syntactic tree from *Charniak parser* [8]), polarity, tense and aspect (using PoS and handcrafted rules). The semantic level features are the semantic role, the governing verb of the current word, role configuration (for governing verbs), lexical semantics (the top four classes from WordNet for each word). This system being the best ranked of the challenge, it was later used as a reference for *TempEval-3*. EDINBURGH [9] relies on text segmentation, rule-based and machine-learning named entity recognition, shallow syntactic analysis and lookup in lexicons compiled from the training data and from *WordNet*. TRIPS parser [1] provides event identification and "TimeML-suggested features", and is semantically motivated. It is based on a proper Logical Form Ontology. TRIOS [26] is based on TRIPS with a Markov Logic Network

(MLN) which is a Statistical Relation Learning Method (SRL). Finally, Ju_cse [12] consists in a very simple and manually designed rule-based method for event extraction, where all the verb PoS tags (from *Stanford* PoS tagger) are annotated as events.

All these systems and their respective performance provide valuable information. Firstly, most of them rely on a classical architecture using machine learning, and CRFs seem to perform well, as they do in many other information extraction tasks. Secondly, the results highlight the necessity of providing semantic information large enough to cover the great number of ways to express events, especially for the nominal events. The systems that we propose in this paper share many points with some of the systems we described here, as they also rely on supervised machine learning, including CRFs, and also make use of lexicons which were in part obtained automatically.

4   Event Detection Systems

The systems proposed in this paper aims at being easily adapted to any new language or text. To do so, as for many state-of-the-art systems, they adopt a supervised machine learning framework: *TimeML* annotated data are provided to train our systems, which are then evaluated on separate test set. The goal of the classifier is to assign each word with a label indicating whether it is an event. Since some events are expressed through multi-word expressions, the IOB annotation scheme is used (B indicates the beginning of an event, I is for inside an event, and O is for outside – if the word does not refer to an event). The training data are excerpts from corpora where each word is annotated with these labels and is described by different features (detailed hereafter). These data are then exploited by machine learning techniques presented in Sub-sects. 4.2 and 4.3. After the training phase, the inferred classifiers can be used to extract the events from unseen texts by assigning the most probable label to each word with respect to its context and features.

4.1   Features

The features used in our systems are simple and easy to extract automatically. They include what we call hereafter internal features: word-form, lemmas and part-of-speech, obtained with TreeTagger[8]). On the other hand, external features bring lexical information coming from existing lexicons, either general or specific to event description:

– for French, a feature indicates for each word whether it belongs to the *VerbAction* [25] and *The Alternative Noun Lexicon* [7] lexicons or not. The former lexicon is a list of verbs and their nominalization describing actions (*e.g.* *enfumage* (act of producing smoke), *réarmement* (rearmament)); the latter is complementary as it records non deverbal event nouns (nouns that are not derived from a verb, eg. *miracle* (miracle), *tempête* (storm)).

---

[8] http://www.ims.uni-stuttgart.de/projekte/corplex/TreeTagger.

– for English, a feature indicates for each word whether it belongs to one of the eight classes of synsets concerned with actions or events, that is *change, communication, competition, consumption, contact, creation, motion, stative.*

We also exploit lexical resources that are automatically built, called *Eventiveness Relative Weight Lexicons* (ERW hereafter), following the seminal work or Arnulphy et al. [3]. These lexicons are lists of words associated with the probability that they express an event. In our case, they are built from newspaper corpora (AFP news wire for French and Wall Street Journal for English). We do not go into further details about the building of ERWs, they may be found in the previously cited reference. It is worth noting that these lexicons bring information on polysemic words. It means that, for instance, most of the entries may express an action, which is then relevant to extract, or the result of an action, which is not wanted (*e.g. enfoncement, décision* in French). Thus, these lexicons are not sufficient by themselves, but they bring valuable information to exploit with more complex method taking the context into account.

4.2   CRF and Decision-Tree Based Systems

We have considered two machine learning techniques usually used for this kind of tasks: conditional random fields (CRFs, for instance used by [16]), and decision trees (DTs) that have shown good performance in previous work [2].

Concerning the DTs, we use the WEKA [10] implementation of C4.5 [22]. The interest of DTs is their ability to handle different types of features: nominal (useful to represent part-of-speech for example), boolean (does a word belong to a lexicon), numeric (ERW values). In order to take into account the sequential aspect of the text, each word is described by its own features (cf. sec. 4.1) and those of the preceding and following words.

CRFs [13] are now a well-established standard tool for annotation tasks. Contrary to DTs, they inherently take into account the sequential dependencies in our textual data. But in contrast, most implementations do not handle numeric features. Thus, the ERW scale of values is splitted into 10 equally large segments and transformed into a 10-value nominal feature. In the experiment reported below, we use WAPITI [15], a fast and robust CRF implementation.

4.3   CRF-kNN Combined System

The two systems described above are quite common for information extraction. We propose here a more original system, still based on CRFs, but aiming at addressing some of their shortcomings. One of them is the fact that CRFs consider the sequential context in a very constrained way. A sequence introducing an event X, as in example 1 below, will be considered as different to example 2 due to the offset caused by the insertion of "*l'événement de*" or "*unexpectedly*". The event Y may thus be undetected, even though example 1, which seems similar, is in the training set.

1. *"c'est à cette occasion que s'est produit X ..."* / at the very moment, X happened
2. *"c'est à cette occasion que s'est produit l'événement de Y ..."* / at the very moment, unexpectedly, X happened

Another issue with CRFs is that available implementations can hardly handle numeric features (like ERW values), or consider sets of synonyms.

To address these different limits, we join a kNN classifier to CRFs to help to label the potential events. CRFs are used as explained in the previous section, but all the possible labels with their probabilities are kept instead of only the most probable label. The kNN then compute a similarity between every candidates (every potential events found by the CRFs, regardless of their probability) and all the training instances.

In our case, this similarity is computed by using n-gram language modeling. It allows us to estimate a probability (written P_{LM}) for a sequence of words. More precisely, for each potential event found by the CRF, its class C^* (event or not) is decided following its probability given by the CRF ($P_{CRF}(C)$), and the probabilities provided by language models on the event itself and on its left and right contexts (resp. candidate, $cont_L$ and $cont_R$). Language models (*i.e.* sets of estimated probabilities) are thus estimated for each class and each position (left or right) from the training data. This is done by counting n-grams occurring at the left and at the right of each event of the training set, and inside the event. These models are denoted $\mathcal{M}_C$, $\mathcal{M}_C^R$ and $\mathcal{M}_C^L$. Finally, the label decision is formalized as:

$$C^* = \operatorname*{argmax}_C P_{CRF}(C) * P_{LM}(cont_L|\mathcal{M}_C^L) * P_{LM}(candidate|\mathcal{M}_C) * P_{LM}(cont_R|\mathcal{M}_C^R)$$

In our experiments, we use bigram models for $\mathcal{M}_C^D$ and $\mathcal{M}_C^G$, and unigram models for $\mathcal{M}_C$; the right and the left context are 5 words long. Based on that, the similarity of the left contexts of examples 1 and 2 would be high enough to detect the event in example 2.

Moreover, one other interest of language models is that it makes it possible to take into account lexical information during the smoothing process. In order to prevent unseen n-grams from generating a 0 probability for a sequence, it is usual to associate a small but non zero probability to them. Several strategies are proposed in the literature [17]. In our case, we use a back-off strategy from unseen bigrams to unigrams and a Laplacian smoothing, as it is easy to implement, for unseen unigrams. One originality of our work is to use also smoothing to exploit the information in our lexicons. Indeed, a word unseen in the training data may be replaced with a seen word belonging to the same lexicon (or synset for WordNet). When several words can be used, the one that maximizes the probability is chosen. In every case, a penalty ($\lambda < 1$) is applied; formally, for a word w unseen in the training data for a model $\mathcal{M}$, we have:

$$P(w|\mathcal{M}) = \lambda * max\{P(w_i|\mathcal{M}) \,|\, w_i, w \text{ is the same lexicon/synset }\}$$

Concerning the ERW values, they give information on the presence of the considered word inside the lexicons, *i.e.* may be interpreted as belonging values (absent words are scored 0) which are used to compute the penalty for the smoothing: the replacement penalty (λ) between one unseen word w with a seen one w_i is proportional to the difference between the values of these two words.

Combining these two systems makes the most of the CRF ability to detect interesting phrases, thanks to a multi-criterion approach (part-of-speech, lemmas), and of the language modeling to consider larger contexts and to integrate lexical information as a smoothing process.

5 Experiments on English

5.1 Settings

To evaluate our systems, the metrics we adopt are the same as for *TempEval-2*: precision (Pr), recall (Rc) and F1-score (F1). They are computed for the whole extraction tasks as well as on a subset of events known to be more difficult, specifically nominal events (events expressed as a noun or a phrase whose head is a noun), and stative nominal events.

Beside the overall performance of the systems, we want to assess the importance of the different features. Here, we report the results for some of the several combinations we tested, according to the type of features: internal and/or external (cf. Sect. 4.1). The configurations tested are:

1. with internal information only: the models only rely on word forms, lemmas and part-of-speech.
2. with both internal and external information;
3. this configuration is a variant of the preceding one, specific to the use of WordNet: the 8 classes of synsets are used as 8 binary features indicating the presence or absence of the word in the synset classes.

5.2 Results

Among all the tested system/feature configurations, Table 1 present the results of the best ones. For comparison purposes, we also report the results of TIPSEM, EDINBURGH, JU_CSE, TRIOS et TRIPS obtained at *TempEval-2*.

On these English data, CRF approaches outperform the ones based on decision trees, especially for the nominal event detection. This is partly due to the fact that nominal events are rare: only 7 % of nouns are events while, for instance, 57.5 % of the verbs are events. This imbalance has a strong impact on DTs while CRFs are less sensitive to that. But more generally, for any system, the performance drops when dealing with nominal events (either with or without states). Here again, this is due to the scarcity of such events, which are therefore less represented in the training data, which in turn causes a low recall. This study also shows that the performances differ depending on the different feature combinations. It shed light on the importance of using lexical information for these tasks, which confirms the state of the art.

Table 1. Performance of the best system/feature combination on the *TempEval-2* English data set.

Type of event	System	Pr	Rc	F1
All events	TIPSEM	0.81	**0.86**	0.83
	EDINBURGH	0.75	0.85	0.80
	JU_CSE	0.48	0.56	0.52
	TRIOS	0.80	0.74	0.77
	TRIPS	0.55	0.88	0.68
	(3) CRF-kNN	**0.86**	**0.86**	**0.86**
	(3) CRF	0.79	0.80	0.79
	(3) DT	0.73	0.71	0.72
Nominal only	(3) CRF-kNN	0.78	0.55	0.65
	(3) CRF	0.72	0.48	0.58
	(2) DT	0.58	0.28	0.38
Nominal without states	(3) CRF-kNN	0.64	0.44	0.52
	(3) CRF	0.53	0.38	0.45
	(3) DT	0.87	0.08	0.15

Last, our CRF-kNN system yields the best results, outperforming CRFs alone, DT or state-of-the-art systems. These results are promising as they only rely on features that are easy to extract from the text (*e.g.* PoS) or publicly available (*e.g.* WordNet). Thus, they are expected to be applicable to any language such as French (cf. next section).

6 Experiments on French

6.1 Dataset and Comparison to English

In contrast to English, few corpora are available to develop, evaluate and compare event extraction systems in French. Among them, the *TempEval-2* French corpus is supposed to be similar to its English counterpart in terms of genre and annotation. As for the English corpus, which was part of the *TimeBank1.2*, this French corpus is a part of the *FR-TimeBank*. In previous work [4], we also proposed an annotated corpus for French. As for *FR-TimeBank*, it is composed of newspaper articles, which makes it comparable in genre to *En-TempEval-2* corpus, but it is only annotated in non-stative nominal events (*TimeML* tag <EVENT class="OCCURRENCE" pos="NOUN">).

Several points are worth mentioning for a fair comparison with English results. Table 2 shows that the proportion of all events is comparable between the French and English *TempEval-2* corpora: about 2.6 by sentence. However a detailed analysis shows that there are more verbal events than nominal ones in *TempEval-2* corpora, but relatively more nominal events in both French corpus

Table 2. Comparison of English (ENG) and French (FRE) corpora with *TimeML* annotations.

		# sentences	# tokens	# events
ENG	*TempEval-2*	2,382	58,299	6,186
FRE	*TempEval-2*	441	9,910	1,150
FRE	corpus of [4]	2,414	54,110	1,863

Table 3. Performance of the best feature/system configurations on the French corpora (*Fr-TempEval-2*, [4] and [18]).

Corpus	Type of event	System	Pr	Rc	F1
TempEval-2	all events	(2) CRF-kNN	**0.87**	**0.79**	**0.83**
français		(2) CRF	0.80	0.76	0.78
		(4) DT	0.78	0.77	0.78
	nominal only	(2) CRF-kNN	**0.69**	0.60	**0.64**
		(2) CRF	0.55	0.52	0.53
		(4) DT	0.58	**0.63**	0.60
	nominal without states	(2) CRF-kNN	**0.65**	**0.52**	**0.58**
		(2) CRF	0.53	0.46	0.50
		(4) DT	0.57	0.49	0.53
Corpus of [4]	nominal without states	(2) CRF-kNN	**0.79**	**0.63**	**0.70**
		(2) CRF	0.76	0.54	0.63
		(4) DT	0.75	0.60	0.67
Corpus of [18]	all events	Parent et al	0.625	0.777	0.693
	nominal only	Parent et al	0.547	0.537	0.542

than for English. Furthermore, the corpus of [4] contains more nominal events than *Fr-TempEval-2*; and about 90 % of nominal events are not states in *Fr-TempEval-2*, versus 80 % in *En-TempEval-2*.

6.2   Results on French

The feature combinations used for English have been tested; Table 3 reports the best performing model/feature configurations. For purposes of comparison, we also implemented a system proposed in a previous work [2] to serve as a baseline, which we note (4). This system also relies on DTs but uses features that are more difficult to obtain and thus less adaptable, namely a deep syntactic analysis, post-edited with manually-built rules. Finally, we also report the results published by [18] on their own corpus.

Overall, the CRF models perform as well as the technique proposed in [2], while using no syntactic information and hand-coded resources. Concerning the non-stative nominal events, the results are significantly better on the corpus

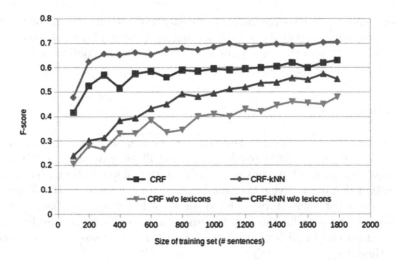

Fig. 1. Performance (F1-score) of CRF-kNN and CRF models with respect to the number of training sentences.

of [4] than on *Fr-TempEval-2* (F1=0.63 *vs.* F1=0.53). This performance gap highlights the above-mentioned intrinsic differences of the two corpora. Finally, even if the comparison is tricky since we deal with different corpora, it is worth noting that our systems outperform the results reported by [18].

French experiments lead to the same observations as for English data: extracting nominal events is more difficult than extracting verbal events. Yet, the difference between nominal and non-stative nominal events is smaller than for English. It may be explained by the proportion of such events which differs, as mentioned in Sect. 6.1. As for English, our system combining CRFs and language-model-based kNNs yields the best overall results. Again, the results obtained with the different sets of features underline the positive impact of lexical information for such extraction tasks.

6.3   Influence of Lexicons and Training Data Size

In order to evaluate the impact of the size of training data on the performance of our CRF-kNN system, we report in Fig. 1 how F1-score evolves according to the number of annotated sentences used for training. For purposes of comparison, we also report the performance of the CRF-alone system in order to shed the light on the contribution of the language models. Two configurations are tested: with and without external lexical information.

First, this figure shows that the interest of combining CRFs with the language-model kNNs is significant, for any size of the training data. Second, the language models improve the CRF performance, whether lexicons are used or not. Obviously, without external lexical information, the F-score progression depends directly on the number of training sentences. In contrast, using lexical resources makes the F1-score increase rapidly with small amount of training data, and then increase again linearly for bigger amount of data. It shows that small training set, and thus small annotation costs, can be considered, provided that lexical resources are available.

7 Conclusion

Extracting events from texts is a keystone for many applications, but definitions of what is an event are often *ad hoc* and difficult to generalize, which makes any comparison impossible. On the other hand, the linguistically motivated and standardized definition given by *TimeML* and implemented in the *TempEval* challenges was not completely explored for some languages such as French. In this paper, we tried to fill this gap by proposing several systems, evaluated on French, but also on English in order to assess their performance with respect to state-of-the-art systems.

The three proposed systems adopt a classical architecture based on supervised machine learning techniques. Yet, one of our contributions is to propose a combination of CRFs and language-model kNNs, which takes advantage of both techniques. In particular, the language model offers a nice way to incorporate lexical information in the event detection process, which has proven to be useful, especially when dealing with few data. This original combination of CRFs and kNNs yields good results on both English and French and outperforms state-of-the-art systems. The good results obtained for English validate our approach and suggest that the performance reported for French may now serve as a reasonable baseline for any further work. Among the perspectives, we will focus on the extraction of the other temporal markers and relations defined in *TimeML*. We also foresee the adaptation of our CRF-kNN method to these tasks as well as other information extraction tasks.

References

1. Allen, J.F., Swift, M., de Beaumont, W.: Deep semantic analysis of text. In: Proceedings of the 2008 Conference on Semantics in Text Processing, STEP 2008, pp. 343–354. Association for Computational Linguistics, Stroudsburg (2008). http://dl.acm.org/citation.cfm?id=1626481.1626508
2. Arnulphy, B.: Désignations nominales des événements: Étude et extraction automatique dans les textes. Ph.D. thesis, Université Paris-Sud - École Doctorale d'Informatique de Paris Sud (EDIPS) / Laboratoire LIMSI (2012)
3. Arnulphy, B., Tannier, X., Vilnat, A.: Automatically generated Noun Lexicons for event extraction. In: Proceedings of the 13th International Conference on Intelligent Text Processing and Computational Linguistics (CicLing 2012), New Delhi, India, March 2012
4. Arnulphy, B., Tannier, X., Vilnat, A.: Event nominals: annotation guidelines and a manually annotated corpus in french. In: Proceedings of the 8th International Conference on Language Resources and Evaluation (LREC 2012), Istanbul, Turkey, May 2012

5. Becker, H., Naaman, M., Gravano, L.: Beyond trending topics: Real-world event identification on twitter. In: Fifth International AAAI Conference on Weblogs and Social Media (2011)
6. Bethard, S., Martin, J.H.: Identification of event mentions and their semantic class. In: Proceedings of the 2006 Conference on Empirical Methods in Natural Language Processing, pp. 146–154. Association for Computational Linguistics, Sydney (2006). http://www.aclweb.org/anthology/W/W06/W06-1618
7. Bittar, A.: Building a TimeBank for French: a reference corpus annotated according to the ISO-TimeML standard. Ph.D. thesis, Université Paris 7 - École doctorale de Sciences du Langage (2010)
8. Charniak, E.: A maximum-entropy-inspired parser. In: 1st Meeting of the North American Chapter of the Association for Computational Linguistics, pp. 132–139 (2000), http://www.aclweb.org/anthology/A00-2018
9. Grover, C., Tobin, R., Alex, B., Byrne, K.: Edinburgh-ltg: Tempeval-2 system description. In: Proceedings of the 5th International Workshop on Semantic Evaluation, pp. 333–336. Association for Computational Linguistics, Uppsala, July 2010. http://www.aclweb.org/anthology/S10-1074
10. Hall, M., Frank, E., Holmes, G., Pfahringer, B., Reutemann, P., Witten, I.H.: The WEKA data mining software: an update. SIGKDD Explor. 11(1), 10–18 (2009)
11. Jean-Louis, L., Besançon, R., Ferret, O.: Text segmentation and graph-based method for template filling in information extraction. In: 5th International Joint Conference on Natural Language Processing (IJCNLP 2011), Chiang Mai, Thailand, pp. 723–731 (2011)
12. Kumar Kolya, A., Ekbal, A., Bandyopadhyay, S.: Ju_cse_temp: A first step towards evaluating events, time expressions and temporal relations. In: Proceedings of the 5th International Workshop on Semantic Evaluation. pp. 345–350. Association for Computational Linguistics, Uppsala, July 2010. http://www.aclweb.org/anthology/S10-1077
13. Lafferty, J., McCallum, A., Pereira, F.: Conditional random fields: probabilistic models for segmenting and labe ling sequence data. In: International Conference on Machine Learning (ICML) (2001)
14. Lanagan, J., Smeaton, A.F.: Using twitter to detect and tag important events in live sports. In: Artificial Intelligence (2011)
15. Lavergne, T., Cappé, O., Yvon, F.: Practical very large scale CRFs. In: Proceedings the 48th Annual Meeting of the Association for Computational Linguistics (ACL), pp. 504–513. Association for Computational Linguistics, July 2010. http://www.aclweb.org/anthology/P10-1052
16. Llorens, H., Saquete, E., Navarro, B.: Tipsem (english and spanish): Evaluating crfs and semantic roles in tempeval-2. In: Proceedings of the 5th International Workshop on Semantic Evaluation, pp. 284–291. Association for Computational Linguistics, Uppsala, July 2010. http://www.aclweb.org/anthology/S10-1063
17. Ney, H., Essen, U., Kneser, R.: On structuring probabilistic dependencies in stochastic language modelling. Comput. Speech Lang. 8, 1–38 (1994)
18. Parent, G., Gagnon, M., Muller, P.: Annotation d'expressions temporelles et d'événements en franqis. In: Béchet, F. (ed.) Traitement Automatique des Langues Naturelles (TALN 2008). Association pour le Traitement Automatique des Langues (ATALA) (2008)
19. Pustejovsky, J., Verhagen, M., Saurí, R., Littman, J., Gaizauskas, R., Katz, G., Mani, I., Knippen, R., Setzer, A.: TimeBank 1.2. Linguistic Data Consortium (2006). http://timeml.org/site/publications/timeMLdocs/timeml_1.2.1.html

20. Pustejovsky, J., Castaño, J., Ingria, R., Saurí R., Gaizauskas, R., Setzer, A., Katz, G.: Timeml: Robust specification of event and temporal expressions in text. In: IWCS-5, Fifth International Workshop on Computational Semantics, Tilburg University (2003)
21. Pustejovsky, J., Lee, K., Bunt, H., Romary, L.: ISO-TimeML: an international standard for semantic annotation. In: Proceedings of the 7th International Conference on Language Resources and Evaluation (LREC 2010). European Language Resources Association (ELRA), Valletta (2010), http://aclweb.org/anthology-new/L/L10/
22. Quinlan, R.: C4.5: Programs for Machine Learning. Morgan Kaufman Publishers (1993)
23. Saurí, R., Knippen, R., Verhagen, M., Pustejovsky, J.: Evita: a robust event recognizer for QA systems. In: Proceedings of the HLT 2005, Vancouver, Canada, October 2005
24. Schmid, H.: Probabilistic part-of-speech tagging using decision trees. In: Proceedings of International Conference on New Methods in Language Processing, Manchester, UK (1994)
25. Tanguy, L., Hathout, N.: Webaffix : un outil d'acquisition morphologique dérivationnelle à partir du Web. In: Pierrel, J.M. (ed.) Actes de Traitement Automatique des Langues Naturelles (TALN 2002), vol. Tome I, pp. 245–254. ATILF, ATALA, Nancy, France, June 2002
26. UzZaman, N., Allen, J.: Trips and trios system for tempeval-2: extracting temporal information from text. In: Proceedings of the 5th International Workshop on Semantic Evaluation, pp. 276–283. Association for Computational Linguistics, Uppsala (2010). http://www.aclweb.org/anthology/S10-1062
27. UzZaman, N., Llorens, H., Derczynski, L., Allen, J., Verhagen, M., Pustejovsky, J.: Semeval-2013 task 1: Tempeval-3: evaluating time expressions, events, and temporal relations. In: Second Joint Conference on Lexical and Computational Semantics (*SEM), Volume 2: Proceedings of the Seventh International Workshop on Semantic Evaluation (SemEval 2013), pp. 1–9. Association for Computational Linguistics, Atlanta, Georgia, USA (2013). http://www.aclweb.org/anthology/S13-2001
28. Verhagen, M., Gaizauskas, R., Schilder, F., Hepple, M., Katz, G., Pustejovsky, J.: Semeval-2007 task 15: tempeval temporal relation identification. In: Proceedings of the SemEval Conference (2007)
29. Verhagen, M., Saurí, R., Caselli, T., Pustejovsky, J.: Semeval-2010 task 13: Tempeval-2. In: Proceedings of the 5th International Workshop on Semantic Evaluation, ACL 2010, Uppsala, Sweden, pp. 57–62 (2010). http://polyu.academia.edu/TommasoCaselli/Papers/1114340/TempEval2_Evaluating_Events_Time_Expressions_and_Temporal_Relations

Distributional Semantics

In Defense of Word Embedding
for Generic Text Representation

Guy Lev, Benjamin Klein, and Lior Wolf[✉]

The Blavatnik School of Computer Science, Tel Aviv University, Tel Aviv, Israel
wolf@cs.tau.ac.il

Abstract. Statistical methods have shown a remarkable ability to capture semantics. The word2vec method is a frequently cited method for capturing meaningful semantic relations between words from a large text corpus. It has the advantage of not requiring any tagging while training. The prevailing view is, however, that it lacks the ability to capture semantics of word sequences and is virtually useless for most purposes, unless combined with heavy machinery. This paper challenges that view, by showing that by augmenting the word2vec representation with one of a few pooling techniques, results are obtained surpassing or comparable with the best literature algorithms. This improved performance is justified by theory and verified by extensive experiments on well studied NLP benchmarks (This work is inspired by [10]).

1 Introduction

Document retrieval and text analytics, in general, benefit from a fixed-size representation of variable sized text. The most basic method in the field, and still highly influential, is the bag-of-words (BOW) method. It has obvious shortcomings, such as uniform distances between the contribution of every two words to the vector representation and invariance to word order. However, these shortcomings can be partially ameliorated by incorporating techniques such as tf-idf and by considering n-grams instead of single words. However, the usage of one dimension per dictionary word leads to a representation that is sparse with respect to the information content and does not capture even the simplest synonyms.

Recently, semantic embeddings of words in vector spaces have gained a renewed interest, especially the word2vec method [19] and related methods. It has been demonstrated that not only are words with similar meanings embedded nearby, but natural word arithmetic can also be convincingly applied. For example, the calculated difference in the embedding vector space between "London" and "England" is similar to the one obtained between "Paris" and "France". Word2vec representations are learned in a very weakly supervised manner from large corpora, and are not explicitly constrained to abide by such regularities.

Despite the apparent ability to capture semantic similarities, and the surprising emergence of semantic regularities that support additivity, word2vec embeddings have been criticized as a tool for higher level NLP. First, the Neural Network employed to learn the word2vec embeddings is a simple "shallow" (not deep)

© Springer International Publishing Switzerland 2015
C. Biemann et al. (Eds.): NLDB 2015, LNCS 9103, pp. 35–50, 2015.
DOI: 10.1007/978-3-319-19581-0_3

network, capable, by common conception, of capturing only low-level information. Taking an analogy from the field of image recognition, where very deep networks are being deployed, word2vec is considered to be a low-level "edge detection" operator, incapable of capturing complex compositional semantics. Second, word2vec has been criticized for being almost equivalent to the much earlier methods of frequency matrix factorization [17]. Third, it has been argued that in order to capture more than single words, mechanisms should be added in order to account for order and hierarchical compositions [32]. The alleged inability of vector embeddings to solve mid- and high-level NLP problems was also demonstrated in various NLP papers, where an average of vector embeddings served as a baseline method.

It is the purpose of this paper to challenge the commonly held view that the word2vec representation is inadequate and markedly inferior to more sophisticated algorithms. The poor performance of the word2vec representation can probably be traced to aggregation techniques that do not take sufficient account of numerical and statistical considerations. It is shown in this paper that proper pooling techniques of the vectors of the text words leads to state of the art or at least very competitive results.

Given a text to represent, we consider it as a multi-set, i.e., as a generalized set in which each element can appear multiple times. We advocate the use of principal component analysis (PCA) or independent component analysis (ICA) as an unsupervised preprocessing step that transforms the semantic vector space into independent semantic channels. For pooling, as shown, the mean vector performs well. In some situations, the more powerful Fisher Vector (FV) [22] representation provides improved results.

Fisher Vectors provide state-of-the-art results on many different applications in the domain of computer vision [7,21,23,29]. In all of these contributions, the FV of a set of local descriptors is obtained as a concatenation of gradients of the log-likelihood of the descriptors in the set with respect to the parameters of a Gaussian Mixture Model (GMM) that was fitted on a training set in an unsupervised manner. In our experiments, we do not observe a clear benefit to GMM over a simple Gaussian Model. Due to the clear disadvantage of the extra parameter (the number of mixture components), we focus on modeling by a unimodal Gaussian. Furthermore, to account for the non-Gaussian nature of the data incurred by the ICA transformation, we propose to use Generalized Gaussian Models. The corresponding Fisher Vectors are derived and formulas are also given to the approximation of the Fisher Information Matrix in order to allow for normalization of the dynamic range of the FV variant presented.

2   Previous Work

Representing text as vectors Word2vec [18,19] is a recently developed technique for building a neural network that maps words to real-number vectors, with the desideratum that words with similar meanings will map to similar vectors. This technique belongs to the class of methods called "neural language models". It uses a scheme that is much simpler than previous work in this domain, where neural networks with many hidden units and several non-linear layers were normally constructed (e.g., [5]), word2vec [18] constructs a simple log-linear classification network [20]. Two such networks are proposed: the Skip-gram and the Continuous Bag-of-words (CBOW) architectures. In our experiments, we employ the Skip-gram architecture, which is considered preferable.

Attention has recently shifted into representing sentences and paragraphs and not just words. The classical method in this domain is Bag of Words [30]. Socher et al. [31] have analyzed sentences using a recursive parse tree. The combination of two subtrees connected at the root, by means of generating a new semantic vector representation based on the vector representations of the two trees, is performed by concatenating their semantic vector representations and multiplying by a matrix of learned parameters. In a recent contribution by Le et al. [15], the neural network learns to predict the following word in a paragraph based on a representation that concatenates the vector representation of the previous text and the vector representations of a few words from the paragraph. This method, called the paragraph vector, achieves state-of-the-art results on the Stanford Sentiment Treebank dataset surpassing a model that averages neural word vectors and ignores word order.

In [40], Yu et al. are using distributed representations that are based on deep learning for the task of identifying sentences that contain the answer to a given question. Given word embeddings, their first model generates the vector representation of a sentence by taking the mean of the word vectors that compose the sentence. Since their first model does not account for word ordering and other structural information, they developed a more complex model that works on the word embedding of the bigrams. Their model matches state of the art performance on the TREC answer selection dataset.

Pooling methods were one of the primary steps in many computer vision pipelines in the era before the advent of Deep Learning. Many different pooling methods were suggested in the last decade, each contributing to the improvement in accuracy on the standard object recognition benchmarks. One of the most known and basic pooling techniques was borrowed from the NLP community when Sivic et al. [30] used clustering over local features of image patches in order to create a bag of words representation for computer vision applications. Richer representations like VLAD [13] and FV [22] were later introduced and were the main contributors to the increasing in accuracy in object recognition benchmarks.

Specifically, the FV representation is today the leading pooling technique in traditional computer vision pipelines and provided state-of-the-art results on many different applications [7,21,23,29]. Although already introduced in 2007, the FV pooling method was able to surpass the bag of words representation only after introducing improvements such as normalization techniques that have dramatically enhanced its performance. Some of the most widely used improvements were introduced by Perronnin et al. [23]. The first improvement is to apply an element-wise power normalization function, $f(z) = \text{sign}(z)|z|^\alpha$ where $0 \leq \alpha \leq 1$ is a parameter of the normalization. The second improvement is to apply a

L2 normalization on the FV after applying the power normalization function. By applying these two operations [23] achieved state-of-the-art accuracy on an image recognition benchmark called CalTech 256 and showed superiority over the traditional Bag of Visual Words model.

3 Pooling

In our approach, a single sentence is represented as a multi-set of word2vec vectors. The notation of a multi-set is used to clarify that the order of the words in a sentence does not affect the final representation and that a vector can appear more than once (if the matching word appears more than once in the sentence). In order to apply machine learning models to the sentences, it is useful to transform this multi-set into a single high dimensional vector with a constant length. This can be achieved by applying pooling.

Since word2vec is already an extremely powerful representation, we find that conventional pooling techniques or their extensions are sufficiently powerful to obtain competitive performance. The pooling methods that are used in this paper are: (1) Mean vector pooling; (2) FV of a single multivariate Gaussian; (3) FV of a single multivariate generalized Gaussian. These are described in the next sections.

3.1 Mean Vector

This pooling technique takes a multiset of vectors, $X = \{x_1, x_2, \ldots, x_N\} \in R^D$, and computes its mean: $v = \frac{1}{N} \sum_{i=1}^{N} x_i$. Therefore, the vector v that results from the pooling is in R^D.

The disadvantage of this method is the blurring of the text's meaning. By adding multiple vectors together, the location obtained – in the semantic embedding space – is somewhere in the convex hull of the words that belong to the multi-set. A better approach might be to allow additivity without interference.

3.2 Fisher Vector of a multivariate Gaussian

Given a multiset of vectors, $X = \{x_1, x_2, \ldots, x_N\} \in R^D$, the standard FV [22] is defined as the gradients of the log-likelihood of X with respect to the parameters of a pre-trained Diagonal Covariance Gaussian Mixture Model. It is common practice to limit the FV representation to the gradients of the means, μ and to the gradients of the standard deviations, σ (the gradients of the mixture weights are ignored).

Since we did not notice a global improvement in accuracy when increasing the number of Gaussian in the mixture, we focus on a single multivariate Gaussian. As a consequence, there are no latent variables in the model and it is, therefore, possible to estimate the parameters $\lambda = \{\mu, \sigma\}$ of this single diagonal covariance Gaussian by using maximum likelihood derivations, instead of using the *EM* algorithm which is usually employed when estimating the parameters of the

Gaussian Mixture Model. Under this simplified version of the FV, the gradients from which the FV is comprised are:

$$\frac{\partial \mathcal{L}(X|\lambda)}{\partial \mu_d} = \sum_{i=1}^{N} \frac{x_{i,d} - \mu_d}{\sigma_d^2} \; ; \quad \frac{\partial \mathcal{L}(X|\lambda)}{\partial \sigma_d} = \sum_{i=1}^{N} \left(\frac{(x_{i,d} - \mu_d)^2}{\sigma_d^3} - \frac{1}{\sigma_{j,d}} \right) \quad (1)$$

and, therefore, the resulting representation is in R^{2D}. Applying PCA and ICA as a preprocessing step is investigated in this work with the purpose of sustaining the diagonal covariance assumption.

As in [22], the diagonal of the Fisher Information Matrix, F, is approximated in order to normalize the dynamic range of the different dimensions of the gradient vectors. For a single Gaussian model, the terms of the approximated diagonal Fisher Information Matrix become: $F_{\mu_d} = \frac{N}{\sigma_{k,d}^2}; \quad F_{\sigma_d} = \frac{2N}{\sigma_{k,d}^2}.$

The FV is the concatenation of two normalized partial derivative vectors: $F_{\mu_d}^{-1/2} \frac{\partial \mathcal{L}(X|\lambda)}{\partial \mu_d}$ and $F_{\sigma_d}^{-1/2} \frac{\partial \mathcal{L}(X|\lambda)}{\partial \sigma_d}$.

It is worth noting the linear structure of the FV pooling, which is apparent from the equations above. Since the likelihood of the multi-set is the multiplication of the likelihoods of the individual elements, the log-likelihood is linear. Therefore, the Fisher Vectors of the individual words can be computed once for each word and then reused. For all of our experiments, the multivariate Guassian (or the generalized Gaussian presented next) is estimated only once, from all word2vec vectors. These vectors are obtained, precomputed on a subset of the Google News dataset, from https://code.google.com/p/word2vec/. Therefore, the encoding is independent of the dataset used in each experiment, is completely generic, and is very efficient to compute as a simple summation of precomputed Fisher Vectors (same runtime complexity as mean pooling).

Following the summation of the Fisher Vectors of the individual words, the Power Normalization and the L2 Normalization that were introduced in [24] (see Sect. 2) are employed, using a constant $a = 1/2$.

3.3 Fisher Vector of a Generalized Multivariate Gaussian

A generalization of the FV that is presented here for the first time, in which the FV is redefined according to a single multivariate generalized Gaussian distribution. The need for this derivation is based on the observation (see below) that word2vec vectors are not distributed in accordance with the multivariate Gaussian distribution.

The generalized Gaussian distribution is, in fact, a parametric family of symmetric distributions and is defined by three parameters: m which is the location parameter and is the mean of the distribution, s the scale parameter and p the shape parameter. The probability density function of the Generalized Gaussian Distribution (GGD) in the univariate case is:

$$ggd(x; m, s, p) = \frac{1}{2sp^{1/p}\Gamma(1 + 1/p)} exp \left(-\frac{|x - m|^p}{ps^p} \right) \quad (2)$$

The estimation of the parameters of a univariate Generalized Gaussian Distribution is done according to [2].

Under the common assumption in the FV that the covariance matrix is diagonal, the multivariate generalized Gaussian distribution is defined:

$$ggd(\boldsymbol{x}; \boldsymbol{m}, \boldsymbol{s}, \boldsymbol{p}) = \prod_{d=1}^{D} \frac{1}{2s_d p_d^{1/p_d} \Gamma(1+1/p_d)} exp\left(-\frac{|x_d - m_d|^{p_d}}{p_d s_d^{p_d}}\right) \tag{3}$$

Since the dimensions of the multivariate GGD are independent, the parameters of the GGD can be estimated dimension-wise.

The FV can now be redefined as the gradients of the log-likelihood of $X = \{x_1, x_2, \ldots, x_N\} \in R^D$ with respect to the parameters of a pre-trained Diagonal Covariance Multivariate Generalized Gaussian Distribution. In practice, the FV is defined in this work only according to the gradients of m and s, since the gradients according to p do not seem to improve the results.

The log likelihood is defined as:

$$\mathcal{L}(\boldsymbol{m}, \boldsymbol{s}, \boldsymbol{p}|X) = \sum_{d=1}^{D}\left[-N\log\left(2s_d p_d^{1/p_d}\Gamma(1+1/p_d)\right) - \frac{\sum_{i=1}^{N}|x_{id}-m_d|^{p_d}}{p_d s_d^{p_d}}\right] \tag{4}$$

The resulting FV in R^{2D} is given by:

$$\frac{\partial \mathcal{L}(\boldsymbol{m}, \boldsymbol{s}, \boldsymbol{p}|X)}{\partial m_d} = s_d^{-p_d}\sum_{i=1}^{N}|x_{id}-m_d|^{p_d-1}\text{sign}(x_{id}-m_d) \tag{5}$$

$$\frac{\partial \mathcal{L}(\boldsymbol{m}, \boldsymbol{s}, \boldsymbol{p}|X)}{\partial s_d} = -N/s_d + s_d^{-p_d-1}\sum_{i=1}^{N}|x_{id}-m_d|^{p_d} \tag{6}$$

The diagonal of Fisher Information Matrix, F, for this distribution is approximated in order to normalize the dynamic range of the different dimensions of the gradient vectors. Let F_{m_d} and F_{s_d} be the terms of diagonal of F that correspond respectively to $\frac{\partial \mathcal{L}(\boldsymbol{m},\boldsymbol{s},\boldsymbol{p}|X)}{\partial m_d}$ and $\frac{\partial \mathcal{L}(\boldsymbol{m},\boldsymbol{s},\boldsymbol{p}|X)}{\partial s_d}$. Then:

$$F_{m_d} = \int_X ggd(X|\lambda)\left[\sum_{i=1}^{N}\frac{\partial \mathcal{L}(x_i|\lambda)}{\partial m_d}\right]^2 dX \tag{7}$$

Where $\lambda = \{m, s, p\}$ Then:

$$F_{m_d} = \sum_{\substack{t=1\ldots N \\ u=1\ldots N \\ t\neq u}}\int_{x_t,x_u}\frac{\partial \mathcal{L}(x_t|\lambda)}{\partial m_d}\frac{\partial \mathcal{L}(x_u|\lambda)}{\partial m_d}ggd(x_t, x_u|\lambda)\,dx_t dx_u$$

$$+\sum_{t=1}^{N}\int_{x_t}\left[\frac{\partial \mathcal{L}(x_t|\lambda)}{\partial m_d}\right]^2 ggd(x_t|\lambda)\,dx_t \tag{8}$$

Since the samples are i.i.d given λ and also the dimensions are independent:

$$\int_{x_t,x_u} \frac{\partial \mathcal{L}(x_t|\lambda)}{\partial m_d} \frac{\partial \mathcal{L}(x_u|\lambda)}{\partial m_d} ggd(x_t, x_u|\lambda) \, dx_t dx_u$$

$$= \int_{x_{t,d}} \frac{\partial \mathcal{L}(x_{t,d}|\lambda)}{\partial m_d} ggd(x_{t,d}|\lambda) \, dx_{t,d} \int_{x_{u,d}} \frac{\partial \mathcal{L}(x_{u,d}|\lambda)}{\partial m_d} ggd(x_{u,d}|\lambda) \, dx_{u,d}$$

Using the fact that $\frac{\partial \mathcal{L}(x_{t,d}|\lambda)}{\partial m_d} = \frac{\partial}{\partial m_d} log(ggd(x_{t,d}|\lambda)) = \frac{\frac{\partial}{\partial m_d} ggd(x_{t,d}|\lambda)}{ggd(x_{t,d}|\lambda)}$:

$$\int_{x_{t,d}} \frac{\partial \mathcal{L}(x_{t,d}|\lambda)}{\partial m_d} ggd(x_{t,d}|\lambda) \, dx_{t,d} = \int_{x_{t,d}} \frac{\partial}{\partial m_d} ggd(x_{t,d}|\lambda) \, dx_t = \frac{\partial}{\partial m_d} \int_{x_{t,d}} ggd(x_{t,d}|\lambda) \, dx_t = 0$$

Therefore, the first expression in the sum of (8) is equal to 0. Assuming that the dimensions are independent, the second expression in the sum of (8) is equal to

$$\sum_{t=1}^{N} \int_{x_{t_d}} \left[\frac{\partial \mathcal{L}(x_{t_d}|\lambda)}{\partial m_d} \right]^2 ggd(x_{t_d}|\lambda) \, dx_{t_d}.$$

Note that $\int_{x_{t_d}} \left[\frac{\partial \mathcal{L}(x_{t_d}|\lambda)}{\partial m_d} \right]^2 ggd(x_{t_d}|\lambda) \, dx_{t_d}$ is the value of the Fisher Information Matrix of a univariate generalized Gaussian distribution for a single sample. Therefore according to [2]:

$$\int_{x_{t_d}} \left[\frac{\partial \mathcal{L}(x_{t_d}|\lambda)}{\partial m_d} \right]^2 ggd(x_{t_d}|\lambda) \, dx_{t_d} = \frac{(p-1)\Gamma\left(\frac{p-1}{p}\right)}{s^2 \Gamma\left(\frac{1}{p}\right) p^{(2-p)/p}} \tag{9}$$

Therefore:

$$F_{m_d} = N \cdot \frac{(p-1)\Gamma\left(\frac{p-1}{p}\right)}{s^2 \Gamma\left(\frac{1}{p}\right) p^{(2-p)/p}} \tag{10}$$

Similarly, since $\int_{x_{t_d}} \left[\frac{\partial \mathcal{L}(x_{t_d}|\lambda)}{\partial s_d} \right]^2 ggd(x_{t_d}|\lambda) \, dx_{t_d} = \frac{p}{s^2}$ according to [2], it can be shown that: $F_{s_d} = N \cdot \frac{p}{s^2}$.

The normalized partial derivatives of the FV are then $F_{m_d}^{-1/2} \frac{\partial \mathcal{L}(X|\lambda)}{\partial m_d}$ and $F_{s_d}^{-1/2} \frac{\partial \mathcal{L}(X|\lambda)}{\partial s_d}$.

In [27], Sanchez et al. state that applying the Principal Components Analysis (PCA) on the data before fitting the GMM is the key to make the FV perform well. In experiments on PASCAL VOC 2007, they show that accuracy does not seem to be overly sensitive to the exact number of PCA components. The explanation is that transforming the descriptors by using PCA is a better fit to the diagonal covariance matrix assumption.

Following this observation, a transformation that will cause the transformed descriptors to be a better fit to the diagonal covariance matrix assumption is sought for the generalized gaussian FV. The optimal transformation will

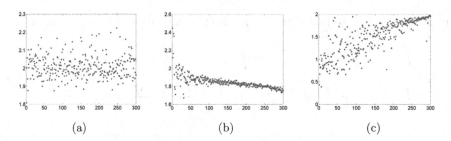

Fig. 1. The shape parameter p of the generalized Gaussian distribution. This parameter was estimated for each dimension of the word2vec representation, based on all word2vec vectors, i.e., a distribution was fit to each coordinate separately. (a) the raw word2vec vectors; (b) after applying PCA, retaining the original dimensionality; (c) after applying ICA. In all three plots, x-axis is the vector coordinate index from 1 to 300, y-axis is the estimated p. Note that the range of the y-axis differs between the plots.

result in transformed descriptors that are dimension independent and are non-Gaussian signals. While PCA suffers from the implicit assumption of an underlying Gaussian distribution [14], the Independent Component Analysis (ICA) [16] explicitly encourages non-Gaussian distributions.

Figure 1 depicts the estimated shape parameters p for each dimension of the word2vec representation, and for all dictionary words used. As can be seen, the shape varies between the dimensions, depending on whether we consider the raw word2vec representation, the representation post-PCA, or that after applying ICA. The baseline distribution is not a Gaussian one, but most shape parameters are between 1.9 and 2.1. Post-PCA, the shape parameters are mostly in a narrow band around 1.9. Post-ICA, the shape parameters follow an almost linear trend between 0.8 and 2.

Finally, The Power Normalization and L2 Normalization are applied using $a = 1/p$ on the resulting FV. While similar to the conventional FV, this constant is not justified directly, we found it experimentally to slightly outperform $a = 1/2$ for this case.

3.4   Classification

The pooled representation of a sentence can be used in combination with any classifier to make predictions based on the sentence. In addition, many of our experiments require the comparison of two sentences. Let u and v be the pooled representations of the two sentences. Our unified representation is given by the concatenation of their difference and their mean: $\begin{bmatrix} |u-v| \\ (u+v)/2 \end{bmatrix}$. This provides information on both the location of the two vectors and the difference between them, in a symmetric manner.

4   Experiments

We perform our experiments on multiple benchmarks: the TREC Answer Selection Dataset, The SemEval-2012 Semantic Sentence Similarity benchmark, and the very recent Yahoo! and AG topic classification benchmarks.

4.1   Answer Selection

The answer sentence selection dataset contains factoid questions each associated with a list of answer sentences. It was created by Wang et al. [36] from the Text REtrieval Conference (TREC) QA track (8–13) dataset, with candidate answers automatically selected from each question's document pool. This selection was based on a combination of overlapping non-stop word counts and pattern matching, and was followed by manual tagging for parts of the dataset. Overall, there are 4718, 1148, and 1517 question-answer pairs in the train, validation, and test set, respectively.

The task is to rank the candidate answers based on their relation to the question. Two standard success metrics are used and in both higher is better: Mean Average Precision (MAP) and Mean Reciprocal Rank (MRR). MRR measures the rank of any correct answer and MAP examines the ranks of all the correct answers and accounts for recall. The two scores are calculated using the official `trec_eval` evaluation scripts.

We compare our results with the state of the art [11,28,35,36,38–40]. Our method employs the concatenated diff+mean vector of Sect. 3.4. Linear SVM is used with a parameter C tuned on the development set.

As can be seen in Table 1, the most basic pooling method of average pooling is already preferable, when applied to word2vec transformed by PCA, to the literature methods. Moreover, when adding FV pooling, the results further improve. Best results are obtained using the ICA + generalized Gaussian FV representation.

It is interesting to compare our method to the method of [40], which also relies on word embedding. While our method employs word2vec, [40] employs the Collobert and Westons neural language model [8] as provided by Turian et al. [33]. The unigram model of [40] is similar to our mean pooling method. However, it uses the classification model of [6]: given vector representations of a question $\mathbf{q}$ and an answer $\mathbf{a}$ (both in $\mathbb{R}^d$), the probability of the answer being correct is $p(y = 1|\mathbf{q}, \mathbf{a}) = \sigma(\mathbf{q}^T \mathbf{M} \mathbf{a} + b)$, where the transformation matrix $\mathbf{M} \in \mathbb{R}^{d \times d}$ and the bias term b are learned model parameters. The bigram model of [40] is a 1D Convolution Neural Network (CNN) with a single convolution layer and a filter size of 2.

The authors of [40] suggest that vector representation based approaches are "not very well equipped for dealing with cardinal numbers and proper nouns, especially considering the small dataset". Therefore, they augment these with two counting based features: word co-occurrence count and word co-occurrence count weighted by idf values. The output of the unigram or bigram model is concatenated in their experiments with these features and then a logistic classifier is applied. In our experiments, we do not observe the need to add such features.

Recently, an extended training set called TRAIN-ALL was proposed [40]. This is a significantly larger training set that was labeled automatically, using pattern matching, and contains many labeling errors. The best result obtained on this dataset [40] has a MAP of 0.711 (MRR 0.785) using the deep learning bigram + count method. Our best result is superior on this training set as well:

Table 1. Experimental results on the TREC Answer Selection benchmark. A long list of literature results are presented, including the state of the art results obtained by Yih et al. [39] and the very recent results of Yu et al. [40]. PCA followed by mean pooling outperform all literature results; ICA + generalized Gaussian FV performs even better. Yu et al. [40] also present results on a larger and noisier training set called TRAIN ALL. Training on this training set (not shown in the table), we obtain a slight improvement only; However, our results are still better than Yu et al. [40]: MAP 0.720 vs. 0.711; MRR 0.824 vs. 0.785.

Method	MAP	MRR
Wang et al. [36]	0.603	0.685
Heilman and Smith [11]	0.609	0.692
Wang and Manning [35]	0.595	0.695
Yao et al. [38]	0.631	0.748
Severyn and Moschitti [28]	0.678	0.736
Baseline: word counts [39]	0.571	0.627
Baseline: tf-idf Word Count [39]	0.596	0.652
Yih et al. LR [39]	0.682	0.762
Yih et al. BDT [39]	0.694	0.789
Yih et al. LCLR [39]	0.709	0.770
Deep learning unigram [40]	0.539	0.628
Deep learning unigram + count [40]	0.689	0.773
Deep learning bigram [40]	0.548	0.644
Deep learning bigram + count [40]	0.706	0.780
Mean pooling	0.665	0.752
PCA + mean pooling	0.710	0.807
ICA + mean pooling	0.679	0.783
Gaussian FV	0.662	0.763
PCA+Gaussian FV	0.621	0.743
ICA + Gaussian FV	0.705	0.810
Generalised Gaussian FV	0.654	0.757
PCA + generalized Gaussian FV	0.623	0.729
ICA + generalized Gaussian FV	**0.719**	**0.824**

MAP of 0.720 (MRR 0.824). Stacking [37], using a fourth linear SVM, all three ICA variants, improves results on TRAIN-ALL to MAP 0.7372 (MRR 0.8511).

4.2 Semantic Sentence Similarity

The task of Semantic Sentence Similarity (STS) has gained considerable attention. Semantic embedding models are at a disadvantage for this task, since the structure of the sentences is complex, and explicit matching between parts of the

Table 2. Results on the STS benchmarks. Our results are shown for PCA followed by mean pooling only, since other pooling options gave almost identical results.

Method	msr-par	msr-vid	smt-eur
ADW [25]	0.694	0.887	0.555
UKP2 [3]	0.683	0.873	0.528
TLsyn [34]	0.698	0.862	0.361
TLsim [34]	0.734	0.880	0.477
VD [12]	-	0.890	-
MTL-GP [26]	0.732	0.888	0.562
DKPro scores [4] (log transformed)	0.734	0.887	0.540
PCA + mean pooling	0.537	0.827	0.513
PCA + mean pooling ∪ DKPro scores	**0.739**	**0.895**	**0.617**

sentence greatly aids the similarity judgment. In our experiments below, we aim to show that word2vec pooling provides a reasonable pipeline, and that when added to a set of literature scores, state of the art results are obtained.

The experimental setup used in the STS task [1] was followed, and for technical reasons (availability of DKPro scores) we employ 3 out of the 5 datasets presented: msr-par, msr-vid, and smt-eur. Each sentence pair in the datasets was given a score from 0 (lowest similarity) to 5 (highest similarity) by human judges. We compare our results to the state of the art results [3,12,25,26,34].

The authors of [3] have released a toolbox called DKPro that contains code for the computation of 75 similarities [4] that is a superset of the 20 similarities used in [3]. Unable to completely identify the 20 similarities, we have rerun the entire set of 75 similarities as an additional pipeline. When taking log scale of the similarities, it seems to outperform [3] on the msr-par benchmark but not on the other two.

We compute the two representations of each pair of sentences and combine them (Sect. 3.4). For the regression problem of the STS benchmarks, we use the effective K-clusters Regression Forests (KRF) [9] method, with the default parameters. Interestingly, on the STS benchmarks the exact combinations of PCA or ICA and pooling method did not show any clear winners. The results of all 9 combinations (including no feature transformation) were almost indistinguishable. We, therefore, present the results of PCA followed by average pooling, which is the most basic method we recommend. We also present results obtained when combining the mean pooling similarity with the DKPro similarities. This is done by the ridge regression method on the 76 similarities, where the regularization parameter was obtained using cross validation on the training set.

The results are presented in Table 2. The results obtained by average pooling would have placed this system as one of the top systems of the SemEval-2012 competition [3,34]. When combined with the DKPro similarities, state-of-the-art results are obtained.

Table 3. Results on the topic classification benchmarks (accuracy). Our word2vec based methods are much better than the word2vec baseline of [41] and nearly as good as the best reported method of [41].

Method	Yahoo!	AG
Large ConvNet + Thesaurus [41]	0.699	**0.916**
Bag of Words [41]	0.666	0.883
word2vec bag-of-centroids [41]	0.588	0.853
PCA + mean pooling + linear SVM	0.688	0.896
PCA + mean pooling + KNN	0.672	0.906
ICA + 3 pooling methods + KNN	**0.703**	0.910

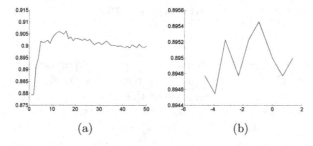

(a)   (b)

Fig. 2. Results on the AG benchmark when varying the papameters of the learning algorithm. (a) varying the parameter k of the KNN algorithm. (b) varying the parameter C of linear SVM (log scale).

4.3   Topic Classification

A week before the submission date, Zhang and LeCun have published a Technical Report presenting topic classification results obtained using deep temporal convolutional networks [41]. The paper presents word2vec as an inferior baseline, performing even worse than the basic bag-of-words method. It is claimed that this might be a result of using the same word2vec representation for all datasets, or "it might also be the case that the hope for linear separability of word2vec is not valid at all". As we show below, this is not the case, and word2vec performs on par with the best results of [41].

Pooling of word2vec in [41] is performed by running k-means on the word vectors (k = 5000), and then using histograms of length 5000 to represent the text, based on nearest centroid association. This is followed by logistic regression. This metod is vastly different from the pooling methods we advocate for.

We performed experiments on two of the datasets used in [41]: Yahoo! and AG. While the exact splits used were not made available yet (personal communication), the protocols for building the benchmarks are available. We verified that different random sampling of train/test have only a minimal effect on the results, with a SD of about 0.005 accuracy. The Yahoo! Answers Topic Classification benchmark is based on the Yahoo! Answers Comprehensive Questions and Answers version 1.0 dataset available through the Yahoo! Webscope program. Topic classification is performed on the 10 largest main categories, where each class contains 140,000 (5,000) random training (testing) samples. Out of all the answers and other meta-information, only the best answer content and the main category information are used for the benchmark. From the AG's corpus of news article http://www.di.unipi.it/~gulli/AG_corpus_of_news_articles. html, the 4 largest categories are used, employing only the title and description fields. From each category, random 40,000 (1,100) samples are taken for training (testing).

Since each vector is classified independently (no pairs), we simply employ linear SVM or the k-nearest neighbor algorithms. The results are depicted in Table 3. As can be seen, our word2vec considerably outperforms the baseline given in [41] and is only slightly worse than the results of the deep networks. Needless to say, the deep networks were completely retrained for each benchmark, and are extremely resource-heavy; A single epoch on the Yahoo! benchmark took a day to train. Also, our system has only the parameters of the classifiers, and as can be seen in Fig. 2, it is insensitive to the choice of these parameter. This, in comparison to the tens of hyperparameters of the deep network solutions.

In this experiment too, the pooling method did almost no difference. For example, for AG KNN classification, all 9 options where at an accuracy level above 0.899. However, by stacking the results obtained, for example, by the three ICA-based pooling methods, performance is slightly improved to 0.910 on this benchmark, and 0.703 on the Yahoo! benchmark.

5 Conclusion

With proper pooling, vector embeddings perform almost as well, if not better, than the best available methods. On the other hand, the proposed pipeline is generic and mostly unsupervised, and only requires a shallow off-the-shelf training in order to adapt to the problem at hand. The Fisher Vector pooling methods share the same runtime complexity as the baseline mean pooling method, and improve results significantly in two out of the three tasks we examined.

Word order is not properly addressed, as is apparent in the STS experiments. We plan to tackle this using a hierarchical pooling scheme that represents text by a list of pooled vectors. In addition, we plan to study pooling of other types of vector embedding such as co-occurance based ones.

Acknowledgments. This research is supported by the Intel Collaborative Research Institute for Computational Intelligence (ICRI-CI).

References

1. Agirre, E., Diab, M., Cer, D., Gonzalez-Agirre, A.: Semeval-2012 task 6: a pilot on semantic textual similarity. In: Proceedings of the First Joint Conference on Lexical and Computational Semantics, SemEval 2012, pp. 385–393. Association for Computational Linguistics, Stroudsburg (2012). http://dl.acm.org/citation.cfm? id=2387636.2387697

2. Agro, G.: Maximum likelihood estimation for the exponential power function parameters. Commun. Stat.-Simul. Comput. **24**(2), 523–536 (1995)
3. Bär, D., Biemann, C., Gurevych, I., Zesch, T.: UKP: computing semantic textual similarity by combining multiple content similarity measures. In: Proceedings of the First Joint Conference on Lexical and Computational Semantics - Volume 1: Proceedings of the Main Conference and the Shared Task, and Volume 2: Proceedings of the Sixth International Workshop on Semantic Evaluation, SemEval 2012, pp. 435–440. Association for Computational Linguistics, Stroudsburg (2012). http://dl.acm.org/citation.cfm?id=2387636.2387707
4. Bär, D., Zesch, T., Gurevych, I.: DKPro similarity: an open source framework for text similarity. In: Proceedings of the 51st Annual Meeting of the Association for Computational Linguistics: System Demonstrations, pp. 121–126. Association for Computational Linguistics (2013). http://aclweb.org/anthology/P13-4021
5. Bengio, Y., Ducharme, R., Vincent, P., Janvin, C.: A neural probabilistic language model. J. Mach. Learn. Res. **3**, 1137–1155 (2003). http://dl.acm.org/citation.cfm?id=944919.944966
6. Bordes, A., Chopra, S., Weston, J.: Question answering with subgraph embeddings. CoRR abs/1406.3676 (2014). http://arxiv.org/abs/1406.3676
7. Chatfield, K., Lempitsky, V., Vedaldi, A., Zisserman, A.: The devil is in the details: an evaluation of recent feature encoding methods. In: British Machine Vision Conference (2011)
8. Collobert, R., Weston, J.: A unified architecture for natural language processing: deep neural networks with multitask learning. In: Proceedings of the 25th International Conference on Machine Learning, ICML 2008, pp. 160–167. ACM, New York (2008). http://doi.acm.org/10.1145/1390156.1390177
9. Hara, K., Chellappa, R.: Growing regression forests by classification: applications to object pose estimation. In: Fleet, D., Pajdla, T., Schiele, B., Tuytelaars, T. (eds.) ECCV 2014, Part II. LNCS, vol. 8690, pp. 552–567. Springer, Heidelberg (2014)
10. Hartley, R.: In defense of the eight-point algorithm. IEEE Trans. Pattern Anal. Mach. Intell. **19**(6), 580–593 (1997)
11. Heilman, M., Smith, N.A.: Tree edit models for recognizing textual entailments, paraphrases, and answers to questions. In: Human Language Technologies: The 2010 Annual Conference of the North American Chapter of the Association for Computational Linguistics, HLT 2010, pp. 1011–1019. Association for Computational Linguistics, Stroudsburg (2010). http://dl.acm.org/citation.cfm?id=1857999.1858143
12. Young, P., Lai, A., Hodosh, M., Hockenmaier, J.: From image descriptions to visual denotations: new similarity metrics for semantic inference over event descriptions. Trans. Assoc. Comput. Linguist. **2**, 67–78 (2014)
13. Jégou, H., Douze, M., Schmid, C., Pérez, P.: Aggregating local descriptors into a compact image representation. In: Proceedings of the IEEE Conference on Computer Vision Pattern Recognition, pp. 3304–3311, June 2010. http://lear.inrialpes.fr/pubs/2010/JDSP10
14. Ke, Y., Sukthankar, R.: Pca-sift: a more distinctive representation for local image descriptors. In: Proceedings of the 2004 IEEE Computer Society Conference on Computer Vision and Pattern Recognition, 2004, CVPR 2004, vol. 2, p. II-506. IEEE (2004)

15. Le, Q.V., Mikolov, T.: Distributed representations of sentences and documents. In: Proceedings of the 31th International Conference on Machine Learning, ICML 2014, Beijing, China, 21–26 June 2014. JMLR Proceedings, vol. 32, pp. 1188–1196. JMLR.org (2014). http://jmlr.org/proceedings/papers/v32/le14.html
16. Lee, T.W.: Independent component analysis: theory and applications [book review]. IEEE Trans. Neural Netw. **10**(4), 982–982 (1999). http://dblp.uni-trier.de/db/journals/tnn/tnn10.html#Lee99a
17. Levy, O., Goldberg, Y.: Neural word embedding as implicit matrix factorization. In: Ghahramani, Z., Welling, M., Cortes, C., Lawrence, N., Weinberger, K. (eds.) Advances in Neural Information Processing Systems, vol. 27, pp. 2177–2185. Curran Associates, Inc. (2014). http://papers.nips.cc/paper/ 5477-neural-word-embedding-as-implicit-matrix-factorization.pdf
18. Mikolov, T., Chen, K., Corrado, G., Dean, J.: Efficient estimation of word representations in vector space. CoRR abs/1301.3781 (2013)
19. Mikolov, T., Sutskever, I., Chen, K., Corrado, G.S., Dean, J.: Distributed representations of words and phrases and their compositionality. In: Advances in Neural Information Processing Systems, pp. 3111–3119 (2013)
20. Mnih, A., Hinton, G.: Three new graphical models for statistical language modelling. In: Proceedings of the 24th International Conference on Machine Learning, ICML 2007, pp. 641–648. ACM, New York (2007). http://doi.acm.org/10.1145/ 1273496.1273577
21. Peng, X., Zou, C., Qiao, Y., Peng, Q.: Action recognition with stacked fisher vectors. In: Fleet, D., Pajdla, T., Schiele, B., Tuytelaars, T. (eds.) ECCV 2014, Part V. LNCS, vol. 8693, pp. 581–595. Springer, Heidelberg (2014)
22. Perronnin, F., Dance, C.: Fisher kernels on visual vocabularies for image categorization. In: IEEE Conference on Computer Vision and Pattern Recognition, 2007, CVPR 2007, pp. 1–8. IEEE (2007)
23. Perronnin, F., Liu, Y., Sánchez, J., Poirier, H.: Large-scale image retrieval with compressed fisher vectors. In: 2010 IEEE Conference on Computer Vision and Pattern Recognition (CVPR), pp. 3384–3391. IEEE (2010)
24. Perronnin, F., Sánchez, J., Mensink, T.: Improving the Fisher kernel for large-scale image classification. In: Daniilidis, K., Maragos, P., Paragios, N. (eds.) ECCV 2010, Part IV. LNCS, vol. 6314, pp. 143–156. Springer, Heidelberg (2010)
25. Pilehvar, T.M., Jurgens, D., Navigli, R.: Align, disambiguate and walk: a unified approach for measuring semantic similarity. In: Proceedings of the 51st Annual Meeting of the Association for Computational Linguistics (vol. 1: Long Papers), pp. 1341–1351. Association for Computational Linguistics (2013). http://aclweb. org/anthology/P13-1132
26. Rios, M., Specia, L.: UoW: multi-task learning gaussian process for semantic textual similarity. In: Proceedings of the 8th International Workshop on Semantic Evaluation, SemEval 2014, pp. 779–784. Association for Computational Linguistics and Dublin City University, Dublin, August 2014. http://www.aclweb.org/ anthology/S14-2138
27. Sánchez, J., Perronnin, F., Mensink, T., Verbeek, J.: Image classification with the fisher vector: theory and practice. Int. J. Comput. Vis. **105**(3), 222–245 (2013)
28. Severyn, A., Moschitti, A.: Automatic feature engineering for answer selection and extraction. In: EMNLP, pp. 458–467. ACL (2013). http://dblp.uni-trier.de/db/ conf/emnlp/emnlp2013.html#SeverynM13
29. Simonyan, K., Parkhi, O.M., Vedaldi, A., Zisserman, A.: Fisher vector faces in the wild. In: Proceedings of BMVC, vol. 1, p. 7 (2013)

30. Sivic, J., Russell, B.C., Efros, A.A., Zisserman, A., Freeman, W.T.: Discovering objects and their location in images. In: Tenth IEEE International Conference on Computer Vision, 2005, ICCV 2005, vol. 1, pp. 370–377. IEEE (2005)

31. Socher, R., Lin, C.C., Ng, A.Y., Manning, C.D.: Parsing natural scenes and natural language with recursive neural networks. In: Proceedings of the 26th International Conference on Machine Learning (ICML) (2011)

32. Socher, R., Perelygin, A., Wu, J., Chuang, J., Manning, C.D., Ng, A., Potts, C.: Recursive deep models for semantic compositionality over a sentiment treebank. In: Proceedings of the 2013 Conference on Empirical Methods in Natural Language Processing, pp. 1631–1642. Association for Computational Linguistics, Seattle, October 2013. http://www.aclweb.org/anthology-new/D/D13/D13-1170.bib

33. Turian, J., Ratinov, L., Bengio, Y.: Word representations: a simple and general method for semi-supervised learning. In: ACL (2010). http://cogcomp.cs.illinois.edu/papers/TurianRaBe2010.pdf

34. Šarić, F., Glavaš, G., Karan, M., Šnajder, J., Bašić, B.D.: Takelab: systems for measuring semantic text similarity. In: Proceedings of the First Joint Conference on Lexical and Computational Semantics - Volume 1: Proceedings of the Main Conference and the Shared Task, and Volume 2: Proceedings of the Sixth International Workshop on Semantic Evaluation, SemEval 2012, pp. 441–448. Association for Computational Linguistics, Stroudsburg (2012). http://dl.acm.org/citation.cfm?id=2387636.2387708

35. Wang, M., Manning, C.D.: Probabilistic tree-edit models with structured latent variables for textual entailment and question answering. In: Proceedings of the 23rd International Conference on Computational Linguistics, COLING 2010, pp. 1164–1172. Association for Computational Linguistics, Stroudsburg (2010). http://dl.acm.org/citation.cfm?id=1873781.1873912

36. Wang, M., Smith, N.A., Mitamura, T.: What is the jeopardy model? a quasi-synchronous grammar for qa. In: Proceedings of the 2007 Joint Conference on Empirical Methods in Natural Language Processing and Computational Natural Language Learning (EMNLP-CoNLL), pp. 22–32. Association for Computational Linguistics, Prague, June 2007. http://www.aclweb.org/anthology/D07-1003

37. Wolpert, D.H.: Stacked generalization. Neural Netw. 5(2), 241–259 (1992). http://dx.doi.org/10.1016/S0893-6080(05)80023-1

38. Yao, X., Van Durme, B., Callison-Burch, C., Clark, P.: Answer extraction as sequence tagging with tree edit distance. In: Proceedings of the 2013 Conference of the North American Chapter of the Association for Computational Linguistics: Human Language Technologies, pp. 858–867. Association for Computational Linguistics, Atlanta, June 2013. http://www.aclweb.org/anthology/N13-1106

39. tau Yih, W., Chang, M.W., Meek, C., Pastusiak, A.: Question answering using enhanced lexical semantic models. In: Proceedings of the 51st Annual Meeting of the Association for Computational Linguistics. ACL Association for Computational Linguistics, August 2013. http://research.microsoft.com/apps/pubs/default.aspx?id=192357

40. Yu, L., Hermann, K.M., Blunsom, P., Pulman, S.: Deep learning for answer sentence selection. In: NIPS Deep Learning Workshop, December 2014. http://arxiv.org/abs/1412.1632

41. Zhang, X., LeCun, Y.: Text Understanding from Scratch. ArXiv e-prints, February 2015

Using Distributed Word Representations and mRMR Discriminant Analysis for Multilingual Text Summarization

Houda Oufaida[1]([⊠]), Philippe Blache[2], and Omar Nouali[3]

[1] Ecole Nationale Supérieure D'Informatique ESI, Algiers, Algeria
h_oufaida@esi.dz
[2] Aix Marseille Université, CNRS, LPL UMR 7309, 13604 Aix En Provence, France
blache@lpl-aix.fr
[3] Centre de Recherche Sur L'Information Scientifique Et Technique CERIST,
Algiers, Algeria
onouali@cerist.dz

Abstract. Multilingual summarization task aims to develop summarization systems that are fully or partly language free. Extractive techniques are at the center of such systems. They use statistical features to score and extract most relevant sentences to form a summary within a size limit. In this paper, we investigate recently released multilingual distributed word representations combined with mRMR discriminant analysis to score terms then sentences. We also propose a novel sentence extraction algorithm to deal with redundancy issue. We present experimental results of our system applied to three languages: English, Arabic and French using the TAC MultiLing 2011 Dataset. Our results demonstrate that word representations enhance the summarization system, MeMoG and ROUGE results are comparable to recent state-of-the-art systems.

Keywords: Multilingual summarization · Distributed word representations · Discriminant analysis · Minimum redundancy · Maximum relevance

1 Introduction

In the past few years, research on summarization systems has received particular attention. Despite of the fact that it is relatively an old field (Luhn, 1958; Edmundson, 1969), the rapid growth of available documents in digital format: web, companies' networks, etc. gave a new boost to the field. With internet expansion in the world, multilingual content is constantly increasing. According to W3Techs study, while English content is still widely used on the internet (55 %), non-English content is expanding constantly. Moreover, non-English usage experiences a real boom. According to the Internet World Stat[1] and based

---

[1] http://www.internetworldstats.com.

© Springer International Publishing Switzerland 2015
C. Biemann et al. (Eds.): NLDB 2015, LNCS 9103, pp. 51–63, 2015.
DOI: 10.1007/978-3-319-19581-0_4

on the most one million visited websites study, English usage increased by around 281 % from 2001 to 2011 which is far less than Spanish usage (743 %), Chinese (1.277 %), Russian (1.826 %) or Arabic (2.501 %) over the same period.

Following multilingual content expansion, new tasks arose in the text summarization field such as cross and multilingual summarization. Cross lingual summarization produce a summary in a different language from the source document language (produce an English summary for an Arabic text for example). Meanwhile in multilingual summarization, we develop systems capable of summarizing documents in different languages where the source and summary share the same language (same system for English source/English summary, French source/French summary ...etc.). The later is the focus of this research.

Supported by two recent workshops: TAC MultiLing 2011 (Giannakopoulos et al. 2011) and ACL MultiLing 2013 (Giannakopoulos, 2013; Kubina et al. 2013), multilingual summarization takes a significant step forward. This comes actually at the opposite of early summarization systems which used heavy natural language processing (NLP) based techniques in a bid to explore in depth the source text and generate new sentences to form an abstract: paraphrasing identification and information fusion for example (Barzilay and McKeown, 2005). In other few researches, NLP techniques were used to identify salient sentences such as the use of rhetorical analysis RST (Marcu, 1997). Indeed, these techniques are closer to a natural human process but are not yet mature; they still require heavy NLP processing often based on limited and language dependent resources.

Currently, developing high level NLP systems for resource poor languages, such as Arabic, is not a feasible option. Recent researches tend to extract multilingual resources in an unsupervised way. They take advantage of huge amounts of raw texts from the web to extract lexical and semantic information. Distributed word representations are an example of unsupervised multilingual resources. Here, words are represented by multi dimensional and real valued vectors. Each dimension hopefully carries syntactic or semantic information of the target word. Experiments on recently released English (Collobert and Weston, 2008; Huang et al. 2012; Mikolov et al. 2013; Socher et al. 2011; Turian et al. 2010) and multilingual distributed word representations (Al-Rfou et al. 2013) are very promising.

In this paper, we propose a novel multilingual summarization system which extracts relevant sentences from single and multiple documents by maintaining minimum redundancy and maximum relevance. We first cluster sentences using semantic information encoded in word representations. To the best of our knowledge, this is the first time that word representations are introduced to sentence semantic relatedness task. We propose a novel sentence semantic relatedness metric and evaluate it using recent multilingual Semantic Textual Similarity STS-2012 and STS-2014 datasets. Second, we use a discriminant analysis method: mRMR (minimum Redundancy Maximum Relevance) (Peng et al. 2005) to score sentences using the discriminant potential of terms. We also propose a novel two speed extraction algorithm depending on the required summary size. We evaluate our summarisation method using TAC Multiling 2011 dataset. This paper is organized as follows: we first introduce a brief review of the related work in Sect. 4. Second, we describe our representations based sentence relatedness metric and present its evaluation on STS-2012 and STS-2014 datasets. Section 4 describes the original mRMR method and our adaptation to the summarization task. Details of our experiments with multilingual multi document summarization are described in Sect. 4. Finally, we conclude our paper with some interesting perspectives.

2   Related work

Multilingual summarization task aims to design systems that are able to summarize texts from different languages. Therefore, the system should use minimum language dependent processing, it is ideally language independent. Whereas in cross lingual summarization, the system produces summaries in a different language from the input documents; producing English summaries for Arabic documents and vice versa is an example.

Recently, Multilingual NLP is the focus of many NLP researches. Particularly, multilingual summarization was the subject of two recent workshops: TAC MultiLing 2011 (Giannakopoulos et al. 2011) and ACL Multiling 2013 (Kubina et al. 2013). The TAC MultiLing 2011 workshop included multi-document multilingual summarization task. Most participating systems adapted existing systems to seven languages: Arabic, English, Czech, French, Greek, Hebrew and Hindi. Solutions range from the use of topic signatures (Conroy et al. 2011), Latent Dirichlet allocation (LDA) topic model (Das and Srihari, 2011) to a multilingual version of MMR (Carbonell and Goldstein, 1998; Hmida and Favre, 2011) to score sentences. Language dependent processing was limited to sentence boundaries detection and basic tokenization.

The ACL MultiLing 2013 workshop added 3 new languages to the 7 other languages: Chinese, Romanian and Spanish. Compared to the first workshop, participating systems proposed more elaborate solutions. Among best participating systems, (Conroy et al. 2013) use different regular expressions to tokenize three language classes: English, non English and Ideographic. Next, three dimensionality reduction techniques are used: LSA, LDA and IBNMF. Four candidate summaries are generated for each text. Among them, the best summary i.e. the one that maximizes coverage and minimizes sentence overlapping is kept. (Li et al. 2013) use hierarchical LDA (hLDA) to define the abstractive level for each sentence. In order to split input text into sentences: the authors used a punctuation based splitter for Chinese, an SVM based splitter for English and French and a Naive Bayes model for the rest of languages. Naturally, sentence splitting and tokenization are required for every extractive summarization system. These two basic tasks are in fact highly language dependent.

In the ACL MultiLang 2013 workshop second task, the corpus consists of 1200 Wikipedia articles: 40 languages with 30 articles per language. Among the four participating systems, only one system was applied to all languages: Conroy et al. 2013). Naturally, only automatic evaluation using ROUGE (Lin, 2004) and MeMoG (Giannakopoulos and Karkaletsis, 2011) metrics was conducted. Moreover, serious questions arose about whether designed systems were over-adapted to summarize news articles.

Until now, introducing high-level syntactic/semantic information to the summarization process often relays on supervised learning and human-made resources (parsers, wordnet, etc.). In a multilingual context, such solutions cannot be applied. Distributed word representations could make up such a constraint in a way that they carry latent syntactic/semantic information. Here, to each word of the vocabulary corresponds a real valued vector of N dimensions so that related words have similar vectors. Extracting relevant word representations is conditioned by having large enough training data (hundreds of millions to billions words raw text datasets). Typically neural networks are used to generate such representations.

Experiments show that resulting vectors capture interesting syntactic and semantic relationships between words (Mikolov et al. 2013). For example, if we use Euclidian distance between word vectors from multilingual Polyglot vectors (Al-Rfou et al. 2013), the five most related words to "king" are mentioned in Table 1.

Table 1. Example of related words extraction using multilingual Polyglot's distributed representations

Language	English	French	Arabic
	King	roi	ملك
1	queen	prince=prince	أمير=prince
2	prince	royaume=kingdom	حاكم=governor
3	princess	Roi=king	زعيم=leader
4	emperor	tsar=tsar	قائد=commander
5	ruler	souverain=sovereign	عرش=throne

Such proprieties are captured in a completely unsupervised way which makes them cheap and easily generated for general or domain specific datasets. Many representations were released for English (Collobert and Weston, 2008; Turian et al. 2010; Huang et al. 2012; Mikolov et al. 2013; Al-Rfou et al. 2013). Unfortunately, only one release offers multilingual representations: Polyglot (Al-Rfou et al. 2013) in which word representations for 137 languages learned from the entire Wikipedia.

Until now, word representations were used as additional features to different NLP syntactic and semantic tasks: singular/plural forms of words, regional spelling and sentiment polarity (Huang et al. 2012; Chen et al. 2013; Mikolov et al. 2013), synonyms/antonyms (Chen et al. 2013), POS tagging (Collobert and Weston, 2008; Al-Rfou et al. 2013), chunking and NER (Collobert and Weston, 2008; Turian et al. 2010), SRL (Collobert and Weston, 2008), syntactic and semantic analogy (Mikolov et al. 2013; Mikolov et al. 2013), word and sentence similarity (Huang et al. 2012; Socher et al. 2011) and word sense desambiguation (Miller et al. 2012).

We believe that semantic information encoded within representations is very interesting especially in a multilingual context. By introducing word representations to a multilingual summarization system, the system remains easily adaptable to a new language.

3 Clustering Sentences Using Multilingual Word Representations

Clustering sentences is very useful to many NLP applications. For text summarization, clustering sentences helps us to detect sub-themes within documents and deal with redundant information. Until now, most of summarization systems use classical cosine similarity to compute sentence similarity. Here we propose a novel sentence similarity metric, it uses proximity between sentence words representations to find words best matches.

Hence, for a couple of sentences (S_1, S_2), we find for each word from one sentence its best match in the second sentence using vectors similarity (cosine similarity for example):

$$Match(w_i) = \arg\max_{w_j \in S_2} Sim(Rep(w_i).Rep(w_j)) \tag{1}$$

Similarly, we find best matches for the second sentence words (which are not necessary the same as the first sentence). We propose to compute sentence relatedness as follows:

$$Sim(S_1.S_2) = \frac{\sum_i Match(w_i) + \sum_j Match(w_j)}{|S_1| + |S_2|} \tag{2}$$

Example

- S_1: "Tea is the most widely consumed beverage in the world"
- S_2: "Coffee is a brewed drink with a distinct aroma and flavour"

Except of stop words, the two sentences do not share any word (Cosine similarity is zero) but they are obviously related since they describe two popular stimulating drinks. Table 2 presents steps of calculating sentence relatedness between the two sentences.

We have evaluated our sentence relatedness formula on English STS-2012 and Spanish STS-2014 datasets. Table 3 displays Pearson correlation scores on both datasets.

We observe that for all datasets, our sentence relatedness metric improve Pearson correlation scores of the classical cosine similarity by a mean of 10.86 % for English and 11.92 % for Spanish. It outperforms best STS 2012 participant for WN and SMTnews English datasets and is very close to best STS 2014 participant. Note that the best STS 2012 participant TakeLab (Saric et al. 2012) use supervised learning over a set of thirteen syntactic and semantic features: n-grams overlap, Wordnet augmented overlap, named entity overlap, ..etc. whereas best Spanish STS 2014 participant UMCC (Chavez et al. 2014) use cross-lingual alignment, Wordnet and supervised learning over STS-2012 and STS-2013 English datasets.

Table 2. Example of a sentence relatedness run

Sentence 1	[tea, widely, consumed, beverage, world]					
Sentence 2	[coffee, brewed, drink, distinct, aroma, flavour]					
Direction 1	$S_1 \rightarrow S_2$			Direction 2	$S_2 \rightarrow S_1$	
	Word	Best Match	Cosine	Word	Best Match	Cosine
	tea	coffee	0.925	coffee	tea	0.925
	widely	distinct	0.309	brewed	consumed	0.688
	consumed	brewed	0.688	drink	beverage	0.757
	beverage	drink	0.757	distinct	world	0.428
	world	drink	0.480	aroma	beverage	0.637
				flavour	beverage	0.716
Relatedness			0.665			

Table 3. STS-2012 English and STS-2014 Spanish sentence relatedness results

	English STS-2012								Spanish STS-2014			
Method	MSR		WN		Europal		SMTnews		News		Wiki	
	750		750		459		399		480		324	
	P.(%)	Cov.(%)	P.(%)	Cov.(%)	P.(%)	Cov.(%)	P.(%)	Cov.(%)	P.(%)	Cov.(%)	P.(%)	Cov.(%)
Polyglot	28.53	95.49	65.16	96.94	48.46	97.39	53.53	95.78	82.47	95.19	74.70	94.17
Huang	45.43	90.07	66.96	97.36	51.82	95.50	50.65	96.02				
Senna	42.78	90.21	68.23	97.51	46.69	96.50	52.80	95.39				
Cosine	30.42		63.29		43.72		38.11		76.71		56.61	
Improvement	15.01		04.94		08.10		15.41		05.76		18.09	
Best STS System	68.30		66.41		52.80		49.37		82.53		78.02	

4   Summarizing Using mRMR Discriminant Analysis

Our goal is to design a multilingual summarizer which includes minimal language dependant processing. Hence, we propose to use a discriminant analysis mRMR (minimum Redundancy and Maximum Relevance) (Peng et al. 2005) to score sentences according to their terms informativeness.

4.1   Step 1: Multilingual Pre-processing

In a multilingual context, language dependency should be minimal. In our method, language dependency is at the pre-treatment step. Our method requires at least proper sentence splitting and basic tokenization. The rest of the summarization steps: terms and sentence scoring are statistical and do not include any language dependant feature.

4.2   Step 2: Sentence Clustering

Grouping similar sentences into clusters is a key element in our summarization method. Here we want to identify different subtopics discussed in the source text. It is particularly useful to summarize multiple source documents in which subtopics may be quite distinct. In a multilingual context, we should compute sentence similarity in a simple but efficient way. Thus we use our representation based similarity metric to compute [Sentence X Sentence] similarity matrix.

Once the similarity matrix computed, we apply a clustering algorithm to group similar sentences in same clusters. K-Medoids and Hierarchical Clustering are among mostly used clustering algorithms.

4.3   Step 3: Scoring Terms

mRMR was originally used in bioinformatics. In our adaptation of mRMR, a terms informativeness increases with its frequency variability among clusters of similar sentences: relevance. At the same time, top terms should not be similar to each other: redundancy. This will put forward most discrminant terms and highlights sentences expressing main ideas.

Term Relevance. Given n clusters of similar sentences and a classification variable h, a terms relevance expresses how much terms mean frequency correlates with h. i.e. the more they are correlated the more the term is discriminant. To compute such a score we use mutual information 3. Such definition aims to maximize coverage of the top n terms.

$$I(X;Y) = \sum_{y \in Y} \sum_{x \in X} p(x,y) \log \frac{p(x,y)}{p(x)p(y)} \tag{3}$$

Thus we need at first, to generate the $[SentencesXTerms]$ matrix M in which each row represents a sentence and each column represents a term. Each value $M[i,j]$ corresponds to the frequency of a term j in a sentence i: $freq_{i,j}$. After sentence clustering, each sentence is attached to a class number; the matrix is then augmented by a new column: h. The use of a good clustering algorithm is critical to the success or failure of mRMR, here we use k-Medoids, of course any good clustering method could be used.

Term Redundancy. However, term relevance is not sufficient, top n terms should be dissimilar. Hence, for each term we compute a redundancy score which is defined in 4 as the mean of all mutual information values between target term and remaining terms.

$$Redundancy(T_i) = \frac{1}{|S|} \sum_{j \in S} I(T_i; T_j) \tag{4}$$

Term Final Weight. Here, we want to sort terms and select those who maximize relevance and minimize redundancy. To combine these two scores, (Peng et al. 2005) propose two possible combinations (see Table 4). Finally, we generate the mRMR vector in which to each term corresponds its mRMR score: $V_{mRMR} = (w_{t1}, w_{t2}, \ldots, w_{tn})$.

Table 4. mRMR relevance and redundancy combinations

MID : Mutual Information Difference	MIQ : Mutual Information Quotient
$MID \equiv max_{t \in T}[Relevance(t) - Redundancy(t)]$	$MIQ \equiv max_{t \in T}[Relevance(t)/Redundancy(t)]$

4.4 Step 4: Sentence Scoring and Extraction

We propose a novel extraction algorithm; it takes into account terms within already selected sentences to compute the score of the next sentence to be included in the summary.

Lets V_{mRMR} be the $mRMR$ vector resulted from Step 3 and $V_s = (s_i, w_i)$, $i \in S$ be the vector of all sentences associated to their initial scores: similarity with mRMR vector. The main idea is to decrease discriminant terms weight along selecting sentences in which they appear. We propose two decreasing speeds: rapid and slow. In the first, rapid decrease, already included terms weights are set to zero. It appears to be suitable for very short summaries and allows us to select the maximum information quickly. The second, low speed, decrease already included mRMR terms weights progressively depending on the weight of the term in the just selected sentence s_j: $weight_j$.

5   Experimentations

5.1   TAC MultiLing 2011 Dataset

TAC MultiLing 2011 dataset (Giannakopoulos et al. 2011) is a parallel multilingual corpus of 7 languages: Arabic, Czech, English, French, Greek, Hebrew and Hindi. The corpus was created by gathering an English corpus which contains 10 document sets of 10 documents with a mean of 246 sentences. The original news articles were extracted from the WikiNews website. Each document set describes one event sequence: 2005 London bombing or Indian Ocean Tsunami, etc. Texts in other languages have been translated by native speakers. For each document set, three model summaries are provided by fluent speakers (native speakers in most cases). Summaries size is between 240 and 250 words.

5.2   Experimental setup

Our main goal is to assess how our summarization method performs in a multilingual context. We generate summaries for three languages: Arabic, English and French (those we more and less master). For each language, we generate three types of peer summaries depending on extraction speed:

– **SUM1:** Baseline, n most relevant sentences with respect to summary size;
– **SUM2:** Rapid decrease extraction;
– **SUM3:** Slow Decrease extraction.

1: **Inputs:**
    $V_{mRMR} = (w_{t1}, w_{t2}, \ldots, w_{tn})$: mRMR weights vector
    $S = \{(s_i, score_i), score_i = sim(V_{s_i}, V_{mRMR}, i \in S)\}$: Initial
    sentence scores
    $Size_R$: Summary size
2: **Initialize:**
    $R = \Phi$
3: Select $s_j, score_j = Max\{score_i, i \in S\}$
4: $R = R \cup \{s_j\}$
5: Update V_{mRMR}
    6: $T = V_{mRMR} \cap T_{s_j}$
    7: **for** $t_k \in T$ **do**
        8: Update weights
        9: **if** *Slow Decrease* **then**
                $w'_k = w_k - weight_j * w_k$
        end
        10: **if** *Rapid Decrease* **then**
                $w'_k = 0$
        end
    end
11: Update S
12: $S = S - s_j$
13: **for** $s_k \in S$ **do**
    14: Update sentence scores
    15: $Score(s'_k) = |Sim(V_{s_k}, NewV_{mRMR}|$
    end
16: **if** $Size(R) < Size_R$ and $\exists w_t > 0$ **then**
    17: **Goto** 3
    end
18: **Return:** Summary R

Algorithm 1. Two speed sentence extraction algorithm

We perform two system runs with 6 and 8 sentence clusters. Our previous experiments with the same dataset showed that best results are obtained with these two configurations. We use, in addition to our multilingual summarization system, two state of the art systems Centroid (Radev et al. 2004) and TextRank (Mihalcea and Tarau, 2004) to generate peer summaries. We also report results of top three TAC Multiling 2011 peer systems.

5.3 Evaluation Metrics

We compare summaries from every run against human made summaries using both ROUGE (Lin, 2004) and MeMoG (Giannakopoulos and Karkaletsis, 2011) metrics. The ROUGE method has been used in DUC conferences. ROUGE (Recall-Oriented Understudy for Gisting Evaluation) counts N-grams matches of model and peer summary. Through DUC and TAC conferences, bi-gram recall ROUGE-2 was best correlated to human judgments compared to other ROUGE

Table 5. ROUGE-2 and MeMoG summarization results for the TAC MultiLing 2011 dataset

Method			English		French		Arabic	
			MeMoG	R-2	MeMoG	R-2	MeMoG	R-2
TextRank	Cosine		0.126	0.116	0.150	0.074	0.113	0.091
	Word representations		0.132	0.093	0.156	0.132	0.129	0.114
mRMR	6 Clusters	SUM1	0.127	0.111	0.141	0.114	0.105	**0.080**
		SUM2	**0.155**	0.089	**0.164**	0.090	**0.117**	0.065
		SUM3	0.131	**0.116**	0.142	0.118	0.101	0.066
	8 Clusters	SUM1	0.127	**0.116**	0.134	0.110	0.105	0.079
		SUM2	0.148	0.071	0.160	0.092	0.114	0.063
		SUM3	0.138	0.113	0.133	**0.119**	0.106	0.077
MultiLing 2011 Peers	CIST		0.152	0.085	0.169	0.099	0.131	0.094
	CLASSY		0.172	0.132	0.176	0.122	0.158	0.140
	JRC		0.172	0.145	0.182	0.149	0.183	0.185
	Baseline_Centroid		0.136	0.076	0.146	0.078	0.124	0.090
	Topline		0.251	0.212	0.266	0.232	0.289	0.281

variants. Note that we have adapted ROUGE script to support Arabic and French accented characters, stemming was also disabled during evaluation.

MeMoG method is based on N-gram graphs. It uses character level n-grams to construct a graph in which every vertex is an N-gram (including spaces, punctuation and so on to avoid any language dependant preprocessing). N-gram vertexes are connected with weighted edges representing number of times the two N-grams appear in the same text window of a certain size. MeMoG merge all N-gram graphs from the human summaries and then compare it to the peer summary graph.

5.4   Results and Discussion

Table 5 presents non-stop ROUGE-2 and MeMog recall scores. Results vary from a language to another. The baseline system: centroid summarizer is outperformed by TextRank and mRMR. TextRank system use sentence relatedness as the sole criteria to extract sentences. It allows us to examine the direct impact of the representations based sentence relatedness compared to the classical cosine similarity. We observe that the use of our sentence relatedness metric improves TextRank scores for all languages and representations (Except for English ROUGE-2 scores). Best improvement is recorded for French ROUGE-2 score: +05.8 %. This confirms the fact that word representations enhance the sentence relatedness assessment.

For our system, mRMR, best MeMoG results are recorded with Rapid Decrease run (SUM2) for all languages. For ROUGE-2 scores, it was actually the opposite: we got the best results with Best n (SUM1) for English and Arabic. Slow decrease run (SUM3) got best results for one English and on French run. MeMoG is less restrictive than ROUGE-2 (multi level character grams against exact word bi-gram matching). We believe that Rapid Decrease strategy selects relatively diversified information compared to information reported in model summaries which explains ROUGE-2 worse results. Compared to Multiling 2011 peer systems results, our system's ROUGE-2 and MeMoG results is just behind top three systems for English and French. Surprisingly, Arabic ROUGE-2 results are week. We believe that the lower coverage and quality of Arabic word representations leaded to this performance drop. Arabic is a highly inflectional language, much more examples of raw text are needed to efficiently train representations. However, Polyglot's representations were generated from all Wikipedia articles, English Wikipedia dataset was 36 times bigger than Arabic dataset. We look forward to use other multilingual representations, once available, to validate/invalidate such conclusions.

6 Conclusion

In this paper, we have presented a novel sentence relatedness metric which makes use of recently released word representations. Evaluation results of English and Spanish sentence relatedness datasets are competitive to recent state of the art systems.

It is currently well established that word representations carry interesting latent syntactic and semantic information. Word representations are induced in a completely unsupervised way if we dispose of large enough raw text data. This makes them particularly attractive especially in a multilingual context. Including such information to a multilingual summarization system is relatively inexpensive, it does not penalize our system which remains easily adaptable to other languages.

We have also proposed a two step summarization method. First, sentences are clustered using our enhanced sentence relatedness metric. Second, the system scores terms and then sentences using mRMR analysis. We also proposed a two speed extraction algorithm, it gives a short or a longer *lifetime* to each discriminant term during the extraction process. We have experimented our system with three languages and three extraction methods using two evaluation metrics. Our experiments revealed that including representations to a state of the art system improves its performance. Automatic evaluation shows that our summarization system is comparable to recent state of the art systems for English and French. However, Arabic week results need further investigations. Is it due to the word representations quality or are there other reasons? This begs an interesting question: why same summarization method leads to different evaluation scores for each language? Further experimentations are needed at this level.

Automatic evaluation results using ROUGE and MeMoG leaded to uncorrelated results. Best ROUGE results were obtained with different configurations than best MeMoG results. Moreover, (Giannakopoulos, 2013) reports that during the ACL MultiLing evaluation, MeMoG metric was better correlated to human evaluation for all languages than ROUGE-2 except for Arabic. Arabic ROUGE-2 and MeMoG results did not correlate with human evaluation at all.

This leaves us perplex, a manual evaluation will help us to better assess our system's performance.

References

Al-Rfou, R., Perozzi, B., Skiena, S.: Polyglot: distributed word representations for multilingual NLP. In: Proceedings of the Seventeenth Conference on Computational Natural Language Learning, pp. 183–192 (2013)

Barzilay, R., McKeown, K.R.: Sentence fusion for multidocument news summarization. J. Computat. Linguist. **31**(3), 297–328 (2005)

Carbonell, J., Goldstein, J.: The use of MMR, diversity-based reranking for reordering documents and producing summaries. IBM J. Res. Dev. **2**, 159–165 (1958)

Chavez, A., Davila, H., Gutierrez, Y., Fernandez-Orquin, A., Montoyo, A., Munoz, R.: UMCC_DLSI_SemSim: Multilingual system for measuring semantic textual similarity. In: Proceedings of the Third Joint Conference on Lexical and Computational Semantics, SemEval 2014, pp. 716–721 (2014)

Chen, Y., Perozzi, B., Al-Rfou, R., Skiena, S.: The expressive power of word embeddings. In: Workshop on Deep Learning for Audio, Speech, and Language Processing, ICML 2013 (2013)

Collobert, R., Weston, J.: A unified architecture for natural language processing: deep neural networks with multitask learning. In: Proceedings of the 25th International Conference on Machine Learning, ICML 2008, pp. 160–167 (2008)

Conroy, J.M., Schlesinger, J.D., Kubina, J., Rankel, P.A., OLeary, D.P.: CLASSY 2011 at TAC: guided and multi-lingual summaries and evaluation metrics. In: Proceedings of the Text Analysis Conference (TAC) (2011)

Conroy, J., Davis, S.T., Kubina, J., Liu, Y.K., O'Leary, D.P., Schlesinger, J.D.: Multilingual summarization: dimensionality reduction and a step towards optimal term coverage. In: Proceedings of the MultiLing 2013 Workshop on Multilingual Multidocument Summarization, ACL 2013, pp. 55–63 (2013)

Das, P., Srihari, R.: Global and local models for multi-document summarization. In: Proceedings of the Text Analysis Conference (TAC) (2011)

Edmundson, H.P.: New methods in automatic extracting. J. ACM **16**(2), 264–285 (1969)

Giannakopoulos, G., El-Haj, M., Favre, B., Litvak, M., Steinberger, J., Varma, V.: TAC 2011 multiling pilot overview. In: Proceedings of the Text Analysis Conference (TAC) (2011)

Giannakopoulos, G., Karkaletsis, V.: AutoSummENG and MeMoG in evaluating guided summaries. In: Proceedings of the Text Analysis Conference (TAC) (2011)

Giannakopoulos, G.: Multi-document multilingual summarization and evaluation tracks. In: ACL 2013 MultiLing Workshop. Proceedings of the MultiLing 2013 Workshop on Multilingual Multi-document Summarization, ACL 2013, pp. 20–28 (2013)

Hmida, F., Favre, B.: LIF at TAC multiling: towards a truly language independent summarizer. In: Proceedings of the Text Analysis Conference (TAC) (2011)

Huang, E.H., Socher, R., Manning, C.D., Ng, A.Y.: Improving word representations via global context and multiple word prototypes. In: Proceedings of the 50th Annual Meeting of the Association for Computational Linguistics: Long Papers - vol. 1, ACL 2012, pp. 873–882 (2012)

Kubina, J., Conroy, J., Schlesinger, J.: ACL 2013 multiling pilot overview. In: Proceedings of the MultiLing 2013 Workshop on Multilingual Multi-document Summarization, ACL 2013, pp. 29–38 (2013)

Li, L., Heng, W., Yu, J., Liu, Y., Wan, S.: CIST system report for acl multiling 2013 track 1: multilingual multi-document summarization. In: Proceedings of the Multi-Ling 2013 Workshop on Multilingual Multi-document Summarization, ACL 2013, pp. 39–44 (2013)

Lin, C.: Rouge: a package for automatic evaluation of summaries. In: Proceedings of the ACL-04 Workshop on Text Summarization Branches Out, pp. 74–81 (2004)

Luhn, H.P.: The automatic creation of literature abstracts. In: Proceedings of the 21st Annual International ACM SIGIR Conference on Research and Development in Information Retrieval, SIGIR 1998, pp. 335–336 (1998)

Marcu, D.: The rhetorical parsing of natural language texts. In: Proceedings of the Eighth Conference on European Chapter of the Association for Computational Linguistics, EACL 1997, pp. 96–103 (1997)

Mihalcea, R., Tarau, P.: TextRank: bringing order into texts. In: Proceedings of the 2004 Conference on Empirical Methods in Natural Language Processing, EMNLP 2004, pp. 404–411 (2004)

Mikolov, T., Chen, K., Corrado, G., Dean, J.: Efficient estimation of word representations in vector space. In: The Computing Research Repository (CoRR) (2013)

Mikolov, T., Yih, W., Zweig, G.: Linguistic regularities in continuous space word representations. In: Proceedings of the HLT-NAACL, pp. 746–751 (2013)

Miller, T., Biemann, C., Zesch, T., Gurevych, I.: Using distributional similarity for lexical expansion in knowledge-based word sense disambiguation. In: Proceedings of COLING, pp. 1781–1796 (2012)

Peng, H., Long, F., Ding, C.: Feature selection based on mutual information criteria of max-dependency, max-relevance, and min-redundancy. IEEE Trans. Pattern Anal. Mach. Intell. J. **27**, 1226–1238 (2005)

Radev, D.R., Jing, H., Stys, M., Tam, D.: Centroid-based summarization of multiple documents. J. Inf. Process. Manage. **40**(6), 919–938 (2004)

Saric, F., Glavas, G., Karan, M., Snajder, J., Basic, B.D.: Takelab: systems for measuring semantic text similarity. Proc. First Jt. Conf. Lexical Comput. Semant. SemEval 2012 **1**, 441–448 (2012)

Socher, R., Huang, E.H., Pennington, J., Ng, A.Y., Manning, C.D.: Dynamic pooling and unfolding recursive autoencoders for paraphrase detection. In: Proceedings of the Advances in Neural Information Processing Systems 24: 25th Annual Conference on Neural Information Processing Systems, pp. 801–809 (2011)

Turian, J., Ratinov, L., Bengio, Y.: Word representations: a simple and general method for semi-supervised learning. In: Proceedings of the 48th Annual Meeting of the Association for Computational Linguistics, ACL 2010, pp. 384–394 (2010)

Combining Pattern-Based and Distributional Similarity for Graph-Based Noun Categorization

Michael Wiegand[1]([⊠]), Benjamin Roth[2], and Dietrich Klakow[1]

[1] Spoken Language Systems, Saarland University, Saarbrücken 66123, Germany
michael.wiegand@lsv.uni-saarland.de
[2] School of Computer Science, University of Massachusetts, Amherst, MA, USA

Abstract. We examine the combination of pattern-based and distributional similarity for the induction of semantic categories. Pattern-based methods are precise and sparse while distributional methods have a higher recall. Given these particular properties we use the prediction of distributional methods as a back-off to pattern-based similarity. Since our pattern-based approach is embedded into a semi-supervised graph clustering algorithm, we also examine how distributional information is best added to that classifier. Our experiments are carried out on 5 different food categorization tasks.

1 Introduction

Automatically inducing semantic categories of nouns from large unlabeled corpora is a pressing problem in natural language processing. Semantic categories are not only needed in order to build lexical ontologies, but they are also vital for relation extraction tasks in order to provide some means of generalization over traditional word-level representations.

With regard to type induction, there are two competing paradigms: *Pattern-based methods* mostly employ few hand-written surface patterns and ensure a high precision while *distributional methods* usually yield a better recall but may be considerably inferior with regard to precision.

In this paper, we examine ways to combine these methods for categorization. We apply them to 5 different tasks in the food domain (3 of which have not been addressed before) providing evidence that a combination works in general. We examine the food domain, since this domain has already been considered for natural language processing tasks [2–5,12]. Moreover, food categories have been shown to substantially improve relation extraction in this domain [23].

2 Data Set and Corpus

Since our task is to induce food categories, we need a food vocabulary as input. We use a proper subset of the food vocabulary employed in [23] where compounds have been removed.[1] It comprises 834 food items. We consider food compounds

---

[1] We remove all food items that contain as a suffix another food item that is also contained in our food vocabulary.

© Springer International Publishing Switzerland 2015
C. Biemann et al. (Eds.): NLDB 2015, LNCS 9103, pp. 64–72, 2015.
DOI: 10.1007/978-3-319-19581-0_5

Table 1. The categorization tasks (*each category is followed by an example and its proportion in the food vocabulary*).

Task	Description	Categories
type	common food categories (inspired by the *Food Guide Pyramid*)	meat/fish (*pork*) 23.9, beverages (*coffee*) 13.9, spices/sauces (*cinnamon*) 12.6, sweets/pastries/snacks (*chocolate*) 12.4, vegetables/salads (*broccoli*) 9.8, starch-based side dishes (*rice*) 7.9, grains/nuts/seeds (*spelt*) 5.9, fruits (*banana*) 5.2, milk products (*cheese*) 4.2, fat (*margarine*) 2.8, eggs (*omelette*) 1.6
dish	compositionality of food items	atom (*apple*) 78.3, dish (*lasagna*) 21.7
taste	predominant taste	umami/salty (*pizza*) 56.7, sweet (*orange*) 25.8, bitter (*beer*) 6.0, sour (*vinegar*) 4.0
temperature	temperature at consumption	cold (*sandwich*) 52.2, warm (*steak*) 41.7
state of matter	state of matter at consumption	solid (*bread*) 76.5, liquid (*remoulade*) 22.5

(e.g. *chocolate-almond cake*) less relevant for our investigation, since one can effectively infer (most) category labels from suffixes/heads as shown in previous work [23].[2] We want to focus on the (sparse) food items that cannot be processed with the help of this linguistic heuristic. This is a more general setting that is also relevant to other domains.

We consider the 5 different categorization tasks summarized in Table 1 addressing different properties of food items. Our food vocabulary has been annotated w.r.t. all of these categories. The first two categorization tasks have already been addressed in previous work [23], however, the remaining three tasks are examined for the first time. In each categorization task, the categories are disjoint.

Our experiments are carried out on German data. Examples are given as English translations. As an unlabeled (domain-specific) corpus from which to induce food categories, we used a crawl of *chefkoch.de* [22] consisting of 418,558 web pages of forum entries.

3   Similarity Types and Categorization

All approaches start with labeled seeds whose category labels are expanded to the remaining unlabeled items with the help of some similarity type.

3.1   Pattern-Based Similarity

For pattern-based similarity, we use the *domain-independent* similarity-patterns from [23]. Each pattern is a lexical sequence that connects the mention of two food items (Table 2). For categorization, the patterns are used to build a similarity graph, where the nodes are the food items and the edges indicate the

---

[2] That is, in order to establish the label of the sparse compound *chocolate-almond cake*, one just considers the label of the suffix/head *cake*. The latter is a more general expression for which a label can be more reliably determined.

Table 2. *Domain-independent* similarity patterns.

Patterns	food_item$_1$ (*or*|*or rather*|*instead of*| "(") food_item$_2$
Example	{*apple: pineapple, pear, fruit, strawberry, kiwi*}
	{*steak: schnitzel, sausage, roast, meat loaf, cutlet*}

Table 3. The 6 most similar food items for two different target food items (<u>underlined</u> items are unintuitive).

pattern-based similarity for		distributional similarity for	
asparagus	*kirsch (brandy)*	*asparagus*	*kirsch (brandy)*
(frequent term)	(rare term)	(frequent term)	(rare term)
vegetable	no matching	*salsify*	*cognac*
mushroom		<u>*salmon*</u>	*calvados*
champignon		<u>*chicken*</u>	*grappa*
salsify		*pasta*	*amaretto*
salad		*savoy*	*liquor*
<u>*fish*</u>		*matjes*	*rum*

occurrences of food items with a similarity pattern (the edge weight is the frequency of the occurrences with these patterns). Then, a semi-supervised graph clustering algorithm (as previously suggested [23]) is applied onto the graph. This requires a set of manually defined seeds for each category to be recognized. The method is a low-resource approach that only requires an unlabeled corpus and a set of seeds.

For all categorization tasks, we always employ the same similarity graph and the same graph clustering method. The only difference is the choice of seeds which represent instances of the respective categories that are to be induced.

3.2   Distributional Similarity

In order to compute distributional similarity, each food item is represented as a feature vector. The components are words that co-occur in a fixed window of 5 words (weighted by *tf-idf*) with mentions of the target food item to be represented. This vector-encoding allows all food items to be compared with each other, using the cosine-similarity. The resulting pair-wise similarities are stored in a similarity matrix (Fig. 1(b)). For classification, a nearest neighbour classifier (using labeled seed food items identical to the ones from Sect. 3.1) is suitable. Such classifier has been found more effective for distributional similarity than graph-based clustering [23].

Unlike in [23], we consider k nearest neighbours rather than just the nearest neighbour. We also extend the vector representation by adding *Brown* clusters [1] of the component words to the vector representation. Brown clusters represent

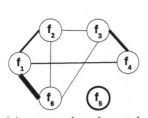 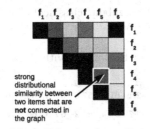 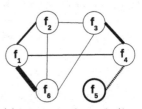

(a) pattern-based graph (*line width of edge indicates similarity*)

(b) distributional similarity matrix (*darkness of cell indicates similarity*)

(c) augmented graph (*line width of edge indicates similarity strength*)

Fig. 1. Combination of pattern-based and distributional similarity (f_i *represents some food item*).

word clusters that are automatically induced. They have been shown to improve named-entity recognition [20] and relation extraction [15].

3.3 Comparing the Two Similarity Types

Pattern-based and distributional methods have complementary properties. This is illustrated by Table 3 which shows the 6 most similar food items to *asparagus* and *kirsch* according to each of the similarity types. *Asparagus* is a frequent food item (31,355 mentions in our corpus) while *kirsch* is rare (34 mentions). As a consequence, none of the similarity patterns are observed with the rare item, hence *kirsch* is an unconnected node in the graph. For unconnected nodes, graph-based clustering is unable to make a prediction. This concerns 15.8 % of the food items in our vocabulary. With distributional similarity, however, we obtain similar food items for *all* food items. But Table 3 also illustrates that the *quality* (precision) of pattern-based similarity is superior to distributional similarity. This is because the similarity patterns are based on *coordination* which is known to ensure semantic coherence [25]. We, therefore, assume that distributional similarity is only helpful when pattern-based similarity provides no prediction.

3.4 Combination Methods

We examine 3 methods to combine distributional and pattern-based similarity. They all use distributional similarity as a back-off to pattern-based similarity. This should primarily mitigate the sparsity in the pattern-based graph caused by food items that are not connected to any other food item (f_5 in Fig. 1(a)). For those food items, some similarity information is obtained by distributional similarity ($edge(f_4, f_5)$ in Fig. 1(b)) and can, for example, be included in the similarity graph (Fig. 1(c)):

– **cascade:** We run graph clustering (on the original pattern-based similarity graph) and the nearest neighbour classifier (using distributional similarity)

Table 4. Varying k in nearest neighbour classification and examining the impact of *Brown* cluster features (*results averaged over tasks*).

	without Brown				with Brown			
k	1	3	5	10	1	3	5	10
Acc	64.9	62.1	61.5	57.8	**67.5**	64.4	64.4	61.3
F	62.3	59.7	58.4	54.2	**64.5**	60.9	60.3	56.9

Table 5. Comparison of combining pattern-based and distributional similarity.

	cascade		$graph-aug_{local}$		$graph-aug_{global}$	
Task	Acc	F	Acc	F	Acc	F
type	78.66	76.96	78.78	76.22	**80.46**	**78.42**
dish	71.34	66.06	**76.74**	69.89	**76.74**	**70.96**
taste	71.47	60.95	73.15	62.73	**74.32**	**63.64**
temperature	77.14	77.07	**78.03**	**78.17**	76.88	76.80
state of matter	81.72	78.32	84.02	80.61	**84.62**	**81.19**
average	76.07	71.87	78.14	73.52	**78.60**	**74.20**

Table 6. Varying the number of edges to be added in $graph-aug_{global}$ (*results averaged over tasks*).

edges	1	2	3	5	10	20
Acc	78.09	**78.60**	78.10	77.74	77.58	75.37
F	74.15	**74.20**	73.81	73.28	73.52	71.85

in parallel; per default the prediction of graph clustering is taken, only if no prediction could be produced by that method; the prediction of the nearest neighbour classifier is used.

- **graph − aug**$_{local}$: Information from the distributional similarity matrix is directly included in the (pattern-based) graph; for each <u>unconnected</u> food item, edges to the n most similar food items according to the distributional similarity matrix are added.
- **graph − aug**$_{global}$: Similar to $graph-aug_{local}$ but for <u>every</u> food item in the food vocabulary, the n most distributionally similar food items are connected by additional edges.

The first method is a naive combination that also keeps pattern-based and distributional similarity separated from each other during training, while the other two methods are integrated solutions. The purpose of the third method is to check whether even beyond food items in the graph that are not connected, additional back-off edges from distributional similarity may help. For both integrated solutions, we employ the distributional similarity score ds as an edge

Table 7. Comparison of different methods.

Task	majority classifier				nearest neighbour (distributional similarity)				graph (pattern - based similarity)				graph − aug_global (combination)			
	Acc	Prec	Rec	F	Acc	Prec	Rec	F	Acc	Prec	Rec	F	Acc	Prec	Rec	F
type	23.9	2.2	9.1	3.5	63.4	64.0	72.2	65.4	74.7	**81.7**	79.9	**79.3**	**80.5**	75.4	**84.3**	78.4
dish	**78.3**	39.2	50.0	43.9	64.2	60.5	65.1	59.1	63.2	68.4	63.9	63.8	76.7	**69.6**	**75.8**	**71.0**
taste	61.4	15.3	25.0	19.0	57.1	49.5	66.8	49.7	64.2	**62.0**	69.9	61.4	**74.3**	59.7	**76.8**	**63.6**
temperature	55.6	27.8	50.0	35.7	75.0	75.0	74.0	74.2	67.0	**79.6**	67.4	72.7	**76.9**	76.9	**77.2**	**76.8**
state of mat.	77.2	38.6	50.0	43.6	78.0	72.8	80.7	74.0	72.6	**81.0**	75.9	76.6	**84.6**	79.0	**87.2**	**81.2**
average	59.3	24.6	36.8	29.1	67.5	64.4	71.7	64.5	68.4	**74.6**	71.4	70.8	**78.6**	72.1	**80.3**	**74.2**

weight in the graph. ds is always in the range $[0; 1[$. It is therefore always smaller than the pattern-based similarity score of observed patterns ps (which denotes the absolute frequency of pattern occurrences), i.e. $ps > ds$ since $ps >= 1$. This encoding should reflect that we consider distributional similarity as a back-off.

4 Experiments

As seeds we randomly sampled for every category of every task (Table 1) 20 seeds. For graph-based clustering, we use the configuration of hyper-parameters from previous work [23]. We induced 1000 Brown clusters from our domain-specific corpus with SRILM [19].

Table 4 shows different configurations for nearest neighbour classification using distributional similarity. Increasing the number of nearest neighbours notably decreases performance. However, using Brown clusters as features is beneficial. Therefore, for all further experiments using a k nearest neighbour classifier, we will always set $k = 1$, however, we include Brown clusters as context features.

Table 5 compares the different methods combining pattern-based and distributional similarity. On average, the naive combination method (i.e. *cascade*) performs worst. The best overall result is obtained by the integrated solution with the global edge extension (i.e. $graph - aug_{global}$).

For the integrated methods in Table 5, we always used the 2 most similar items from the distributional similarity matrix. Table 6 shows that for this value we obtained maximum performance.

Table 7 compares the best combination method against the original graph clustering, nearest neighbour and majority-class classifier. For most tasks, the combination outperforms the best individual classifier (*nearest neighbour/graph*).

The improvement in F-score by combining pattern-based and distributional similarity is most notably caused by raising recall. The combined approach largely outperforms the majority-class classifier w.r.t. F-score. (In terms of accuracy, there is only one task, i.e. *dish*, in which that baseline is not beaten.) The proposed method also produces reasonable results on the new categorization tasks not previously examined (i.e. *taste, temperature* and *state of matter*).

5   Related Work

The types of categorizations we present in this paper are typical instances of noun classification. For that task, both distributional methods [7,10,11,16,18,21,24] and pattern-based methods [6,8,9,14] have been explored. The complementarity of those methods has only been examined for textual entailment [13] and categorization of *raw semantic classes* [17]. While our paper is the first work that combines these methods in the context of graph-based clustering, those previous publications consider different classification methods, i.e. supervised learning and query set expansion, that require a different combination.

This work also extends the types of categorizations applied on the food domain addressing *taste*, *state of matter* and *temperature* for the first time.

6   Conclusion

We presented a combined approach for the induction of noun categories using pattern-based and distributional similarity. We considered various food categorization tasks, including three novel tasks. The best combination is a clustering approach on a pattern-based graph that also includes for each food item edges to the two most similar food items according to distributional similarity. This method outperforms both mere pattern-based and distributional methods.

Acknowledgements. This work was supported, in part, by the German Federal Ministry of Education and Research (BMBF) under grant no. 01IC12SO1X and the Information Extraction and Synthesis Lab at the University of Massachusetts. The authors would like to thank Stephanie Köser for annotating the dataset presented in this paper.

References

1. Brown, P.F., deSouza, P.V., Mercer, R.L., Pietra, V.J.D., Lai, J.C.: Class-based n-gram models of natural language. Comput. Linguist. **18**(4), 467–479 (1992)
2. Chahuneau, V., Gimpel, K., Routledge, B.R., Scherlis, L., Smith, N.A.: Word salad: relating food prices and descriptions. In: Proceedings of the Joint Conference on Empirical Methods in Natural Language Processing and Computational Natural Language Learning (EMNLP/CoNLL), Jeju Island, Korea, pp. 1357–1367 (2012)
3. Druck, G., Pang, B.: Spice it up? mining refinements to online instructions from user generated content. In: Proceedings of the Annual Meeting of the Association for Computational Linguistics (ACL), Jeju, Republic of Korea, pp. 545–553 (2012)
4. van Hage, W.R., Katrenko, S., Schreiber, G.: A method to combine linguistic ontology-mapping techniques. In: Gil, Y., Motta, E., Benjamins, V.R., Musen, M.A. (eds.) ISWC 2005. LNCS, vol. 3729, pp. 732–744. Springer, Heidelberg (2005)
5. van Hage, W.R., Kolb, H., Schreiber, G.: A method for learning part-whole relations. In: Cruz, I., Decker, S., Allemang, D., Preist, C., Schwabe, D., Mika, P., Uschold, M., Aroyo, L.M. (eds.) ISWC 2006. LNCS, vol. 4273, pp. 723–735. Springer, Heidelberg (2006)

6. Hearst, M.A.: Automatic acquisition of hyponyms from large text corpora. In: Proceedings of the International Conference on Computational Linguistics (COLING), Nantes, France, pp. 539–545 (1992)
7. Huang, R., Riloff, E.: Inducing domain-specific semantic class taggers from (almost) nothing. In: Proceedings of the Annual Meeting of the Association for Computational Linguistics (ACL), Uppsala, Sweden, pp. 275–285 (2010)
8. Kozareva, Z., Hovy, E.: Semi-supervised method to learn and construct taxonomies using the web. In: Proceedings of the Conference on Empirical Methods in Natural Language Processing (EMNLP), Cambridge, MA, USA, pp. 1110–1118 (2010)
9. Kozareva, Z., Riloff, E., Hovy, E.: Semantic class learning from the web with hyponym pattern linkage graphs. In: Proceedings of the Annual Meeting of the Association for Computational Linguistics (ACL), Columbus, OH, USA, pp. 1048–1056 (2008)
10. Lenci, A., Benotto, G.: Identifying hypernyms in distributional semantic spaces. In: Proceedings of the Joint Conference on Lexical and Computational Semantics (*SEM), Montréal, Quebec, Canada, pp. 75–79 (2012)
11. Lin, D.: Automatic retrieval and clustering of similar words. In: Proceedings of the Annual Meeting of the Association for Computational Linguistics and International Conference on Computational Linguistics (ACL/COLING), Montreal, Quebec, Canada, pp. 768–774 (1998)
12. Miao, Q., Zhang, S., Zhang, B., Meng, Y., Yu, H.: Extracting and visualizing semantic relationships from chinese biomedical text. In: Proceedings of the Pacific Asia Conference on Language, Information and Compuation (PACLIC), Bali, Indonesia, pp. 99–107 (2012)
13. Mirkin, S., Dagan, I., Geffet, M.: Integrating pattern-based and distributional similarity methods for lexical entailment acquisition. In: Proceedings of the International Conference on Computational Linguistics and Annual Meeting of the Association for Computational Linguistics (COLING/ACL), Sydney, Australia, pp. 579–586 (2006)
14. Pantel, P., Ravichandran, D., Hovy, E.: Towards terascale knowledge acquisition. In: Proceedings of the International Conference on Computational Linguistics (COLING), Geneva, Switzerland, pp. 771–777 (2004)
15. Plank, B., Moschitti, A.: Embedding semantic similarity in tree kernels for domain adapation of relation extraction. In: Proceedings of the Annual Meeting of the Association for Computational Linguistics (ACL), Sofia, Bulgaria, pp. 1498–1507 (2013)
16. Riloff, E., Shepherd, J.: A corpus-based approach for building semantic lexicons. In: Proceedings of the Conference on Empirical Methods in Natural Language Processing (EMNLP), Providence, RI, USA, pp. 117–124 (1997)
17. Shi, S., Zhang, H., Yuan, X., Wen, J.R.: Corpus-based semantic class mining: distributional vs. pattern-based approaches. In: Proceedings of the International Conference on Computational Linguistics (COLING), Beijing, China, pp. 993–1001 (2010)
18. Snow, R., Jurafsky, D., Ng, A.Y.: Learning syntactic patterns for automatic hypernym discovery. In: Advances in Neural Information Processing Systems (NIPS), Vancouver, British Columbia, Canada (2004)
19. Stolcke, A.: SRILM - an extensible language modeling toolkit. In: Proceedings of the International Conference on Spoken Language Processing (ICSLP), Denver, CO, USA, pp. 901–904 (2002)

20. Turian, J., Ratinov, L., Bengio, Y.: Word representations: a simple and general method for semi-supervised learning. In: Proceedings of the Annual Meeting of the Association for Computational Linguistics (ACL), Uppsala, Sweden, pp. 384–394 (2010)
21. Weeds, J., Weir, D., McCarthy, D.: Characterising measures of lexical distributional similarity. In: Proceedings of the International Conference on Computational Linguistics (COLING), Geneva, Switzerland, pp. 1015–1021 (2004)
22. Wiegand, M., Roth, B., Klakow, D.: Web-based relation extraction for the food domain. In: Bouma, G., Ittoo, A., Métais, E., Wortmann, H. (eds.) NLDB 2012. LNCS, vol. 7337, pp. 222–227. Springer, Heidelberg (2012)
23. Wiegand, M., Roth, B., Klakow, D.: Automatic food categorization from large unlabeled corpora and its impact on relation extraction. In: Proceedings of the Conference on European Chapter of the Association for Computational Linguistics (EACL), Gothenburg, Sweden, pp. 673–682 (2014)
24. Yamada, I., Torisawa, K., Kazama, J., Kuroda, K., Murata, M., Saeger, S.D., Bond, F., Sumida, A.: Hypernym discovery based on distributional similarity and hierarchical structures. In: Proceedings of the Conference on Empirical Methods in Natural Language Processing (EMNLP), Singapore, pp. 929–927 (2009)
25. Ziering, P., van der Plas, L., Schuetze, H.: Bootstrapping semantic lexicons for technical domains. In: Proceedings of the International Joint Conference on Natural Language Processing (IJCNLP), Nagoya, Japan, pp. 1321–1329 (2013)

Acquiring a Large Scale Polarity Lexicon Through Unsupervised Distributional Methods

Giuseppe Castellucci[1]([⊠]), Danilo Croce[2], and Roberto Basili[2]

[1] Department of Electronic Engineering, University of Roma Tor Vergata,
Via Del Politecnico 1, 00133 Roma, Italy
castellucci@ing.uniroma2.it
[2] Department of Enterprise Engineering, University of Roma Tor Vergata,
Via Del Politecnico 1, 00133 Roma, Italy
{croce,basili}@info.uniroma2.it

Abstract. The recent interests in Sentiment Analysis systems brought the attention on the definition of effective methods to detect opinions and sentiments in texts with a good accuracy. Many approaches that can be found in literature are based on hand-coded resources that model the *prior* polarity of words or multi-word expressions. The construction of such resources is in general expensive and coverage issues arise with respect to the multiplicity of linguistic phenomena of sentiment expressions. This paper presents an automatic method for deriving a large-scale polarity lexicon based on Distributional Models of lexical semantics. Given a set of sentences annotated with polarity, we transfer the sentiment information from sentences to words. The set of annotated examples is derived from Twitter and the polarity assignment to sentences is derived by simple heuristics. The approach is mostly unsupervised, and the experimental evaluation carried out on two Sentiment Analysis tasks shows the benefits of the generated resource.

Keywords: Polarity lexicon generation · Distributional semantics

1 Introduction

Opinion Mining [17] aims at tracking the opinions expressed in texts with respect to specific topics, e.g. products or people. Sentiment Analysis (SA) deals with the problem of deciding whether a piece of text, e.g. a sentence or a phrase, is expressing some particular sentiment, e.g. positivity or negativity. Recently, SA has been at the very center of many researches, as demonstrated by the growing number of works and evaluation campaigns in the Natural Language Processing (NLP) area (see [16,21] or [18]).

Social media data are often called at measuring the sentiment expressed in the Web with respect to specific topics. For example, Twitter[1] messages are often used by companies or politicians to measure their approval in the Web. The interest in this analysis brought to the definition of highly participated

---

[1] http://www.twitter.com.

© Springer International Publishing Switzerland 2015
C. Biemann et al. (Eds.): NLDB 2015, LNCS 9103, pp. 73–86, 2015.
DOI: 10.1007/978-3-319-19581-0_6

challenges in the research community, as the recent SemEval tasks of Sentiment Analysis (SA) ([16,21]). Many systems that were proposed in literature analyze short messages by adopting supervised Machine Learning (ML) techniques in conjunction with specific data representations (see for example [3] or [10])). In these works, data are usually modeled by adopting hand-coded resources that define the prior polarity of words or terms, e.g. *good* can be associated to a `positive` attitude in contrast to *sad* that can be defined as `negative`. Compiling by hand these resources, as [25] or [9], can be very expensive, while the resulting coverage of linguistic phenomena can be limited. These resources exist for the English language although they are poorer for other languages. Moreover, sentiment expressions are often topic dependent, e.g. occurrences of the word *mouse* are mostly neutral in the consumer electronics domain, while this phenomenon can be highly biased towards negativity in others (e.g. restaurant recommending systems). Accounting for topic specific phenomena in hand-coded lexicon would require manual revisions.

In this paper, we propose an automatic and efficient methodology to derive large-scale polarity lexicons without human supervision, but mainly deriving sentiment information by observing the usage of words in the Web. In particular, we aim at observing social media data, which are often characterized by extra-linguistic information that can be adopted to classify a text as belonging to a specific sentiment class. For example, the presence of a *happy* emoticon, e.g. *:)*, provides evidence for a positive attitude in a message. The approach is based on Distributional Models of lexical semantics that are broadly used in NLP to derive shallow semantic relations between words through the distributional analysis of large-scale corpora. Our proposal is to exploit the equivalence in the representation that some distributional models allow for words and sentences (e.g. the dual LSA space for words and texts introduced in [12]). As sentences can be related to a given polarity, a classifier can be trained in such spaces to transfer the sentiment information from sentences to words. Specifically, we train polarity classifiers by observing sentences and we generate a polarity lexicon by classifying words. Annotated messages are derived from Twitter and their polarity assignment is determined by simple heuristics. It means that words in specific domains can be related to sentiment classes by exploiting their occurrences in annotated sentences, i.e. by observing the usage of words in texts. The resulting methodology is highly applicable as the distributional model can be acquired without any supervision and the provided heuristics do not have any specific bias with respect to languages or domains.

In our evaluations, we generated a large-scale polarity lexicon that is made available to the research community[2]. The resource is measured against two diversified sentiment analysis tasks, i.e. Twitter Sentiment Analysis and Aspect Based Sentiment Analysis. Experiments show the contribution of the generated resource, and a comparison with a well-known subjectivity lexicon is provided.

In the rest of the paper, Sect. 2 presents the proposed methodology for lexicon generation. Section 3 describes the experimental evaluations. Finally, a brief survey of the related methods for polarity lexicon generation is provided in Sect. 4.

---

[2] http://sag.art.uniroma2.it/distributional-polarity-lexicon.

2   A Distributional Approach to the Generation of Polarity Lexicon

In a nutshell, the approach for lexicon generation is based on representing both sentences and words similarly, in order to transfer sentiment information from sentences to individual words. In this Section, we first describe how sentences and words can be represented by exploiting Distributional Models of lexical semantics (Sect. 2.1). Then, we describe a classification-based approach, to transfer the sentiment information from sentences to words (Sect. 2.2). Finally, we describe a heuristic to generate a dataset of sentences annotated with sentiment information (Sect. 2.3).

2.1   Distributional Models

In order to have a similar representation for words and sentences, Distributional Models (DM) of lexical semantics are exploited. DMs are intended to acquire semantic relationships between words, mainly by looking at the words usage. The foundation for these models is the *Distributional Hypothesis* [8], i.e. words that are used and occur in the same contexts tend to purport similar meanings. In recent years, DMs have been at the basis of many advances in NLP, and different methods have been proposed to derive them in efficient ways.

DM approaches characterize semantic relationships in terms of vector similarity. Different relationships can be modeled, e.g. *topical* similarities if vectors are built considering the occurrence of a word in documents or *paradigmatic* similarities if vectors are built considering the occurrence of a word in the context of other words [22]. In such models, words like *run* and *walk* are close in the space, while *run* and *read* are projected in different subspaces. These representations can be derived mainly in two ways: *counting* the co-occurrences between words, e.g. [12], or *predicting* word representations in a supervised setting. In particular, in [13] a simple Recursive Neural Network architecture is exploited to derive such representations. These show linguistic regularities at syntactic and semantic levels that allow to reason about analogy tasks, e.g. judging whether *king:man* $\sim$ *queen:woman*. In [13] simple algebraic operations between word vectors are proposed to model semantic relations such as *king : man* $\sim$ *X : woman*, X is found as the element closer to *king - man + woman*. Roughly speaking, these regularities are reflected in the specific subspaces, i.e. specific dimensions, of the generated vectors. While specific algorithms can be used for the space acquisition, these approaches allow to derive a projection function $\Phi(\cdot)$ of words into a geometrical space, so that the vector representation for a word $w_k \in \mathbb{W}$ is obtained as $\boldsymbol{w_k} = \Phi(w_k)$. Regularities existing in the acquired spaces will be exploited to determine the prior sentiment for words, i.e. our assumption is that polarized words lie in specific subspaces. However, in DMs opposite polarity words are often similar, mainly because they share the same contexts. In the next Section, we discuss how we can capture the differences in sentiment by representing both words and sentences in the same space, and how we can transfer known sentence polarity to individual words by exploiting the sentiment related subspaces.

2.2   A Classification Strategy for Lexicon Generation

The semantic similarity (closeness) established by traditional DMs is not equivalent to emotional similarity. A method to capture the sentiment differences between words in order to derive a representation with respect to the expressed sentiment is needed.

The approach we propose is based on the way a linear classifier derives a discriminant function between two or more classes. Let us consider a space $\mathbb{R}^d$ where some geometrical representation of a set of annotated examples can be derived. In general, a linear classifier can be thought as a separating hyperplane $\theta \in \mathbb{R}^d$ that is then used to classify a new example represented in the same space. Each θ_i corresponds to a specific dimension, or feature i, that has been extracted from the annotated examples. Roughly speaking, after a learning stage, the magnitude of each θ_i reflects the importance of the feature i with respect to a target phenomenon. In this sense, when applied to distributional vectors of word semantics, linear classifiers are expected to learn those regions useful to discriminate examples with respect to the target classes. If these classes reflect the sentiment expressed by words, a classifier trained on distributional vectors should determine those subspaces better correlating examples with the sentiment classes.

In this way, given a set of words $w_i \in \mathbb{W}$ associated with their prior polarity, it should be possible to learn a classifier on its distributed representation derived from a Distributional Models. In order to derive a polarity lexicon, we could define a set of seed words $w_k^{seed} \in \mathbb{W}$, whose elements are words whose prior polarity is known. In this way, we could project seed words through the Word Space model ($\boldsymbol{w}_k^{seed} = \Phi(w_k^{seed})$) in a vector space $\mathbb{R}^d$. Then, we could learn a linear classifier and find what dimensions of $\mathbb{R}^d$ are related to the different polarities. Transferring the information from seed words to others can be accomplished through a classification step. However, the definition of a set of seed words could be expensive and it could not be applicable to several languages. Moreover, assigning a polarity class to individual words out of any context can be difficult. In fact, the polarity of a word is clearly dependent on the sentence where the word appears in. For this reason, we propose a methodology that avoid a seed selection and that stress the idea of distributional models by deriving sentiment related subspaces by observing words usages in full sentences. We propose to select a collection of sentences labeled with the underlying sentiment. We represent both words (i.e. candidate entries for the polarity lexicon) and sentences into the same space. Being able to project sentences and semantically related words into similar subspaces, we would be able to acquire a classifier by observing sentences, i.e. determine those subspaces more related to sentiment in sentences, and transfer this information to words through the classifier training.

In details, we have words $w_k \in \mathbb{W}$ and their relative vector representation $\boldsymbol{w}_k \in \mathbb{R}^d$ obtained by projecting them in a Word Space, i.e. $\boldsymbol{w}_k = \Phi(w_k)$.

We also have a training set $\mathbb{T}$, composed by sentences associated to polarity. In order to project an entire sentence in the same space, we apply a simple but effective linear combination operator. For each sentence $t \in \mathbb{T}$, we derive the vector representation $t \in \mathbb{R}^d$ by summing all the word vectors composing the sentence, i.e. $t = \sum_{w_i \in t} \Phi(w_i)$. It is one of the simpler, but still expressive, method that is used to derive a representation that accounts for the underlying meaning of a sentence, as discussed in [12]. Having projected an entire sentence in the space, we can look at all the dimensions of the space that are related to a sentiment class. The sentence representations are fed to a linear learning algorithm that will induce a discriminant function f, which is expected to capture the sentiment related subspaces by properly weighting each dimension i of the original space. The lexicon is generated by applying f to the words in $\mathbb{W}$. As we deal with multiple sentiment classes, f can be seen as m distinct binary function $(f_1, f_2, \ldots, f_m)$ one for each sentiment class, similarly to a OneVsAll approach [20]. Each function is used to classify a word $w_k \in \mathbb{W}$, thus deriving m distinct numerical scores s_i^k, each reflecting the confidence the classifier has in deciding whether w_k belongs to class i. Each score s_i^k is normalized through a softmax-like function[3], obtaining the normalized polarity score o_i^k: each word w_k in $\mathbb{W}$ can be represented both with its distributional representation, i.e. $\boldsymbol{w_k} = \Phi(w_k)$, and with its sentiment representation, i.e. $\boldsymbol{o^k}$.

2.3   Generating a Dataset Through Emoticons

As discussed in Sect. 2.2, an annotated dataset of sentences $\mathbb{T}$ is needed to acquire a linear classifier that emphasizes specific subspaces. Although different dataset of such kind exists, our aim is to use a general methodology that can enable the use of this technique in different scenarios. In fact, if many dataset exists for English, this is not true for other languages. For this reason, we are going to use a heuristic to generate a training set by exploring Twitter messages and the emoticons that can be found in them. The method is based on a Distant Supervision approach applied on Twitter [7].

In order to derive messages that belongs to the positive and negative classes, we select Twitter messages whose last token is a smile either positive, e.g. :), :-) or :D or negative, e.g. :(or :-(. Neutral messages are selected by filtering messages that end with an url, as in many cases these are written by newspaper accounts and they use mainly non-polar words to announce an article. From this dataset, we further filter out those messages that contain one element of another class, i.e. if a message ends with a positive smile and it contains either a negative smile or a link it will be discarded. This is useful to have a more accurate dataset with respect to the involved classes. It is worth nothing that if one have at one's disposal a more fine-grained emoticons classification, it is possible to derive a dataset composed by more than 3 groups of data, i.e. deriving a lexicon with more emotions.

---

[3] $o_i^k = e^{s_i^k} / \sum_{j=1}^m e^{s_j^k}$.

Table 1. Example of polarity lexicon terms and relative sentiment scores (English language).

Term	Positivity	Negativity	Neutrality
good::j	0.74	0.11	0.15
:)	0.86	0.04	0.10
bad::j	0.12	0.80	0.08
pained::v	0.13	0.74	0.13
#apple::h	0.14	0.16	0.70
article::n	0.16	0.09	0.75

Table 2. Example of polarity lexicon terms and relative sentiment scores (Italian language).

Term	Positivity	Negativity	Neutrality
ottimo::j	0.77	0.08	0.15
:)	0.73	0.08	0.19
sofferenza::n	0.16	0.58	0.26
soffrire::v	0.08	0.65	0.27
#apple::h	0.17	0.12	0.71
articolo::n	0.19	0.05	0.76

3   Evaluating an Automatically Generated Polarity Lexicon

In this Section, the experimental setting used to acquire the polarity lexicon is discussed. Then, two Sentiment Analysis tasks are addressed to verify the benefits of this resource.

Word Vector Generation. As discussed in Sect. 2.2 distributional representation for words are needed. We generate word vectors according to a Skip-gram model [13] through the word2vec[4] tool. In particular, we derive 250 dimensional word vectors, by using a corpus of more than 20 million tweets downloaded during the last months of the 2014 year[5]. We process each tweet by applying a custom tweet version of the Chaos parser [2]. Tokenization, morphological analysis and part-of-speech (pos) tagging are applied to derive the input for the word vector generation, i.e. we use the lemma and the part-of-speech tag (e.g. *lemma::pos*) for word2vec. We obtained 188,635 words that will be classified to generate the polarity lexicon, i.e. it will be the final lexicon size.

Dataset Generation. We applied the heuristic described in Sect. 2.3 to derive a labeled dataset to train a classifier. We filtered the data used for the word vectors generation obtaining 352,620 positive tweets, 74,166 negative tweets and 5,280,738 neutral tweets. We finally generated the training set $\mathbb{T}$ by randomly selecting 7,000 tweets for each class.

Linear Classifier. Support Vector Machines (SVM) [27] is one of the most effective classifier that have been used in many different domains. In NLP, they have been used for their capability to learn both linear and non-linear (exploiting the notion of kernel function [24]) classifiers. In this paper, we learn a linear function that can separate data between the three sentiment classes of interest. We adopted the LibLinear [6] formulation of SVM that can be found in KeLP[6], a Java machine learning platform developed at the Enterprise Engineering Department of the University of Roma Tor Vergata.

---

[4] https://code.google.com/p/word2vec/.
[5] word2vec settings are: *min-count=50*, *window=5*, *iter=10* and *negative=10*.
[6] http://sag.art.uniroma2.it/demo-and-software/kelp.

Distributional Polarity Lexicon (DPL) Generation. In order to generate a polarity lexicon, we build a training set for the LibLinear algorithm with the distributional representations of sentences in $\mathbb{T}$. The feature representation is obtained by linearly combining word vectors[7] in a tweet considering only verbs, nouns, adjectives and adverbs. The combination should capture the meaning of the message, and the learning phase should emphasize specific subspaces, i.e. those related to sentiment classes.

As we are dealing with three sentiment classes (i.e. `positive`, `negative` and `neutral`) a One-Vs-All (OVA) strategy [20] is used to derive the optimal classifiers. That is, first we learn m distinct classifiers, where m is the number of involved classes, i.e. in our case $m = 3$ that are able to classify a tweet with respect to the involved classes. In the OVA strategy the classification decision for a test example is made up as the class the received the maximal classification score. A classifier tuning phase is pursued on an 80/20 split of the training data $\mathbb{T}$ and the accuracy of classifications is optimized, i.e. the percentage of correctly classified examples. The lexicon is obtained by generating a test set of examples by considering all the words represented though the distributional model. We classify each word with the three classifiers, and polarity scores are obtained with the softmax normalization, as described in Sect. 2.2.

In Table 1 examples of English words and their generated scores can be found. As it can be seen, the approach here proposed is able to recognize the polarity of words, given the classifiers that have been trained on full sentences. Qualitatively, it seems that polar words lie in specific subspaces and that these can be captured through a linear classification strategy. In order to demonstrate the generality of the approach, we generated also an Italian lexicon by applying the same methodology[8]. In Table 2 examples of Italian words and their generated scores can be found. Again, the proposed methodology seems able to capture sentiment information in sentences and transfer it to single words even for another language.

Tasks Settings. Experiments reported in the next Sections are all performed as classification tasks, in particular by exploiting the kernelized formulation of the SVM algorithm [27] that can be found in the `KeLP` framework. Kernel functions [24] have been shown to achieve state-of-the-art results in many NLP tasks (see for example [4]). Kernels provide an efficient way to represent data at an abstract level, while their computation still look at their intrinsic and informative properties. For instance, a *polynomial* kernel applied on a Bag-Of-Words (BOW) representation can look both to single or to word pairs without ever computing the explicit space of bigrams. Moreover, kernel functions can be combined, e.g. the contribution of kernels can be summed up. It allows having different kernels operating on specific representations of the data, each emphasizing some

---

[7] In order to not have a bias over the query terms, the last token is not used in the combination.

[8] The Distributional Model for Italian is acquired with `word2vec` on 2 million Italian tweet; we used exactly the same emoticons to select the messages and the same learning strategy.

characteristics. We address the tasks below by combining multiple kernels, thus verifying the contribution of each representation, and, in particular, the contribution of the automatically generated polarity lexicon.

As presented in Sect. 2, the proposed approach outputs a m-dimensional vector representation o^k for each word in the vocabulary, i.e. in our setting $m = 3$ and each dimension o_i^k respectively represents the positivity, negativity and neutrality of a word w_k. In order to use the lexicon to represent an entire sentence t inside SVM, we propose to generate a very simple feature representation by summing up all the polarity lexicon vectors o^k corresponding to the words w_k in t[9], i.e. $t = \sum_{w_k \in t} o^k$. This should be able to capture when many words agree with respect to the polarity; the dimension associated to a particular sentiment should have a higher score. Obviously, this representation has some limitations, e.g. it doesn't consider the scope of a negation.

3.1 Sentiment Analysis in Twitter

Web 2.0 technologies allow users to generate new contents on blogs, forums or social networks. In recent years, the interest in mining this information is growing as people often write opinions about facts or events. In fact, different Twitter based challenges have been proposed in the research community of computational linguistics. We want to apply the lexicon in two tasks that have been proposed in 2013 and 2014, [16,21]. In both cases, the task concerns with the assignment of a sentiment class to a tweet. For example, the tweet *"Porto amazing as the sun sets...* http://bit.ly/c28w" should be recognized as positive, while *"@knickfan82 Nooo; (they delayed the knicks game until Monday!"* should be recognized as negative.

We model the task with a multiple kernel approach by exploiting two vector representations: Bag-Of-Word (BOW) and Word Space (WS). The former aims at capturing pure lexical information, where each dimension reflects the presence or not of a particular word in a sentence. Two messages can be thus compared through the lexical overlap by adopting the dot product between their BOW vector representations. The latter tries to smooth the lexical overlap between messages by relying on a Word Space model that emphasizes paradigmatic relations among words, i.e. when a word w_i can be substituted with w_j in a sentence s without changing the overall meaning of s, as discussed in [22]. The geometrical space behind the WS representation is the same used for the generation of the lexicon, and the sum of verbs, nouns, adjectives and adverbs is considered as the final representation. In the SVM learning algorithm, a linear kernel is applied on each representation, and their sum is considered as the final kernel function. We further augment these representations with the one derived from the Distributional Polarity Lexicon (DPL), i.e. we add a linear kernel applied on the polarity lexicon representation. Again, this is obtained by considering only the verbs, nouns, adjectives and adverbs in a message. In order to verify the effectiveness of DPL, we compare it with a well-known lexicon, i.e. the Subjectivity

---

[9] We apply a normalization on the resulting vector t so it has norm 1.

Table 3. Twitter sentiment analysis 2013 results. *Best-system* refers to the top scoring system in SemEval 2013.

Kernel	Pn-F1	Pnn-F1
BOW	59.72	63.53
BOW+SUBJL	61.46	64.95
BOW+DPL	60.78	64.09
BOW+WS	66.12	68.56
BOW+WS+SUBJL	65.20	67.93
BOW+WS+DPL	66.40	68.68
Best-system	69.02	-

Table 4. Twitter sentiment analysis 2014 results. *Best-system* refers to the top scoring system in SemEval 2014.

Kernel	Pn-F1	Pnn-F1
BOW	58.74	61.38
BOW+SUBJL	60.82	62.85
BOW+DPL	62.49	64.01
BOW+WS	65.20	66.35
BOW+WS+SUBJL	64.29	66.13
BOW+WS+DPL	66.11	67.07
Best-system	70.96	-

Lexicon [28], in the following evaluations. It is composed by a set of $15,991$ manual annotated subjective expressions each associated to a polarity (positive, negative, neutral) and a strength (weak or strong) value. For each tweet we generate a new feature representation (SUBJL) where each dimension refers to a polarity value with its relative strength, as found in the message. For example, the SUBJL representation of *"Getting better!"* is a feature vector whose the only non-zero element is the feature strong_pos. Again, a linear kernel is applied and it is combined to the BOW and WS representations. We expect that DPL is able to capture more phenomena, as it should cover more expressions than SUBJL.

In Tables 3 and 4 the experimental outcomes for the 2013 and 2014 datasets are reported. Performance measures are the *Pn-F1* and the *Pnn-F1*. The former is the arithmetic mean between the F1 measures of the positive and negative classes. This was the official score used in the SemEval challenges. The latter is the arithmetic mean between the F1 measures of the positive, negative and neutral classes. In Table 3 results are shown for the 2013 test dataset, which is composed by 3814 examples. First, the achievable performance with a linear kernel applied on the simple BOW representation is shown. Further results combine other representations to the BOW one. When applying the WS representation, an improvement can be noticed. It means that distributional representations of word semantics are useful to capture semantic phenomena behind sentiment related expressions. When combining also DPL further improvements are obtained for both performance measures. It seems that DPL is effectively able to smooth the contribution of the pure lexical semantics representation (WS). It is noticeable that the *BOW+WS+DPL* system would have ranked in 2nd position in the 2013 ranking.

Same trends are observable for the 2014 test set, as shown in Table 4. However, in this case we were not able to have the complete SemEval test set due to the unavailability of the data. In fact, given Twitter restriction policies for downloading data, the 2014 task organizers were able to release only the IDs of the messages. At the time of this experimentation, some of the messages were no longer available for download. Thus, our evaluation is carried out on the 1562 test examples we were able to download, while the full test was composed by 1853. It makes a direct comparison between our approach and the ones in SemEval 2014 impossible, but it still can give an idea of the performances of the generated lexicon. Again, we report the performance measures with the BOW and WS representation combined with the representations obtained from the Subjectivity Lexicon (SUBJL) and with the automatically generated lexicon (DPL). As it can be noticed, even in this scenario the use of distributed word representation is beneficial, as demonstrated by the *BOW+WS* row of Table 4. Again, when adopting the lexicon with an improvement in both the performance scores is measured.

3.2 Aspect Based Sentiment Analysis

The second task where we want to test the Distributional Polarity Lexicon is the task of Aspect Based Sentiment Analysis (ABSA) applied over restaurant reviews. This experiment is intended to verify the applicability of the lexicon on a very different domain with respect to the one it was acquired from. The domain is different mainly with respect to the language used to convey messages (reviews vs. tweets) and also with respect the topic (restaurants vs. general topics found in tweets). An ABSA system aims at identifying the underlying discussed aspects and the sentiment expressed towards them. A typical approach to ABSA consists of a cascade of four different sub-tasks. For example, consider the following review: *"The bartender on my recent visit was so incredibly rude that I will never go back!"*. The four subtasks are: (i) Aspect Term Extraction (ATE), i.e. the identification of the aspect terms (*bartender*); (ii) Aspect Term Polarity (ATP), i.e. the assignment of a polarity class to the aspect terms (*negative*); (iii) Aspect Category Detection (ACD), i.e. identifying the aspect categories, from a predefined set of them (*service*); (iv) Aspect Category Polarity (ACP), i.e. the assignment of a polarity class (positive, negative, neutral or conflict) to each aspect category (*negative*).

Here, we focus on the Aspect Category Polarity (ACP) task by adopting the SemEval 2014 Task 4 [18] corpus consisting of customer reviews of restaurants, where the mentioned aspect categories can be *price, food, service, ambience, miscellaneous*. The corpus is composed by 3,041 reviews in training data and 800 reviews in the testing data. The task is threaded as a multi-classification task with respect to the *positive, negative, neutral* and *conflict* classes. Again, considering a review and its category, the classifier predicts a sentiment label as the one that maximizes the prediction score.

Bag-of-Word (BOW) and two different Word Space representations[10] are adopted. The first Wordspace is obtained from a corpus of Wikipedia English documents (WS_{UK}) [1]; the latter is obtained from the TripAdvisor dataset[11] (WS_{TRIP}), an in-domain corpus consisting of more than 40 million words of the restaurant and hotel domain. Again, word2vec with the same settings used for the generation of the Twitter based Word Space is exploited. All the sentences

---

[10] We tested the Twitter Word Space, but the results were unsatisfactory as of the noise of tweets.

[11] http://sifaka.cs.uiuc.edu/~wang296/Data/index.html.

for the ABSA task have been analyzed through the Stanford Parser [11] and each WS_* representation is derived considering only verbs, nouns, adjectives and adverbs. Again, we verified the contribution of the polarity lexicon by adding a new representation and by applying a linear kernel on it.

Table 5. Aspect Category Polarity (ACP) results in aspect based sentiment analysis

Kernel	Accuracy
BOW	70.04
BOW+SUBJL	71.31
BOW+DPL	72.68
BOW+WS$_{UK}$+WS$_{TRIP}$	75.21
BOW+WS$_{UK}$+WS$_{TRIP}$+SUBJL	74.53
BOW+WS$_{UK}$+WS$_{TRIP}$+DPL	76.19

In Table 5 results for the Aspect Category Polarity task, in terms of accuracy, i.e. the percentage of correctly classified instances, are reported. Again, we compare the proposed lexicon (DPL) with the subjectivity lexicon [28] (SUBJL). In the results, we report the performance measure obtained by using or not the representation that consider the automatically generated lexicon. Even in this case, when using our lexicon improvements are achieved. It is remarkable considering that the lexicon was acquired on very different data, i.e. Twitter messages, which are characterized by many misspellings and slangs that can affect the quality of the inference. The best system here would have been ranked in 4th position. It is a straightforward result considering that only very simple features are used here, and this method is fully portable.

4   Related Work

Polarity lexicon generation has been tackled in many researches and three main areas can be pointed out.

Manually Annotated Lexicons. Earlier works are based on manual annotations of terms with respect to polarity (or emotional) categories. For example, in [25] sentiment labels are manually associated to 3600 English terms. In [9] a list of positive and negative words are manually extracted from customer reviews. The MPQA Subjectivity Lexicon [28] contains words, each with its prior polarity (positive or negative) and discrete strength (strong or weak). The NRC Emotion Lexicon [15] is composed by frequent English nouns, verbs, adjectives, and adverbs annotated through Amazon Mechanical Turk with respect to eight emotions (e.g. joy, sadness, trust) and positive or negative sentiment. The main drawback of manual approaches is that they cover a limited number of sentiment related phenomena. The method here proposed is mostly unsupervised and it allows to derive sentiment information for a higher number of terms.

Lexicons Acquired Over Graphs. Graph based approaches exploit an under-lying semantic structure that can be built upon words. In [5] the WordNet [14] synset glosses are exploited to derive three scores describing the positivity, neg-ativity and neutrality of the synsets. The work in [19] generates a lexicon as graph label propagation problem. Each node in the graph represents a word. Each weighted edge encodes a relation between words derived from WordNet [14]. The graph is constructed starting from a set of manually defined seeds. The labels for the other words, i.e. the polarity, are generated by exploiting some well known graph-based methods. [23] similarly exploits the WordNet resource by constructing a graph starting from 14 manually defined seed words, and applying the Personalized Page Rank algorithm to propagate sentiment labels.

Corpus-Based Lexicons. Statistics based approaches are more general as they mainly exploit corpus processing techniques. For example, [26] proposed a minimally supervised approach to associate a polarity tendency to a word by determining if it is co-occurring more positive words than negative words. More recently, [29] proposed a semi-supervised framework for generating a domain-specific sentiment lexicon. Their system is initialized with a small set of labeled reviews, from which segments whose polarity is known are extracted. It exploits the relationships between consecutive segments to automatically generate a domain-specific sentiment lexicon. In [10] They compute a numerical score, reflecting the polarity of each word, through a point-wise mutual information based measure between a word and an emotion. This work is close to [10] where a minimally-supervised approach based on social media data is proposed. They use hashtags or emoticons that are related to emotions, e.g., #happy, #sad, :) or :(, to retrieve tweet messages. They count if a word co-occurs with differ-ent emotion-related words, and compute a Point-wise Mutual Information score reflecting the association to specific emotions: 0 means that a word is not associ-ated to an emotion while higher values mean that the word is highly associated. Differently from [10], we exploit distributional models to represent the meaning of whole messages. We induce a classifier able to induce the sentiment underlying a message by observing its all meaning. We do not account for simple and local patterns in the acquisition of the classifier. The induced classification function is thus used to relate both messages and words to emotions. Moreover, our app-roach assigns a polarity distribution to each word, thus making clear those terms having different polarity shades.

5   Conclusion

In this paper, a new and effective unsupervised methodology to generate large-scale polarity lexicon is presented. It is based on representing both sentences and words in the same space defined by a Distributional Model. Once sentence and words lie in the same space, a sentiment transfer from sentences to words is exploited by learning a linear classifier derived by observing sentences annotated with sentiment classes. The classifier is expected to find what are the sentiment related subspaces, and it is then used to classify a single word, i.e. emphasizing such dimensions. The method can be considered general, as it does not rely on any hand-coded resource, but mainly uses information that can be derived from emoticons in order to generate a labeled set of sentences. Moreover, distributional models are derived entirely in an unsupervised manner by only exploiting large-scale unlabeled corpora. It has been shown that when applied on two very diverse sentiment analysis tasks, the generated lexicon is always beneficial. However, the usage of the proposed distributional polarity lexicon does not solve traditional limitations of existing lexicons, e.g. negation effects or ironic phenomena in texts. Dealing with these problems with automatically generated lexicon is a future research direction. Moreover, as for now, we deal only with single words, while many sentiment carrying expressions are multi-word, e.g. *give up* can be considered negative. As the generation of the lexicon depends only on emoticons, which can be considered cross-lingual anchors, evaluations in different languages can be pursued. Moreover, as sentiment expressions can be topic dependent, e.g. *mouse* can be `neutral` when opinionating of the electronic domain while it can be `negative` in the restaurant domain. Thus, a more accurate selection of training sentences based not only on emoticons, but also on the topic, could lead to domain specific polarity lexicon.

References

1. Baroni, M., Bernardini, S., Ferraresi, A., Zanchetta, E.: The wacky wide web: a collection of very large linguistically processed web-crawled corpora. LRE **43**(3), 209–226 (2009)
2. Basili, R., Pazienza, M.T., Zanzotto, F.M.: Efficient parsing for information extraction. In: ECAI, pp. 135–139 (1998)
3. Castellucci, G., Filice, S., Croce, D., Basili, R.: Unitor: combining syntactic and semantic kernels for twitter sentiment analysis. In: Proceedings of SemEval, 2nd Joint Conference on Lexical and Computational Semantics (*SEM), pp. 369–374. ACL, Atlanta (2013)
4. Croce, D., Moschitti, A., Basili, R.: Structured lexical similarity via convolution kernels on dependency trees. In: Proceedings of EMNLP, Edinburgh, Scotland, UK (2011)
5. Esuli, A., Sebastiani, F.: Sentiwordnet: a publicly available lexical resource for opinion mining. In: Proceedings of 5th Language Resources and Evaluation Conference, pp. 417–422 (2006)
6. Fan, R.E., Chang, K.W., Hsieh, C.J., Wang, X.R., Lin, C.J.: Liblinear: a library for large linear classification. J. Mach. Learn. Res. **9**, 1871–1874 (2008)
7. Go, A., Bhayani, R., Huang, L.: Twitter sentiment classification using distant supervision. Technical report. Stanford University (2009). https://sites.google.com/site/twittersentimenthelp/home
8. Harris, Z.: Distributional structure. In: Katz, J.J., Fodor, J.A. (eds.) The Philosophy of Linguistics. Oxford University Press, Oxford (1964)
9. Hu, M., Liu, B.: Mining and summarizing customer reviews. In: Proceedings of 10th International Conference on Knowledge Discovery and Data Mining, pp. 168–177. ACM (2004)
10. Kiritchenko, S., Zhu, X., Mohammad, S.M.: Sentiment analysis of short informal texts. J. Artif. Intell. Res. **50**, 723–762 (2014)

11. Klein, D., Manning, C.D.: Accurate unlexicalized parsing. In: Proceedings of ACL 2003, pp. 423–430 (2003)
12. Landauer, T., Dumais, S.: A solution to plato's problem: the latent semantic analysis theory of acquisition, induction and representation of knowledge. Psychological Review **104**(2), 211 (1997)
13. Mikolov, T., Chen, K., Corrado, G., Dean, J.: Efficient estimation of word representations in vector space. CoRR abs/1301.3781 (2013). http://arxiv.org/abs/1301.3781
14. Miller, G.A.: Wordnet: a lexical database for english. Commun. ACM **38**(11), 39–41 (1995)
15. Mohammad, S.M., Turney, P.D.: Emotions evoked by common words and phrases: using mechanical turk to create an emotion lexicon. In: Proceedings of NAACL 2010 Workshop on Computational Approaches to Analysis and Generation of Emotion in Text. ACL (2010)
16. Nakov, P., Rosenthal, S., Kozareva, Z., Stoyanov, V., Ritter, A., Wilson, T.: Semeval-2013 task 2: sentiment analysis in twitter. In: Proceedings of SemEval, 2nd Joint Conference on Lexical and Computational Semantics (*SEM), pp. 312–320. ACL, Atlanta, June 2013
17. Pang, B., Lee, L.: Opinion mining and sentiment analysis. Found. Trends Inf. Retr. **2**(1–2), 1–135 (2008)
18. Pontiki, M., Galanis, D., Pavlopoulos, J., Papageorgiou, H., Androutsopoulos, I., Manandhar, S.: Semeval-2014 task 4: aspect based sentiment analysis. In: Proceedings of SemEval (2014)
19. Rao, D., Ravichandran, D.: Semi-supervised polarity lexicon induction. In: Proceedings of the EACL, pp. 675–682. ACL (2009)
20. Rifkin, R., Klautau, A.: In defense of one-vs-all classification. J. Mach. Learn. Res. **5**, 101–141 (2004)
21. Rosenthal, S., Ritter, A., Nakov, P., Stoyanov, V.: Semeval-2014 task 9: sentiment analysis in twitter. In: Proceedings SemEval. ACL and Dublin City University (2014)
22. Sahlgren, M.: The Word-Space Model. Ph.D. thesis, Stockholm University (2006)
23. San Vicente, I., Agerri, R., Rigau, G.: Simple, robust and (almost) unsupervised generation of polarity lexicons for multiple languages. In: Proceedings of the 14th EACL. ACL (2014)
24. Shawe-Taylor, J., Cristianini, N.: Kernel Methods for Pattern Analysis. Cambridge University Press, Cambridge (2004)
25. Stone, P.J., Dunphy, D.C., Smith, M.S., Ogilvie, D.M.: The General Inquirer: A Computer Approach to Content Analysis. MIT Press, Cambridge (1966)
26. Turney, P.D., Littman, M.L.: Measuring praise and criticism: Inference of semantic orientation from association. ACM Trans. Inf. Syst. **21**(4), 315–346 (2003)
27. Vapnik, V.N.: Statistical Learning Theory. Wiley-Interscience, Wiley (1998)
28. Wilson, T., Wiebe, J., Hoffmann, P.: Recognizing contextual polarity in phrase-level sentiment analysis. In: Proceedings of EMNLP. ACL (2005)
29. Zhang, Z., Singh, P.M.: Renew: a semi-supervised framework for generating domain-specific lexicons and sentiment analysis. In: Proceedings of ACL, pp. 542–551. ACL (2014)

Querying and Question Answering Systems

Query Refinement Using Conversational Context: A Method and an Evaluation Resource

Maryam Habibi[✉] and Andrei Popescu-Belis

Idiap Research Institute and École Polytechnique Fédérale de Lausanne (EPFL),
Rue Marconi 19, 1920 Martigny, Switzerland
{maryam.habibi,andrei.popescu-belis}@idiap.ch

Abstract. This paper introduces a query refinement method applied to queries asked by users during a meeting or a conversation. The proposed method does not require further clarifications from users, to avoid distracting them from their conversation, but leverages instead the local context of the conversation. The method first represents the local context by extracting keywords from the transcript of the conversation. It then expands the queries with keywords that best represent the topic of the query, i.e. expansion keywords accompanied by weights indicating their topical similarity to the query. Moreover, we present a dataset called AREX and an evaluation metric based on relevance judgments collected in a crowdsourcing experiment. We compare our query expansion approach with other methods, over queries extracted from the AREX dataset, showing the superiority of our method when either manual or automatic transcripts of the AMI Meeting Corpus are used.

Keywords: Query refinement · Speech-based information retrieval · Crowdsourcing · Evaluation

1 Introduction

We introduce a query refinement technique for explicit queries addressed by users to a system during a conversation. Retrieval based on these queries can be erroneous, due to their inherent ambiguity. The proposed technique uses the local context of the conversation to properly answer the users' information needs, without the need for explicit query refinement, which would interrupt users from their discussion. For instance, in the example discussed throughout the paper (see Sect. 5.4 and the Appendix), people are talking about the design of a remote control, and a participant needs more information about the acronym "LCD". Our goal is to find the most helpful Wikipedia pages to answer users' information needs in the context of designing a remote control.

Previous query refinement techniques enrich queries either interactively, or automatically, by adding relevant specifiers obtained from an external data source. However, interacting with users for query refinement may distract them from their current conversation, while using an external data source outside the

© Springer International Publishing Switzerland 2015
C. Biemann et al. (Eds.): NLDB 2015, LNCS 9103, pp. 89–102, 2015.
DOI: 10.1007/978-3-319-19581-0_7

users' local context may cause misinterpretations. For example, the acronym "LCD" can be interpreted as the 'lowest common denominator' or the 'Lesotho Congress for Democracy', in addition to 'liquid-crystal display', which is the correct interpretation in this case. To address this issue, several techniques have attempted to use the local context of users' activities, without requiring user interaction [1,8]. However, as we will show, they are not entirely suitable for a conversational environment, because of the nature of the vocabulary and the errors introduced by the ASR, such as 'recap' in the dialogue example of the paper.

In this paper, the local context of an explicit query is represented by a keyword set that is automatically obtained from the conversation fragment preceding each query as in [15,16]. We assign a weight value to each keyword, based on its topical similarity to the explicit query, to reduce the effect of the ASR noise, and to recognize appropriate interpretations of the query. In order to evaluate the improvement brought by this method, we constructed the AREX dataset (AMI Requests for Explanations and Relevance Judgments for their Answers, now publicly available). This dataset contains a set of explicit queries inserted in several conversations of the AMI Meeting Corpus [9], along with a set of human relevance judgments over sample retrieval results from Wikipedia for each query; it is accompanied by an automatic evaluation metric based on Mean Average Precision (MAP). The results show the superiority of our technique over previous ones and its robustness against unrelated keywords or ASR noise.

The paper is organized as follows. In Sect. 2, we review existing methods for query refinement. In Sect. 3, we describe the proposed query refinement method using conversational context. Section 4 explains how the AREX dataset was constructed and specifies the evaluation metric. Section 5 presents and discusses the experimental results obtained both with ASR output and with human-made transcripts of the AMI Meeting Corpus.

2   Related Work

Several methods for the refinement of explicit queries asked by users have been proposed in the field of information retrieval, and are often classified into query expansion techniques and relevance feedback ones [11]. Query expansion generates one or more hypotheses for query refinement by recognizing possible interpretations of a query, based on knowledge coming either directly from the document corpus over which retrieval is performed [2,3,10,24,29] or from Web data or personal profiles in the case of Web search [12,13,21,30]. Query expansion techniques select suggestions for query refinement either interactively or automatically [11]. For instance, relevance feedback gathers judgments obtained from the users on sample results obtained from an initial query [19,25,26].

These methods are not ideal for refinement of explicit queries asked during a conversation, because they require users to interrupt their conversation. On the contrary, our overall goal is to estimate users' information needs from their explicit queries with as little intrusion as possible. Moreover, using the local context for query refinement instead of external, non-contextual resources has the potential to improve retrieval results [8].

To the best of our knowledge, two previous systems have utilized the local context for the augmentation of explicit queries. The JIT-MobIR system for mobile devices [1] used contextual features from the physical and the human environment, but the content of the activities itself was not used as a feature. The WATSON system [8] refined explicit queries by concatenating them with keywords extracted from the documents being edited or viewed by the user. However, in order to apply this method to a retrieval system for which the local context is a conversation, the keyword lists must avoid considering irrelevant topics from ASR errors. Moreover, unlike written documents which follow generally a planned and focused structured, in a conversation users often turn from one topic to another, and adding such a variety of keywords to a query might deteriorate the retrieval results [4,11].

3   Content-Based Query Refinement

The system that we have been building is the Automatic Content Linking Device [22,23], which monitors a conversation between its users, such as a business meeting, and makes spontaneous recommendations of relevant documents, but also allows the users to formulate explicit spoken queries to retrieve documents. In this paper, our focus is the second functionality. The documents can be retrieved from the Web or a specific repository: in the experiments presented here, this repository is always the English Wikipedia obtained using the Freebase Wikipedia Extraction (WEX) dataset[1] from Metaweb Technologies (version dated 2009-06-16).

The users can simply address the system by using a pre-defined unambiguous name, which is robustly recognized by the real-time ASR component of the ACLD [14]. More sophisticated strategies for addressing a system in a multi-party dialogue context have been studied [6,28], but they are beyond the scope of this paper, which is concerned with processing the query itself. Once the results are generated by the system, they are displayed on a shared projection screen or on each user's device.

To answer an explicit query Q, the process of query refinement starts by modeling the local context using the transcript of the conversation fragment preceding the query. We use the same fixed length for all the fragments, though more sophisticated strategies are under consideration too. From the local context, we extract a set of keywords C using a diverse keyword extraction technique that we previously proposed [15,16], which maximizes the coverage of the fragment's topics with keywords. We then weigh the extracted keywords by using a filter that assigns a weight m_i, with $0 \leq m_i < 1$, to each keyword $kw_i \in C \setminus Q$ based on the normalized topical similarity of the keyword to the explicit query, as formulated in the following equation:

$$m_i = \frac{\sum_{z \in Z} p(z|Q)p(z|kw_i)}{\sqrt{\sum_{z \in Z} p(z|kw_i)^2}\sqrt{\sum_{z \in Z} p(z|Q)^2}} \tag{1}$$

---

[1] See http://download.freebase.com/wex.

In this equation, Z is the set of abstract topics which correspond to latent variables inferred using a topic modeling technique over a large collection of documents, and $p(z|kw_i)$ is the distribution of topic z in relation to the keyword kw_i. Similarly, $p(z|Q) = (\sum_{q \in Q} p(z|q))/|Q|$ is the averaged distribution of topic z in relation to the query Q made of query words q.

The topic distributions are created using the LDA topic modeling technique [5], implemented in the Mallet toolkit [20]. The topic models are learned over a large subset of the English Wikipedia with around 125,000 randomly sampled documents [18]. Following several previous studies, we fixed the number of topics at 100 [7,18].

Each query Q is thus refined by adding additional keywords extracted from the fragment, with a certain weight. Note that we do not weigh all the words of the fragment, but only those selected as keywords, in order to avoid expanding the query with words that are relevant to one of the query aspects but not to the main topics of the fragment. We obtain a parametrized refined query $RQ(\lambda)$ which is a set of weighted keywords, i.e. pairs of (word, weight):

$$RQ(\lambda) = \{(q_1, 1), \dots, (q_{|Q|}, 1), (kw_1, m_1^\lambda), \dots, (kw_{|C|}, m_{|C|}^\lambda)\} \qquad (2)$$

In other words, the refined query contains the words from the explicit query with weight 1, and the expansion keywords with a weight proportional to their topic similarity to the query.

The λ parameter has the following role. If $\lambda = \infty$, the refined query is the same as the initial explicit query (with no refinement) because $0 \leq m_i < 1$. By setting λ to 0, the query is like the one used in the Watson system [8], giving the same weight to the query words and to the keywords representing the local context. Because the keywords are related to topics that have various relevance values to the explicit query, we will set the intermediate value $\lambda = 1$ in our experiments, to weigh each keyword based on its relevance to the topics of the query. The value of λ could be optimized if more training data were available.

4   Dataset and Evaluation Method

Our experiments are conducted on the AREX dataset ("AMI Requests for Explanations and Relevance Judgments for their Answers") which we constructed and made publicly available at http://www.idiap.ch/dataset/arex. The dataset contains a set of explicit queries, inserted at various locations of the conversations in the AMI Meeting Corpus [9], as explained in Sect. 4.1. The dataset also includes relevance judgments gathered using a crowdsourcing platform over the documents retrieved for four queries prepared by the four different methods described in Sects. 4.2 and 5. These judgments can be used as ground truth to evaluate a retrieval system automatically.

4.1   Explicit Queries in the Dataset

The AMI Meeting Corpus contains conversations about designing remote controls, in series of four scenario-based meetings each, for a total of 138 meetings.

Our dataset is made of a set of explicit queries with the time of their occurrence in the AMI Corpus. Since the number of naturally-occurring queries in the corpus is insufficient for evaluating our system, we artificially generated and inserted a number of queries, using the following procedure.

Initially, utterances containing an acronym X are automatically detected, for two reasons. First, acronyms are one of the typical items which are likely to require explanations because of their potential ambiguity. Second, several acronyms already appear in explicit queries that occurred naturally in the AMI Corpus. Nevertheless, our query expansion technique is applicable to any explicit query.

We formulate explicit queries such as "I need more information about X", and insert them after the utterances containing the acronym (see for instance the example in the Appendix). Seven acronyms, all-but-one related to the domain of remote controls, are considered: LCD (liquid-crystal display), VCR (videocassette recorder), PCB (printed circuit board), TFT (thin-film-transistor liquid-crystal display), $NTSC$ (National Television System Committee), IC (integrated circuit), and RSI (repetitive strain injury). These acronyms occur 74 times in the scenario-based meetings of the AMI Corpus and are accompanied by 74 different conversation fragments in the AREX dataset.

We used both manual and ASR transcripts of the fragments from the AMI Corpus in our experiments. The ASR transcripts were generated by the AMI real-time ASR system for meetings [14], with an average word error rate (WER) of 36 %. In addition, for experimenting with a variable range of WER values, we have simulated the potential speech recognition mistakes as in [16], by applying to the manual transcripts of these conversation fragments three different types of ASR noise: deletion, insertion and substitution. In a systematic manner, i.e. altering all occurrences of a word type, we randomly selected the conversation words, as well as the words to be inserted, from the vocabulary of the English Wikipedia. The percentage of simulated ASR noise varied from 10 % to 30 %, as the best recognition accuracy reaches around 70 % in conversational environments [17]. However, noise was never applied to the explicit query itself.

4.2 Evaluation Using the Dataset

Ground Truth Relevance Judgments. Following a classical approach for evaluating information retrieval [27], we build a reference set of retrieval results by merging the lists of the top 10 results from four different query expansion methods used to answer users' explicit queries. The retrieval results are obtained by the Apache Lucene search engine over the English Wikipedia. Three of the methods are listed in Sects. 3 and 5, and the last one builds a query which consists of only the keywords extracted from conversation fragments, with no words from the queries. We found that each explicit query had at least 31 different results for all the 74 fragments, and we decided to limit the reference set to 31 documents for each query.

Each fragment is about 400 words long, for the following reason. We computed the sum of the weights assigned to the keywords extracted from each fragment by $RQ(1)$ which weighs keywords based on their relevance to the query topics. Then we averaged them over 25 queries, which were randomly selected from the AREX dataset to serve as a development set for tuning our hyper-parameters. The values obtained from five repetitions of the experiment with the fragment lengths varying from 100 to 500 words in increments of 100 were, respectively: 2.14, 2.32, 2.08, 2.08, and 2.08. Since there is no variation in these values for the last three values, we set fragment size to 400 words. We have also limited the weighting to the first 10 keywords extracted from each fragment, following several previous studies [11], thus speeding up the query processing.

We designed a set of tasks to gather relevance judgments for the reference set from human subjects. We showed to the subjects the transcript of the conversation fragment ending with the query: "I need more information about X" with 'X' being one of the acronyms considered here. This was followed by a control question about the content of the conversation, and then by the list of 31 documents from the reference set. The subjects had to decide on the relevance value of each document by selecting one of the three options among 'irrelevant', 'somewhat relevant' and 'relevant' (noted below as $A = \{a_0, a_1, a_2\}$).

We collected judgments for the 74 queries of our dataset from 10 subjects per query. The tasks were crowdsourced via Amazon's Mechanical Turk, each judgment becoming a "human intelligence task" (HIT). The average time spent per HIT was around 2 min. For qualification control, we only accepted subjects with greater than 95 % approval rate and with more than 1000 previously approved HITs, and we only kept answers from the subjects who answered correctly the control questions. We applied furthermore a qualification control factor to the human judgments, in order to reduce the impact of "undecided" cases, inferred from the low agreement of the subjects. We compute the following measure of the uncertainty of subjects regarding the relevance of document j: $H_{tj} = -\sum_{a \in A}(s_{tj}(a) \ln(s_{tj}(a))/\ln|A|)$, where $s_{tj}(a)$ is the proportion in which the 10 subjects have selected each of the allowed options $a \in A$ for the document j and the conversation fragment t. Then, the relevance value assigned to each option a is computed as $s'_{tj}(a) = s_{tj}(a) \cdot (1 - H_{tj})$, i.e. the raw score weighted by the subjects' uncertainty.

Scoring a List of Documents. Using the ground truth relevance of each document in the reference set, weighted by the subjects' uncertainty, we will measure the MAP score at rank n of a candidate document result list. We start by computing gr_{tj}, the global relevance value for the conversation fragment t and the document j by giving a weight of 2 for each "relevant" answer (a_2) and 1 for each "somewhat relevant" answer (a_1).

$$gr_{tj} = \frac{s'_{tj}(a_1) + 2s'_{tj}(a_2)}{s'_{tj}(a_0) + s'_{tj}(a_1) + 2s'_{tj}(a_2)} \tag{3}$$

Then we calculate $AveP_{tk}(n)$ the Average Precision at rank n for the conversation fragment t and the candidate list of results of a system k as follows:

$$AveP_{tk}(n) = \sum_{i=1}^{n} P_{tk}(i)\triangle r_{tk}(i) \tag{4}$$

where $P_{tk}(i) = \sum_{c=1}^{i} gr_{tl_{tk}(c)}/i$ is the precision at cut-off i in the list of results l_{tk}, $\triangle r_{tk}(i) = gr_{tl_{tk}(i)}/\sum_{j \in l_t} gr_{tj}$ is the change in recall from document in rank $i-1$ to rank i over the list l_{tk}, and l_t is the reference set for fragment t.

Finally, we compute $MAP_k(n)$, the MAP score at rank n for a system k by averaging the Average Precisions of all the queries at rank n as follows, where $|T|$ is the number of queries.

$$MAP_k(n) = \sum_{t=1}^{|T|} \frac{AveP_{tk}(n)}{|T|} \tag{5}$$

Comparing Two Lists of Documents. We compare two lists of documents obtained by two systems k_1 and k_2 through the percentage of the relative MAP at rank n improvement, defined as follows:

$$\%RelativeScore_{k_1,k_2}(n) = \frac{MAP_{k_1}(n) - MAP_{k_2}(n)}{MAP_{k_2}(n)} \times 100. \tag{6}$$

5   Experimental Results

We defined in Sect. 3 three methods for expanding queries based on the values of λ in Eq. 2. The first method has $\lambda = \infty$ and is therefore noted $RQ(\infty)$ – it only uses explicit query keywords, with no refinement. The second one refines explicit queries using the method of the Watson system [8], with $\lambda = 0$, hence noted $RQ(0)$. The third method has $\lambda = 1$ and is noted $RQ(1)$ – this is the novel method proposed here, which expands the query with keywords from the conversation fragment based on their topical similarity to the query. Comparisons are performed over the human-made transcripts and the ASR output, using as a test set the remaining 49 queries not used for development.

5.1   Variation of Fragment Length

We study first the effect of the fragment length on the retrieval results of the three methods, $RQ(1)$, $RQ(\infty)$, and $RQ(0)$. Keyword sets used for expansion are extracted here from the manual transcript of the conversation fragments preceding the 49 queries of the testset. The fragments have a fixed-length per experiment, but we ran our experiments over lengths from 100 to 500 words.

The relative MAP scores of $RQ(1)$ over $RQ(\infty)$ for different ranks n from $n=1$ to $n=4$ are provided in Fig. 1a, demonstrating the superiority of $RQ(\infty)$

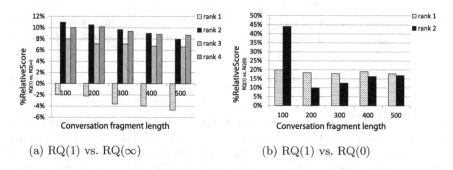

(a) RQ(1) vs. RQ(∞)   (b) RQ(1) vs. RQ(0)

Fig. 1. Relative MAP scores of $RQ(1)$ against $RQ(\infty)$ up to rank 4 (a), and against $RQ(0)$ up to rank 2 (b). The scores were obtained using manual transcripts with fragment lengths of 100, 200, 300, 400 and 500 words. $RQ(1)$ outperforms the other two methods, except for $RQ(\infty)$ at rank $n = 1$.

at $n = 1$. However, $RQ(1)$ surpasses $RQ(\infty)$ for ranks 2, 3 and 4. The improvement over $RQ(\infty)$ slightly decreases by increasing the conversation fragment length, likely because of the topic drift in longer fragments. Indeed, when increasing the fragment length, the proposed method $RQ(1)$ behaves more similarly to $RQ(\infty)$ by assigning small weight values (close to zero) to the candidate expansion keywords.

The relative MAP scores of $RQ(1)$ over $RQ(0)$ are reported at ranks $n = 1$ and $n = 2$ in Fig. 1b. We do not report values for higher ranks, because of the lack of enough judgments for the retrieval results of RQ(0) among the reference set. The improvements over $RQ(0)$ at rank $n = 1$ are approximately the same for different fragment lengths. They, nevertheless, vary a lot with the length of fragments when looking at rank $n = 2$. The improvement is minimum at length 200 words, likely due to more relevant candidate expansion keywords at this length compared to the others. As shown above, the average sum of the weights of the expansion keywords is maximized by our method, $RQ(1)$, at length 200 words. When the length decreased or increased from 200 words, the query topics are not completely covered, or the topics are changed respectively. Therefore, the improvement over $RQ(0)$ is increased by decreasing or increasing the length from 200 words at rank $n = 2$, thus showing that $RQ(1)$ is more robust to out-of-topic keywords than $RQ(0)$.

5.2   Comparisons on Manual Transcripts

We now compare the proposed method $RQ(1)$ with two methods, $RQ(0)$ and $RQ(\infty)$ over the manual transcripts of the 49 conversation fragments, for ranks n from $n = 1$ to $n = 8$, with fragments of 400 words preceding each query. The improvements obtained by $RQ(1)$ over the two others are represented in Fig. 2 (the results for 400 words from Fig. 1 are reused in this figure).

The relative MAP scores of $RQ(1)$ over $RQ(\infty)$, except at rank $n = 1$, demonstrate the significant superiority of $RQ(1)$ over $RQ(\infty)$ (between 7 % to 11 %)

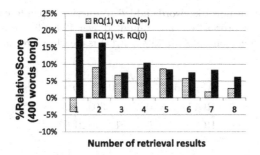

Fig. 2. Relative MAP scores of $RQ(1)$ over the two baseline methods $RQ(\infty)$ and $RQ(0)$ up to rank 8, obtained over the manual transcript of the 49 fragments of 400 words. $RQ(1)$ surpasses both methods for ranks 2 to 8.

up to rank $n = 6$ on average. There are also on average small improvements around 2 % over $RQ(\infty)$ at ranks $n = 7$ and 8, because of retrieving the documents which are relevant to both the queries and the fragments by $RQ(\infty)$ (which does not disambiguate the query) at ranks $n = 1, 7$ and 8.

The relative MAP scores of $RQ(1)$ over $RQ(0)$ show significant improvements of more than 15 % for ranks $n = 1$ and $n = 2$. Although the scores decrease from rank 2, they remain considerably high at around 7 %.

5.3   Comparisons on ASR Transcripts

We applied the explicit query expansion methods to our dataset using the ASR transcripts of the conversations, in order to consider the effect of ASR noise on the retrieval results of the expanded queries. We experimented with real ASR transcripts with an average word error rate of 36 % and with simulated ones with a noise level varying from 10 % to 30 %. We computed the average of the scores over five repetitions of the experiment with simulated ASR transcripts, which are randomly generated, and provide below the relative MAP scores of $RQ(1)$ over $RQ(\infty)$ up to rank 3, and over $RQ(0)$ up to rank 2. Moreover, upon manual inspection, we found that there are many relevant documents retrieved in the presence of ASR noise, which have no judgment in the AREX dataset, because they do not overlap with the 31 documents obtained by pooling four methods.

First we compared the two contextual expansion methods, $RQ(0)$ and $RQ(1)$, in terms of the proportion of noisy keywords that each method added to the refined queries. This proportion was computed by summing up the weight value of the keywords used for query refinement that were in fact ASR errors (their set is noted N_j), normalized by the sum of the weight value of all keywords used for the refinement of the query j, as follows:

$$pn_j = \frac{\sum_{kw_i \in (C_j \cap N_j)} m_i^\lambda}{\sum_{kw_i \in C_j} m_i^\lambda} \times 100\,\% \qquad (7)$$

Table 1. Proportion of noisy keywords added to queries depending on ASR noise on $RQ(1)$ and $RQ(0)$. The proportions are computed over 49 explicit queries from AREX, for a noise level varying from 10 % to 30 %. $RQ(1)$ is clearly more robust to noise than $RQ(0)$.

ASR noise	10 %	20 %	30 %
$RQ(1)$	0.78	1.30	2.27
$RQ(0)$	5.64	12.07	21.07

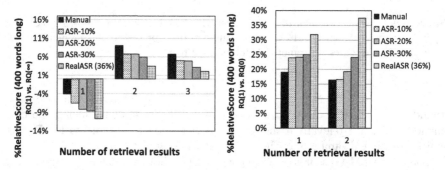

Fig. 3. Relative MAP scores of $RQ(1)$ against $RQ(\infty)$ up to rank 3 (a), and against $RQ(0)$ up to rank 2 (b), obtained over the real or simulated ASR transcripts. The results show that $RQ(1)$ outperforms the other two methods.

We averaged these values over the 49 explicit queries and the five experimental runs with different random ASR errors. The results shown in Table 1 reveal that the proposed method, $RQ(1)$, is more robust to the ASR noise than $RQ(0)$.

We also represent the relative scores of $RQ(1)$ over $RQ(0)$ in Fig. 3b. The improvement over $RQ(0)$ increases when the percentage of noise added to the fragments increases, and shows that our method exceeds $RQ(0)$ considerably. Moreover, we compare the retrieval results of $RQ(1)$ and $RQ(\infty)$ (which does not consider context) in noisy conditions, in Fig. 3a. Although the improvement over $RQ(\infty)$ slightly decreases with the noise level, $RQ(1)$ still outperforms $RQ(\infty)$ in terms of relevance, and is generally more robust to ASR noise.

5.4   Examples of Expanded Queries and Retrieval Results

To illustrate how $RQ(1)$ surpasses the other techniques, we consider an example from one of the queries of our dataset, using the ASR transcript of the conversation fragment given in Appendix of this paper. The query is: "I need more information about LCD". So the query bears on the acronym "LCD". The list of keywords extracted for this fragment is the following, where three keywords ('recap', 'sleek', and 'snowman') are in fact ASR noise: $C = \{$'interface', 'design', 'decision', 'recap', 'user', 'control', 'final', 'remote', 'discuss', 'sleek', 'snowman'$\}$.

The proposed method $RQ(1)$ assigns, in this particular example, a weight of zero to keywords from ASR noise and to those unrelated to the conversation

Table 2. Examples of retrieved Wikipedia pages (ranked lists) using three methods. Results of $RQ(1)$ are more relevant to the query and conversation topics.

RQ(1)	RQ(∞)	RQ(0)
Liquid-crystal display	Liquid-crystal display	User interface
Backlight	Backlight	X Window System
Liquid-crystal display television	Liquid-crystal display television	Usability
Thin-film transistor	Lowest common denominator	Wii Remote
LCD projector	LCD Soundsystem	Walkman
LG Display	LCD projector	Information hiding
LCD shutter glasses	Pakalitha Mosisili	Screensaver
Universal remote	LG Display	Apple IIc

topics. So its corresponding expanded query is: $RQ(1) = \{(\text{lcd},1.0), (\text{control},0.7), (\text{remote},0.4), (\text{design},0.1), (\text{interface},0.1), (\text{user},0.1)\}$.

$RQ(0)$ assigns a weight 1 to each keyword of the list C and uses all of them for expansion, regardless of their importance to the query. Therefore, the expanded query contains many more irrelevant words. Finally, $RQ(\infty)$ does not expand the query so it considers only 'lcd'.

The retrieval results up to rank 8 obtained for the three methods are displayed in Table 2. All the results of $RQ(1)$ are related to 'liquid-crystal display', which is the correct interpretation of the query, while $RQ(\infty)$ provides three irrelevant documents: 'lowest common denominator' (a mathematic function), 'LCD Soundsystem' (an American dance band), and 'Pakalitha Mosisili' (a politician at Lesotho Congress for Democracy). None of the results provided by $RQ(0)$ addresses 'liquid-crystal display' directly, due to irrelevant keywords added to the query from topics unrelated to the conversation or from ASR noise.

6 Conclusion

The best method for contextual query refinement appears to be the proposed method $RQ(1)$ over both manual and ASR transcripts. Although, $RQ(\infty)$ outperforms $RQ(1)$ at rank $n = 1$, the scores of $RQ(1)$ show a significant improvement up to rank $n = 8$ over manual transcripts and up to rank $n = 3$ over ASR ones. Moreover, $RQ(1)$ outperforms $RQ(0)$ on both manually-made and ASR transcripts. The scores also demonstrate that the proposed method $RQ(1)$ is robust to various ASR noise levels and to the length of the conversation fragment used for expansion. The dataset accompanying these experiments, AREX, is public and can be used for future comparisons of conversational query-based retrieval systems.

In future work, we plan to setup experiments with human subjects in a scenario that encourages them to use spoken queries during a task-oriented conversation, and confirm the superiority of our proposal with respect to the state-of-the-art through evaluation on a deployed system.

Acknowledgments. The authors are grateful to the Swiss National Science Foundation (SNSF) for its financial support through the IM2 NCCR on Interactive Multimodal Information Management (see www.im2.ch), as well as to the Hasler Foundation for the REMUS project (n. 13067, Re-ranking Multiple Search Results for Just-in-Time Document Recommendation).

Appendix: Transcript of a Conversation Fragment from the AMI Meeting Corpus

We provide here a 150-word fragment of the ASR from a conversation of the AMI Corpus (segmented by the ASR into utterances), which was used as an example in this paper. The discussion is about designing a remote control, and a query was introduced at the end of the fragment for the AREX dataset. The document results retrieved for this query by three methods are given in Table 2.

A: Okay well .. All sacked .. Right .. Oh i see a kind of detailed design meeting .. Um .. We're gonna discuss the the look-and-feel design user interface design and .. We're gonna evaluate the product .. And .. For .. The end result of this meeting has to be a decision on the details of this remote control like a sleek final decision .. Uh-huh .. Um i'm then i'm gonna have to specify the final design .. In the final report ..

B: Yeah .. So um just from from last time .. To recap .. So we're gonna have a snowman shaped remote control with no LCD display new need for tap bracket so if you're gonna be kinetic power and battery .. Uh with rubber buttons maybe park lighting the buttons with um .. Internal LEDs to shine through the casing .. Um hopefully a job down and incorporating the slogan somewhere as well I think i missed .. Okey .. Um so .. Uhuh .. If you want to present your prototype .. Go ahead ..

C [inserted]: I need more information about LCD.

References

1. Alidin, A.A., Crestani, F.: Context modelling for just-in-time mobile information retrieval (JIT-MobIR). Pertanika J. Sci. Technol. **21**(1), 227–238 (2013)
2. Attar, R., Fraenkel, A.S.: Local feedback in full-text retrieval systems. J. ACM (JACM) **24**(3), 397–417 (1977)
3. Bai, J., Song, D., Bruza, P., Nie, J.Y., Cao, G.: Query expansion using term relationships in language models for information retrieval. In: Proceedings of the 14th ACM International Conference on Information and Knowledge Management, pp. 688–695 (2005)

4. Bhogal, J., Macfarlane, A., Smith, P.: A review of ontology based query expansion. Inf. Process. Manage. **43**(4), 866–886 (2007)
5. Blei, D.M., Ng, A.Y., Jordan, M.I.: Latent dirichlet allocation. J. Mach. Learn. Res. **3**, 993–1022 (2003)
6. Bohus, D., Horvitz, E.: Models for multiparty engagement in open-world dialog. In: Proceedings of the SIGDIAL 2009 Conference: The 10th Annual Meeting of the Special Interest Group on Discourse and Dialogue, pp. 225–234 (2009)
7. Boyd-Graber, J., Chang, J., Gerrish, S., Wang, C., Blei, D.: Reading tea leaves: how humans interpret topic models. In: Proceedings of 23rd Annual Conference on Neural Information Processing Systems, pp. 288–296 (2009)
8. Budzik, J., Hammond, K.J.: User interactions with everyday applications as context for just-in-time information access. In: Proceedings of the 5th International Conference on Intelligent User Interfaces, pp. 44–51 (2000)
9. Carletta, J.: Unleashing the killer corpus: experiences in creating the multi-everything AMI meeting corpus. Lang. Resour. Eval. J. **41**(2), 181–190 (2007)
10. Carpineto, C., De Mori, R., Romano, G., Bigi, B.: An information-theoretic approach to automatic query expansion. ACM Trans. Inf. Syst. (TOIS) **19**(1), 1–27 (2001)
11. Carpineto, C., Romano, G.: A survey of automatic query expansion in information retrieval. ACM Comput. Surv. (CSUR) **44**(1), 1–50 (2012)
12. Chirita, P.A., Firan, C.S., Nejdl, W.: Personalized query expansion for the web. In: Proceedings of 30th Annual International ACM SIGIR Conference on Research and Development in IR, pp. 7–14 (2007)
13. Diaz, F., Metzler, D.: Improving the estimation of relevance models using large external corpora. In: Proceedings of 29th Annual International ACM SIGIR Conference on Research and Development in IR, pp. 154–161 (2006)
14. Garner, P.N., Dines, J., Hain, T., El Hannani, A., Karafiat, M., Korchagin, D., Lincoln, M., Wan, V., Zhang, L.: Real-time ASR from meetings. In: Proceedings of the 10th Annual Conference of the International Speech Communication Association, pp. 2119–2122 (2009)
15. Habibi, M., Popescu-Belis, A.: Diverse keyword extraction from conversations. In: Proceedings of the 51st Annual Meeting of the Association for Computational Linguistics, pp. 651–657 (2013)
16. Habibi, M., Popescu-Belis, A.: Keyword extraction and clustering for document recommendation in conversations. IEEE/ACM Trans. Audio Speech Lang. Process. **23**(4), 746–759 (2015)
17. Hain, T., Burget, L., Dines, J., Garner, P.N., El Hannani, A., Huijbregts, M., Karafiat, M., Lincoln, M., Wan, V.: The AMIDA 2009 meeting transcription system. In: Proceedings of INTERSPEECH, pp. 358–361 (2010)
18. Hoffman, M.D., Blei, D.M., Bach, F.: Online learning for Latent Dirichlet Allocation. In: Proceedings of 24th Annual Conference on Neural Information Processing Systems (NIPS), pp. 856–864 (2010)
19. Lavrenko, V., Croft, W.B.: Relevance based language models. In: Proceedings of 24th Annual International ACM SIGIR Conference on Research and Development in IR, pp. 120–127 (2001)
20. McCallum, A.K.: MALLET: A machine learning for language toolkit (2002). http://mallet.cs.umass.edu
21. Park, L.A.F.: Query expansion using a collection dependent probabilistic latent semantic thesaurus. In: Zhou, Z.-H., Li, H., Yang, Q. (eds.) PAKDD 2007. LNCS (LNAI), vol. 4426, pp. 224–235. Springer, Heidelberg (2007)

22. Popescu-Belis, A., Yazdani, M., Nanchen, A., Garner, P.N.: A speech-based just-in-time retrieval system using semantic search. In: Proceedings of the 49th Annual Meeting of the ACL, Demonstrations, pp. 80–85 (2011)
23. Popescu-Belis, A., Boertjes, E.M., Kilgour, J., Poller, P., Castronovo, S., Wilson, T., Jaimes, A., Carletta, J.E.: The AMIDA automatic content linking device: just-in-time document retrieval in meetings. In: Popescu-Belis, A., Stiefelhagen, R. (eds.) MLMI 2008. LNCS, vol. 5237, pp. 272–283. Springer, Heidelberg (2008)
24. Robertson, S.E., Walker, S., Beaulieu, M., Willett, P.: Okapi at TREC-7: automatic ad hoc, filtering, VLC and interactive track. NIST Special Publication SP, pp. 253–264 (1999)
25. Rocchio, J.J.: Relevance feedback in information retrieval. In: Salton, G. (ed.) The SMART Retrieval System: Experiments in Automatic Document Processing. ch. 14, pp. 313–323. Prentice-Hall, Englewood Cliffs (1971)
26. Salton, G., Buckley, C.: Improving retrieval performance by relevance feedback. Readings in Information Retrieval 24, 5 (1997)
27. Voorhees, E.M., Harman, D.K. (eds.): TREC: Experiment and Evaluation in Information Retrieval. MIT Press, Cambridge (2005)
28. Wang, D., Hakkani-Tur, D., Tur, G.: Understanding computer-directed utterances in multi-user dialog systems. In: Proceedings of the 2013 IEEE International Conference on Acoustics, Speech and Signal Processing (ICASSP), pp. 8377–8381 (2013)
29. Xu, J., Croft, W.B.: Query expansion using local and global document analysis. In: Proceedings of 19th Annual International ACM SIGIR Conference on Research and Development in IR, pp. 4–11 (1996)
30. Xu, J., Croft, W.B.: Improving the effectiveness of information retrieval with local context analysis. ACM Trans. on Inf. Syst. (TOIS) 18(1), 79–112 (2000)

Applying Semantic Parsing to Question Answering Over Linked Data: Addressing the Lexical Gap

Sherzod Hakimov[✉], Christina Unger, Sebastian Walter, and Philipp Cimiano

Semantic Computing Group Cognitive Interaction Technology – Center of Excellence (CITEC), Bielefeld University, Bielefeld 33615, Germany
{shakimov,cunger,swalter,cimiano}@cit-ec.uni-bielefeld.de

Abstract. Question answering over linked data has emerged in the past years as an important topic of research in order to provide natural language access to a growing body of linked open data on the Web. In this paper we focus on analyzing the lexical gap that arises as a challenge for any such question answering system. The lexical gap refers to the mismatch between the vocabulary used in a user question and the vocabulary used in the relevant dataset. We implement a semantic parsing approach and evaluate it on the QALD-4 benchmark, showing that the performance of such an approach suffers from training data sparseness. Its performance can, however, be substantially improved if the right lexical knowledge is available. To show this, we model a set of lexical entries by hand to quantify the number of entries that would be needed. Further, we analyze if a state-of-the-art tool for inducing ontology lexica from corpora can derive these lexical entries automatically. We conclude that further research and investments are needed to derive such lexical knowledge automatically or semi-automatically.

1 Introduction

The topic of question answering over linked data has started to receive substantial attention in the Semantic Web community [8], and benchmarking campaigns such as QALD[1] [7] have been organized in order to support the systematic comparison of different approaches on the same task, on a shared dataset and using the same evaluation protocol.

The main task in question answering over linked data can be framed as finding a mapping of natural language questions to SPARQL[2] queries which can then be executed over an RDF dataset. As an example, consider the question in 1 together with the given SPARQL query that can be executed over DBpedia in order to retrieve the answer.

1. Who was the first to climb Mount Everest?

---

[1] http://www.sc.cit-ec.uni-bielefeld.de/qald/.
[2] http://www.w3.org/TR/sparql11-query/.

© Springer International Publishing Switzerland 2015
C. Biemann et al. (Eds.): NLDB 2015, LNCS 9103, pp. 103–109, 2015.
DOI: 10.1007/978-3-319-19581-0_8

```
PREFIX res: <http://dbpedia.org/resource/>
PREFIX dbo: <http://dbpedia.org/ontology/>
SELECT DISTINCT ?uri
WHERE {
        res:Mount_Everest dbo:firstAscentPerson ?uri .
}
```

The benchmarking challenges organized so far identified the *lexical gap* as one of the main problems in developing question answering approaches to linked data. The lexical gap refers to the problem that the vocabulary used by a user and the vocabulary used to formally represent the data can differ substantially. In the above example, for instance, the natural language question uses the expression *the first to climb*, while the corresponding property in the SPARQL query is called `firstAscentPerson`.

In order to develop a system that is successful and robust in mapping natural language questions to corresponding SPARQL queries, substantial lexical knowledge is needed, such as the knowledge that the property `firstAscentPerson` can be expressed as *the first to climb*. In this paper we analyze what lexical knowledge is needed for a question answering approach over linked data in order to be able to correctly interpret questions such as the above one. To this end, we have implemented the semantic parsing approach proposed by Zettlemoyer and Collins [14] and adapted it for the task of question answering over linked data.

As first contribution, we show that a vanilla implementation of this approach achieves poor results on the task. The main reason for this is that it was designed for a scenario in which the vocabulary used in the training data largely overlaps with the vocabulary used in the test data. This assumption, however, does not hold in open ended question answering systems over linked data in which the questions on which the system is trained can be rather different to the questions that actual users will ask, with respect to both wording and structure.

As second contribution, we investigate how much lexical knowledge would need to be added so that a semantic parsing approach can perform well on unseen data. We manually add a set of lexical entries on the basis of analyzing the test portion of the QALD-4 dataset. Further, we analyze if a state-of-the-art tool for inducing ontology lexica from corpora can derive these lexical entries automatically.

2   Semantic Parsing for Question Answering Over Linked Data

In order to apply semantic parsing for question answering over linked data, we adapt Zettlemoyer and Collins' approach [14] (ZC05). This approach relies on Combinatory Categorial Grammar (CCG) [9,10] for consituent-based syntactic representations, and typed-lambda calculus expressions [3] for semantic representations. A simple example of a CCG lexicon for the sentence *Barack Obama is married to Michelle Obama* is given in Table 1. The forward and

Table 1. Example CCG lexicon.

Lexical item	Syntactic category	Semantic representation
Barack Obama	NP	`Barack_Obama`
is	(S\NP)/(S\NP)	$\lambda f.\lambda x.f(x)$
married to	(S\NP)/NP	$\lambda y.\lambda x.\mathtt{spouse}(x, y)$
Michelle Obama	NP	`Michelle_Obama`

backward application rules of CCG are applied to these lexical items in order to construct the parse tree of the sentence and its semantic representation spouse(`Barack_Obama`, `Michelle_Obama`).

Input to the ZC05 algorithm is a set of training examples (S_i, L_i) with $i = 1 \ldots n$, where each S_i is a sentence and each L_i is a corresponding semantic representation (*logical form*). The output is a pair (ϕ, θ), where ϕ is a set of features and θ is a vector of weights for those features.

At the heart of the algorithm is the method GENLEX(S,L). It takes as input a sentence S and a corresponding logical form L, and generates a set of potential lexical items with syntactic categories and semantics, and finally pairs them with all possible substrings of S using rules defined in [14]. The resulting lexical items are then used in the actual semantic parsing step, together with initially defined lexical items for domain-independent expressions, such as wh-words, prepositions, determiners, etc.

The actual semantic parsing step returns the highest scoring parses that derive the expected logical form L using all possible lexical items. Parsing itself is an iterative process: The first step uses all possible lexical items generated by GENLEX, and only those lexical items that were used in the successful parses are then passed to the second step of parsing, with newly estimated parameter values.

We re-implemented the algorithm following the description in [14], using CKY-style parsing and a stack decoder, and changing the parameter estimation step into perceptron updates as in [15]. In Table 2 we show the updated GENLEX rules to apply ZC05 semantic parsing approach. Newly added input triggers are highlighted in boldface. Domain-independent expressions were specified manually, based on the domain-independent expressions used in [14]. These expressions and 200 training examples from QALD-4 [11], used as input to the ZC05 algorithm, can be found at http://pub.uni-bielefeld.de/data/2715997.

In order to evaluate semantic parsing on the QALD-4 dataset, the provided SPARQL queries are automatically converted to semantic representations using the following translation rules:

- Every resource in the query is translated into a constant.
- Every property in the query is translated into a binary function.
- Every COUNT solution modifier is translated into the function constant *count*.

Table 2. GENLEX rules from Zettlemoyer and Collins [14] adapted to question answering over linked data.

Input Trigger	Output Category and Example
Constant c	NP : c
	NP : dbr:Brooklyn_Bridge
Arity-two predicate p	(S\NP)/NP : $\lambda x.\lambda y.p(y,x)$
	(S\NP)/NP : $\lambda x\lambda y.$dbo:author(y,x)
Arity-two predicate p	(S\NP)/NP : $\lambda x.\lambda y.p(x,y)$
	(S\NP)/NP : $\lambda x.\lambda y.$dbo:starring(x,y)
Arity-two predicate p	(S\NP)/NP : $\lambda g.\lambda x.\lambda y.p(y,x) \wedge g(y)$
	(S\NP)/NP : $\lambda g.\lambda x.\lambda y.$dbo:crosses$(x,y) \wedge g(y)$
Arity-two predicate p	N/NP : $\lambda x.\lambda y.p(x,y)$
	N/NP : $\lambda x.\lambda y.$ dbo:officialColor(x,y)
Arity-two predicate p	N/NP : $\lambda g.\lambda x.\lambda y.p(y,x) \wedge g(y)$
	N/NP : $\lambda g.\lambda x.\lambda y.$dbo:capital$(y,x) \wedge g(y)$
Arity-two predicate p	N : $\lambda x.p(x,c)$
and constant c	N : $\lambda x.$rdf:type $(x,$dbo:River$)$
Arity-two predicate p	(N\N)/NP : $\lambda x.\lambda g.\lambda y.p(y,x) \wedge g(y)$
	(N\N)/NP : $\lambda x.\lambda g.\lambda y.$dbo:crosses$(y,x) \wedge g(y)$
Arity-two predicate p	N/N : $\lambda g.\lambda y.p(y,c) \wedge g(y)$
and constant c	N/N : $\lambda x.$dbo:country $(x,$ dbr:Germany$) \wedge g(x)$
argmax/min with second	NP/N : $\lambda g.\lambda x.$ *argmax/min* $(g(x), f(x))$
argument arity-two function f	NP/N : $\lambda g.\lambda x.argmax(g(x), \lambda d.$ dbo:birthDate$(x,d))$

3 Evaluation

After having trained the Zettlemoyer and Collins algorithm on the QALD-4 training set, the learned model was tested on the QALD-4 test set, comprising 50 questions. We excluded questions that require YAGO classes, UNIONs, ORDER BY statements and FILTERs, leaving 37 questions with respect to which the results produced by the semantic parsing approach were compared to the QALD-4 gold standard results. For each question q, precision and recall were computed as follows:

$$Recall(q) = \frac{\text{number of correct system answers for } q}{\text{number of gold standard answers for } q}$$

$$Precision(q) = \frac{\text{number of correct system answers for } q}{\text{number of system answers for } q}$$

In addition, F1-measure is computed as the harmonic mean of precision and recall. Since the QALD-4 training queries cover only a small part of the DBpedia vocabulary, we decided to increase lexical coverage of the system by adding a lexical item for each DBpedia predicate and class on the basis of their label, according to the GENLEX rules in Table 2.

Table 3. Results on the QALD-4 test dataset in terms of precision, recall and F-measure, together with the number of correctly answered questions (out of 37).

	Precision	Recall	F1	Correct
Learned lexicon + ontology labels	0.66	0.05	0.09	2
Learned lexicon + ontology labels + handcrafted items	0.93	0.70	0.80	26
Learned lexicon + ontology labels + M-ATOLL	0.70	0.18	0.30	7

Table 4. Example lexical items created for the QALD-4 test data.

Expression	Syntax	Semantics
first to climb	N/NP	$\lambda x \lambda y.$ dbo:firstAscentPerson(x, y)
artistic movement	N	$\lambda x \lambda y.$ dbo:movement(x, y)
launched from	(S\NP)/NP	$\lambda x \lambda y.$ dbo:launchPad(y, x)
extinct	N	$\lambda x.$ dbo:conservationStatus$(x,$'EX'$)$
German	N/N	$\lambda g \lambda x. g(x) \wedge$ dbo:country$(x,$ dbr:Germany$))$
taikonauts	N	$\lambda x.$ rdf:type$(x,$dbo:Astronaut$)$ $\wedge$ dbo:nationality$(x,$ dbr:China$)$

The test results are given in the first row of Table 3, where *correct* specifies the number of correctly answered questions (out of 37). Most prominently, recall turns out to be very low. This is because most of the expressions in the test questions appear neither in the training data nor among the DBpedia labels. Thus, the system lacks a great deal of lexical knowledge of expressions that were not seen during training.

For example, to answer the question *Who was the first to climb Mount Everest*, the system would need a lexical item such as shown in the first row of Table 4. Such an item is not present in the induced lexicon, neither is it contained among the ontology labels. In such cases we therefore need external lexical resources to bridge the lexical gap. In order to test how much additional lexical knowledge is needed, we manually handcrafted lexical items for the test data. Some examples are given in Table 4.

In total we created 54 lexical items. The results using those additional lexical items are presented in the second row in Table 3, showing that recall significantly increased, from 5 % to 70 %.

The system thus shows remarkable improvements by using the handcrafted lexical items. However, for large domains the required manual effort is not always feasible. Therefore we ran M-ATOLL [12,13], a system that automatically extracts lexicalizations for ontology elements from a text corpus, on the predicates used in the training dataset. It managed to find 10 of the required 54 lexical items. Results using lexical items per predicate that were automatically extracted by M-ATOLL are shown in the third row in Table 3.

Despite the range of automatically and manually created lexical items, the system still failed to answer questions such as *What was Brazil's lowest rank in the FIFA World Ranking*. This is mainly due to the n-gram size used to match vocabulary elements with expressions occuring in the natural language question. Currently we consider only 4-grams, in order restrict the number of parse trees produced during semantic parsing, whereas 7-grams would be needed to map *lowest rank in the FIFA World Ranking* to the corresponding property `fifaMin`.

4   Related Work

A very prominent work on learning grammars for semantic parsing is Zettlemoyer and Collins [14], who proposed lexical induction and parameter estimation using pairs of questions and logical forms. Our learning algorithm is based on their approach but differs in the parameter estimation step, using perceptron-style updates (as in [15]) instead of gradient updates. Kwiatkowski et al. [5] proposed an approach for lexicon induction without using handcrafted domain-independent lexical items. The approach is based on an iterative splitting of the sentence and the logical form, such that the approach learns which splitting operation produces the most accurate lexical items. Preceding work by Kwiatkowski et al. [6] leverages the same splitting strategy but generalizes better by using templates for lexical items. Other work on semantic parsing with CCG is Artzi and Zettlemoyer [1,2], and Krishnamurthy et al. [4] who apply semantic parsing to open-domain question answering.

Research on applying semantic parsing to Freebase has also gained a lot of attention, examples are Cai and Yates (2013); Kwiatkowski et al. (2013); Berant et al. (2013); Berant and Liang (2014); Reddy et al. (2014). Like our system, these systems need external lexical knowledge to parse unseen expressions during the test phase.

5   Conclusion

We have implemented the semantic parsing approach by Zettlemoyer and Collins [14] and adapted it to question answering over linked data. In order to quantify the effort needed to address the lexical gap, we have analyzed the amount of entries that would be needed in order to get acceptable results on the QALD-4 benchmark. By manually adding 54 lexical entries to the seed lexicon of the semantic parser we achieve a precision of 93 % and a recall of 70 %. We have further analyzed whether these lexical entries can be induced automatically from a corpus using the state-of-the-art ontology induction system M-ATOLL. While these preliminary results bear some promise, they also clearly show that automatic methods still leave a large part of the lexical gap open, that until now can only be filled manually, and that further research and investments are needed in techniques that induce lexical entries from corpora or by crowd-sourcing in order to build successful question answering systems over linked data.

References

1. Artzi, Y., Zettlemoyer, L.: Bootstrapping semantic parsers from conversations. In: Proceedings of the Conference on Empirical Methods in Natural Language Processing, pp. 421–432. Association for Computational Linguistics (2011)
2. Artzi, Y., Zettlemoyer, L.: Weakly supervised learning of semantic parsers for mapping instructions to actions. TACL **1**, 49–62 (2013)
3. Carpenter, B.: Type-Logical Semantics. MIT Press, Cambridge (1997)
4. Krishnamurthy, J., Mitchell, M.T.: Joint syntactic and semantic parsing with combinatory categorial grammar. In: Proceedings of the 52nd Annual Meeting of the Association for Computational Linguistics, Long Papers, vol. 1, pp. 1188–1198 (2014)
5. Kwiatkowski, T., Zettlemoyer, L., Goldwater, S., Steedman, M.: Inducing probabilistic CCG grammars from logical form with higher-order unification. In: Proceedings of the 2010 Conference on Empirical Methods in Natural Language Processing, pp. 1223–1233. Association for Computational Linguistics (2010)
6. Kwiatkowski, T., Zettlemoyer, L., Goldwater, S., Steedman, M.: Lexical generalization in CCG grammar induction for semantic parsing. In: Proceedings of the Conference on Empirical Methods in Natural Language Processing. pp. 1512–1523. Association for Computational Linguistics (2011)
7. Lopez, V., Unger, C., Cimiano, P., Motta, E.: Evaluating question answering over linked data. Web Semant. Sci. Serv. Agents World Wide Web **21**, 3–13 (2013)
8. Lopez, V., Uren, V., Sabou, M., Motta, E.: Is Question Answering fit for the Semantic Web? A Survey. Semant. Web **2**, 125–155 (2011)
9. Steedman, M.: Surface Structure and Interpretation. MIT Press, Cambridge (1996)
10. Steedman, M.: The Syntactic Process, vol. 35. MIT Press, Cambridge (2000)
11. Unger, C., Forascu, C., Lopez, V., Ngonga Ngomo, A.C., Cabrio, E., Cimiano, P., Walter, S.: Question Answering over Linked Data (QALD-4). In: Cappellato, L., Ferro, N., Halvey, M., Kraaij, W. (eds.) Working Notes for CLEF 2014 Conference (2014)
12. Walter, S., Unger, C., Cimiano, P.: ATOLL - a framework for the automatic induction of ontology lexica. Data Knowl. Eng. **94**, 148–162 (2014)
13. Walter, S., Unger, C., Cimiano, P.: M-ATOLL: a framework for the lexicalization of ontologies in multiple languages. In: Mika, P., Tudorache, T., Bernstein, A., Welty, C., Knoblock, C., Vrandečić, D., Groth, P., Noy, N., Janowicz, K., Goble, C. (eds.) ISWC 2014, Part I. LNCS, vol. 8796, pp. 472–486. Springer, Heidelberg (2014)
14. Zettlemoyer, L.S., Collins, M.: Learning to map sentences to logical form: structured classification with probabilistic categorial grammars. arXiv preprint (2005). arXiv:1207.1420
15. Zettlemoyer, L.S., Collins, M.: Online learning of relaxed CCG grammars for parsing to logical form. In: Proceedings of the 2007 Joint Conference on Empirical Methods in Natural Language Processing and Computational Natural Language Learning (EMNLP-CoNLL-2007. Citeseer (2007)

Pragmatic Query Answering: Results from a Quantitative Evaluation

Jon Scott Stevens[1]([✉]), Anton Benz[1], Sebastian Reuße[2], Ralf Klabunde[2], and Lisa Raithel[3]

[1] Center for General Linguistics, Berlin, Germany
stevens@zas.gwz-berlin.de
[2] Ruhr University, Bochum, Germany
[3] University of Potsdam, Potsdam, Germany

Abstract. This paper reports on an implementation of methods for generating indirect responses in question-answering dialogue based on domain-level strategic reasoning. User's questions are interpreted as reflexes of underlying *user requirements* which are potentially satisfied by information beyond what is directly asked about. We find that the algorithms that reason about user requirements yield significantly shorter dialogues than a simpler baseline, and that users are able to interact with these systems in a pragmatically natural way.

1 Introduction

The issue of how a user-supplied question may be transformed into a database query and then used to supply an appropriate answer has been a lively topic of research in recent years (Hoffart et al. 2011; Yao et al. 2014). In the present paper, we offer a particular pragmatic perspective on processing natural language queries in the context of automatic question-answering (QA) systems. Rather than specifying a one-to-one translation procedure from question to database query, we investigate whether considering possible unobserved *user requirements* as possible motivators of the observed question, can increase the efficiency and naturalness of a user's interaction with the system. We explore an approach which permits us to address questions *indirectly* in cases where it is in both the user's and the system's best interest to supply information that was not directly asked about. We implement the strategic reasoning model of Stevens et al. (2014) to generate such responses, and report results from a comparison of this model with two baseline models—a simplified algorithm for generating indirect answers, and a baseline producing only direct answers—in terms of dialogue length and judged coherence.

Automated methods of generating truthful and contextually appropriate responses to questions have often been confined to direct modes of answering (Moriceau 2006). In contrast, it has been noted that an analysis of questions used in naturally occurring discourse must take a variety of factors into account, ranging from considerations of cooperative behavior (Potts 2008), to probabilistic reasoning based on world-knowledge (de Marneffe et al. 2009) and discourse

© Springer International Publishing Switzerland 2015
C. Biemann et al. (Eds.): NLDB 2015, LNCS 9103, pp. 110–123, 2015.
DOI: 10.1007/978-3-319-19581-0_9

structure (Bertomeu et al. 2006; Chai and Jin 2004). These results indicate that questions are not isolated instructions akin to a database query clad in natural language; rather, pragmatic factors are involved in addressing questions.

Furthermore, even in the case when a system only deals with a *semantically tractable* subset of expressions (Popescu et al. 2003), these superficially 'easy questions' are nonetheless sensitive to the interactive context of a conversation. This is apparent even in one-off exchanges of a kind one would encounter in the scope of an automated QA system. For example, consider the following exchange between a prospective buyer of an apartment and an automated realtor system.

(1)   Q.   Does the apartment have a large garden out back?
       A.   There is a nice little park right next to the building.

From (1-A), a user will likely infer his original question has received a negative answer. That is, the system's indirect answer carries the implicature that there is no large garden. Because of this implicature, the realtor can fulfill his discourse obligation, both (indirectly) answering the user's question while also supplying additional information which may be to the user's benefit. The potential benefit of the system's response stems from the possibility that the user's original question was prompted by a more complex underlying requirement—perhaps for a place to grow flowers, or for a scenic place to walk her dog—rather than by a fixed interest in the attribute the original question aimed at. In other words, questions serve not only to elicit information, but also to help solve problems in the real world.

The current work aims to assess the effect on dialogue of allowing a system to supply valid indirect answers of the kind shown in 1 given the required domain-specific knowledge. Where previous approaches to the automatic generation of indirect answers have utilized so-called *stimulus conditions* which may license surplus information to be included with the supplied answer (Green and Carberry 1994), it has also been proposed that indirect answers may arise directly from domain-level constraints and a strategic model of the interaction (Benz et al. 2011).

In order to further explore this latter approach, we present evaluation data for three alternative content determination methods deployed within the scope of a QA system which is intended to function as a virtual realtor for prospective apartment tenants. Specifically, we begin with a simple baseline algorithm meant to supply only literal yes/no answers, against which we evaluate the strategic answer generation algorithm proposed by (Stevens et al. 2014), which is based on a game-theoretic model of strategic rationality, and a variant of intermediate complexity, which does not implement the complete strategic reasoning procedure, but nonetheless represents possible unobserved user requirements. By observing how users, instructed to identify flats fulfilling certain criteria of our choosing, proceed in their task, we are able to evaluate the potential benefits of the pragmatic approach in a quantitative way. To this end, Sect. 2 presents the three candidate answer generation algorithms, which we deployed in the context of the QA system described in Sect. 3. In Sect. 4, we report on the evaluation method and results, which we subsequently discuss.

2   Three Answer Selection Algorithms

In the following, we present three alternative models of selecting an answer to a user's polar (yes/no) question. We assume that the state of the underlying system is tracked in four variables which are accessible by the answer selection algorithm:

- KB refers to a domain knowledge base in a first-order language. In the following, we refer to unary predicates within KB as *attributes*.
- FUD tracks the entity from KB considered to be the current *flat under discussion*.
- QUERY contains the question's predicate and is assumed to be defined within KB.
- HIST contains all predicates which have so far been asserted about the current FUD.

Upon completion, the selected answer is returned in the form of an ASSERT or DENY move, both of which take the asserted/negated predicate as an argument.

2.1   Polar Answers

The polar model of answer selection, laid out in Algorithm 1, supplies a truthful yes/no answer to the original question.

Algorithm 1. Polar answer selection.

---
    function SELECTANSWER(KB, FUD, QUERY)
        if KB ⊢ QUERY(FUD) **then**
            return ASSERT(QUERY)
        else
            return DENY(QUERY)
        end if
    end function

2.2   Answering with Alternatives

More involved computational models of discourse proceed from the assumption that interaction is driven by a process of inferring the agent's *underlying plan* or *goal*, and aiding him in realizing it ((Allen and Perrault 1980; Gaasterland et al. 1994; Green and Carberry 1992). Indirect answers may thus be generated in a cooperative fashion by providing information which, while it does not logically entail an answer to the original question, enables the questioner to achieve his underlying goal.

We assume here that queries are motivated by unobserved *user requirements*, where a user requirement is formalized as a set of satisfying conditions, such that a specific user desire is fulfilled if any one of those conditions holds. Where plan inference attempts to infer an intended sequence of future actions from a single action, our approach holds that in many scenarios, user questions are motivated by simple unordered sets of disjunctive requirements which could potentially be satisfied either by a direct answer to the question, or else by a single alternative piece of information. For our purposes, a satisfying condition is taken to be an attribute statement of the form *the flat under discussion has attribute a*, for which we simply write a. If we let A be the set of all possible attributes a, then a user requirement is taken to be a subset of A.

To give a concrete example, a customer in dialogue with a real estate agent might require that the apartment she is interested in have a place outside to grow flowers. This is obtained if the apartment in question has either the attribute *balcony* or else the attribute *garden*. We can thus represent a requirement {*balcony, garden*}, where either member of this set serves to satisfy the customer's need. If the customer asks whether there is a balcony, in the case that the user's question is motivated by this requirement, then offering the information that there is a garden is more helpful than simply saying there is no balcony. Moreover, the user should be able to pragmatically infer from this alternative alone that there is no balcony, because if there were, a more straightforward direct answer would have been at least as helpful to the customer. This approach has its roots in decision- and game-theoretic analyses of question answering (Benz and van Rooij 2003; van Rooij 2007).

We now introduce the answer generation algorithm of (Stevens et al. 2014), which we dub the *strategic* model of generation. It proceeds from a model of goal-driven rationality, by assuming that a customer chooses her question in a principled way. Specifically, the customer chooses her question such that a direct answer to that question could determine whether an underlying requirement is met. The sales agent knows this to be the case, and, drawing upon a known space of plausible requirements, can choose to provide an indirect answer that might equally address the customer's motivating requirement. This strategy can be shown to be part of an equilibrium in a signaling game between a seller and a buyer (Stevens et al. 2014). The answer selection algorithm derived from this observation implements the seller's in-equilibrium behavior and allows the system to reason about likely underlying requirements based on the customer's chosen predicate.

Given a set of attributes, A, let the set of possible user requirements be $D \subset \mathcal{P}(A)$.[1] Given a requirement $d \in D$, we assume uniform prior probability over D, i.e.,

$$p(d) := \frac{1}{|D|}$$

Moreover, we assume that each attribute a in a requirement set d is an equally likely option to find out whether d is satisfied. Accordingly, we define:

---

[1] (Stevens et al. 2014) use "$d \in D$" to denote requirement sets, since in their framework these sets are shorthand representations of *decision problems*, which are formally more elaborate and do not need to be elucidated here. While we refer to these sets simply as *requirements*, we carry their notation over here.

$$p(a|d) := \frac{1}{|d|} \text{ if } a \in d \text{ and } 0 \text{ otherwise}$$

Using Bayes' rule, it follows that:

$$p(d|a) = \frac{p(a|d) \times p(d)}{p(a)}$$

where $p(a) = \sum_{d' \in D} p(a|d') \times p(d')$.

Given some query, we can now assign a score, p_{comp}, encoding the degree of "compatibility" between the observed query and an alternate answer which addresses a different question, query', based on how likely it is we could have observed query' from someone who issued the original query.

$$p_{comp}(\text{query}', \text{query}) := \sum_{\substack{d \in D \\ \text{query}' \in d}} p(d|\text{query})$$

It follows from a game-theoretic model that the interaction between system and user is in equilibrium when the user chooses his queries so that they are relevant to his requirements, and when the system assumes the user to behave in this fashion and maximizes p_{comp}. Such a system is realized in Algorithm 2.

Algorithm 2. Strategic answering with alternatives.

---

function SELECTANSWER(KB, FUD, QUERY, HIST)

    ANSWER ← $\underset{\substack{a \in A \setminus \text{HIST} \\ \text{KB} \vdash a(\text{FUD})}}{\text{argmax}} p_{comp}(a, \text{QUERY})$

    ▷ Return optimal answer, but don't repeat alternatives already in HIST.

    if $p_{comp}(\text{ANSWER}, \text{QUERY}) = 0$ **then**

        ▷ No true alternatives were found.

        return DENY(QUERY)

    else

        return ASSERT(ANSWER)

    end if

end function

This algorithm will return a "yes" answer to a query if it is true, since that answer is necessarily compatible with any user requirement which could have prompted the query, thus $p_{comp}(\cdot, \text{query}) = 1$. On the other end of the spectrum, $p_{comp}(\cdot, \text{query})$ will be zero for all answers that do not overlap with a known user requirement that contains the query. For example, if there is no plausible requirement containing both *garden* and *basement*, then $p_{comp}(basement, garden)$ cannot be positive.

Illustration. Consider Table 1a, which defines two plausible unobserved user requirements a prospective buyer of an apartment might wish to address.

Table 1. An example domain listing user requirements and the derived p_{comp} values.

(a) A set of two possible user requirements in the real-estate domain. Columns show desires a customer might have; rows list possible satisfying conditions.

Flat has...	Grow flowers	Walk dog
Balcony	👍	👎
Large garden	👍	👍
Nearby park	👎	👍
Community garden	👍	👎

(b) p_{comp} values resulting from the example domain defined in Table 1a.

Observed query	p_{comp}			
	Balcony	Garden	Park	Comm. garden
Balcony	1	1	0	1
Garden	.4	1	.6	.4
Nearby park	0	1	1	0
Comm. garden	1	1	0	1

Assuming the domain shown in Table 1a, Table 1b lists all p_{comp} values given the set of possible queries. Suppose the flat under discussion features all attributes except a garden. In that case, Algorithm 2 will preferentially answer a "garden" query by referring to the presence of a nearby park, as it is maximally likely that this answer, above all other true alternatives, is compatible with the user's needs.

2.3 Alternatives Without Probabilities

Although the probability ranking function in 2 provides a concise way of representing the validity of potential positive answers to yes/no questions, we wonder whether in some simple contexts a similar output could be generated by a simpler variant where potential alternatives are weighted equally. As a point of comparison, we implement such a variant as follows. Given some attribute $a \in A$, we can define the function:

$$\text{relevantReqs}(a) := \{d | a \in d, d \in D\}$$

From this we can calculate a set of possible alternatives to a "no" answer to a question as follows, which forms the basis of Algorithm 3.

$$\text{possibleAlts}(a) := \cup \text{relevantReqs}(a) \setminus \{a\}$$

3   Implementation

Our three candidate algorithms were deployed within a dialogue system based on the information state update (ISU) approach to dialogue modeling (Traum and Larsson 2003). Under the ISU approach, dialogue is modeled by means of a formal description of an agent's *information state* (such as his private beliefs,

---

Algorithm 3. Non-probabilistic answering with alternatives.

---

function SELECTANSWER(KB, FUD, QUERY, HIST)
   if KB ⊢ QUERY(FUD) **then**
      return ASSERT(QUERY)
   else
      ALTERNATIVES ←
            {$a|a \in$ possibleAlts(QUERY)
                KB ⊢ a(FUD)}
      ALTERNATIVES′ ← ALTERNATIVES \ HIST
      ▷ Disprefer repetition.
      if ALTERNATIVES′ $\in \emptyset$ **then**
         CHOICE ← random(ALTERNATIVES′)
         return ASSERT(CHOICE)
      else
         return DENY(QUERY)
      ▷ Fall back to direct answer.
      end if
   end if
end function

---

intentions and plans, as well as shared beliefs such as the common ground), and a set of *dialogue moves*, which may trigger updates to the information state (e.g., asserting a proposition, making a request, etc.). Implementing an ISU-based model entails creating one or more *dialogue move engines* (DME; Larsson and Traum 2000), which perform the aforementioned updates to the information state, as well as a *control algorithm*, which manages the application of the system's other modules, such as input and output, and sequences these with the DME's update procedures.

A simple dialogue system intended to perform the task of a cooperative real-estate agent was built using a modified version of the PyTrindiKit toolkit.[2] The system's information state was defined as a record consisting of a single data-base object considered the flat under discussion (FUD), as well as a history of previously discussed objects and prior assertions made by the system. System input and output, in terms of dialogue move types, and their respective informa-tion state update effects, are listed in Table 2. As a control algorithm, a simple sequential scheme was employed, where input from the user would alternate with system output. This is outlined in more detail in Sect. 4.

Each selection algorithm was implemented as an update rule triggered by a user's YNQUESTION. The selection algorithm would then produce a DENY or an ASSERT, where the asserted predicate was selected by the algorithm.

The knowledge base back end was implemented using an adapted version of PyKE, a Horn logic theorem prover[3].

---

[2] https://code.google.com/p/py-trindikit.
[3] http://pyke.sourceforge.net/.

Table 2. Move types and corresponding update effects defined by the system. Move contents specify information associated with move tokens; this information may then be accessed by update rules.

Move type	Generated by	Move contents	Update effect
ACCEPT	User	None	Dialogue is ended
REJECT	User	None	The currently focused database object is cleared
YNQUESTION	User	An attribute	A token is added to the system's agenda. This token subsequently causes the system to provide an answer during its turn
ASSERT	System	An attribute	None
DENY	System	An attribute	None
FLATPROPOSAL	System	A flat ID	The current flat under discussion is set

We deployed our system in the domain of *real-estate sales talk*. The system was set up to emulate the behavior of a real-estate agent tasked with answering customers' (polar) questions for a range of attributes pertaining to individual flats. A set of 12 predicates was chosen to model the attributes of objects within our domain.

The pragmatic systems proceeded from the assumption that user queries are motivated by one of twelve possible user requirements, which were defined as disjunctions over sets of attributes. The chosen requirements were defined so as to be partially overlapping; i.e., a single attribute was relevant to more than one. Additionally, some of these possible requirements were singletons, i.e., did not allow for any helpful alternative solutions.

4   Evaluation

Subjects interacted with our system by means of an online interface accessible remotely through a web browser. At the outset of the experiment, subjects were tasked with identifying, among a sequence of presented flats, a flat which would satisfy a set of supplied requirements. One out of a total of four lists, each containing three requirements, was assigned to subjects at random. The requirements were constructed by the researchers to be plausible desiderata for users looking for an apartment to rent or buy (e.g., connection to public transit, which could be satisfied either by a nearby bus stop, or by a nearby train station).

Subjects' interaction with our system proceeded sequentially. After an initial offer on the part of the system, users were able to issue a polar query for the presence of each of the attributes relevant to their issued tasks. Once subjects had determined with certainty whether the discussed flat matched their issued goals, they could either accept the flat, ending the conversation, or reject it, triggering a new proposal by the system. Referents in flat proposals were introduced using fake, non-evocative address descriptions.

The flats presented by the system were individually generated for each participant type, such that the knowledge base would contain one satisfying instance for each possible combination of the three conditions issued to subjects, plus two additional flats that satisfied two of the conditions (i.e., $2^3 + 2 = 10$ flats overall). For each flat, if a condition was not satisfied, an alternative attribute (e.g., a garden, if there is no balcony) was added with 50 % probability in order to create more opportunities for indirect answers. Crucially, the sequence in which flats were presented was fixed so that the last flat offered would be the sole object satisfying all of the desired criteria; this allows us to obtain an apples-to-apples comparison of complete dialogues with respect to their length. If subjects failed to accept the optimal solution, the interaction was terminated.

Each subject was randomly assigned one of the three question-answering systems: either the *literal* system (Algorithm 1), which produces simple yes/no answers, the *strategic* system (Algorithm 2), or the intermediate *random alternative* system (Algorithm 3). After completing interaction with our system, subjects were asked to complete a short survey, asking them to rate the perceived coherence of the system's answers (i.e. relatedness/relevance to question, helpfulness, evasiveness, and whether answers were left open) on a seven-point Likert scale.

We predict that the random alternative and strategic models, which we dub the *pragmatic* models, will improve overall efficiency of dialogue above that of the literal system by both (i)offering helpful alternative solutions to the customer's problem, and (ii) allowing customers to infer implicit "no" answers from indirect answers, leading to rejections of sub-optimal flats. Due to the simplified nature of the dialogue task—there are only three given requirements per customer, with a maximum of three satisfying conditions per problem—we do not predict any large difference between the two pragmatic models. If, contrary to our hypothesis, subjects fail to draw implicatures from indirect answers, then we expect subjects to repeat questions (in order to obtain a direct answer) at a rate proportional to the number of unhelpful alternatives given (i.e., alternatives which are not an element of the user's requirement set).

With respect to the questionnaire items, the literal system is predicted to be judged maximally coherent, since only straightforward yes/no answers are offered. The question is whether the pragmatic systems also allow for coherent dialogue. If subjects judge indirect answers to be incoherent, then we expect the difference in average Likert scale ratings between pragmatic and literal systems to reflect the proportion of indirect answers given by the pragmatic systems.

4.1   Results

We obtained 99 subjects via Amazon Mechanical Turk, of which 73 completed the dialogue task. We found no meaningful effect of model type on whether the task was completed. Four outliers—two from the random alternative condition and one each from the literal and strategic conditions—were excluded for being

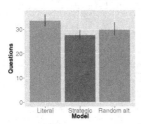

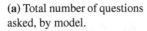

(a) Total number of questions asked, by model.

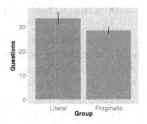

(b) Total number of questions asked (pragmatic models vs. literal model).

Fig. 1. Results on dialogue length.

more than 1.5 interquartile ranges above the third quartile with regard to total number of questions.[4] There were no outliers below the first quartile. This left us with 69 complete dialogues in total, 26 from the literal condition, 22 from the strategic condition and 21 from the random alternative condition. The two pragmatic conditions both yielded about 19 % indirect answers, with no meaningful difference between the two.

We begin by analyzing the number of questions asked during dialogues in which the optimal solution was found. This is used as a proxy for dialogue length, as each question received a single answer, and each complete dialogue contained discussion of exactly 10 database objects. Figure 1a breaks down average number of questions by model. The strategic model yields significantly fewer questions than the literal model (27.6 vs. 33.7). The random alternative model yields a somewhat higher average than the strategic model (29.9), but the difference is not statistically significant. Our hypothesis predicts that the difference in mean for the two pragmatic models grouped together (28.7) compared to the literal model (33.7) will be statistically significant. Three statistical tests bear this out. First, 95 % confidence intervals for the two groups (obtained via bootstrap resampling) are non-overlapping (see Fig.1b). Second, a one-way analysis of variance (ANOVA) testing the effect of group choice (pragmatic vs. literal) on number of questions yields significance ($F = 6.15$, $p = 0.002$). Finally, to allow for the possibility of random item effects (the items in this case being the set of requirements randomly assigned to the subject), we tested a mixed effects regression model with a fixed effect of group choice and random effect of item. Again, number of questions is predicted by whether the model is pragmatic ($t = -3.14$, $p = 0.003$).

We now ask whether the observed effect is due only to the presence of helpful alternatives, or whether the presence of implicatures contributes to dialogue efficiency. The example in Fig. 2, taken from a dialogue with the strategic system,

---

[4] These subjects simply asked all possible questions for most flats, failing to reject even if it was immediately clear that the flat under discussion was sub-optimal.

```
S: I can also offer you an apartment in the Brownstone
   district.
U: Does the apartment have a train station nearby?
S: Well, there is a bus stop nearby.
U: Does the apartment have a pub nearby?
S: There is a pub near the apartment.
U: Does the apartment have a balcony?
S: Well, it has a garden.
U: [Rejects]
```

Fig. 2. Part of an unfolding interaction between system (S) and user (U). The user was instructed to find a flat for a friend with a balcony, a nearby place to get a drink (either café, pub or restaurant) and a nearby public transit connection (either bus or train).

exemplifies both effects.[5] By offering the information that there is a bus stop, the user's underlying problem motivating that question (requiring close public transit) is shown to be solved, allowing the user to move on. Then, when asked whether there is a balcony, the system offers the garden attribute as an alternative. Presumably drawing from this the implicature that there is in fact no balcony, the user immediately rejects.

If we take the null hypothesis to be that subjects do *not* draw such negative implicatures, then we expect a certain proportion of questions asked to the pragmatic systems about a particular flat to be repeated. Subjects are allowed to ask questions multiple times, and alternatives are never presented twice, such that repeating questions will ultimately lead to a direct yes/no answer. We do see some instances of this behavior in the dialogues. If this is indicative of an overall difficulty in drawing pragmatic inferences from an online dialogue system, then we expect the number of such repetitions to reflect the number of unhelpful alternatives (such as garden for balcony in Fig. 2) that are offered. We find no evidence of this. Proportion of unhelpful alternatives does not correlate in any significant way with proportion of repeated questions. Moreover, we can estimate the expected proportion repeated questions if the null hypothesis is true by: (i) taking the proportion of unhelpful answers (which should prompt at least one repetition of the question) and (ii) adding the proportion of repeated questions in the literal condition, taken to be an estimate of the baseline level

---

[5] Early feedback on the system indicated that the indirect answers were more natural when preceded by a discourse connective like "well", which we included for our evaluations. Shortly before the time of writing we ran the same experiment again but without the discourse connective. Looking only at those subjects who finished the task (50 total), the effects reported here were replicated; however, in contrast with the first experiment, when we looked at all subjects who asked more than a single question (72 total), we found a large group effect (a 69 % decrease for the pragmatic group) on whether the subjects successfully completed the task (a mixed effects binomial regression yields significance, with $z = -2.19$, $p = 0.03$). This is consistent with initial intuitions that the discourse marker makes the answers more natural, thus making the task easier and/or more pleasant.

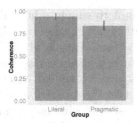

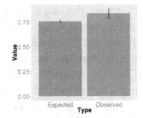

(a) Coherence score (pragmatic models vs. literal model).

(b) Coherence score across the two pragmatically motivated models.

Fig. 3. Results on perceived coherence.

of question repetition due to noise (since there are no indirect answers in that condition). By this metric we would expect somewhere around 6.4 % redundancy in the pragmatic conditions. The observed value is less than one percent.

Finally, a rating of perceived coherence was prompted by several questionnaire items. We obtained a composite measure, the *coherence score*, from all coherence-related items on the seven point Likert scale by summing all per-item scores for each subject and normalizing them to a unit interval, where 1 signified the upper bound of perceived coherence. Figure 3a shows, as one might expect, that the pragmatic models show lower scores than the literal model. The difference is significant (one-way ANOVA yields $F = 7.52$, $p = 0.008$, mixed effects model yields $t = -2.74$, $p = 0.008$). While this suggests room for improvement vis-à-vis the naturalness of the indirect answers provided, we emphasize that the ratings given are nonetheless quite high. At about 84 % coherence on average, the pragmatic models exhibit higher scores than would be expected if indirect answers contributed no coherence (i.e., if the answers were categorically judged to be non-helpful, evasive, or not answering or relating to the question). If we take the 0.94 composite rating of the literal condition to be the expected maximum value, we can calculate expected coherence given the null hypothesis that coherence correlates negatively with proportion of indirect answers. We simply multiply the expected maximum value by the number of direct answers given by the pragmatic models. Figure 3b compares the mean expected values with the mean observed values. Observed values are significantly higher (one-way ANOVA yields $F = 8.11$, $p = 0.006$).

5   Discussion

We implemented two variants of a pragmatically motivated answer-generating system in a sales domain, based on the notion that unobserved user requirements provide an underlying motivation for questions in this kind of dialogue. A user requirement is taken to be a set of potential conditions under which a particular need is met. In both variants, the system is supplied with a space of possible customer requirements, and when asked a simple yes/no query about an attribute which the current flat does not have, it will attempt to supply a helpful alternative by considering what positive attribute statements might solve whatever problem motivated the question. The first variant, from Stevens et al. (2014), maximizes a probability measure corresponding to how likely a response is to be helpful to the customer. The second variant achieves a similar outcome by constructing a set of candidate requirements based on the question asked, then selecting a response from a set of alternatives consistent with one or more of those candidates. These were evaluated against a baseline which produces only direct answers.

The indirect ("pragmatic") models allow for more efficient dialogue. Similarly to a plan inference system, providing alternate ways of satisfying requirements can allow the user to avoid follow-up questions by letting them know that a particular need is met. Moreover, if the customer's immediate need is not met, we have found that they are able to pragmatically infer from an indirect answer that this is the case, which further contributes to the efficiency of the dialogue. Participants rated the dialogues along a number of dimensions relating to dialogue coherence. While the pragmatic models were judged slightly less coherent than the baseline, the ratings were nonetheless reasonably high, and we found no evidence that the number of indirect answers given correlated inversely with perceived dialogue coherence.

This represents a further step in the direction of pragmatically competent dialogue systems. Indirect answers and the implicatures which they carry are a natural part of dialogue in many domains, and the current work suggests that notions from decision theory and game theory can be useful theoretical tools from which to draw inspiration for work on dialogue systems. Current research is extending the practicality of this approach, e.g. by considering how a space of possible user requirements can be learned rather than being pre-supplied.

References

Allen, J.F., Perrault, C.R.: Analyzing intention in utterances. Artif. Intell. **15**(3), 143–178 (1980)

Benz, A., Bertomeu, N., Strekalova, A.: A decision-theoretic approach to finding optimal responses to over-constrained queries in a conceptual search space. In: Proceedings of the 15th Workshop on the Semantics and Pragmatics of Dialogue, pp. 37–46 (2011)

Benz, A., van Rooij, R.: Optimal assertions, and what they implicate: a uniform game theoretic approach. Topoi **26**(1), 63–78 (2007)

Bertomeu, N., Uszkoreit, H., Frank, A., Krieger, H.-U., Jörg, B.: Contextual phenomena and thematic relations in database QA dialogues: results from a wizard-of-oz experiment. In: Proceedings of the Interactive Question Answering Workshop at HLT-NAACL, pp. 1–8 (2006)

Chai, J.Y. Jin, R.: Discourse structure for context question answering. In: Proceedings of the Workshop on Pragmatics of Question Answering at HLT-NAACL, pp. 23–30 (2004)

de Marneffe, M.-C., Grimm, S., Potts, C.: Not a simple yes or no: uncertainty in indirect answers. In: Proceedings of the SIGDIAL 2009 Conference: The 10th Annual Meeting of the Special Interest Group on Discourse and Dialogue, pp. 136–143 (2009)

Gaasterland, T., Godfrey, P., Minker, J.: An overview of cooperative answering. J. Intell. Inf. Syst. **1**(2), 123–157 (1992)

Green, N., Carberry, S.: Generating indirect answers to yes-no questions. In: Proceedings of the 7th International Workshop on Natural Language Generation, pp. 189–198 (1994)

Hoffart, J., Suchanek, F.M., Berberich, K., Lewis-Kelham, E., De Melo, G., Weikum, G.: YAGO2: exploring and querying world knowledge in time, space, context, and many languages. In: Proceedings of the 20th International Conference Companion on World Wide Web, pp. 229–232. Association for Computational Linguistics (2011)

Larsson, S., Traum, D.R.: Information state and dialogue management in the TRINDI dialogue move engine toolkit. Nat. Lang. Eng. **6**(3&4), 323–340 (2000)

Moriceau, V.: Generating intelligent numerical answers in a question-answering system. In: Proceedings of the 4th International Natural Language Generation Conference, pp. 103–110. Association for Computational Linguistics (2006)

Popescu, A.-M., Etzioni, O., Kautz, H.: Towards a theory of natural language interfaces to databases. In: Proceedings of the 8th International Conference on Intelligent User Interfaces, pp. 149–157. ACM (2003)

Potts, C.: Indirect answers and cooperation: on asher and lascarides's making the right commitments in dialogue. http://www.stanford.edu/cgpotts/commentaries/potts-umich08-cmts-on-asherlascarides.pdf (2008). Accessed 2 Sept 2013

Stevens, J.S., Benz, A., Reusse, S., Laarmann-Quante, R., Klabunde, R.: Indirect answers as potential solutions to decision problems. In: Proceedings of the 18th Workshop on the Semantics and Pragmatics of Dialogue, pp. 145–153 (2014)

Traum, D.R., Larsson, S.: The information state approach to dialogue management. In: van Kuppevelt, J., Smith, R.W. (eds.) Current and New Directions in Discourse and Dialogue. Text, Speech and Language Technology, pp. 325–353. Springer, Netherlands (2003)

van Rooij, R.: Questioning to resolve decision problems. Linguist. Philos. **26**(6), 727–763 (2003)

Yao, X., Berant, J., Van Durme, B.: Freebase QA: information extraction or semantic parsing?. In: Proceedings of the ACL 2014 Workshop on Semantic Parsing, pp. 82–86. Association for Computational Linguistics (2014)

What was the Query? Generating Queries for Document Sets with Applications in Cluster Labeling

Matthias Hagen[(✉)], Maximilian Michel, and Benno Stein

Bauhaus-Universität Weimar, Weimar, Germany
{matthias.hagen,maximilian.michel,benno.stein}@uni-weimar.de

Abstract. We deal with the task of generating a query that retrieves a given set of documents. In its abstract form, this can be seen as a "compression" of the document set to a short query. But the task also has a real-world application: cluster labeling (e.g., for faceted search). Our solution to cluster labeling is the usage of queries that approximately retrieve a cluster's documents. To be generalizable, our approach does not require access to a search index but only a public interface like an API. This way, our approach can also be implemented at client side.

In an experimental evaluation, a basic version of our approach using a simple retrieval model is on par with standard cluster labeling techniques. A further user study reveals that queries as labels are often preferred when they are not too long.

1 Introduction

In this paper, we study the problem of generating a query that would retrieve a given set of documents from some search interface. At first glance, the problem itself seems rather abstract and only of theoretical interest. However, we suggest it as a means to identify good human-understandable labels for document clusters. The labels should tell the users something about the contained documents. In our opinion, many users nowadays conceptually connect search queries with document sets—the returned results. We thus exploit this connection by using as cluster labels such queries that approximately retrieve the documents from one cluster but not from the others.

Our approach does not require full access to some search index; a public interface like an API is sufficient. This way, our approach is applicable even at client side. However, the full potential can be utilized at search engine side when for instance generating search result facets that provide some clues on what the results are about. As facets are only useful with good labels, we propose to cluster the original query's result set and to provide other search queries as labels for the different clusters/facets. In this way, facets could work similar to query suggestions. By clicking on a facet, the user implicitly submits the label as a search query and is provided with a set of results—as accepted and expected by many users.

© Springer International Publishing Switzerland 2015
C. Biemann et al. (Eds.): NLDB 2015, LNCS 9103, pp. 124–133, 2015.
DOI: 10.1007/978-3-319-19581-0_10

In a user side scenario, the constructed queries can also be seen as a way of "compressing" a document set using the search engine as the "compression" algorithm. Instead of the whole document set, just the query could be stored. Against some retrieval system that does not change too frequently (e.g., some research search engines but probably not the big commercial search engines), the query in some sense contains all the information necessary to retrieve the document set again. However, the main use case of our approach at user side is that of labeling small to medium sized document clusterings. Our algorithm can derive queries for each cluster that approximately retrieve the documents from the respective cluster. To this end, it is not even necessary to build a fully-fledged search engine for the whole clustering but some on-the-fly computations of retrieval scores would suffice.

We envision the usage of queries as cluster labels as particularly promising due to the nature of queries. Traditional cluster labeling approaches often purely rely on text statistics. However, many users accept queries as the dominant way of retrieving a set of documents from a larger collection. Using queries as labels, we are able to go beyond the simple text statistics model of traditional cluster labeling such that we can exploit all the tools developed for effective document retrieval and make them applicable to cluster labeling itself.

In an empirical evaluation, we show our query-based labels to be on par with standard approaches. We examine the label quality with classic measures of similarity to human-generated labels (i.e., Jaccard index or F-measure) and we also develop a new semantics-aware quality measure based on ESA. Additionally, we conduct a user study to manually assess the usefulness of the generated query labels. In all experiments it turns out that queries are a good means of labeling when they are not too long.

2 Related Work

Query Formulation. Fuhr et al. suggest an optimum clustering framework based on vectors of document-query similarities [7] that inspired our idea. One way of storing such important queries for a document is the reverted index [17] that we will also employ. For deriving queries for a *single* document, several strategies from the literature [3,5,24] were shown not to perform as well as the approach by Hagen and Stein [11] that also inspired our idea. However, contrary to the above single-document query formulation approaches, our scenario requires queries that retrieve complete document *sets*. This problem was first examined by Jordan at al. [13] who used language models based on full access to corpus statistics. Instead, we are focusing on a black-box scenario where we just apply the public search engine interface. Bonchi et al. [4] deal with a scenario very similar to ours. For a given result set of a query, they want to find queries in a query log that "cover" the result set in a set-cover manner. We generalize their setting by not requiring any log information but simply relying on public interfaces as in the maximum query setting [10].

Cluster Labeling. We suggest to use queries as a new approach to cluster labeling. In general, there are two different strategies applied to cluster labeling: differential cluster labeling and cluster-internal labeling [14]. Differential cluster labeling compares term distributions within a cluster to the distributions of other clusters. A very effective such approach is based on the χ^2-test yielding labels of k terms that have a high weight according to their presence within the cluster and their "absence" outside of the cluster [6]. The cluster-internal labeling methods instead simply construct labels from the terms appearing within a cluster's *centroid* document—a prominent example being the weighted centroid approach (WCC) [21] identified as a simple yet very effective technique based on $tf \cdot idf$ weights in a recent cluster labeling comparison [15]. Our own approach will be a mixture of both general strategies: we also exploit the centroid document to identify candidate terms as a form of cluster-internal labeling but then derive queries by paying attention to the result set in comparison to the whole clustering as a form of differential cluster labeling. A drawback for both approaches (WCC and χ^2-test) is that the size k of the label (number of desired terms) has to be pre-determined whereas in our scenario it is automatically derived. Whenever the query is not descriptive enough, another term is added. We compare our query-based labels to WCC and the χ^2-test on the AMBIENT dataset that has been applied in different clustering studies [16, 22, 23].

3 Approach

We first describe our basic approach of generating a query for a given document set against a search engine interface. In the second part, we apply this approach to cluster labeling.

3.1 Generating Queries for Document Sets

The goal of generating a query for a given document set is to find a keyword (or keyphrase) combination that approximately returns the given document set from a search engine interface but not too many other documents. In a web search scenario this setting may seem rather artificial. It becomes more applicable and tractable when in the use case of cluster labeling the search engine is set up only for the documents in the clustering (typically much smaller than the web). Still, also against some web search engine, our approach is able to "compress" a given document set to a short query. In both settings, we treat the retrieval system as a black box. Thus, no real information about the employed retrieval model or about the index structure can be used. Similar to other approaches [2, 12], only the public black-box search interface needs to be available.

Reverted Index. To store some information about the to-be-retrieved document set, we employ a reverted index [17]. Instead of mapping document IDs to index terms as in the traditional inverted index, the reverted index stores for each document the queries that return that document. Pickens et al. [17] originally

Input: document set D, RevertedIndex
Output: query term candidates W_{cand}
1: $Map \leftarrow \emptyset$
2: **for all** $d \in D$ **do**
3:   $W_d \leftarrow$ RevertedIndex(d)
4:   **for all** $w \in W_d$ **do**
5:     $Map(w) \leftarrow Map(w) + 1$
6: $W_{cand} \leftarrow \emptyset$
7: **for all** $w \in Map$ **do**
8:   $\#d \leftarrow Map(w)$
9:   $weight \leftarrow \#d/|D|$
10:   $W_{cand} \leftarrow W_{cand} \cup \{\langle w, weight\rangle\}$
11: Sort W_{cand} by decreasing weight
12: **return** W_{cand}

Input: D, W_{cand}, threshold k
Output: query q with D in top-k results
1: $v \leftarrow 0$
2: $q \leftarrow \emptyset$
3: **for all** $w \in W_{cand}$ **do**
4:   $q \leftarrow q \cup \{w\}$
5:   $D_{top\text{-}n} \leftarrow$ top-k results of q
6:   $v' \leftarrow |D_{top\text{-}n} \cap D|/|D|$
7:   **if** $v' \geq v$ **then**
8:     $v \leftarrow v'$
9:   **else**
10:     $q \leftarrow q \setminus \{w\}$
11:     **break**
12: **return** q

Fig. 1. Left: Identifying candidate terms. Right: Greedy combination of candidate terms.

suggest to use query logs or frequent terms as the basis queries to automatically populate the reverted index. Each returned document in the top-k results of some basis query (e.g., the top-1000 results) becomes a key for some postlist in the reverted index. The postlist contains the queries that return the document weighted by the rank at which the document appears (i.e., the first queries rank the document higher than later queries in a postlist). Note however that query logs are not always available and that using frequent terms may result in problems of retrievability [1].

Constructing the Basis Queries. Since we do not have up-to-date query logs at our disposal, we can only employ Pickens et al.'s suggestion of using frequent terms as the basis queries [17] but will adapt it to the use case of cluster labeling. Given a document set, we first automatically construct its centroid document. To this end, the documents are represented as *tf*-vectors (stopwords removed) and the centroid document is the arithmetic midpoint of the resulting vector space. One can think of the terms in the centroid document as the ones that on average appear at least once in each document. One crucial point is that in an online scenario of generating a good query for a given document set, each of the basis queries needs processing time when automatically submitted to a search engine. For a faster response time, we propose to have a cut-off value of using at most n terms for the basis queries. In a pilot study on the AMBIENT dataset (also used in our evaluation), the centroid document on average contained about 90 terms which we choose as the cut-off value for n.

Query Generation with the Reverted Index. The query generation using the reverted index runs in three phases: (1) constructing the reverted index on the fly for the given document set, (2) identifying candidate terms, and (3) composition of a good query from the candidates.

To construct the reverted index, we submit the centroid document's terms as basis queries. Having the reverted index at hand, we assign weights to the terms in the index according to the number of documents they retrieve from the document set and return the terms by decreasing weight. The respective algorithm is given in the left part of Fig. 1.

We can then combine the candidate terms to a final query in a third phase. The goal is to find a query that returns as many of the documents from the given document set as possible. To this end, we propose a greedy strategy (cf. the right part of Fig. 1). The algorithm adds terms from the candidate list to a query q. Whenever the returned result list does not get worse (i.e., does not return less of the documents from the given document set), the term is added to the query. Otherwise, it is dropped and the combination process stops since we expect the remaining terms to be of even worse quality given their smaller weight. If time is not an issue, the combination could also proceed in a backtracking manner and test several queries from which the shortest or otherwise best might be chosen.

3.2 Application to Cluster Labeling

The described query formulation approach can be easily transferred to the task of cluster labeling. The research question then is whether queries can serve as promising cluster labels.

Query formulation in the context of cluster labeling can be seen as a mixture of cluster-internal and differential labeling. The first phase of term selection is completely internal based on the cluster's centroid document. However, when weighting the terms and combining them to a query, the information of how many documents from different clusters are retrieved, is exploited. The constructed query for one cluster should return as many documents of that cluster but as few documents as possible from other clusters.

We view each of the candidate terms as a classifier that selects documents from the desired cluster and documents from the other clusters. As a weighting scheme, we propose the F-Measure derived from the recall of documents from the desired cluster and the precision in form of the retrieval of only few documents from other clusters. Note that these values can also be computed on the reverted index when constructed for the whole clustering. The set of documents that ideally should not be contained in the retrieved results forms a slight difference to the general query formulation from above. But apart from that slight difference (adding F-Measure weighting), the greedy combination works as described before.

4   Evaluation

We compare our new query-based cluster labeling approach to standard approaches from differential and cluster-internal labeling: the χ^2-test labeling [14] and weighted centroid covering [21]. Both performed very well in a recent cluster labeling study [15].

Our evaluation is divided into two parts. First, we compare the labels with traditional measures: Jaccard index and cosine similarity to reference labels. As a new measure taking also semantic similarity into account, we also propose an ESA-based similarity [8] of a generated and a reference label. This newly proposed measure is also a contribution in itself to cluster labeling evaluation.

Table 1. Average label quality (791 AMBIENT subtopics with Wikipedia disambiguation description as the reference label). The computed labels' quality is measured by the traditional measures F-Measure (precision and recall of the computed label terms against the reference), Jaccard index (overlap of computed and reference terms), and cosine similarity of the tf-weighted term vectors of the computed and the reference labels, as well as the newly proposed ESA similarity between the computed and the reference label. Bold font depicts the best approach in a row.

	Query generation	χ^2	Weighted centroid covering
F-Measure	0.103	**0.137**	0.056
Jaccard index	0.051	**0.068**	0.028
Cosine similarity	**0.367**	0.352	0.188
ESA similarity	**0.443**	0.434	0.311

Second, to complement the machine-computable measures, we also conduct a small-scale user study on the quality of the derived labels.

4.1 Evaluation Corpus

Our evaluation corpus is based on the AMBIENT dataset[1] often used in cluster evaluation [16,22,23]. The dataset contains 44 topics referring to ambiguous terms with a Wikipedia disambiguation page. The short descriptions of the 791 subtopics in the disambiguation pages form the reference labels. The original corpus contains documents obtained by submitting the 44 topics to a commercial search engine. However, since only the top-100 documents for each of the 44 topics were fetched and some topics contain as many as 37 subtopics, there are a lot of subtopics with only very few or no assigned documents. To enlarge the corpus, we submitted all the 791 subtopics as search queries to the Bing API and fetched the top-50 results for each query. Note that in the evaluation, we do not run a clustering algorithm but use the "correct" clustering given by the enlarged AMBIENT subtopics' document sets as the reference—a standard procedure in evaluating cluster labeling.

We set up a BM25F index [19,20] for the enlarged AMBIENT corpus. To simulate web-scale search, queries against this small index are also submitted to the BM25F-based ChatNoir search engine [18] for the ClueWeb09. The results of our local AMBIENT search and the accompanying ChatNoir search are always merged using the BM25F-scores.

4.2 Automatic Label Evaluation

For each of the 791 subtopics, the three cluster labeling approaches χ^2-test, weighted centroid covering, and our newly proposed query-based method are run. In a first evaluation phase, we employ the standard evaluation scheme

---

[1] http://credo.fub.it/ambient/, last accessed: May 20, 2014.

Table 2. User study results for the query-based labels ("Query"), the χ^2-based labels, and the weighted centroid covering ("WCC"). Shown are the absolute and relative number of votes from our 29 participants on the 100 sampled subtopics. The last two columns show for how many of the subtopics an approach received the most votes ("Winner") and the absolute majority of votes.

Approach	User votes (absolute)	User votes (relative)	Winner	Absolute majority
Query	1276	0.44	43	31
χ^2	1160	0.40	33	16
WCC	463	0.16	4	2
Total	2900	1.00	80	49

of comparing the reference labels in the AMBIENT dataset (the disambiguation descriptions) to the computed labels. Standard measures of similarity are F-Measure (precision and recall of the computed compared to the reference label terms), Jaccard index (overlap of computed and reference terms), and cosine similarity of the tf-weighted term vectors of the computed and the reference labels. Since these measures are only able to capture lexical similarity, we also propose to use a semantics-aware measure in form of the ESA-similarity [8]. In this case, also semantically related terms that have no or only a very low lexical similarity are counted as "correct." The background collection for the ESA-similarity is formed by a random sample of 100,000 English Wikipedia articles. Note that the usage of ESA as a cluster labeling quality measure is novel and a contribution in itself. Before, only lexical similarity was measured.

The results can be found in Table 1. For evaluation, we set the label length $k = 5$ for the approaches χ^2-test and weighted centroid covering since this is the average length of the query generation labels. Interestingly, the measures that simply evaluate the term overlap with the reference label (F-Measure and Jaccard) favor the χ^2-labels while the more advanced ESA similarity favors the query labels. Thus, depending on the used evaluation measure, our new query generated labels are somewhat on par with the standard χ^2 labeling approach and clearly improve upon the weighted centroid covering.

4.3   User Study

Complementing the automatic evaluation of similarity to reference labels, we also conduct a user study in which human participants should select the best label from the three approaches according to their personal perceived similarity to the also displayed reference label.

For the user study, we sampled 100 of the 791 subtopics that had to be evaluated by each of our participants. The study was conducted online with a short introduction to the idea of cluster labeling. To ensure a meaningful word order of the generated cluster labels (remember that χ^2 and weighted centroid covering just present labels composed of 5 single words), we post-processed the labels to find frequent word n-grams in the cluster's documents and in the Google

n-grams. The label terms were re-ordered whenever a frequent n-gram like new york was identified and the ordering in the original label was york new. This improves the labels' readability for our human participants and could possibly be a useful post-processing step in any labeling approach working with single words.

In our study, 29 subjects each spent 15–30 min on their judgments. Table 2 shows the aggregated results. According to the number of votes, the users favor query- and χ^2-based labels. However, the situation changes when looking at the number of topics where one approach received the most votes (column "Winner"; for 20 subtopics there was a tie) and where one approach got an absolute majority of at least 6 out of 10 votes (no such majority for 51 topics). Here, the users clearly favor the query-based labels. However, a general critique amongst our users that could also be observed from the votes was the label length. Whenever the query labels are longer than 5 terms (the threshold for the other two approaches), the users often favored the χ^2 labels or even the weighted centroid covering.

4.4 Discussion

The traditional automatic evaluation of similarity against the reference labels results in a tie between the query-based and the χ^2 labels. But our user study indicates the promising potential of query-based labels since many users favor them and if they do not, the query labels often are almost as popular as the χ^2 labels.

5 Conclusion and Outlook

We have presented a solution to the abstract problem of automatically formulating a query that retrieves a given document set. This abstract problem has an interesting use case in the scenario of cluster labeling where the task is to generate good labels for the individual clusters that "tell" the user something about the contained documents. Our idea of using queries as the labels (derived by solving the abstract query formulation problem) has shown promising performance when compared against standard cluster labeling approaches. Using traditional and our newly proposed ESA-based evaluation measure, our query-based cluster labels are on par with the standard methods. A further user study showed a clear tendency that users prefer the idea of queries as cluster labels over the standard methods.

As for future research, the full potential of our query-based cluster labeling idea should be exploited by enhancing the currently used rather basic BM25F retrieval model. Including for instance synonyms and putting more emphasis on keyphrases as the basis queries, we envision an even better quality of queries as cluster labels. It also would be very interesting to examine the usage of queries itself to guide the whole clustering process by for instance using a document's keyqueries [9] as the clustering features. The queries used for clustering would then directly form appropriate labels at no additional costs.

References

1. Azzopardi, L., Vinay, V.: Retrievability: an evaluation measure for higher order information access tasks. In: Proceedings of the 17th ACM conference on Information and knowledge management (CIKM 2008), pp. 561–570, ACM, New York (2008)
2. Bar-Yossef, Z., Gurevich, M.: Random sampling from a search engine's index. J. ACM **55**(5), 1–74 (2008)
3. Bendersky, M., Croft, W.B.: Finding text reuse on the web. In: Proceedings of the Second ACM International Conference on Web Search and Data Mining (WSDM 2009), pp. 262–271, ACM, New York (2009)
4. Bonchi, F., Castillo, C., Donato, D., Gionis, A.: Topical query decomposition. In: Proceedings of the 14th ACM SIGKDD International Conference on Knowledge Discovery and Data Mining (KDD 2008), pp. 52–60, ACM, New York (2008)
5. Dasdan, A., D'Alberto, P., Kolay, S., Drome, C.: Automatic retrieval of similar content using search engine query interface. In: Proceedings of the 18th ACM Conference on Information and Knowledge Management (CIKM 2009), pp. 701–710, ACM, New York (2009)
6. Fuglede, B., Topsøe, F.: Jensen-Shannon divergence and Hilbert space embedding. In: Proceedings of International Symposium on Information Theory (ISIT 2004), pp. 31, IEEE, Piscataway (2004)
7. Fuhr, N., Lechtenfeld, M., Stein, B., Gollub, T.: The optimum clustering framework: implementing the cluster hypothesis. Inf. Retrieval **15**(2), 93–115 (2011)
8. Gabrilovich, E., Markovitch, S.: Computing semantic relatedness using wikipedia-based explicit semantic analysis. In: Proceedings of the 20th International Joint Conference on Artifical Intelligence (IJCAI 2007), pp. 1606–1611, Morgan Kaufmann Publishers Inc, San Francisco (2007)
9. Gollub, T., Hagen, M., Völske, M., Stein, B.: From keywords to keyqueries: content descriptors for the web. In: Proceeding of the 36th International ACM SIGIR Conference on Research and Development in Information Retrieval (SIGIR 2013), pp. 981–984, ACM, New York (2013)
10. Hagen, M., Stein, B.: Search strategies for keyword-based queries. In: 7th International Workshop on Text-Based Information Retrieval (TIR 2010), pp. 37–41, Piscataway, IEEE (2010)
11. Hagen, M., Stein, B.: Candidate document retrieval for web-scale text reuse detection. In: Grossi, R., Sebastiani, F., Silvestri, F. (eds.) SPIRE 2011. LNCS, vol. 7024, pp. 356–367. Springer, Heidelberg (2011)
12. Huston, S., Croft, W.B.: Evaluating verbose query processing techniques. In: Proceeding of the 33rd International ACM SIGIR Conference on Research and Development in Information Retrieval (SIGIR 2010), pp. 291–298, ACM, New York (2010)
13. Jordan, C., Watters, C., Gao, Q.: Using controlled query generation to evaluate blind relevance feedback algorithms. In: Proceedings of the 6th ACM/IEEE-CS Joint Conference on Digital Libraries (JCDL 2006), pp. 286–295, ACM, New York (2006)
14. Manning, C.D., Raghavan, P., Schütze, H.: Introduction to Information Retrieval. Cambridge University Press, New York (2008)
15. Muhr, M., Kern, R., Granitzer, M.: Analysis of structural relationships for hierarchical cluster labeling. In: Proceeding of the 33rd International ACM SIGIR Conference on Research and Development in Information Retrieval (SIGIR 2010), pp. 178–185, ACM, New York (2010)

16. Navigli, R., Crisafulli, G.: Inducing word senses to improve web search result clustering. In: Proceedings of the 2010 Conference on Empirical Methods in Natural Language Processing (EMNLP 2010), pp. 116–126, Stroudsburg (2010) (Association for Computational Linguistics)

17. Pickens, J., Cooper, M., Golovchinsky, G.: Reverted indexing for feedback and expansion. In: Proceedings of the 19th ACM international conference on Information and knowledge management (CIKM 2010), pp. 1049–1058, ACM, New York (2010)

18. Potthast, M., Hagen, M., Stein, B., Graßegger, J., Michel, M., Tippmann, M., Welsch, C.: ChatNoir: A search engine for the ClueWeb09 corpus. In: The 35th International ACM SIGIR Conference on Research and Development in Information Retrieval (SIGIR 2012), pp. 1004, ACM, New York (2012)

19. Robertson, S.E., Zaragoza, H.: The probabilistic relevance framework: BM25 and beyond. Found. Trends Inf. Retrieval **3**(4), 333–389 (2009)

20. Robertson, S.E., Zaragoza, H., Taylor, M.J.: Simple BM25 extension to multiple weighted fields. In: Proceedings of the 13th ACM International Conference on Information and Knowledge Management (CIKM 2004), pp. 42–49, ACM, New York (2004)

21. Stein, B., zu Eißen, S.M.: Topic identification: framework and application. In: Proceedings of the 4th International Conference on Knowledge Management (I-KNOW 2004), Journal of Universal Computer Science, Know-Center, pp. 353–360, Graz (2004)

22. Stein, B., Gollub, T., Hoppe, D.: Beyond precision@10: Clustering the long tail of web search results. In: Proceedings of the 20th ACM International Conference on Information and Knowledge Management (CIKM 2011), pp. 2141–2144, ACM, New York (2011)

23. Turel, A., Can, F.: A new approach to search result clustering and labeling. In: Salem, M.V.M., Shaalan, K., Oroumchian, F., Shakery, A., Khelalfa, H. (eds.) AIRS 2011. LNCS, vol. 7097, pp. 283–292. Springer, Heidelberg (2011)

24. Yang, Y., Bansal, N., Dakka, W., Ipeirotis, P., Koudas, N., Papadias, D.: Query by document. In: Proceedings of the Second ACM International Conference on Web Search and Data Mining (WSDM 2009), pp. 34–43, ACM, New York (2009)

Context-Aware NLP

Using Context-Aware and Semantic Similarity Based Model to Enrich Ontology Concepts

Zenun Kastrati[(✉)], Sule Yildirim Yayilgan, and Ali Shariq Imran

Faculty of Computer Science and Media Technology,
Gjøvik University College, Gjovik, Norway
{zenun.kastrati,sule.yayilgan,ali.imran}@hig.no

Abstract. Domain ontologies are a good starting point to model in a formal way the basic vocabulary of a given domain. However, in order for an ontology to be usable in real applications, it has to be supplemented with lexical resources of this particular domain. The learning process of enriching domain ontologies with new lexical resources employed in the existing approaches takes into account only the contextual aspects of terms and does not consider their semantics. Therefore, this paper proposes a new objective metric namely SEMCON which combines contextual as well as semantic information of terms to enriching the domain ontology with new concepts. The SEMCON defines the context by first computing an observation matrix which exploits the statistical features such as frequency of the occurrence of a term, term's font type and font size. The semantics is then incorporated by computing a semantic similarity score using lexical database WordNet. Subjective and objective experiments are conducted and results show an improved performance of SEMCON compared with *tf*idf* and χ^2.

Keywords: Domain ontology · Context aware · Semantic similarity · Concept

1 Introduction

A domain ontology represents the knowledge of a given domain in a principled way but in order to be implemented in real applications, an ontology has to be enriched with new lexical resources of this particular domain. This process, known as onto-terminology [1], populates the existing ontology with new concepts without considering ontological types and relations of these concepts. Therefore, the structure of existing ontology remains unchanged.

Recently, the population of the ontology with lexical data has been subject of research by various authors. In this regard, authors in [2] proposed a new approach named *Synopsis* to automatically building a lexicon for each specific term called criterion. The lexicon built is then used to populate the ontology. An adaptation of *Synopsis* approach is presented by researchers in [3]. They used the same methodology but rather than building terms lexicon, they built the lexicon of an ontology concepts. In order to do this, they built an information retrieval

© Springer International Publishing Switzerland 2015
C. Biemann et al. (Eds.): NLDB 2015, LNCS 9103, pp. 137–143, 2015.
DOI: 10.1007/978-3-319-19581-0_11

system called *CoLexIR* which automatically identifies all parts of a document that are related to a given concept.

The learning process of enriching ontology concepts employed in these approaches uses only contextual aspects of terms while they lack to consider the semantic information of these terms. Therefore, this paper proposes a new approach namely SEMCON, which combines the contextual and semantic information through its learning process of enriching the ontology concepts. Besides using contextual information, new statistical features such as term's font size and term's font type are also considered.

The rest of the paper is organized as follows. Section 2 describes our proposed method in detail. In Sect. 3 we describe the subjective experiment, while Sect. 4 describes the objective evaluation of proposed method. Lastly, Sect. 5 concludes the paper.

2   Proposed Model

The proposed model, shown in Fig. 1, initially partitions a document into subsets of text known as passages. After these passages are partitioned, each passage is treated as an independent document. More concrete, each passage represented by a presentation slide is considered as an independent document.

The next step is a morpho-syntatic analysis using TreeTagger [4] where the partitioned passages are tokenized and lemmatized. As a result, a list of potential terms which can either be a noun, verb, adverb or adjective is obtained. Finally, only nouns are filtered out as the most meaningful terms in a document [5].

Computation of the observation matrix is the next step in the proposed model. Observation matrix is a matrix where the rows represent the terms extracted from a document, and the columns are the possible passages from that particular document. Each entry of the observation matrix represents the observed values for a term, namely term frequency, term font size and term font type in each of the corresponding passages, as shown in Eq. 1. Introducing terms' font type and terms' font size, as the very important factors in the information finding process [6], is inspired from the representation of tags in the tag cloud [7].

$$O_{i,j} = \sum_{i \in t} \sum_{j \in p} (Freq_{i,j} + FT_{i,j} + FS_{i,j}) \tag{1}$$

where, t and p show the set of terms and passages, respectively. $Freq_{i,j}$ denotes the frequency of occurrences of a term t_i in passage p_j. $FT_{i,j}$ and $FS_{i,j}$ show font type and font size of a term t_i in passage p_j, respectively.

We adopt a linear increase for different font types and font sizes, varied in the range between 0 and 1. More formally, font type of a term t is found using Eq. 2, while font size is found using Eq. 3.

$$FT(t) = 0.75 * B + 0.5 * U + 0.25 * I \tag{2}$$

$$FS(t) = 1.0 * T + 0.75 * L_1 + 0.50 * L_2 + 0.25 * L_3 \tag{3}$$

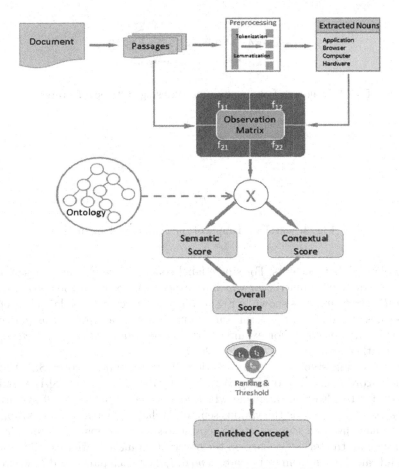

Fig. 1. Block diagram of SEMCON model

where, B, U and I denote bold font type, underlined font type and italic font type, respectively. Similarly, T, L_1, L_2 and L_3 represent title font size, level 1 font size, level 2 font size and level 3 font size, respectively.

A concrete example of building of observation matrix using statistical features is represented in Fig. 2. It can be seen from the illustration that term *Web* occurs 4 times in the slide 2, where, 2 times it appears as level 1 font size and as bold font type and 2 times it appears as level 2 font size.

The next step is finding of term to term contextual score (S_{con}) which is calculated using the cosine similarity metric with respect to the passages, and it is given in Eq. 4.

$$S_{con}(t_i, t_j) = \frac{t_i \cdot t_j}{\| t_i \| \| t_j \|} \tag{4}$$

Further, we extract and use a subset of the terms in order to extend the concept list of ontology. There maybe single label concepts in an ontology as well

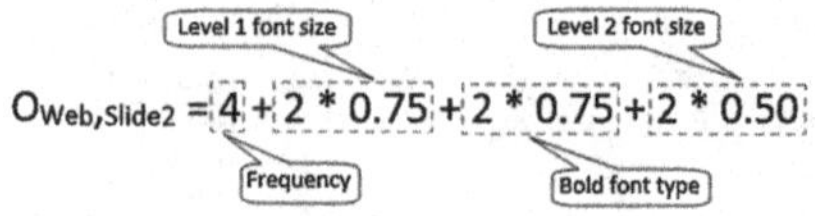

Fig. 2. Building of observation matrix using statistical features

Fig. 3. Ontology sample of the computer domain

as compound label concepts. For single label concepts, we use only those terms from the term square matrix for which an exact term exists in the ontology, i.e. *Application* or *Storage*, as shown in Fig. 3. Whereas, for compound label concepts, we use those terms from the term square matrix which are present as part of a concept in the ontology. For example, for concept *InputAndOutputDevices*, we consider either term *Input, Output* or *Device*.

The following step is the computation of the semantic score (S_{sem}). The semantic score is computed using WordNet database. WordNet [8] is a lexical database for the English language which groups words into sets of synonyms called synsets and records the various semantic relations between these synonym sets. We have used all the synsets to represent specific terms being considered. Go through all the terms we have on the observation matrix, we take all possible pairs and calculate the semantic score, $sem(t_i, t_j)$, for each pair t_i and t_i, where t_i, $t_j \in O$ and O is the observation matrix. After calculating the semantic score for all pair of terms, we generate a table for each term and the most similar terms set to be the synonyms for that term. The Wu&Palmer algorithm [9] is used to find the semantic score which is implemented in a freely available software package WordNet::Similarity [10] and the score is computed using Eq. 5.

$$S_{sem}(t_i, t_j) = \frac{2 * depth(lcs)}{depth(t_i) + depth(t_j)} \tag{5}$$

where, *depth(lcs)* indicates least common subsumer of terms t_i, t_j and *depth(t_i)* and *depth(t_j)* indicate the path's depth of term t_i and t_j in the WordNet::similarity.

The overall correlation between two terms, t_i, t_j, is found using the contextual and semantic score. Mathematically, the overall score is given in Eq. 6.

$$S_{overall}(t_i, t_j) = w * S_{con}(t_i, t_j) + (1 - w) * S_{sem}(t_i, t_j) \tag{6}$$

where w is a parameter with value set as 0.5 based on the empirical analysis from the PowerPoint presentation data set. The overall score is in the range (0,1]. The overall score is 1 if two terms are the same.

Table 1. Top 10 close related terms of 'Application' concept

Concept	Top 10 terms obtained by SEMCON model
Application	Apps, Application, Software, Program, Control, Task, Part, Master, Operation, Function

Table 2. Borda count of subjects' responses for the 'Application' concept.

Rank	Term	Borda count	Rank	Term	Borda count
1	Apps	50	6	Task	11
2	Application	40	7	Browser	10
3	Software	34	8	Function	9
4	Program	21	9	User	9
5	Windows	14	10	Process	7

Finally, to obtain terms which are more closely related to a given term, a rank cut-off method is applied using a specified threshold. Terms which are above the threshold are considered to be the relevant terms for enriching the concepts.

A simple example of the SEMCON output, given in Table 1, shows the top 10 terms obtained as the most related terms of *Application* concept. 6 of these terms, namely *Application, Program, Apps, Function, Task* and *Software* are amongst the top 10 terms selected by subjects as the closest terms with concept *Application* performed in the subjective experiment given in Sect. 3.

3   Subjective Evaluation

To evaluate the performance of the SEMCON, we have used PowerPoint presentations dataset from 5 different domains. The dataset consists of 39 slides which cover 369 terms and 41 concepts.

A subjective survey was also carried out by publishing an online questionnaire to 15 subjects. They were asked to select 5 closely related terms from a list of terms for each given concept. From the subjective survey, we found then the most related terms selected by subjects for a given concept using Borda Count method [11]. Mathematically, Borda count method is defined in Eq. 7.

$$BordaCount(t) = \sum_{i=1}^{m}[(m+1-i)*freq_i(t)] \tag{7}$$

where, $freq_i(t)$ is the frequency of term t chosen at Position i, and m is the total number of possible positions, in our case 5.

The scores from the Borda count are then sorted to obtain the top 'n' terms, giving us the refined list of the most relevant terms selected by subjects. For our experiment, we set n = 10, and this gives us the top 10 terms. Table 2 shows the top 10 terms selected by subjects as the closest terms of concept *Application*.

Table 3. The performance of objective methods

Concept	tf*idf (%)			χ^2 (%)			SEMCON (%)		
	P	R	F1	P	R	F1	P	R	F1
Computer	20.0	30.0	24.0	20.0	30.0	24.0	26.7	40.0	32.0
Software	46.7	70.0	56.0	40.0	60.0	48.0	46.7	70.0	56.0
Hardware	26.7	40.0	32.0	33.3	50.0	40.0	33.3	50.0	40.0
Web	26.7	40.0	32.0	26.7	40.0	32.0	46.7	70.0	56.0
Storage	53.3	80.0	64.0	46.7	70.0	56.0	46.7	70.0	56.0
Microprocessor	40.0	60.0	48.0	33.3	50.0	40.0	40.0	60.0	48.0
I&ODevices	26.7	40.0	32.0	20.0	30.0	24.0	33.3	50.0	40.0
Application	46.7	70.0	56.0	46.7	70.0	56.0	53.3	80.0	64.0
Windows	46.7	70.0	56.0	40.0	60.0	48.0	46.7	70.0	56.0
Average	**37.0**	**55.6**	**44.4**	**34.1**	**51.1**	**40.9**	**41.5**	**62.2**	**49.8**

4 Objective Evaluation

In order to validate the SEMCON model, we have also performed an objective evaluation where the results obtained from the SEMCON are compared with the results obtained from the *tf*idf* [12] and χ^2 [13] methods. To evaluate the effectiveness of objective metrics, we employed the standard information retrieval measures such as Precision, Recall and F1 [12].

We evaluated the performance of objective methods on how well these methods score the top subjective terms. In order to do this, scores for the 10 top subjective terms are taken as the ground truth. The score of these terms obtained using the objective methods are then evaluated. Under this light, the most related terms of computer concepts are observed and the comparison, in terms of precision, recall and F1, is shown in Table 3. The comparison shows that the SEMCON has achieved an improvement on finding the new terms to enrich the concepts of computer domain ontology. More precisely, it achieved the average improvement of F1 of 12.0 % over the *tf*idf* and 21.7 % over the χ^2.

5 Conclusion and Future Work

In this paper, we proposed a new approach to enriching the domain ontologies with new concepts by combining contextual as well as semantics of terms extracted from the accompanying documents. The proposed approach is a generic model which can be applied to any existing domain ontology for extending it with new concepts. The model defines the context using new statistical features such as term's frequency, term's font size and font type. The semantics is then incorporated by computing a semantic similarity score using lexical database WordNet. The experimental results show an improved performance of SEMCON compared with *tf*idf* and χ^2. In future work we plan to investigate into how the combination of contextual and semantic components contributes to the overall task of ontology concepts enrichment.

References

1. Roche, C., Calberg-Challot, M., Damas, L., Rouard, P.: Ontoterminology - a new paradigm for terminology. In: Dietz, J.L.G. (ed.) KEOD 2009 - Proceedings of the International Conference on Knowledge Engineering and Ontology Development, Portugal (2009)
2. Duthil, B., Trousset, F., Roche, M., Dray, G., Plantié, M., Montmain, J., Poncelet, P.: Towards an automatic characterization of criteria. In: Hameurlain, A., Liddle, S.W., Schewe, K.-D., Zhou, X. (eds.) DEXA 2011, Part I. LNCS, vol. 6860, pp. 457–465. Springer, Heidelberg (2011)
3. Ranwez, S., Duthil, B., Sy, M.F., Montmain, J., Augereau, P., Ranwez, V., Hovy, E.: How ontology based information retrieval systems may benefit from lexical text analysis. In: Oltramari, A., Vossen, P., Qin, L., Hovy, E. (eds.) New Trends of Research in Ontologies and Lexical Resources. Ideas, Projects, Systems. Theory and Applications of Natural Language Processing, pp. 209–228. Springer, Heidelberg (2013)
4. Schmid, H.: Probabilistic part-of-speech tagging using decision trees. In: Proceedings of International Conference on New Methods in Language Processing (1994)
5. Li, H., Tian, Y., Ye, B., Cai, Q.: Comparison of current semantic similarity methods in WordNet. In: Computer Application and System Modeling, International Conference, vol. 4, pp. 4008-4011 (2010)
6. Halvey, M.J., Keane, M.T.: An assessment of tag presentation techniques. In: Proceedings of the 16th International Conference on World Wide Web, USA, pp. 1313–1314. ACM (2007)
7. Bateman, S., Gutwin, C., Nacenta, M.: Seeing things in the clouds: the effect of visual features on tag cloud selections. In: Proceedings ACM Conference on Hypertext and Hypermedia, HT 2008, pp. 193–202 (2008)
8. Fellbaum, C.: WordNet: An Electronic Lexical Database. The MIT Press, Cambridge (1998)
9. Wu, Z., Palmer, M.: Verb semantics and lexical selection. In: Proceedings of the 32nd Annual Meeting of the Associations for Computational Linguistics, pp. 133–138 (1994)
10. Pedersen, T., Patwardhan, S., Michelizzi, J.: WordNet: similarity - measuring the relatedness of concepts. In: Proceedings of 19th National Conference on Artificial Intelligence, pp. 1024–1025 (2004)
11. Young, P.: Optimal voting rules. J. Econ. Perspect. **9**, 51–64 (1995)
12. Sebastiani, F.: Machine learning in automated text categorization. ACM Comput. Surv. (CSUR) **34**(1), 1–47 (2002)
13. Liu, J.N.K., He, Y.-L., Lim, E.H.Y., Wang, X.-Z.: A new method for knowledge and information management domain ontology graph model. IEEE Trans. Syst. Man Cybern.: Syst. **43**, 115–127 (2013)

NADIA: A Simplified Approach Towards the Development of Natural Dialogue Systems

Markus M. Berg[1,2](✉)

[1] Institute of Computer Science, University of Kiel, Kiel, Germany
mail@mmberg.net
[2] Department of Electrical Engineering and Computer Science,
University of Wismar, Wismar, Germany

Abstract. Spoken Dialogue Systems have enormously improved during the last couple of years and gave rise to voice-controlled mobile assistants. While the abilities of these systems are very sophisticated, there is a lack of tools enabling us to easily describe a natural dialogue that can afterwards be processed by a dialogue engine without having to programme the engine itself. In this paper we present NADIA, a dialogue engine that can process an easy to define XML-based dialogue description.

Keywords: Spoken Dialogue Systems · Dialogue Modelling · Natural Language Generation · Voice User Interface

1 Introduction

Since the fading of *Interactive Voice Response Systems* due to insufficient control over the dialogue and an unnatural loop of endlessly selecting options from a menu, *Spoken Dialogue Systems* (SDS) start to get interesting again for the community of standard users. With the release of *Apple Siri* or *Google Now* almost everybody already got in touch with speech-controlled mobile assistants. However, proper environments for an easy development of SDS are still missing. Without suitable tool support, it seems impossible to teach students all the theoretical fundamentals within one semester as part of a standard computer science class and at the same time enable them to develop their own spoken dialogue system that – according to the students' expectations – should be on a comparable level to the aforementioned mobile assistants.

Therefore, in this work we present how we have combined the results of our previous research to build the Natural Dialogue System (NADIA) that facilitates the development of natural dialogues. After an overview of related work, we first introduce the basic concepts and then show an example of how to create a simple dialogue.

2 Related Work

The following section provides a short overview of languages that can be used to create a dialogue system without having to develop the engine itself.

© Springer International Publishing Switzerland 2015
C. Biemann et al. (Eds.): NLDB 2015, LNCS 9103, pp. 144–150, 2015.
DOI: 10.1007/978-3-319-19581-0_12

Voice XML [9] is a markup language that allows the definition of speech dialogues. It is mostly used in connection with interactive voice response systems and integrates well with many media platforms. The development of Voixe XML dialogues requires the definition of grammars and mostly leads to a system-directed dialogue although a basic form-filling approach can be used to create mixed-initiative dialogues. The specification of system utterances is realised with simple texts and has to be done for any variation of the utterance in any language. The dialogue flow is explicitly defined in a declarative way as well. One of the biggest challenges is the interpretation of user utterances. It is not possible to integrate any form of parsing, chunking, named entity recognition or anaphora resolution. Another aspect is the missing connection to databases or web services. In order to facilitate the creation of Voice XML documents, IDEs have been developed that allow the graphical, flow-based creation of dialogues (e.g. *Voxeo CXP*). These tools also enable the interaction with back-ends.

The Artificial Intelligence Markup Language [11] has been released by Richard S. Wallace in 2001 and has been the basis for famous chatbots like *A.L.I.C.E.* The intention of the XML-based AIML is not the creation of IVR systems, but instead the development of text-based dialogue systems. The idea resembles Joseph Weizenbaum's *ELIZA* [12] that consists of decomposition templates and reassembly rules. The creation of mixed-initiative dialogues is only possible in a very limited way. Moreover, an easy integration of back-ends like databases or web services is missing at the moment, but part of the new AIML 2.0 draft. However, despite the simplicity of the approach, AIML has proven to be a good tool to create convincing chatbots as several Loebner prices indicate.

TrindiKit [8] is a Prolog toolkit to develop a dialogue manager that is based on the information state theory. A dialogue move engine updates the information state (the information stored by the dialogue system) based on observed dialogue moves (inferred from the user's utterances) and selects appropriate moves that are performed in order to react on the user's inputs. Apart from the dialogue move engine itself, also interpretation, generation, input and output modules are provided. Any of these modules is able to read from and write to the information state with the help of update rules [8]. TrindiKit is a good choice to create dialogue managers based on the information state. It is much more generic and supports more complex scenarios than VoiceXML or AIML. However, it focusses on the dialogue manager and the development is more complicated and requires more theoretical knowledge, even to create a simple dialogue system.

3   Goals and Characteristics of NADIA

After having introduced some approaches to creating dialogue systems, we now want to highlight the features that we expect from a modern dialogue system and that are addressed by *NADIA* (NAtural DIAlogue System) [6].

Separation of Dialogue Description and Engine. In simple approaches, people tend to directly connect the dialogue behaviour with the actual domain knowledge and speech recognition. This style of development makes changes or adaptations to other domains hard or even impossible. Hence, it is important to separate the actual (domain dependent) dialogue description from the (generic) algorithm that executes the dialogue.

Declarative Setting of the Dialogue Behaviour. For demonstrating the differences between dialogue strategies (e.g. system vs. mixed initiative), a declarative approach is useful as it allows to run the same dialogue description with a different behaviour and outcome without changing the code.

Avoidance of Grammar Definition. Because of enormous improvements in open-domain speech recognition in dictation scenarios, we do not need to specify grammars anymore. Instead of locally installed speech recognisers, we can use cloud-based services like offered by *Google* [10]. These are optimised for a grammar-less recognition of search queries and text messages.

Reuse of Interpretation Modules. Many dialogue systems share the same types of questions like yes-no-questions, questions for a date or a city. The interpretation of such recurring questions should be handed over to the dialogue engine and not be part of the dialogue description.

True Mixed Initiative and Subdialogues. The most important aspect of mixed initiative is characterised by the ability to influence the dialogue flow by changing the topic, by answering questions in a different order, or by providing more information than has been asked for. The selection of the best matching entity of the dialogue is done by the engine based on the provided dialogue description and the current context. Often, the user needs to get further information before he is able to answer a question. Therefore, the use of subdialogues improves the overall usability of a dialogue system.

Basic Support of Dialogue Acts. Dialogue acts infer the user's intention from the surface form and can help to disambiguate user utterances. Especially in simple dialogue systems without a deep grammatical analysis of the input, they can help to support basic approaches like keyword spotting. The categorisation of user utterances also helps to generalise the meaning and to facilitate further processing.

Easy and Adaptive Prompt Definition. Many dialogues suffer from the fact that users memorise the exact phrasing after a few times of using the system. It also feels strange if the user addresses the system in a very informal way but the system gives very formal answers. Since the manual definition of adaptive and varying utterances is effortful, concept-to-text approaches contribute to adaptive, more natural dialogue systems.

Easy Integration with External Systems. A drawback of many dialogue systems is the lack of generic connectors to external systems like web services or databases.

Web-based Access and Easy Installation. In restricted environments it is often not possible to install software. Also, depending on the programming language, portings for different operating systems may be necessary. Here, a web-based approach simplifies the access to the dialogue system.

4   Components of the Dialogue Engine

We now give a short overview of NADIA's functionality. After the user input has been externally recognised, the text is sent to the dialogue system. In the first step, the utterance is categorised into one of three different dialogue acts that describe the intention of the user. Afterwards, depending on the dialogue act, the current context and the dialogue description, the system tries to find questions that the user utterance could be the answer to. This does not necessarily have to be the question that the system just asked, as the user may try to change the dialogue flow. Also, the user might give more information than the system has asked for. The system should process this additional information instead of asking for it again in a later step. Afterwards, the extracted information is stored as a frame and the system asks the next question. It makes use of language generation in order to be able to adapt to a certain level of politeness, formality, and language. Once all information has been gathered, the system will connect to the specified type of back-end, execute an action and return to the user. The resulting utterance is synthesised by an external service again. Some of the techniques the engine makes use of will be shortly summarised now.

System-oriented Dialogue Acts: Dialogue acts base on speech acts [1] and capture the type of an utterance in a dialogue, be it a greeting, a command, a wish, a question or an apology. For a simple dialogue system we do not need the full range of complexity of dialogue acts. We make use of the concept to categorise an utterance according to its intention. When the user wants the system to do something, we speak of an *action-requesting act*. *Information-seeking acts* comprise all sorts of formulations that have the aim of getting information from the system (e.g. questions or commands). *Information-providing acts* provide the system with more information. This can be any answer during the information collection phase of a dialogue. With the help of these *System-oriented Dialogue Acts* [2,4], it is possible to use simple keyword-spotting approaches and still be able to infer the correct action, as the following example shows: The requests *"Is the light turned on?"* and *"Please turn the light on!"* may result in the extracted information light on. Only with the help of the dialogue act it is possible to perform the correct action and not to switch on the light in the basement when the user only wants to know if it is switched on.

Linguistic Data Types Instead of Grammars: In order to simplify the specification of possible answers we use the *Abstract Question Description* (AQD) [3]. It's based on the idea that a question can be described by the type of answer that it expects. A question like *"Where do you want to go?"* expects a city whereas *"When do you want to leave?"* awaits a date. This can be passed like *"24/12/2014"* or as a paraphrase like *"next Wednesday"* or *"on Christmas Eve"*. Instead of specifying grammars as part of the dialogue description (like in Voice XML), we propose the use of 'linguistic' data types like *date, city* or *name*. The analysis whether an utterance matches the type is handled by the dialogue engine with arbitrarily sophisticated approaches. This supports reusability and keeps the dialogue description short and clear.

We can see that there is a close connection between question and answer: An answer is only processable in context of the question, i.e. we can only understand the utterance *"Two"* if we know that the question was *"With how many persons do you want to travel?"*. We call this combined unit *Information Transfer Object* (ITO). The most important part of an ITO is the AQD.

Prevent Static Formulations: We make use of Natural Language Generation techniques to create adaptive system prompts based on parameters like politeness, formality and language. The questions *"Where do you want to start?"* and *"Where do you want to go"* both expect the answer type *location*, which is sufficient for answer processing. But in order to be able to choose the correct words for language generation we extend the AQD by a context layer [7]. The first question is a question for a date with a reference to a *trip* and the pragmatic aspect *begin*. We describe this as the role *begin of trip*. The second question is described with the role *end of trip*. Beyond the specification of the requirements for the question, we need to annotate a lexicon with the word meaning. The word *go* can be used in a temporal (date → when) or spatial (city → where) dimension in the context of the beginning of a trip. Eventually, the AQD consists of three layers now: type (what the question asks for), form (how the question is asked) and context (what the question is about).

5   Dialogue Model

After having described some of the engine's foundations, we now briefly sketch the structure of the dialogue description. We consider a dialogue as the whole conversation between a user and the system across different topics. It reflects the domain knowledge of the system, i.e. what the system is able to talk about and what information it needs for executing an action. This can roughly be compared to a Voice XML or AIML file.

Any dialogue consists of at least one task, e.g. finding a hotel, getting weather information or switching a lamp. Every task consists of a number of pieces of information that are required to fulfil the goal. In the first case, the system needs to know the city, the number of persons and the travel dates before it is able to send a query to a booking system. Every question is represented by an ITO as we have described in the last section. Every ITO contains exactly one AQD that gives an abstract description of the question and the valid answers. After all information has been gathered from the user, the system will execute a piece of code that is specified as an *action*. This can be a request to a web service or the execution of a Groovy script. On the dialogue level, it is possible to set global properties like the style of initiative, whether to use dialogue acts, politeness and formality scores, or the language that should be used for NLG.

Let's have a look at a source code extract of a very simple dialogue for getting the current temperature. Because a dialogue consists of different tasks, every task needs a *task selector* for enabling the dialogue system to decide which task to activate. We use a very simple bag of words selector that just looks for keywords, e.g. *weather* or *temperature*.

```
Dialogue d = new Dialogue();
d.setAllowSwitchTasks=true;
Task task=new Task("getWeatherInformation");
task.setSelector(new BagOfWordsTaskSelector("weather"...));
```

As we only need to know the city, we only have one ITO, containing the AQD that we need for the understanding of the utterance (NLU).

```
ito=new ITO("getWeatherCity", "For which city do you...");
aqd=new AQD().setType(new AQDType("fact.named_entity.
  non_animated.location.city"));
```

Finally, we add the action that performs an HTTP request to the *Openweathermap* API. We use *XPath* to extract the result.

```
ha=new HTTPAction("... in %getWeatherCity is #result ...");
ha.setUrl("http://api.openweathermap.org/...");
ha.setParams("q=%getWeatherCity&mode=xml...");
ha.setXpath("/current/temperature/@value");
```

Now we can add the task to the dialogue and store it as an XML file. The resulting dialogue supports mixed initiative and over-informative answers, so that the following two dialogues are possible:

- System: **For which city do you want to know the weather?**
- User: Glasgow.
- System: **The temperature in Glasgow is 18°C.**

- User: I'd like to know the weather in Glasgow.
- System: **The temperature in Glasgow is 18°C.**

A more detailed description can be found in the manual [5] and in the thesis [6]. This also shows how to u se language generation and dialogue acts, how to create open-ended questions and how to realise more complex dialogues.

6   Summary and Evaluation

We have presented a system that allows the easy definition of information-seeking dialogues. Although we are aware of the fact that this system still has limitations and does not cover the full range of computational linguistic theories, we regard it as a valuable tool for teaching, rapid-prototyping and as a test bed for new ideas. We don't know of any other dialogue system that uses an NLG approach to generate system prompts in information-seeking dialogue systems. Also, the usage of dialogue acts, reusable interpretation modules, and declarative control of the overall dialogue strategy does exist in research prototypes, but it is not part of a dialogue description language yet. NADIA has been tested by 7 experts and 6 test users who confirm the usefulness of this approach and confirm that it facilitates the development of natural dialogues.

References

1. Austin, J.L.: How to Do Things with Words. Oxford University Press, Oxford (1962)
2. Berg, M.: Towards the modelling of backend functionalities in task-oriented dialogue systems. In: 5th Language & Technology Conference: Human Language Technologies as a Challenge for Computer Science and Linguistics, pp. 274–278. Poznań (Poland) (2011)
3. Berg, M.M., Düsterhöft, A., Thalheim, B.: Towards interrogative types in task-oriented dialogue systems. In: Bouma, G., Ittoo, A., Métais, E., Wortmann, H. (eds.) NLDB 2012. LNCS, vol. 7337, pp. 302–307. Springer, Heidelberg (2012)
4. Berg, M., Thalheim, B., Düsterhöft, A.: Dialog acts from the processing perspective in task oriented dialog systems. In: Proceedings of the 15th Workshop on the Semantics and Pragmatics of Dialogue (SemDial 2011). Los Angeles (USA) (2011)
5. Berg, M.M.: NADIA manual (2013). https://github.com/mmberg/nadia/wiki
6. Berg, M.M.: Modelling of Natural Dialogues in the Context of Speech-Based Information and Control Systems. Ph.D. thesis, University of Kiel, Germany (2014)
7. Berg, M.M., Isard, A., Moore, J.D.: An OpenCCG-based approach to question generation from concepts. In: Métais, E., Meziane, F., Saraee, M., Sugumaran, V., Vadera, S. (eds.) NLDB 2013. LNCS, vol. 7934, pp. 38–52. Springer, Heidelberg (2013)
8. Larsson, S., Berman, A., Bos, J., Grönqvist, L., Ljunglöf, P., Traum, D.: TrindiKit 2.0 Manual. Technical report, University of Gothenburg (2000)
9. McGlashan, S., Burnett, D.C., Carter, J., Danielsen, P., Ferrans, J., Hunt, A., Lucas, B., Porter, B., Rehor, K., Tryphonas, S.: Voice Extensible Markup Language (VoiceXML) Version 2.0 (2004). http://www.w3.org/TR/voicexml20
10. Schalkwyk, J., Beeferman, D., Beaufays, F., Byrne, B., Chelba, C., Cohen, M., Kamvar, M., Strope, B.: Your word is my command: google search by voice: a case study. In: Neustein, A. (ed.) Advances in Speech Recognition: Mobile Environments, Call Centers and Clinics, pp. 61–90. Springer, USA (2010)
11. Wallace, R.S.: AIML overview. http://www.pandorabots.com/pandora/pics/wallaceaimltutorial.html
12. Weizenbaum, J.: ELIZA - a computer program for the study of natural language communication between man and machine. Commun. ACM **9**(1), 36–45 (1966)

Cognitive and Semantic Computing

How to Talk to a Cognitive Computer

Csaba Veres[(✉)]

The University of Bergen, Bergen, Norway
`csaba.veres@uib.no`

Abstract. *Cognitive Computing* is becoming a catchphrase in the technology world, with the promise of new smart services offered by industry giants like IBM and Google. We observe that the latest technologies do not represent a major departure from previous achievements in Artificial Intelligence. An example from language processing demonstrates that present day Cognitive Computing still struggles with long-standing issues in AI. We conclude that in the absence of fundamental breakthroughs, it might be more fruitful to follow Licklider's lead and adopt *Symbiotic Computing* as a metaphor for designing software programs that enhance human cognitive performance.

Keywords: Cognitive computing · Cognition · AI · Symbiosis · Language

1 Introduction

The Gartner Hype Cycle for Smart Machines, 2014 names *Cognitive Computing* as a technology that is on the rise.[1] The IEEE Technical Activity for Cognitive Computing defines it as "an interdisciplinary research and application field" ... which ... "uses methods from psychology, biology, signal processing, physics, information theory, mathematics, and statistics" ... in an attempt to construct ... "machines that will have reasoning abilities analogous to a human brain".

The IBM corporation has invested heavily in bringing Cognitive Computing to the commercial world, starting perhaps with the computer Deep Blue which for the first time in history, on May 11, 1997, beat the world chess champion after a six-game match[2]. They then developed the computer 'Watson' which could process and reason about natural language, and learn from documents without supervision. In February 2011 Watson beat two previous champions in the "Jeopardy!" quiz show, demonstrating its ability to understand natural language questions, search its database of knowledge for relevant facts, and compose a natural language response with the correct answer. John Kelly, director of IBM Research, claims that "The very first cognitive system, I would say, is the Watson computer that competed on Jeopardy!"[3]. Kelly continues that cognitive systems

---

[1] https://www.gartner.com/doc/2802717/hype-cycle-smart-machines-.

[2] http://www-03.ibm.com/ibm/history/ibm100/us/en/icons/deepblue/.

[3] http://www.scientificamerican.com/article/will-ibms-watson-usher-in-cognitive-computing/.

© Springer International Publishing Switzerland 2015
C. Biemann et al. (Eds.): NLDB 2015, LNCS 9103, pp. 153–159, 2015.
DOI: 10.1007/978-3-319-19581-0_13

can "understand our human language, they recognize our behaviours and they fit more seamlessly into our worklife balance. We can talk to them, they will understand our mannerisms, our behaviours - and that will shift dramatically how humans and computers interact."

Google inc. has also been heavily involved in commercializing cognitive technologies, particularly deep learning[4], an evolution of neural networks with many hidden layers [1] which are particularly good at image recognition tasks. Google demonstrated GoogLeNet, the winning application at the 2014 ImageNet Large-Scale Visual Recognition Challenge [2].

IBMs public promotional materials boldly state that "cognitive computers can process natural language and unstructured data and learn by experience, much in the same way humans do" and "interact naturally with people to extend what either humans or machine could do on their own."[5] John Searle coined the term 'strong AI' to describe systems which process information "in the same way humans do". Strong AI holds that "the appropriately programmed computer literally has cognitive states and that the programs thereby explain human cognition", which is on opposition to 'weak AI' where the computer merely "enables us to formulate and test hypotheses in a more rigorous and precise fashion" [3]. Searle argues against the possibility of strong AI with his famous Chinese room scenario, where he argues that an ungrounded symbol manipulation system lacks, in principle, the capacity for human understanding. The history of cognitive computing is sprinkled with claims approaching strong AI.

The term 'cognitive computing' has been in use since the 1980s, as can be seen in the Google Ngram Viewer. The use of the term was associated with a strong growth in neural network computing following a joint US-Japan conference on Cooperative/Competitive Neural Networks in 1982[6]. In 1986 the backpropagation algorithm was detailed in the two volume manifesto: "Parallel distributed processing: Explorations in the microstructure of cognition" [4], which made neural network modeling much more versatile and accessible to researchers, and resulted in a plethora of new research programs exploiting the connectionist paradigm.

The advances in neural network computing also helped revive research in Fuzzy Logic with the emergence of neuro-fuzzy systems which could learn parameters in a fuzzy system, leading to a set of methodologies that could perform imprecise reasoning, or *soft computing* [6]. Finally, the mid-1980s also saw the advent of genetic algorithms which could be used to avoid local minima in learning systems [5]. In 1993 the state of the art could be summarized by: "Cognitive computing denotes an emerging family of problem-solving methods that mimic the intelligence found in nature" ... "all three core cognitive computing technologies neural-, fuzzy- and genetic-based derive their generality by interpolating the solutions to problems with which they have not previously been faced from the solutions to ones with which they are familiar" [7].

---

[4] http://deeplearning.net.

[5] http://www.research.ibm.com/cognitive-computing/#fbid=GZ_iDrBgajZ.

[6] http://cs.stanford.edu/people/eroberts/courses/soco/projects/neural-networks/ History/history2.html.

While none of these technologies could decisively meet Searle's challenge for strong AI, it was pretty clear that the claimed biological plausibility of neural networks was to take us in that direction. Similarly, neuro-fuzzy systems were supposed to operate in ways analogous to human cognition: "In the final analysis, the role model for soft computing is the human mind." [6]. These technologies offered themselves as the foundation of programs that could indeed mimic human cognition.

Thirty years earlier Licklider was already contemplating a future with computers capable of thought like behaviour [13]. He imagined that the emergence of something like strong AI was not imminent, and there would be an interim period of "between 10 and 500 years" in which humans and computers would exist in a symbiotic relationship which would "bring computing machines effectively into the processes of thinking". He argued that for many years computer programs would not be able to mimic human thought processes, but instead work with humans as "dissimilar organisms living together in intimate association", enhancing the weaker parts of human cognition. We must understand how humans solve problems so that we can design programs that can take over those aspects of problem solving that are most mundane or difficult, and not so that we can design programs that mimic human reasoning. The principles of human cognition must be well understood even if it can't be implemented, so computer programs can be written to fit precisely where they are needed.

In the remainder of this paper we argue that modern Cognitive Computing still falls short of realizing human-like thought. We suggest that Symbiotic Computing is a better metaphor since it maximizes the usefulness of programmable systems without trying to force a singularity with human cognition, or degrading them to a set of 'merely useful technologies'.

2   Cognitive Computing?

While the popular discourse about Cognitive Computing emphasize their human-like characteristics, the scientific publications on the inner workings of Watson clearly show the many non human-like aspects of the implementation. For example, during the primary search phase Watson retrieves a large amount of potentially relevant data through a number of different techniques including the use of an inverted index in the Lucene search engine, and SPARQL queries to retrieve RDF triples from a triplestore [8]. This retrieves a huge volume of potentially relevant facts which are then further processed, often with statistical techniques. It is very unlikely that human reasoning would follow a similar process.

Noam Chomsky at the MIT symposium on "Brains, Minds and Machines" held in May 2011[7] took modern AI to task more generally, voicing the opinion that the statistical learning techniques cannot reveal causal principles about the nature of cognition in general, and language in particular. They are useful engineering tools which can perform very useful tasks, but they will not give insight into cognitive processes, and will not operate by the same principles as those processes.

---

[7] http://mit150.mit.edu/symposia/brains-minds-machines.

Peter Norvig, a fellow speaker at the symposium and director of research at Google argues, in a long essay on his web site, that this is a false dichotomy and that Chomsky's proposed explanatory variables in linguistic knowledge are a fiction[8]. In his opinion predictive statistical models based on vast quantities of data are simply all there is to natural language cognition. Scientific progress is to be made not by postulating hypothetical causal mental states and testing their consequences through intuition in the form of grammaticality judgement, but by collecting vast quantities of language data and finding statistical models that best fit the data. If Norvig is correct then the current optimism about the possibilities of statistical models for cognitive computing are perhaps justified (and some of Watson's heuristics could be considered 'cognitive'), but if Chomsky is correct, then we might expect the current approaches to run into difficulties when human and machine processes differ. Our position is that if such differences are inevitable then it would be an advantage to know about them in advance to design solutions.

3   Which Theory of Language?

Chomsky believes in a distinction between linguistic *competence*, the tacit, internalised knowledge of language, and *performance* which is the observable manifestation of the former (speech acts, written texts, etc.). However, performance data is not a pure reflection of competence since linguistic productions are riddled with errors due to attention shifts, memory limitations and environmental factors. Chomsky therefore eschews corpus data as evidence for theory building, preferring instead grammaticality judgements which are elicited in response to sentences constructed to test a certain theory about competence.

Norvig defends the use of corpora, rejecting the use of grammaticality judgement as a form of linguistic evidence since it does not accurately reflect real language use. He cites the famous example from Chomsky [9] who claims that neither sentence 1 or 2 (or any part of the sentences) has ever appeared in the English language, and therefore any statistical model of grammaticalness will rule them as being equally remote from English. Yet it is clear to humans that 1 but not 2 is a grammatical sentence of English:

1. Colourless green ideas sleep furiously.
2. Furiously sleep ideas green colourless.

Pereira [10] demonstrates that modern statistical models of language prove Chomsky wrong. In fact, 1 is 200,000 times more probable than 2 in a large corpus of newspaper text. In his essay Norvig discusses a replication of the experiment on a different corpus "to prove that this was not the result of Chomsky's sentence itself sneaking into newspaper text", which corroborates the result. In addition, he finds that both sentences are much less probable than a normal grammatical sentence. Thus not only is Chomsky wrong about the statistical facts, but he is also

---

[8] http://norvig.com/chomsky.html.

wrong about the categorical distinction between grammatical/ungrammatical: 1 is more grammatical than 2, but less grammatical than ordinary sentences, according to Norvig.

We disagree with the conclusions, and argue that this experiment in fact supports Chomsky's view. Suppose Norvig's concerns about the proliferation of Chomsky's sentence turns out to be true, but it is true about 2 rather than 1. Perhaps a fundamentalist Chomskian government assumes power in the future and enforces a rule that every written text must be headed by Chomsky's "Furiously sleep ideas green colourless." to remind writers to use only grammatical sentences. Before long, the probability of 2 will exceed that of 1. But will 2 become more grammatical than 1, or will it just become annoyingly omnipresent? We think the latter, in which case the statistical theory would make the wrong prediction. To deny grammaticality judgement as a source of linguistic evidence in favour of corpora seems mistaken. There must be a principled criterion for what sort of observed strings should be counted as linguistic evidence.

One important use of statistical methods is for lexical disambiguation, as summarized in [10] "the co-occurrence of the words 'stocks','bonds' and 'bank' in the same passage is potentially indicative of a financial subject matter, and thus tends to disambiguate those word occurrences, reducing the likelihood that the 'bank' is a river bank, that the 'bonds' are chemical bonds, or that the 'stocks' are an ancient punishment device". Norvig points out that 100 % of the top contenders at the 2010 SemEval-2 completion used statistical techniques. However, the limitations of the approach can be easily demonstrated. Consider the following examples involving the ambiguous word bank.

3. I will go to the river bank this afternoon, and have a picnic by the water.
4. I will go to the local bank this afternoon, and afterwards have a picnic by the nearby water feature.

The word 'bank' in sentence 3 is clearly about "the land alongside or sloping down to a river or lake" (Oxford English Dictionary), while 4 is more difficult to interpret, but appears to be about the 'financial' interpretation of 'bank'. Both 3 and 4 contain words that are likely to co-occur with the 'sloping land' interpretation of 'bank' (i.e. picnic, water), which makes 4 misleading. But 4 also contains 'local' which is more likely to co-occur with the 'financial' interpretation, especially when they are strictly adjacent as in 'the local bank'. Note, however, that 'local river bank' is not an entirely uncommon phrase, and could refer to the 'sloping land' sense. The interpretation of 'local bank' is something like 'the local branch of the bank', which is a sensible interpretation if the mental representation of the financial sense of 'bank' includes the fact that banks have 'branch offices'. We suggest that the resolution of ambiguity requires a suitable theory of compositional lexical semantics (e.g. [11]) rather than statistical models. In fact, even Watson uses a structured lexicon in question analysis and candidate generation [12]. Statistical techniques are easier to implement, but are limited when compared to human cognition.

We can push the example in sentence 4 a little further, by swapping the word 'local' with 'river':

5. I will go to the river bank this afternoon, and afterwards have a picnic by the nearby water feature.

On first reading this might seem odd, but suppose one was given as context that the person who uttered the sentence lived in a city which recently developed the previously neglected riverside into a business hub, and several banks were opened. With such knowledge the 'financial' reading of 'bank' becomes instantly clear, without an a priori change in statistical distributions. As more people started talking and writing about the river branch of their bank then no doubt over time the statistical facts would come to reflect this usage. Statistical models completely miss the causal explanation for the change in the observed facts. Statistics does not drive interpretation: interpretation drives statistics. And interpretation, it seems, requires a model of compositional semantics.

4   Towards Human and Machine Symbiosis

We adopt the working assumption that the differences between human and machine 'cognition' explain why simple tasks for humans, like resolving ambiguity, can be difficult problems for machines. To achieve symbiosis, humans can help by providing information that programs can't easily infer. A simple possibility is to develop applications that elicit and use human input as part of their normal operation in areas where computers have difficulty, as in the following two examples.

LexiTags [14,15] is a social semantic bookmarking service in which users can save URLs of interest and tag them with disambiguated tags that are either WordNet senses or DBPedia identifiers[9]. LexiTags provides ambiguity resolution as well as the identification of key topics in a document. While sophisticated statistical algorithms exist for topic analysis (e.g. [16]), the problem of allocating the most contextually significant topic(s) or tags to documents is more difficult because it relies on the subjective goals and beliefs of the tagger. MaDaME [17] is a tool for web developers who wish to mark up their sites with schema.org properties[10]. The tool allows users to highlight key words in their web site, and disambiguate them by selecting a sense from WordNet or DBPedia. The tool then automatically infers the most appropriate schema.org concepts and generates markup that adds schema.org as well as WordNet and SUMO identifiers to the HTML web page. Both of these tools generate curated metadata about web resources which can subsequently be used to automatically infer generalizations about, and relationships between web resources. The little human sourced semantics can go a long way in facilitating automated reasoning about the resources.

In conclusion, we argue that the foreseeable future will see advances in Symbiotic Computing rather than Cognitive Computing. To maximize the benefits

---

[9] http://lexitags.org.
[10] http://mobilesemantics.dyndns.org:3000.

and reduce the intellectual friction, we should acknowledge the shared contribution of cognitive theories as well as engineering solutions in programming smart machines, and not oversell the successes of the machines.

References

1. Hinton, G.E., Osindero, S., Teh, Y.-W.: A fast learning algorithm for deep belief nets. Neural Comput. **18**(7), 1527–1554 (2006). doi:10.1162/neco.2006.18.7.1527
2. Szegedy, C., Liu, W., Jia, Y., Sermanet, P., Reed, S., Anguelov, D., et al.: Going Deeper with Convolutions. Google Technical report (2014). arXiv.org
3. Searle, J.R.: Minds, brains, and programs. Behav. Brain Sci. **3**(03), 417–424 (1980). doi:10.1017/S0140525X00005756
4. Rumelhart, D.E., McClelland, J.L., CORPORATE PDP Research Group, (eds.): Parallel Distributed Processing: Explorations in the Microstructure of Cognition, Vol. 1: Foundations. MIT Press, Cambridge (1986)
5. Yen, J.: Fuzzy logic-a modern perspective. IEEE Trans. Knowl. Data Eng. **11**(1), 153–165 (1999). doi:10.1109/69.755624
6. Zadeh, L.A.: Soft computing and fuzzy logic. IEEE Softw. **11**(6), 48–56 (1994). doi:10.1109/52.329401
7. Johnson, R.C.: What is cognitive computing? Dr. Dobb's J. **18**(2), 18–24 (1993)
8. Ferrucci, D., Brown, E., Chu-Carroll, J., Fan, J., Gondek, D., Kalyanpur, A.A., et al.: Building watson: an overview of the DeepQA project. AI Mag. **31**(3), 59–79 (2010). doi:10.1609/aimag.v31i3.2303
9. Chomsky, N.: Syntactic Structures. Mouton, The Hague (1957)
10. Pereira, F.: Formal grammar and information theory: together again? Philos. Trans. R. Soc. Lond. Math. Phys. Eng. Sci. **358**(1769), 1239–1253 (2000). doi:10.1098/rsta.2000.0583
11. Pustejovsky, J.: The Generative Lexicon. MIT Press, Cambridge (1995)
12. Deep parsing in Watson. IBM J. Res. Dev. 56(3.4), 3:1–3:15 (2012). doi:10.1147/JRD.2012.2185409
13. Licklider, J.C.R.: Man-computer symbiosis. IRE Trans. Hum. Factors Electron. **1**, 411 (1960). doi:10.1109/THFE2.1960.4503259
14. Veres, C.: LexiTags: an interlingua for the social semantic web. Social Data on the Web (SDoW) Workshop at the 10th International Semantic Web Conference, 23–27 October (2011)
15. Veres, C.: Crowdsourced semantics with semantic tagging: "Dont just tag it, LexiTag it!". In: Acosta, M., Aroyo, L., Bernstein, A., Lehmann, J., Noy, N.F, Simperl, E. (eds.) Proceedings of the 1st International Workshop on Crowd- sourcing the Semantic Web, Sydney, Australia, 19 October 2013. CEUR Workshop Proceedings (2013)
16. Blei, D.M.: Probabilistic topic models. Commun. ACM **55**(4), 77–84 (2012). doi:10.1145/2133806.2133826
17. Veres, C.: Schema.org for the Semantic Web with MaDaME. In: Lohmann, S. (ed.) Proceedings of the I-SEMANTICS 2013 Posters & Demonstrations Track. CEUR-WS.org, Graz, Austria, 4–6 September 2013, vol. 1026. CEUR Workshop Proceedings (2013)

Comparing Recursive Autoencoder
and Convolutional Network for Phrase-Level
Sentiment Polarity Classification

Johannes Jurgovsky$^{(\boxtimes)}$ and Michael Granitzer

University of Passau, 94032 Passau, Germany
{Johannes.Jurgovsky,Michael.Granitzer}@uni-passau.de
http://www.fim.uni-passau.de/medieninformatik/

Abstract. We present a comparative evaluation of two neural network
architectures, which can be used to compute representations of phrases or
sentences. The *Semi-Supervised Recursive Autoencoder* (SRAE) and the
Convolutional Neural Network (CNN) are both methods that directly
operate on sequences of words represented via word embeddings and
jointly model the syntactic and semantic peculiarities of phrases. We
compare both models with respect to their classification accuracy on
the task of binary sentiment polarity classification. Our evaluation shows
that a single-layer CNN produces equally accurate phrase representations
and that both methods profit from the initialization with word embeddings trained by a language model. We observe that the initialization
with domain specific word embeddings has no significant effect on the
accuracy of both phrase models. A pruning experiment revealed that
up to 95 % of the parameters used to train the CNN could be removed
afterwards without affecting the model's accuracy.

Keywords: Natural language processing · Deep learning · Artificial
neural networks · Recursive autoencoder · Convolutional neural network

1 Introduction

When applying deep learning methods to natural language processing, in particular for sentiment polarity classification, there are currently two main approaches
that map both the meaning and the structure of a variable-length sentence to a
fixed-dimensional representation.

Foremost, there are recursive neural networks that exploit the properties of
compositionality present in natural language. The principle of compositionality
states that the meaning of an expression is determined by its structure and the
meanings of its constituents. Recursive architectures comprise the compositional
properties of a sentence globally over all its constituents. They are inherently
deep architectures, as they recursively apply the same transformation over the
sentence structure. In natural language processing, they have been successful

© Springer International Publishing Switzerland 2015
C. Biemann et al. (Eds.): NLDB 2015, LNCS 9103, pp. 160–166, 2015.
DOI: 10.1007/978-3-319-19581-0_14

in learning sequence and tree structures, mainly continuous sentence representations based on continuous word embeddings. These sentence representations have been shown to retain enough information of the sentence to be used as features in a simple linear classifier for sentiment polarity classification [1–3].

In contrast, convolutional neural networks learn local feature detectors indicating the presence or absence of word sequences within a sentence. In these networks, the composition function is not learned but given a-priori as an architecture consisting of alternating convolution and pooling layers. By pooling over the output of a convolution layer one can obtain a translation invariant representation of the input in terms of local feature activations. Convolutional neural networks have been employed in a variety of tasks in natural language processing for many years. Similarly to recursive architectures, they have also been used to derive sentence representations from sequences of word embeddings for classification purposes [4,5].

In this paper we directly compare the sentiment polarity classification accuracy of phrase representations obtained from either a single-layer *Convolutional Neural Network* (CNN) or the *Semi-Supervised Recursive Autoencoder* (SRAE) [2], when initialized with word embeddings that were pre-trained on different corpora. We show that

(i) both models give equal accuracy on the movie reviews dataset.
(ii) independent of the method used, domain specific language corpora, i.e. sentiment corpora in our case, are not necessary for obtaining accurate phrase models. Contrary, good phrase models for sentiment polarity analysis can be estimated from general purpose corpora like Wikipedia.
(iii) both phrase models can be represented with a fraction of the parameters used to train them. In case of the CNN, by pruning 95 % of the parameters, the accuracy remains the same.

2   Estimating Sentence Representations with Neural Networks

When solving NLP tasks with neural networks using, it has become a common practice to incorporate word embeddings. In contrast to one-hot coded words, word embeddings are distributed representations that exhibit different notions of word similarities within a language. They can be obtained from language models that were trained on large text corpora. Since word embeddings are usually implemented as vectors, all words together form an embedding matrix. Thus, we can represent a sequence of words as a sequence of word vectors by using the embedding matrix as a look-up table.

In [2], the authors adopted the approach of recursive auto-associative memories [6] and included a modification so that the model can both learn to predict an approximate composition structure and compute the corresponding sentence representation. The SRAE model computes composite parent representations from word representations by recursively applying an autoencoder on pairs of neighbouring words. Through pairwise composition the model builds a binary tree in which leaf nodes correspond to single word representations and inner nodes correspond to multi-word phrase representations. The representation induced at each inner node captures the semantics of the multi-word phrase spanned by the sub-tree of the node. The parent representation in the root node of the tree is considered to be representative of the whole sentence.

In contrast to the SRAE, our CNN has a simple feed-forward architecture without recurrence. We constructed a strongly simplified convolutional network with only one convolutional and pooling layer and without additional hidden layers.

The network takes as input a sequence of word vectors and learns to compute a vector of feature activations. The feature detectors in the convolutional layer, each span a window of five input words. The detectors share the same parametrization for all contiguous regions of the input sentence. Together, all the region-specific responses of a detector form a feature map, on which we apply a max-pooling operation to select the response of maximum value. The pooled feature detectors are considered to be representative of the phrase.

In both the SRAE and the CNN architecture, we stack a softmax output-layer on top of the last layer that computes the representation, in order to learn from positive/negative sentiment label information during training. We include the embedding matrix as additional parameters in each model. Thus, during training, the word vectors can be fine-tuned to capture sentiment information induced by the target labels.

3   Experimental Setup

We compared the learned sentence representations from either models with respect to the usefulness of their features in a sentiment polarity classification task. We employed the movie reviews sentiment polarity dataset, which consists of 10'662 (5331 positive, 5331 negative) labelled sentences collected from an online rating platform[1] and were provided by Pang et al. [7].

We employed a set of simple regular expressions to split sentences into word tokens. Punctuation marks were considered as individual tokens and words containing apostrophes were split into two tokens. Only those tokens that occurred more than once were included in the vocabulary, giving a total vocabulary size of 10'046 words. Any other word was mapped to a special *UNKNOWN* token. The corresponding word vectors were initialized either with small random values or with pre-trained values computed by two types of language models. We obtained word embeddings from the *Skip-Gram Model* [8] which can be computed very efficiently with the *word2vec*-tool[2]. We trained the Skip-Gram model on about one billion tokens extracted from English Wikipedia articles[3] and, for a matter of evaluating the impact of domain specific word embeddings, we trained a second instance on one billion tokens from Amazon product reviews[4].

---

[1] http://www.rottentomatoes.com/.
[2] https://code.google.com/p/word2vec/.
[3] http://dumps.wikimedia.org/enwiki/latest/.
[4] http://snap.stanford.edu/data/web-Amazon.html.

We implemented both neural networks in the programming language Python with the use of the symbolic computation library Theano[5]. In order to reproduce the results reported in the SRAE-paper, we also ran L-BFGS for a maximum of 80 iterations over the complete training data in batch mode to minimize the objective function. Regarding parametrization, we adopted the hyper-parameter settings reported in the paper. In particular, we set the dimension of word vectors and the number of feature detectors in the CNN to 100, such that the dimension of the computed sentence representation vectors is the same (100) for both models. The CNN's objective function was minimized via stochastic gradient descent with mini-batches (20) and a linearly decreasing learning rate over a total of 15 epochs.

We evaluated the performance of both models via 10-fold cross validation. For each train/test split, we trained one of the two phrase models on the training set in a (semi-)supervised setting. After convergence, we used the trained model to extract phrase representations from all sentences in the dataset. We trained a binary Logistic Regression classifier on the training set phrase representations and evaluated its sentiment polarity accuracy on the test set representations.

4   Results

4.1   Word Embeddings

As Mikolov et al. [9] demonstrated, the Skip-Gram word vectors encode various relational similarities, which can be recovered using vector arithmetic and used to solve word analogy tasks. Therefore, we evaluated them on the Google analogical reasoning dataset.[6]

With regard to the embeddings obtained with the Skip-Gram architecture, we observe that the Wikipedia(W) embeddings (64.2 %) achieve an overall better accuracy than the Amazon(A) embeddings (41.7 %) for almost all relation types; with the only exception being the *present-participle* category (A:65.0 %, W:57.4 %). However, we see comparable accuracies for relation types that are commonly used in colloquial and informal language, like *third-person singular* (A:52.4 %, W:54.1 %), *comparative* (A:69.1 %, W:76.2 %) and *plurals* (A:60.4 %, W:71.7 %). From these results we conclude that well-formed texts, which consistently follow the grammar of a language, yield better word representations.

4.2   Sentiment Polarity Classification

The results of our implementation of the SRAE suggest that we correctly re-implemented the SRAE model in Python. In case of initializing the word vectors with small random values, our implementation achieves 76.8 % accuracy on average over the 10 train/test splits. This result is consistent with the result originally reported in their paper. Our simple version of a convolutional network could also

---

[5] http://deeplearning.net/software/theano/.

[6] code.google.com/p/word2vec/source/browse/trunk/questions-words.txt.

Table 1. Mean accuracy and standard deviation of 10-fold cross validation for both models with different word vector initializations.

Initialization	SRAE (Socher et al.)	SRAE (our impl.)	CNN (our impl.)
Random	76.8	76.8 ± 1.75	76.0 ± 1.17
Skip-Gram (Wikipedia)	-	79.0 ± 1.17	78.4 ± 1.26
Skip-Gram (Amazon)	-	78.6 ± 0.84	79.5 ± 1.35

achieve 76.0 %. The Wilcoxon rank sum test revealed ($p = 0.29$) that the difference is not significant at the 5 % significance level, due to large variations in the individual results per split (Table 1).

Regarding our experiments, it seems that training the Skip-Gram model on the Amazon text corpus, which mainly consists of subjective personal opinions, does not significantly increase the utility of word vectors as compared to those trained on a general purpose corpus like Wikipedia. This effect can be observed for both the SRAE ($p_{RAE} = 0.53$) and the CNN ($p_{CNN} = 0.23$) representations. A pairwise comparison of the SRAE and the CNN performance for each word vector initialization mode reveals that both models yield about the same ($p_{SGa} = 0.53$, $p_{SGw} = 0.23$) mean accuracy.

4.3 Pruning

To investigate the extent to which the phrase models could make use of their parameters, we conducted a pruning experiment. First, we trained the model on a particular train/test split to make the parameters best fit the training data. Then we set a certain percentage (pruning level) of the model's smallest parameter values to zero. After this pruning step, we let the model compute phrase representations for all examples in the training set and test set. Again, we trained a Logistic Regression classifier on the training set representations and evaluated its performance on the test set representations. We repeated this process for several percentages of zeroed values.

We applied this pruning strategy to each parameter matrix of a model individually. The SRAE's parameters $\Theta_{SRAE} = (W^{(1)}, W^{(2)}, \mathbf{b}^{(1)}, \mathbf{b}^{(2)}, L)$ comprise a total of 1'044'800 values and the CNN's parameters $\Theta_{CNN} = (W, \mathbf{b}, L)$ a total of 1'054'700. We determine individual threshold values for W, $\mathbf{b}$ and L, such that all values below these thresholds are set to zero and thus do not contribute to the construction of the phrase representation.

Figure 1 shows the accuracy of the Logistic Regression classifier after training it on the pruned SRAE and CNN sentence representations with random word vector initialization. In both models, we could prune up to 81 % of the parameters without observing any major impact on the classification accuracy induced by the modified phrase representations (not shown in the figure). In case of the CNN, even if we removed up to 95 % of all model parameters, the representations seem to preserve enough information about the inputs to be classified correctly with 76 % accuracy.

Small weight values only
have modest impact on the net
input of neurons in the network. Accordingly, a neuron's
output does not change much
in its activation with respect
to changes in the inputs that
are received from low-weight
connections. In a neural network there are typically many
settings of weights that could
potentially model the dataset
quite well - especially for small
datasets and lots of parameters,
like in our case.

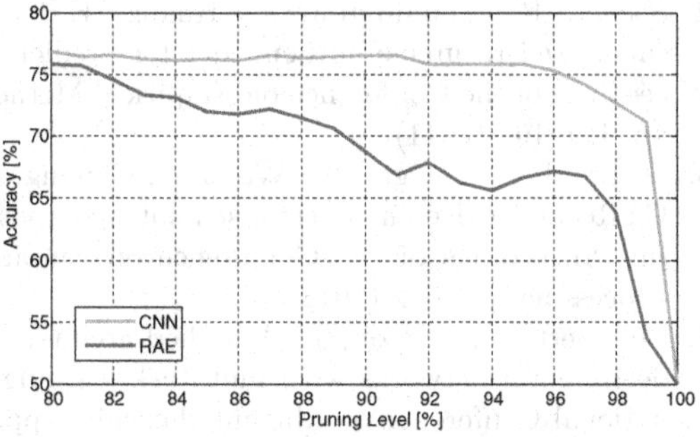

Fig. 1. Binary sentiment polarity classification
accuracy of Logistic Regression. Underlying sentence representations were extracted with pruned
versions of CNN or RAE at different pruning
levels.

The drop in the CNN's accu-
racy is rather small for pruning levels below 95 %, when it starts decreasing
rapidly. With max-pooling, one essentially samples a particular instance from
the set of potential network architectures which have the same weights (the one
where only the neuron with maximum activation is included in the computation
graph). Since the input weights of neurons in a convolutional layer are shared,
pruning small weights, in many cases, does not change the selection of a neuron after applying the max-operation. Thus, the CNN's particular architecture
to which the model converged to after training, is mostly unaffected by many
settings of pruned weights.

5 Conclusion

We presented a comparison of the Semi-Supervised Recursive Autoencoder and
a Convolutional Neural Network and evaluated the capability of both networks
to extract sentence representations. From our experiments we conclude that a
very simple CNN architecture with only one convolution and pooling layer can
be as effective as the SRAE for sentiment polarity classification of sentences. We
showed that word embeddings trained with the Skip-Gram language model on
a corpus of subjective text does not significantly improve classification performance on this sentiment analysis task. A pruning experiment revealed that in
both neural phrase models up to 81 % of all parameters can be omitted without
causing a major impact on the learned phrase representations. This might point
towards more efficient training methods for neural networks.

References

1. Socher, R., Manning, C.D., Ng, A.Y.: Learning continuous phrase representations
   and syntactic parsing with recursive neural networks. In: Proceedings of the NIPS-
   2010 Deep Learning and Unsupervised Feature Learning Workshop, pp. 1–9 (2010)

2. Socher, R., Pennington, J., Huang, E.H., Ng, A.Y., Manning, C.D.: Semi-supervised recursive autoencoder for predicting sentiment distributions. In: Proceedings of the Conference on Empirical Methods in Natural Language Processing, pp. 151–161 (2011)
3. Socher, R., Perelygin, A., Wu, J.Y., Chuang, J., Manning, C.D., Ng, A.Y., Potts, C.: Recursive deep models for semantic compositionality over a sentiment treebank. In: Proceedings of the Conference on Empirical Methods in Natural Language Processing, p. 1642 (2013b)
4. Collobert, R., Weston, J.: A unified architecture for natural language processing: deep neural networks with multitask learning. In: Proceedings of the 25th International Conference on Machine Learning, pp. 160–167 (2008)
5. dos Santos, C.N., Gatti, M.: Deep convolutional neural networks for sentiment analysis of short texts. In: Proceedings of the 25th International Conference on Computational Linguistics (COLING) (2014)
6. Pollack, J.B.: Recursive distributed representations. Artif. Intell. **46**(1), 77–105 (1990)
7. Pang, B., Lee, L.: Seeing stars: exploiting class relationships for sentiment categorization with respect to rating scales. In: Proceedings of the 43rd Annual Meeting on Association Computational Linguistics, pp. 115–124 (2005)
8. Mikolov, T., Sutskever, I., Chen, K., Corrado, G.S., Dean, J.: Distributed representations of words and phrases and their compositionality. Adv. Neural Inf. Proc. Syst. **26**, 3111–3119 (2013)
9. Mikolov, T., Chen, K., Corrado, G., Dean, J.: Efficient estimation of word representations in vector space. arXiv preprint arXiv:1301.3781 (2013)

The Interplay of Language Processing, Reasoning and Decision-Making in Cognitive Computing

Sergei Nirenburg[✉] and Marjorie McShane

Rensselaer Polytechnic Institute, Troy, NY 12180, USA
{zavedomo,margemc34}@gmail.com

Abstract. Integrating language processing, reasoning and decision making is a prerequisite to any breakthroughs in cognitive computing. This paper discusses historical attitudes that have worked against such integration, then describes a cognitive architecture called OntoAgent that illustrates both the feasibility and the payoffs of pursuing integration. Examples are drawn from the Maryland Virtual Patient prototype application, which offers medical trainees the opportunity to diagnose and treat a cohort of cognitively modeled virtual patients that are capable of language processing, reasoning, learning, decision making and simulated action.

Keywords: Natural language processing · Intelligent agents · Reasoning · Cognitive architecture

1 The Analytic Approach and Its Consequences

Since at least the times of Descartes, the scientific method has become more or less synonymous with the analytical approach, whereby a phenomenon or process is decomposed into contributing facets or components. The general idea is that, after each such component has been sufficiently studied independently, there would follow a synthesis step that would result in a comprehensive explanation of the phenomenon or process. A well-known example of the application of the analytical approach is the tenet of the autonomy of syntax in theoretical linguistics, which has been widely adopted by – and has strongly influenced – the field of computational linguistics.

The analytical approach makes good sense because it is well nigh impossible to expect to account for all the facets of a complex phenomenon simultaneously and at a consistent grain size of description. But it comes with a cost: it artificially constrains the purview of theories and the scope of models, and it has unwittingly fostered indefinite postponement of the all-important synthesis step.

Within the field of cognitive modeling, the analytical approach is evident in the traditionally defined pipeline of agent functionalities: *perception, reasoning, action*. Under this view, natural language understanding is subsumed under

© Springer International Publishing Switzerland 2015
C. Biemann et al. (Eds.): NLDB 2015, LNCS 9103, pp. 167–179, 2015.
DOI: 10.1007/978-3-319-19581-0_15

perception and natural language generation is subsumed under *action*. And, although it is well understood that human-level natural language processing will require extensive reasoning, decision-making and learning, these *language-oriented* manifestations have not been on the agenda of the respective research communities.

This paper explores our efforts at synthesizing, within the OntoAgent cognitive architecture, the treatment of natural language understanding with the treatment of reasoning, decision-making, learning, and non-linguistic perception. After introducing the historical landscape and briefly describing our cognitive architecture, we will illustrate this synthesis using the following four simple premises as motivation: (1) While natural language understanding provides input to reasoning, reasoning, in turn, supports the process of natural language understanding. (2) While natural language understanding provides input to decision-making, decision-making, in turn, is an integral part of the process of natural language understanding. (3) Not only does NLP support an agent's learning about the language and world, learning about language and the world is needed for the agent to engage in human-level language understanding. (4) Agent memory can be populated not only by results of natural language understanding; it be populated by output from other channels of perception.[1] Although we have been working toward actualizing these aspects of integration for some time, this paper represents our first attempt to generalize about how this effort contributes to the goal of *synthesis at an early stage* in the development of human-inspired intelligent agents.

1.1 Natural Language Processing Within Cognitive Modeling, Historically

Mainstream NLP focuses predominantly on the shallow analysis of uninterpreted text strings (see [7] for a balanced overview of the state of the art), avoiding language problems that require anything beyond statistically-oriented computation. Automatic reasoning systems, for their part, typically start with high-quality knowledge structures and pay relatively little attention to the provenance of these inputs, in expectations that language problems will eventually be solved by external processors (see, e.g., [10]). This division between NLP and reasoning was recognized already in the 1950s by Bar Hillel:

> "...The evaluation of arguments presented in a natural language should have been one of the major worries of logic since its beginnings. However, [...] the actual development of formal logic took a different course. It seems that [...] the almost general attitude of all formal logicians was to regard such an evaluation process as a two-stage affair. In the first stage, the original language formulation had to be rephrased, without loss, in a normalized idiom, while in the second stage, these normalized

---

[1] Results of reasoning can also populate the agent memory. We do not develop this topic in this paper.

formulations would be put through the grindstone of the formal logic evaluator. [...] Without substantial progress in the first stage even the incredible progress made by mathematical logic in our time will not help us much in solving our total problem." ([2], pp. 202–203).

If we substitute the more modern terms "knowledge representation language" for "normalized idiom," and "reasoning system" for "formal logic evaluator," it becomes clear that the state of affairs described by Bar Hillel has only modestly changed over the last half-century. That is, in spite of the important work on semantically-oriented NLP by groups led by Schank [24], Wilks [29], Woods [30], Allen [1], Schubert [27] and others, the "evaluation of arguments presented in a natural language" has yet to garner intensive, widespread attention. Instead, the center of gravity for mainstream NLP has been on statistical tools applied to text strings in large text corpora.

As mentioned earlier, in the study of cognitive architectures, it is customary to modularize agent functionalities. The most coarse-grained categorization was presented above: *perception, reasoning, action.* Various more fine-grained classifications have also been put forth, such as the one found in Langley et al.'s 2009 survey article [9]. They describe nine capabilities that any good cognitive system must have: (1) recognition and categorization; (2) decision making and choice; (3) perception and situation assessment; (4) prediction and monitoring; (5) problem solving and planning; (6) reasoning and belief maintenance; (7) execution and action; (8) interaction and communication; and (9) remembering, reflection and learning. The authors primarily subsume NLP under "interaction and communication" but acknowledge that it involves other aspects of cognition as well. They recognize the lack of fundamental integration of NLP with other aspects of agent cognition, stating, "Although natural language processing has been demonstrated within some architectures, few intelligent systems have combined this with the ability to communicate about their own decisions, plans, and other cognitive activities in a general manner." Indeed, of the 18 representative architectures briefly described in the Appendix, only two – SOAR [12] and GLAIR [26] – are overtly credited with involving NLP; and one, ACT-R, is credited indirectly by reference to applied work on tutoring [8] within its framework. Of all of the cross-modular influences, one that has been particularly well explored is the interaction between NLP and planning: e.g., the pioneering work of Cohen, Levesque and Perrault (see [5] and [22]) demonstrated the utility of approaching NLP tasks in terms of AI-style planning, and planning is a first-order concern in the field natural language generation [23].

The reason why NLP has been addressed rather peripherally and not in depth in the field of cognitive modeling is because it is a very hard problem; more precisely, it is a collection of many hard problems. However, satisfactory solutions to these problems can only be expected if they are tackled as part of overall agent functioning. In other words, at least outside the realm of "low-hanging fruit" applications, NLP must be integrated with other aspects of simulated agent cognition. This is the hypothesis pursued in the OntoAgent program of research and development, to which we now turn.

2   OntoAgent and the Maryland Virtual Patient Application

OntoAgent can be described as both a cognitive architecture and a knowledge environment: it has the same goals as traditional cognitive architectures, but, unlike many other architectures, it also stresses descriptive and system-building work aimed at creating a non-toy knowledge substrate to support the functioning of agents in applications [15]. OntoAgent has three core static knowledge resources: the ontology, the fact repository and the lexicon. The **ontology** is a knowledge base of descriptions that contains knowledge about types of objects, events, relations that link them and attributes that describe them. It also contains the scripts that support agent simulation. The current version of the OntoAgent ontology contains about 9,000 concepts described by an average of 16 properties each. Some agents are not endowed with the full ontology, largely to simulate individual differences in knowledge of specialized domains. The **fact repository** contains remembered instances of ontological concepts and their property values. Naturally, each agent has its own fact repository to reflect its individual simulated experience. The **lexicon** describes approximately 30,000 word senses of English, both syntactically and semantically, with semantic descriptions written in the ontological metalanguage.

Language understanding in OntoAgent means automatically translating ambiguous and often elliptical natural language inputs into an unambiguous, ontologically-grounded metalanguage suited for reasoning. This translation is carried out using heuristic evidence that relies primarily on information in the lexicon and ontology (for a description of the analysis process, see [20]). The analyzer produces text meaning representations like the one shown in Table 1, which represents the meaning of the English sentence *Dolores has severe chest pain.* Concepts are shown in small caps, and numerical suffixes differentiate instances. The similarity of concept names to English words is solely to support manual knowledge acquisition: the meaning of a concept is defined as its inventory of property values.[2]

Ontologically-grounded text meaning representations facilitate expectation-driven reasoning during both language processing and decision-making. This is because the ontology contains more information about each *type* of object and event than is known about each textually attested *instance*. For example, whereas the concept instance PAIN-23 in our example includes fillers for the properties INTENSITY and LOCATION, the ontology includes many more properties of PAIN. One is PAIN-CAUSE-TYPE, whose literal fillers are *nociceptive, neuropathic* and *psychogenic*. When the concept PAIN is activated (i.e., used in a text meaning representation), the agent may expect further dialog to include references to as-yet "unused" properties. So, given the subsequent input, *The pain was due to tissue damage.*, the polysemous phrases *due to* and *tissue damage* will be analyzed as the highly specific property-value pair PAIN-CAUSE-TYPE NOCICEPTIVE using expectation-driven reasoning.

---

[2] The same ontology can be used for representing the meaning of utterances in any language, given an ontological semantic lexicon for that language.

Table 1. Text meaning representation for *Dolores has severe chest pain.*

PAIN-23		
	EXPERIENCER	HUMAN-37
	INTENSITY	.8
	LOCATION	CHEST-BODY-PART-14
	textstring	"has"
	from-sense	have-v18 (a phrasal entry for *have...pain*)
HUMAN-37		
	EXPERIENCER-OF	PAIN-23
	HAS-NAME	"Dolores"
	HAS-GENDER	female
	textstring	"Dolores"
	from-sense	*personal-name*
CHEST-BODY-PART-14		
	LOCATION-OF	PAIN-23
	textstring	"chest"
	from-sense	chest-n1

The application that has served as the substrate for validating our approach to agent modeling is Maryland Virtual Patient (MVP). MVP is a prototype clinician training application that features a cohort of cognitively modeled virtual patients that can be diagnosed and treated by human trainees in open-ended simulations [14, 19]. Virtual patients are comprised of linked physiological and cognitive simulations. Physiologically, virtual patients change over time and in response to interventions by the user. Cognitively, virtual patients can engage in dialog with the user, make decisions about their health care and lifestyle, learn and remember new information, and carry out simulated action. The bridge between physiology and cognition is *interoception*, defined as the perception of one's bodily signals. We model interoception as one of two channels of perception, the other being natural language understanding. Both interoception and natural language understanding generate identical meaning representations that are used to populate the agent's fact repository (cf. Sect. 3.4).

Along with virtual patients, the MVP environment features a tutoring agent, which can provide context-sensitive guidance to the trainee. As described in [16] and [17], the same knowledge substrate used by the tutor could be used to support the functioning of an advisor to practicing clinicians.

3   Four Examples of Integrating NLP with Other Agent Functionalties

This section briefly describes the previously introduced four points of integration of NLP with other aspects of agent cognition in OntoAgent.

3.1   NLU Supports Reasoning and Reasoning Supports NLU

The fact that natural language understanding (NLU) supports reasoning is well-attested in the cognitive systems literature: after all, NLU is typically viewed as a type of perception, which precedes reasoning and action. For example, in the MVP application, once a virtual patient has understood the meaning of (i.e., generated a text meaning representation for) "What brings you here?" it must:

1. Detect the goal(s) that the interlocutor is pursuing in uttering the dialog turn.
2. Integrate the results of its analysis of text meaning and speaker goal into its memory.
3. Decide to generate an instance of a "Be-a-Cooperative-Conversationalist" goal and add it to its active goal agenda.
4. Prioritize goal instances on the agenda (in this case, the above instance will be prioritized).
5. Select a plan to pursue to attain this goal (in this case, the plan will be to carry out a verbal action).
6. Decide on content of the verbal action to be produced (in this case, this will involve checking its memory for recent symptoms).
7. Generate an English sentence that realizes the above content by outputting, for example, "I've been having difficulty swallowing."

In short, natural language understanding launches a cascade of other agent reasoning functions.

But just as NLU supports reasoning, so must reasoning be brought to bear for NLU, since the challenges presented by natural language are formidable: lexical ambiguity, referential ambiguity, idiomaticity, ellipsis, indirect speech acts, non-literal language, unexpected input and more. These challenges have been underplayed in the past 20 years, as mainstream NLP has chosen to focus on those linguistic phenomena that are most amenable to supervised machine learning. For example, there have been significant efforts toward detecting (but not semantically interpreting) multi-word expressions [25]; resolving the simpler cases of textual coreference [11]; and selecting which relations hold between uninterpreted nominals in nominal compounds [28]. However, as long as NLP is approached as the manipulation of uninterpreted textual strings, its results will not be sufficient to support human-level reasoning by intelligent agents.

Within OntoAgent, by contrast, we do not shun the difficult problems posed by natural language. Addressing such problems naturally requires knowledge-based, reasoning-intensive methods [13,20,21]. Of course, the goal of fully understanding all language phenomena in open text will not be achieved overnight – all of the contributing algorithms and the knowledge bases they rely on require long-term, iterative improvement. However, even setting the goal of language *understanding* shifts the perspective away from isolated NLP modules and toward the integration of natural language into overall agent cognition.

3.2 NLU Supports Decision-Making and Decision-Making is Needed for NLU

As we just saw, NLU supports reasoning and reasoning is a prerequisite to decision-making, so the statement "NLU supports decision-making" should be self-evident. However, the reverse dependency – i.e., decision-making in support of NLU – has received little attention. This is in large part because NLU has been recently treated as a monolithic process that results in a singularly right or wrong answer. However, this orientation fails to account for the fact that normal, unedited, natural language use does not consist of exclusively "clean" utterances that can be understood – even by people – with 100 % precision. Instead, language is littered with false starts, infelicitous ellipses, intentional and unintentional vagueness, unnecessary detail, incomprehensibly formulated thoughts, and so on. For this reason, NLU is better modeled as a multi-stage process after each stage of which the agent asks itself *Have I understood enough to proceed to reasoning (and action)?* Decisions about "enough" will spare agents from endlessly pursuing ever deeper language analysis.

Let us consider a few examples in which "enough" is achieved at different stages of processing.

– *The basic text meaning representation is sufficient.* The basic text meaning representation is sufficient to support reasoning when an agent is faced with a direct question (*Do you have chest pain?*) or a direct command (*Please tell me your symptoms.*). In these cases, the basic text meaning representation includes an instance of a REQUEST-INFO (request information) or REQUEST-ACTION event, which is sufficient for the agent to generate a verbal action in response.
– *Indirect speech act detection is needed.* When the basic text meaning representation does not include a direct request for information or action, the agent attempts to determine whether an indirect speech act was used, as is the case in the following: *I'd like to know if you ever have chest pain* (indirect question), *I think that surgery is your best option* (indirect request for action). In such instances, the result of the agent's reasoning is recorded in a so-called *extended* text meaning representation, which includes the initially masked REQUEST-INFO or REQUEST-ACTION concept.
– *Not all input needs to be fully understood.* As mentioned above, it is not uncommon for natural language input to include words, phrases and even whole sentences that are functionally superfluous. Such is the case, for example, when the speaker precedes a request for information or action by a long preface: *I know we've talked about a lot of things related to your past and current symptom profile, but what I'd really like to know at this point in time is, do you have chest pain?* Even if the agent cannot not confidently disambiguate every lexeme preceding the question, it can still answer the question and hold up its end of the dialog interaction. We not suggesting that it is optimal for an agent to fail to fully understanding something; however, we are suggesting that if the goal is to build useful intelligent agents in the near- and mid-term, teaching them to focus on actionable aspects of utterances is well-motivated.

- *Clarification from the human collaborator is needed.* In some cases, the agent might fail to understand a necessary portion of an input and therefore be blocked from subsequent reasoning. This can happen, for example, if an agent is asked a question about an unknown word/concept: *Do you feel pain in the area of your lower esophageal sphincter?* In such cases, the agent must make the decision to pursue learning during language processing itself, as described furher in Sect. 3.3.
- *The agent decides reason about the speaker's goal.* In many cases, people respond not only to the direct meaning of a question or request, but to their understanding of the speaker's goal in uttering it. For example, the following dialog turns are quite natural: (a) *"Where are your keys?" "You can't borrow my car."* (b) *"I have a stomache ache." "You're going to school."* (c) *"Can you run and fetch me a screwdriver?" "This knife will work just as well."* In each of these cases, responding to the direct meaning would have also been appropriate, but the interlocutor considers these responses more efficient. Whether a speaker will respond to the direct meaning of an utterance or to his/her interpretation of the speaker's goal is a function of the interlocutor's personality traits, understanding of the situation, the relationship between the speaker and interlocutor, and so on.

Consider an example in which five different virtual patients (VPs), who present to the doctor with the symptom of coughing, respond to the question *"Have you been traveling lately?"* in different ways for different reasons.[3]

- VP1: No, I haven't been anywhere that might have made me sick.
- VP2: Yes, I was on a crowded plane last week.
- VP3: No.
- VP4: Yes, I drove to Washington to visit my sister.
- VP5: No. Why are you asking?

VP1 and VP2 have an inventory of personality traits and physical and mental states that compel them to hypothesize about the goal the speaker is pursuing. Since they are reporting to the doctor with a complaint, and since they have not yet been diagnosed, they hypothesize that the doctor's goal is diagnosis. They try to figure out – using ontological search – how COUGH is connected with TRAVEL-EVENT. The ontologies of these two VPs are the same, and they include the information that COUGH can be CAUSED-BY INFLUENZA, that INFLUENZA is a COMMUNICABLE-DISEASE, and that COMMUNICABLE-DISEASE can be CAUSED-BY AIRPLANE-TRAVEL, BUS-TRAVEL, TRAIN-TRAVEL (the latter is a simplification of a much longer causal chain that involves being located in crowded spaces). So, they understand that their cough might be caused by something they encountered during these types of travel. The two patients differ, however, with respect to their fact repositories (i.e., memories of past experiences). VP1 does not have any recorded memories of relevant travel events – i.e., travel in an airplane, bus or train – so it responds 'no'. The remainder of the utterance serves as a trace that it is responding to the goal behind the question, not to its literal meaning. VP2, by contrast, has a different fact repository: it recently traveled on a plane. So it answers positively,

---

[3] See [18] for further discussion of agent parameterization.

and the elaboration of its response serves as a trace that it responded to the speaker's goal as well.

VP3, for its part, decides to bypass goal-oriented mindreading and to respond to the direct meaning of the question.[4] The decision not to read extra meaning into the question is quite natural for this agent since it has no recent travel events in its fact repository – i.e., nothing to trigger decision-making about whether or not an event is relevant.

As for VP4, it is unclear from its utterance whether or not it attempted to reason about the speaker's goal. Either it did attempt to and failed, or it simply answered the surface meaning of the question. If it failed, then the likely reason is that its ontology lacks the necessary link between coughing and certain types of travel.

Finally, VP5 decides not to mindread but attempts to learn what goal the user is pursuing in asking this question, reflecting a knowledge-seeking character trait.

In sum, it would be counterproductive to assume that all agents must at all times understand every input to the same depth and with the same precision. Some inputs defy precise analysis, even by humans; some inputs provide more detail than is needed for the sufficient functioning of an agent in an application; and many inputs can be analyzed and responded to differently by agents with different knowledge, histories, traits and states. All of this variability is best supported by opening up the many stages of the language understanding process and giving the agent an option to make decisions about how to proceed following each of those stages.

3.3   NLP Supports Learning and Learning Supports NLP

Learning in OntoAgent is understood as the agent's ability to populate, modify and use its knowledge bases – ontology, lexicon and fact/belief repository – over the period of its functioning (its "lifespan"). As an illustration of agent learning, let us look at the MVP dialog excerpt shown in Table 2. At the time point we observe, the human agent is interviewing the virtual patient and recommends that she have an EGD. This particular virtual patient does not know the word EGD or the concept EGD – they are absent from her lexicon and ontology. So, she must engage in further dialog to learn the needed information (NLP supports learning), and she must learn the information in order to carry out further dialog on this topic (learning supports NLP). Traces of system functioning (which can be seen in dynamic "under the hood" panels of the interface) are presented in italics. The fact that the patient asks questions about the procedure is due to its character traits: it requires extensive information before making a decision.

---

[4] We can only know for sure that this agent did not undertake goal-related mindreading by looking at the trace of system processing. VP1 could also have generated just 'No' as its response.

Table 2. Dialog between the doctor (MD) and the virtual patient (VP). Dialog turns are in boldface, whereas traces of VP decision-making are in italics.

MD:	**I suggest having an EGD, which is a diagnostic procedure**
VP:	*I record* EGD *as a child of* DIAGNOSTIC-PROCEDURE *in my ontology*
	I record EGD as a noun in my lexicon, mapped to EGD
	I don't know enough about its properties to agree
	I need to ask questions about risk and pain
	How risky is it?
MD:	**It's not risky at all**
VP:	*Add* RISK 0 *to ontological specification of* EGD
	Is it painful?
MD:	**It's only a little uncomfortable**
VP:	*Add* PAIN .1 *to ontological specification of* EGD
	I can tolerate that amount of risk and pain
	OK, I'll agree to that

3.4 Memory Population from Natural Language Understanding or Other Perception Modes

Perception is the primary source of agent learning. As mentioned above, OntoAgents have two channels of perception: language understanding and simulated interoception. Perception via both of these channels results in identical meaning representations, formulated in the ontological metalanguage. Earlier (Table 1) we saw how text analysis results in such a meaning representation; let us now consider the analogous process of interoception.

During physiological simulation, when certain property values reach a given threshold, they trigger the instantiation of a symptom. For example, when a virtual patient's esophagus is sufficiently inflamed, this triggers the symptom of heartburn. Table 3 shows how the ontological representation of a mild esophageal inflammation, which is generated by the physiological simulation of gastroesophageal reflux disease, automatically generates the symptom *mild, occasional heartburn*. What is noteworthy for this discussion is that the ontological representation of this perceived symptom is the same as the meaning representation for the sentence, "I am experiencing mild, occasional heartburn." So, whether the agent experiences this symptom or is told it has this symptom (for whatever reason the latter might happen), it will be stored to memory in the same form and can support the same subsequent reasoning.

4 Final Thoughts

We have selectively illustrated the tight integration of language processing with reasoning, decision making, learning and alternative channels of perception in

Table 3. Ontological representation for esophageal inflammation generating the symptom of mild, occasional chest pain. Intensity and frequency of chest pain are measured in the abstract scale 0,1.

INFLAMMATION-1		
	LOCATION	ESOPHAGUS-1
	EFFECT	CHEST-PAIN-1
ESOPHAGUS-1		
	LOCATION-OF	INFLAMMATION-1
	PART-OF-OBJECT	PATIENT-1
CHEST-PAIN-1		
	INTENSITY	.1
	FREQUENCY	.1
	EXPERIENCER	PATIENT-1
	CAUSED-BY	INFLAMMATION-1
PATIENT-1		
	EXPERIENCER-OF	CHEST-PAIN-1
	HAS-OBJECT-AS-PART	ESOPHAGUS-1

the OntoAgent environment. Addressing this integration is, we believe, a prerequisite to any breakthroughs in semantic and cognitive computing. In our opinion, the most crucial of these is sufficiently fine-grained knowledge resources, such ontologies and lexicons. Therefore, R&D in knowledge acquisition must assume an ever growing importance.

The current trend in computational linguistics is "supply-side" – pursuing broad coverage at the expense of the depth of analysis. By contrast, the long-continuing trend in the fields of cognitive architectures and application systems that deploy language capabilities is still predominantly "demand-side" – devoting just as much effort to resource development as is minimally needed, with "minimally needed" being defined differently in the different paradigms. In cognitive architectures, language capabilities must be minimally sufficient to test the reasoning algorithms and representational formalisms, whereas in application systems, language capabilities must be minimally sufficient to achieve user acceptance.

In the framework of OntoAgent – and work that preceded it in the paradigm of Ontological Semantics – we have been developing a hybrid approach that seeks a better balance between supply-side and demand-side approaches, as well as between the breadth and depth of acquired knowledge. It is our hope that recent and current knowledge acquisition efforts – e.g., [3,4,6] – bring in new insights and results that will permit the research community to succeed in building adequate knowledge resources for cognitive computing.

Acknowledgments. This research was supported in part by Grant N00014-09-1-1029 from the U.S. Office of Naval Research. All opinions and findings expressed in this material are those of the authors and do not necessarily reflect the views of the Office of Naval Research.

References

1. Allen, J., Ferguson, G., Stent, A.: An architecture for more realistic conversational systems. In: Proceedings of the Conference on Intelligent User Interfaces, pp. 1–8 (2001)
2. Bar Hillel, Y.: Aspects of Language. Magnes, Jerusalem (1970)
3. Bobrow, D., Condoravdi, C., Karttunen, L, Zaenen, A.: Learning by reading: normalizing complex linguistic structures onto a knowledge representation. In: Proceedings of the AAAI Spring Symposium on Learning by Reading and Learning to Read (2009)
4. Clark, P., Harrison, P.: Large-scale extraction and use of knowledge from text. In: Proceedings of the Fifth International Conference on Knowledge Capture, pp. 153–160. ACM (2009)
5. Cohen, P.R., Levesque, H.J.: Rational interaction as the basis for communication. In: Cohen, P.R., Morgan, J., Pollack, M.E. (eds.) Intentions in Communication. MIT Press, Cambridge (1990)
6. Hahn, U., Marko, K.G.: Ontology and lexicon evolution by text understanding. In: Proceedings of the ECAI 2002 Workshop on Machine Learning and Natural Language Processing for Ontology Engineering (2002)
7. Jurafsky, D., Martin, J.H.: Speech and Language Processing: An Introduction to Natural Language Processing, Speech Recognition, and Computational Linguistics, 2nd edn. Prentice-Hall, Upper Saddle River (2009)
8. Koedinger, K.R., Anderson, J.R., Hadley, W.H., Mark, M.A.: Intelligent tutoring goes to school in the big city. Int. J. Artif. Intell. Educ. **8**, 30–43 (1997)
9. Langley, P., Laird, J.E., Rogers, S.: Cognitive architectures: research issues and challenges. Cogn. Syst. Res. **10**, 141–160 (2009)
10. Langley, P., Meadows, B., Gabaldon, A., Heald, R.: Abductive understanding of dialogues about joint activities. Interact. Stud. **15**(3), 426–454 (2014)
11. Lee, H., Chang, A., Peirsman, Y., Chambers, N., Surdeanu, M., Jurafsky, D.: Deterministic coreference resolution based on entity-centric, precision-ranked rules. Comput. Linguist. **39**(4), 885–916 (2013)
12. Lewis, R.: An architecturally-based theory of human sentence comprehension. Ph.D. Thesis. Carnegie Mellon University. CMU-CS-93-226 (1993)
13. McShane, M.: Reference resolution challenges for an intelligent agent: the need for knowledge. IEEE Intell. Syst. **24**(4), 47–58 (2009)
14. McShane, M., Jarrell, B., Fantry, G., Nirenburg, S., Beale, S., Johnson, B.: Revealing the conceptual substrate of biomedical cognitive models to the wider community. In: Westwood, J.D., Haluck, R.S., et al. (eds.) Medicine Meets Virtual Reality 16, pp. 281–286. IOS Press, Amsterdam (2008)
15. McShane, M., Nirenburg, S.: A knowledge representation language for natural language processing, simulation and reasoning. Int. J. Seman. Comput. **6** (2012)
16. McShane, M., Beale, S., Nirenburg, S., Jarrell, B., Fantry, G.: Inconsistency as diagnostic tool in a society of intelligent agents. Artif. Intell. Med. (AIIM) **55**(3), 137–148 (2012)

17. McShane, M., Nirenburg, S., Jarrell, B.: Modeling decision-making biases. Biologically-Inspired Cogn. Archit. (BICA) J. **3**, 39–50 (2013)
18. McShane, M.: Parameterizing mental model ascription across intelligent agents. Interact. Stud. **15**(3), 404–425 (2014)
19. Nirenburg, S., McShane, M., Beale, S.: A simulated physiological/cognitive "double agent". In: Beal, J., Bello, P., Cassimatis, N., Coen, M., Winston, P. (eds.), Papers from the AAAI Fall Symposium, Naturally Inspired Cognitive Architectures, Washington, D.C., Nov. 7–9. AAAI Technical report FS-08-06, Menlo Park, CA: AAAI Press (2008)
20. McShane, M., Nirenburg, S., Beale, S.: Language understanding with ontological semantics. Adv. Cogn. Syst. (forthcoming)
21. McShane, M., Beale, S., Babkin, P.: Nominal compound interpretation by intelligent agents. Linguist. Issues Lang. Technol. (LiLT) **10** (2014)
22. Perrault, C.R., Allen, J.F.: A plan-based analysis of indirect speech acts. Am. J. Comput. Linguist. **6**, 167–182 (1980)
23. Reiter, E.: Natural language generation. In: Clark., A., Fox, C., Lappin, S. (eds.) The Handbook of Computational Linguistics and Natural Language Processing, pp. 574–598. Wiley-Blackwell (2010)
24. Schank, R., Riesbeck, C.: Inside Computer Understanding. Erlbaum, Hillsdale (1981)
25. Schone, P., Jurafsky, D.: Is knowledge-free induction of multiword unit dictionary headwords a solved problem? In: Proceedings of Empirical Methods in Natural Language Processing, Pittsburgh, PA (2001)
26. Shapiro, S.C., Ismail, H.O.: Anchoring in a grounded layered architecture with integrated reasoning. Robot. Auton. Syst. **43**(2–3), 97–108 (2003)
27. Schubert, L., Hwang, C.H.: Episodic logic meets little red riding hood: a comprehensive, natural representation for language understanding. In: Iwanska, L., Shapiro, S. (eds.) Natural Language Processing and Knowledge Representation: Language for Knowledge and Knowledge for Language. MIT/AAAI Press, Menlo Park (2000)
28. Tratz, S., Hovy, E.: A taxonomy, dataset, and classifier for automatic noun compound interpretation. In: Association for Computational Linguistics (2010)
29. Wilks, Y., Fass, D.: Preference semantics: a family history. Comput. Math. Appl. **23**(2) (1992)
30. Woods, W.A.: Procedural semantics as a theory of meaning. Research Report No. 4627. Cambridge, MA: BBN (1981)

Towards Benevolent Sales Assistants in Retailing Scenarios

Sabine Janzen$^{(\boxtimes)}$ and Wolfgang Maass

Saarland University, Saarbruecken, Germany
{sabine.janzen,wolfgang.maass}@iss.uni-saarland.de

Abstract. Non-collaborative dialogues like sales dialogues are characterized by congruent intentions, i.e. intentions that are agreed by dialogue partners, and conflicting intentions. We will refer to these intentional structures as mixed intention sets. In this paper, we will investigate dialogue systems that are benevolent towards a dialogue partner, i.e. benevolent agents try to find a fair balance between partner intentions and agent intentions in particular with respect to conflicting intentions. For the class of question-answering dialogues, we propose a model for the intelligent generation of answers considering mixed intention sets and demonstrate its application in the retailing domain in form of a benevolent sales assistant (BSA). BSA processes mixed intention sets in a strategic way by means of a game-theoretical equilibrium approach to find a fair balance between intentions of dialogue partners. We evaluated the BSA by a run-time analysis of 500 simulated sales dialogues between customers and retailers and show how the sales assistant strategically generates answers considering mixed intention sets in retailing scenarios.

1 Introduction

It has long been thought that planning of system-supported collaborative dialogues with congruent participants' intentions, i.e. intentions that are agreed by dialogue partners, is preferential because these kinds of dialogues represent the main part of daily conversations [15,18,22,24,26]. But, do you remember your last dialogue? Perhaps, this dialogue was of collaborative nature, i.e., all dialogue partners had congruent intentions for participating in the dialogue, e.g., when cooking dinner together. More likely, your intentions for participating in this dialogue were not congruent with those of your dialogue partner(s), for instance, in a meeting, during shopping, in training or private life in general. Then, you were part of a non-collaborative dialogue characterized by congruent as well as conflicting intentions of dialogue partners. Reflecting on the number of dialogues of this category occurring in everyday life, it can be stated that such non-collaborative dialogues with congruent as well as conflicting intentions of dialogue partners represent the main part of everyday dialogues and dominate the daily communication between people. Nonetheless, dialogues with congruent and conflicting intentions of dialogue partners were rarely considered in the

© Springer International Publishing Switzerland 2015
C. Biemann et al. (Eds.): NLDB 2015, LNCS 9103, pp. 180–193, 2015.
DOI: 10.1007/978-3-319-19581-0_16

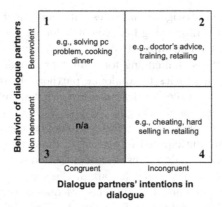

Fig. 1. Categorization of dialogues regarding benevolence of dialogue partners and congruence of dialogue partners' intentions

context of dialogue planning [15,18,22,24,26]. Regarding the processing of congruent and conflicting intentions targeting all involved dialogue partners, existing approaches rather focus on intentions of single actors or on shared respectively joint intentions [2,9,14,35,36]. References [31,32] focus on dialogues possessing a shared discursive goal among the participants. In the field of multi-agent systems, approaches for non-collaborative interaction are considered but without giving insights into their application in dialogue scenarios or the processing of the mixture of conflicting and congruent intentions of dialogue partners [13]. Figure 1 clarifies the distinction between dialogues covering exclusively congruent intentions of dialogue partners and dialogues where intentions of dialogue partners are congruent as well as incongruent, i.e. conflicting. In the following, we will refer to these intentional structures as mixed intention set. Additionally, Fig. 1 indicates a second dimension of categorization: the level of benevolence of dialogue partners' behavior in the sense of good will to find a compromise between conflicting intentions and to create a fair dialogue. So, we distinguish four categories, with category 1 representing dialogue situations, where all dialogue partners have congruent intentions – also known as joint intention – for participating in the dialogue and behave benevolently, e.g., solving a pc problem together. In category 2, the dialogue partners behave benevolently despite of their incongruent intentions, e.g., a doctor's advice where the patient argues that he needs some pills because of stomach pain whereas the doctor diagnoses stress-related illness that induces a more healthy lifestyle in the future. Category 3 is not applicable because dialogues with congruent dialogue partners' intentions but non-benevolent behavior does not exist. Last, in category 4 the class of dialogues with incongruent intentions but non-benevolent behavior of dialogue partners is represented, e.g., cheating when selling a car. Whereas the buyer wants to buy a proper car for a fair price, the seller wants to cheat his counterpart and sell a faulty car for an overcharged price. So, the seller reveals non-benevolent behavior.

In our research, we investigate dialogues with congruent as well as conflicting intentions of two actors showing benevolent behavior exemplified in retailing (cf. category 2 in Fig. 1). The support of such dialogues by dialogue systems in real-world environments is challenging for a number of reasons. First, dialogue systems take the role of proxies for dialogue partners. In retailing, a dialogue systems can represent a retailer's sales assistant. That means dialogue systems adopt intentions of represented dialogue partners. Second, for creating benevolent behavior despite of conflicting intentions, dialogue systems need to be able to balance adopted and anticipated intentions of dialogue partners during dialogue [19, pp. 176] [32,33]. This means, mixed intention sets are satisfied [23] for gaining dialogues that are perceived as fair in the sense of a balance between intentions of dialogue partners. For the class of question-answering dialogues, we propose a model for satisficing mixed intention sets in dialogues based on game-theoretic approaches exemplified by a benevolent sales assistant (BSA).

Next, we illuminate the role of intelligent assistants in the retailing domain as well as related requirements. In Sect. 3, the model for satisficing mixed intention sets is introduced. Afterwards, a case study is presented consisting of (1) a prototypical implementation of a BSA, (2) an exemplary course of satisficing mixed intention sets by the BSA, (3) a run-time analysis, as well as (4) a discussion of results (Sect. 4). Last, we close with a summary (Sect. 5).

2   Intelligent Assistants in Retailing

Applying natural language technologies in intelligent assistants is a long-standing vision [38, 41, 43] because it enables familiar user interaction with intuitive access to information. In recent years, natural language technologies are becoming sophisticated while the support of dialogues. Existing sales assistants in eCommerce cope with conflicting intentions of retailers and customers by applying simple strategies. Initial efforts were made to integrate natural language technologies into sales dialogues of eCommerce scenarios [20]. However, support of dialogues and consideration of conflicting intentions in retailing scenarios are still a key problem. As shown in Fig. 1, dialogues with mixed intention sets occur in numerous domains, e.g., healthcare, sport, retailing. For exemplifying the model for satisficing mixed intention sets, we selected the retailing domain for a number of reasons. First, sales dialogues in retailing follow well-studied phases (e.g., [1, 4, 5]). Second, sales dialogues have clear results, i.e., purchase decisions. Third, intentions in sales dialogues can be concretely specified. From the customers perspective, a sales dialogue has the goal to reduce uncertainty related to buying decisions [11] by considering product characteristics (e.g., price), alternative products, and customer reviews [27]. Also product search efforts shall be reduced [17]. By contrast, the goal of retailers is to increase revenue by diverse strategies, for instance product and price differentiation [8,39] or pushing slow-sellers [12]. Contrary to the aforementioned example of doctors consultation, it could be supposed that the retailer – as provider of the sales assistant – is not interested in revealing benevolent behavior for finding a fair balance between

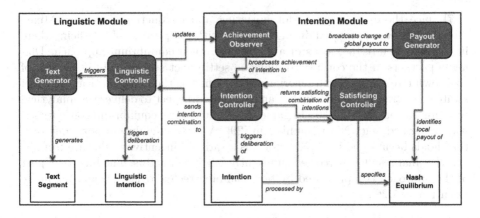

Fig. 2. Model for satisficing mixed intention sets in dialogues

his own intentions and those of customers. But, according to interviews with sales trainers and literature review, such behavior is described as hard selling that is beyond the scope of the retailing scenarios considered in this paper. We assume benevolent sales assistants that act in nearly symmetric dialogue situations comparable to doctors consultations. Taking up this position when creating sales assistants, customer relationship is improved that helps generating trust in technology (e.g., eTRUST).

3   Model for Satisficing Mixed Intention Sets

Before giving formal definitions, let us start by describing an exemplary dialogue situation between a customer and a sales assistant in an online-shopping scenario. The customer searches for a body lotion, so she asks the sales assistant: *"In which fragrances is Sunshine body lotion available?"* The sales assistant gives the following answer: *"Sunshine body lotion is available in the following 3 fragrances: Orange, Water lily and Spring. Today, there is a 7% discount on Sunshine body lotion"*. The customer intends to get comprehensive product information regarding available fragrances whereas the sales assistant has the intention to increase revenue. A fair balance between the conflicting intentions is found by giving information regarding the available fragrances followed by a discount offer for the requested body lotion. In order to model this kind of benevolent behavior of an natural language assistant in a dialogue situation, a model for satisficing mixed intention sets was defined (cf. Fig. 2). We addressed the specific conversational setting of the considered dialogue type (cf. category 2 in Fig. 1) by separating linguistics from the conceptual intention module. Due to the fact, that dialogues between multiple actors are considered, intentions represent the set of all – congruent and incongruent – intentions of dialogue partners in the dialogue situation [6], i.e., a mixed intention set. We concentrate primarily on future-directed intentions [10, 37, 44] in the sense of pro-attitudes

[6,7] since those are required for initiating and conducting dialogues. Mixed intention sets are processed during dialogue with the purpose of satisficing them in a strategic way by means of a game-theoretical equilibrium approach. Dialogue partners in the considered dialogue setting act strategically in pursuit of their own intentions. Game theory can be defined as the mathematical theory of strategic interaction and as such is an adequate prospect to deliver the analytical tools for our purpose [28,30]. In game theory literature, equilibrium concepts are widely applied, e.g., Nash equilibrium [29]. A Nash equilibrium is an outcome that holds because no involved actor has a rational incentive to deviate from it, i.e., the final result is perceived sufficiently fair by all actors. The linguistic part of the approach is represented by text planning technologies [21] and linguistic intentions [15,26].

3.1   Intention Module

Intentions represent the core of the model (cf. Fig. 2). Intentions of users of dialogue system are distinguished from intentions brought into the dialogue by represented dialogue partners, e.g., sales persons in retailing scenarios. A dialogue system that instantiates the model regards these intentions as predefined intentions of the system. According to [40], intentions indirectly trigger situated dialogue actions, e.g., the generation of answers. We assume that intentions are deduced from desires. Since we consider question-answering settings in this contribution, beliefs and desires of actors [6] are not taken into account. This will be part of future work. *Intentions* in the model represent combinations of deliberative and non-deliberative intentions [6]. They are formed earlier and retain all during the course of the dialogue. But, dialogue partners that are internally represented by abstract players deliberate about real-valued weights of these *Intentions* continuously during dialogue.

$$Intentions = \{INT_1 \ldots INT_n\} \tag{1}$$

$$Players = \{P_1 \ldots P_n\}$$

$$Weight_{INTn} = \{Weight_{INTn,P1} \ldots Weight_{INTn,Pn}\}$$

At cold start, all *Intentions* relevant in the domain-specific dialogue situation are initialized with default weights for each player respectively dialogue partner. We assume, that *Intentions* - more precisely their weights by players - persist until they are successfully achieved [44], i.e. dialogue counterparts showed the intended reaction. In the context of the model, this means that answers were given that contributed to the achievement of dialogue partners' intentions.

Satisficing mixed intention sets is considered as multi-player nonzero-sum game. The game is played for infinitely many rounds, more precisely infinitely many dialogue acts each consisting of question and answer. Players have intentions they wish to satisfy. The game has an initial state and the values that are given to the variables, i.e., weights that are given to intentions at each dialogue act determine the next state of the game. In each round of the game, it has to be decided which intentions are selected as trigger for generating an answer that supports the creation of dialogues perceived as sufficiently fair by all actors. Therefore, intentions of all actors are processed by the *Satisficing Controller*. Internally, the *Satisficing Controller* generates strategy profiles that can be played by dialogue partners, i.e. their abstract players. Strategy profiles consist of strategies that are specified by generating the power set of all intentions relevant from a linguistic point of view within this round of the game. This subset of intentions is called salience set ($SalSet_{INT}$). Each strategy (1) represents a combination of intentions of the salience set, and (2) is rated by a local payout (*LocPayout*) that can be expected by players when playing this strategy (S).

$$SalSet_{INT} = \{INT_1 \dots INT_n\} \tag{2}$$

$$StratProfile_{SalSet} = \{S_1 \dots S_n\}$$

$$LocPayout_{Sn,Pn} = \| \sum_{\substack{i=1 \\ j=\#SalSet}}^{j} Weight_{INTi,Pn} \|$$

Satisfaction of players depends on outcomes of the game, i.e., local payouts of strategies. Players prefer those strategies over others that provide high local payouts:

$$\forall S_i \in StratProfile_i : LocPayout(S_i^*, S_{-i}) \geq LocPayout(S_i, S_{-i}) \tag{3}$$

Intuitively, a *Nash Equilibrium* formalizes the idea that no player can be better off given that all other players do not change their strategies [16]. "Thus each player's strategy is optimal against those of the others". Reference [29, p. 287] That means, a combination of strategies by all players is a *Nash Equilibrium* if for every player P_i and strategy S_i we have:

$$\forall P_i, S_i \in StratProfile_i : LocPayout(S_i^*, S_{-i}^*) \geq LocPayout(S_i, S_{-i}^*) \tag{4}$$

Specified by the *Satisficing Controller*, the *Nash Equilibrium* $NE_{Si,-i}$ represents a satisficing combination of intentions at a particular time in the dialogue, that is good enough for each player [34] in the sense of a "win-win-situation" (cf. Fig. 2). The *Satisficing Controller* passes this satisficing combination of intentions to the *Intention Controller* that represents the interface to the linguistic module. The *Payout Generator* identifies local payouts of the *Nash Equilibrium* and adds them to global payouts of players that evolve during dialogue. Changes in global payouts are broadcasted to the *Intention Controller* that trigger a deliberation of all intentions, more precisely their weights (cf. Fig. 2). This is due to the fact, that based on gains and losses in global payouts, intentions are considered differently by players [3, 42]. That means the perspective of players changes when evaluating intentions in the sense of available options. Instead of gaining high global payouts, the objective of the model is to balance payouts of all players during dialogue or to approximate them in case of drifting apart. We assume that similar global payouts of all players can be regarded as evidence for satisficed mixed intention sets.

3.2 Linguistic Module

Receiving a satisficing combination of intentions from the *Intention Controller*, the *Linguistic Controller* has to select appropriate linguistic resources, i.e., *Text Segments* that contribute to the achievement of these intentions of dialogue partners in dialogue. According to [15, 26], a dialogue model must include information about the intended effect of individual text segments of the generated answers on dialogue partners. These *Text Segments* (e.g., phrases) fulfill certain functions regarding the overall dialogue; they satisfy *Linguistic Intentions*. Therefore, the model distinguishes *Intentions* from *Linguistic Intentions* (cf. Fig. 2). To be clear, the selection of specific *Intentions* by the intention module does not trigger the *Text Generator* directly, but leads to an activation of *Linguistic Intentions* by the *Linguistic Controller* which in turn triggers the generation of specific *Text Segments* by the *Text Generator*. Intuitively, a *Linguistic Intention* represents the ability of an answer text segment to contribute to the achievement of overall intentions of dialogue partners [15]. Each intention is supported by a set of *Linguistic Intentions* that contribute to the achievement of this intention ($SupportSet_{\text{INT}}$). On the other hand, each *Linguistic Intention* can contribute to the achievement of several intentions. The ability of a *Linguistic Intentions* to support the achievement of intentions is expressed by linguistic weights for each player. These are initialized and updated by propagation of weights by intentions across their set of supporting *Linguistic Intentions* (LI).

$$SupportSet_{\text{INT}} = \{LI_1 \dots LI_n\} \tag{5}$$

$$LingWeight_{\text{LIn,Pn}} = \left(\frac{|Weight_{\text{INTn,Pn}}|}{\#SupportSet_{\text{INT}}} \right)$$

One of the reasons for separating *Intentions* from *Linguistic Intentions* is to handle the m:n mapping between (1) *Intentions* and (2) plain text segments that contribute to an achievement of these intentions [26]. Last, the *Linguistic Controller* continuously updates the *Achievement Observer* of the intention module with respect to satisfied *Linguistic Intentions*. In case, *Intentions* of dialogue partners are regarded as achieved, this information is broadcasted to the *Intention Controller* that triggers the deliberation of all intentions. As mentioned before, the model uses text planning technologies [21] for generating answers during dialogue. The relationship between *Linguistic Intentions* and text plans is part of other contributions and beyond the scope of this paper.

4 Case Study

We implemented a text-based QA system *Satisficing Dialogue Engine (SDE)* that is derived from the proposed model (cf. Fig. 2) and provides a BSA in an online shopping scenario[1]. Users are able to construct questions term-by-term. SDE only suggests those question terms that result in meaningful questions, for

---

[1] cf. http://redqueen.iss.uni-saarland.de/satin/.

instance *Is – Apple iPod Nano – up-to-date?*. Having tapped the last term of a question, the complete question is presented together with the answer, e.g., *"There are MP3 players that are more up-to-date, i.e. Sony NWZ-E585B Walkman by Sony. Currently, there is a 4.0 % discount on Sony NWZ-E585B Walkman. So, the total comes to 104.54 EUR"*.

Technically, SDE was implemented as web application. We applied a pool of text plans as well as linguistic intentions that were defined based on a text corpus of sales conversations that was imposed empirically. Based on reviewing 19 scientific journals of retailing and consumer research[2], customer and retailer intentions in sales dialogues were derived and classified as well as prioritized by means of item sorting (n=11) [25]. Combinations of the two top-ranked retailer and customer intentions were analyzed in 12 simulated sales conversations in the domain of electronic products between 3 real retailers and 12 subjects acting as customers. Retailers and customers got personal scenarios describing their intentions in the dialogue and were advised to act according to these incongruent and partially conflicting intentions. All 12 conversations were recorded as video files and validated in a web-based user study with 120 subjects. In this web-based study also the identified intentions were validated by 120 subjects concerning their relevance (i.e. weight) in real-world sales conversations. Furthermore, the 120 participants evaluated the sales conversations shown in video clips regarding (1) their naturalness and (2) their consistency with the given intentions of the involved dialogue partners. The best-ranked 8 sales conversations were transcribed, aggregated to a text corpus and analyzed regarding (1) questions and answers that occurred frequently and relations between them, and (2) the composition of answer parts and the rhetorical relations between them. Last, results of this analysis were transferred to 72 text schemata [24], 31 plan operators [21], 33 rhetorical relations [18,22] and 18 linguistic intentions [26] and represented semantically as linguistic knowledge base of SDE.

4.1 Exemplary Course of Satisficing Mixed Intention Sets in Retailing Scenario

We will illustrate our approach by an exemplary course of satisficing mixed intention sets in a retailing scenario supported by SDE. Customer and sales assistant respectively SDE are represented by player A (P_a) and B (P_b). Internally, intentions of customers are distinguished from intentions brought into the system by the retailer. SDE regards these intentions as own intentions and represents them during dialogue. Therefore, we have system intentions adopted from a retailer (e.g., intention to increase sale of product bundles) and anticipated intentions of a customer (e.g., intention to purchase products at the best price attainable). As mentioned before, intentions were identified by means of literature review before validating them by a web-based study with 120 subjects and deriving default weights.

Imagine, the customer posed a question concerning the difference between two products: **Q:** *"Where is the difference between body lotion BodyCocoon and*

---

[2] 1971–2008.

Table 1. Identification of nash equilibria

Player B°

LocPayout (Sx,Sy)

		⊘	INT $_E$	INT $_{IR}$	INT $_{IT}$	INT $_{CR}$	INT $_{BA}$
	⊘	S_1 & S_1 (0.00,0.00)	S_1 & S_2 (0.00,0.14°)	S_1 & S_3 (0.00*,0.14°)	S_1 & S_4 (0.00,0.06)	S_1 & S_5 (0.00,0.14°)	S_1 & S_6 (0.00,0.11)
	INT $_E$	S_2 & S_1 (0.15,0.00)	S_2 & S_2 (0.15, 0.14)	S_7 & S_7 (-0.23,0.29°)	S_8 & S_8 (0.54*,0.20)	S_9 & S_9 (0.15,0.29°)	S_{12} & S_{12} (-0.07,0.26)
	INT $_{IR}$	S_3 & S_1 (-0.38,0.00)	S_7 & S_7 (-0.23,0.29°)	S_3 & S_3 (-0.38, 0.15)	S_{16} & S_{16} (0.00,0.20)	S_{10} & S_{10} (-0.38,0.29°)	S_{13} & S_{13} (-0.62,0.26)
	INT $_{IT}$	S_4 & S_1 (0.38*,0.00)	S_8 & S_8 (0.54*, 0.20°)	S_{16} & S_{16} (0.00*,0.20°)	S_4 & S_4 (0.38,0.06)	S_{11} & S_{11} (0.38*,0.20°)	S_{14} & S_{14} (0.15*,0.17)
	INT $_{CR}$	S_5 & S_1 (0.00,0.00)	S_9 & S_9 (0.15, 0.29°)	S_{10} & S_{10} (-0.38,0.29°)	S_{11} & S_{11} (0.38,0.20)	S_5 & S_5 (0.00, 0.14)	S_{15} & S_{15} (-0.23,0.26)
	INT $_{BA}$	S_6 & S_1 (-0.23,0.00)	S_{12} & S_{12} (-0.07,0.26°)	S_{13} & S_{13} (-0.62,0.26°)	S_{14} & S_{14} (0.15,0.17)	S_{15} & S_{15} (-0.23,0.26°)	S_6 & S_6 (-0.23,0.11)

Player A* — LocPayout (Sx,Sy)

body lotion BodySplash?" Thereupon, SDE processes a text plan for generating an appropriate answer. Each text plan consists of (1) an obligatory part that allocates information requested by the user by posing the question, and (2) several optional text segments. These optional text segments provide additional information and are linked with linguistic intentions. In this example, the optional text segments of the exemplary text plan are related to the following linguistic intentions that force the integration of specific information into the answer:

- *Additive Product (AP)*: information about additive products
- *Pure Information (PI)*: product information in general
- *External Review (ER)*: reviews by product owners
- *External History (EH)*: products considered by other customers

Based on these linguistic intentions, the supported intentions *Extravagance, Increasing Revenue, Information Transparency, Customer Relationship* and *Behavior Analysis* are identified and assigned to the salience set, i.e. the set of intentions relevant from a linguistic point of view within the current dialogue act.

$$SalSet_{INT} = \{INT_E, INT_{IR}, INT_{IT}, INT_{CR}, INT_{BA}\} \qquad (6)$$

This mixed intention set offered by the salience set is weighted differently by player A and B. Some intentions (e.g., INT_E) can be considered as congruent intentions whereas several intentions (e.g., INT_{IR}) represent conflicting intentions. Next, the Satisficing Controller generates the strategy profile, i.e. the power set of the salience set restricted to all possible dual intention combinations. These 16 strategies, e.g., S_1, are measured by normalized local payouts for player A and B that base on the summed weights of involved intentions. For simplification, the following equation shows only an extract of the strategy profile as well as local payouts for S_4:

$$StratProfile_{\text{SalSet}} = \{\ldots S_1(\emptyset), S_2(INT_E),$$
$$S_3(INT_{IR}), S_4(INT_{IT})\ldots S_{14}(INT_{IT}; INT_{BA})\ldots\} \tag{7}$$

$$LocPayout_{S4,\text{Pa}} = 0.38; LocPayout_{S4,\text{Pb}} = 0.06$$

Next, strategies of both players are processed to identify the Nash Equilibrium. Therefore, local payouts are compared by means of a matrix. For each player, the best answer in the sense of the highest local payout is identified within the table rows and columns respectively. In Table 1, these best answers are labeled with * for P_A and ° for P_B. The table fields marked grey show the best answers of P_A according to the strategy played by P_B and vice versa. Therefore, four equilibria are identified. The Satisficing Controller chooses the equilibrium with the smallest difference in local payouts, i.e., $NE_{S1,S7}$, instead of selecting the pareto-dominant option. This is due to the main purpose of our approach of gaining similar global payouts of all players during dialogue for satisficing mixed intention sets in the sense of benevolent behavior. The gained local payouts $(0.00, 0.14)$ change the existing global payout of each player that lead to a deliberation of all intentions, i.e. a revalidation of weights. Changes in weights are propagated across the set of supporting linguistic intentions by each intention.

$$GlobPayout_{\text{Pa}} = (0.10)_{\text{existing}} + 0.00 = 0.10;$$
$$GlobPayout_{\text{Pb}} = (0.01)_{\text{existing}} + 0.14 = 0.15 \tag{8}$$

The strategies marked by the Nash equilibrium cover (1) zero intentions, (2) a single intention, or (3) a combination of intentions. In case of zero intentions left, none of the optional answer text segments is integrated into the answer. Otherwise, by comparing the support set of the intention(s) left with the set of linguistic intentions triggered by the text plan, an intersection is defined that represents the set of resulting and relevant linguistic intentions.

$$NE_{S1,S7} = \{\emptyset, INT_{IR}\} \tag{9}$$

$$SupportSet_{\text{INTIR}} = \{LI_{EQ}, LI_{AV}, LI_{AP}\}$$

This intersection contains the linguistic intention $\{LI_{AP}\}$ that forces the integration of information about additive products into the answer. So, in this exemplary answer generation, only this optional text segment of the text plan is considered for integration in the answer. The remaining optional text segments are not integrated into the answer. Finally, the linguistic intention LI_{AP} is satisfied and supports the achievement of the intention *"Increasing Revenue"* specified by the Nash equilibrium. The final text plan is processed to generate the following exemplary answer: **A:** *"The products BodyCocoon and BodySplash differ in fragrance (almond and apricot), price ($5.49 and $2.99) and in the focused skin type (dry skin and normal skin). Together with the corresponding shower creme, BodyCocoon is available for $6.99"*.

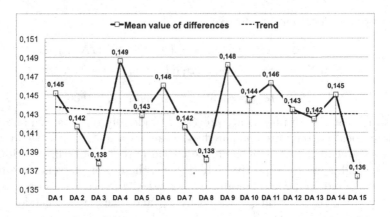

Fig. 3. Mean values of differences in global payouts in 500 simulated sales dialogues consisting of 15 dialogue acts (DA = dialogue act)

Last, linguistic weights of satisfied linguistic intentions contribute to the achievement of intentions they do support. As INT_{IR} is supported by the satisfied linguistic intention LI_{AP}, its linguistic weight contributes to the achievement of the intention. An intention is regarded as achieved if the corresponding achievement is equal to or greater than its weight, i.e. answers were given that satisfied the intention. Then, the weight is reduced by the distance measure between its weight and 1.00. Here, the achievement of INT_{IR} becomes greater than its weight of (4.86) regarding player B. The Achievement Observer informs the Intention Controller about the achievement of INT_{IR} which leads to a deliberation of intentions. This means the weight of INT_{IR} is reduced and the resulting difference is propagated across the whole set of intentions, i.e. all weights are adjusted.

$$Achievement_{INTIR,Pb} = (3.80)_{existing} + (1.67)_{LI\ AP} = 5.47 \qquad (10)$$

$$Weight_{INTIR,Pb} = 4.86 - Dist(4.86, 1.00) = 1.00$$

$$Weight_{INT2...n,Pb}+ = \left(\frac{Dist(4.86, 1.00)}{\#SupportSet_{INT\ IR} - 1} \right)$$

4.2 Discussion

The model (cf. Fig. 2) was evaluated by means of a run-time analysis of simulated sales dialogues between customers and retailers supported by SDE. Subject of the run-time analysis was the conduction of 500 simulated sales dialogues each consisting of 15 dialogue acts. In this context, a dialogue act constitutes a combination of a supposed arbitrary question posed by a simulated customer and an answer generated by a randomly selected text plan. When triggering a new simulated sales dialogue, global payouts were reset to the initial value (0.00).

Objective was the validation of the model concerning its performance in satisficing mixed intention sets in non-collaborative dialogues as exemplified before. The extent of satisficing mixed intention sets was measured by means of differences between global payouts of players. Figure 3 illustrates the evolvement of global payouts in 500 simulated sales dialogues based on the mean value of differences in global payouts for dialogue act #1–15. It can be observed that steep ascents in payout differences (e.g., in dialogue act #9) are mitigated by a subsequent stabilization and minimization (e.g., in dialogue act #13–15). The slope of the trend function in Fig. 3 as well as the fact that on average mean values of differences account for 0.143 indicate balanced global payouts of players and can be regarded as evidence for satisfied mixed intention sets and thus benevolent behavior of the dialogue system. In future work, an empirical user study will be conducted as lab experiment. Users will get personal scenarios describing their intentions in the dialogue before interacting with SDE, i.e. posing questions to the system and evaluating given answers. Beside a questionnaire that focuses on the perceived fairness of the dialogue by the user amongst other aspects, we will use biofeedback technologies (skin conductance) during users interaction with SDE for completing and crosschecking user statements. From a conceptual perspective, we intend to process data gained by the dynamic evolvement of intentions, respectively their weights during dialogue by means of supervised learning techniques. This offers the possibility for predicting default weights and building up customer models for upcoming dialogues.

5 Conclusion

Non-collaborative dialogues are characterized by congruent intentions, i.e. intentions that are agreed by dialogue partners, and conflicting intentions. These intentional structures are defined as mixed intention sets. In this paper, we investigated dialogue systems that are benevolent towards a dialogue partner, i.e. benevolent agents try to find a fair balance between partner intentions and agent intentions in particular with respect to conflicting intentions. For the class of question-answering dialogues, we proposed a model for the intelligent generation of answers considering mixed intention sets and demonstrated its application in the retailing domain in form of a benevolent sales assistant (BSA). BSA processes mixed intention sets in a strategic way by means of a game-theoretical equilibrium approach to find a fair balance between intentions of dialogue partners. We evaluated the BSA by a run-time analysis of 500 simulated sales dialogues between customers and retailers and show how the sales assistant strategically generates answers considering mixed intention sets in retailing scenarios.

Acknowlegdement. This work was partially funded by the German Federal Ministry for Education and Research (BMBF) under the contract 01IS12030.

References

1. Ajzen, I.: The theory of planned behavior. Organ. Behav. Hum. Decis. Process. **50**(2), 179–211 (1991)
2. André, E., Rist, T., Mulken, S.V., Klesen, M., Baldes, S.: The automated design of believable dialogues for animated presentation teams. In: Embodied Conversational Agents, pp. 220–255. MIT Press (2000)
3. Bellman, R.: Dynamic Programming. Princeton University Press, 1 edn. (1957)
4. Bettman, J.R., Luce, M.F., Payne, J.W.: Constructive consumer choice processes. J. Consum. Res. **25**(3), 187–217 (1998)
5. Blackwell, R.D., Miniard, P.W., Engel, F.: Consum. Behav. Harcourt, USA (2001)
6. Bratman, M.: Intention, Plans, and Practical Reason. Center for the Study of Language and Information. Harvard University Press, Cambridge (1987)
7. Bratman, M.: What is intention? In: Cohen, P.R., Morgan, J. (eds.) Intentions in Communication. MIT Press, Cambridge (1990)
8. Choudhary, V., Ghose, A., Mukhopadhyay, T., Rajan, U.: Personalized pricing and quality differentiation. Manage. Sci. **51**(7), 1120–1130 (2005)
9. Chu-Carroll, J., Carberry, S.: Conflict resolution in collaborative planning dialogs. Int. J. Hum. Comput. Stud. **53**, 969–1015 (2000)
10. Cohen, P.R., Levesque, H.J.: Intention is choice with commitment. Artif. Intell. **42**(2–3), 213–261 (1990)
11. Daft, R.L., Lengel, R.H.: Organizational information requirements, media richness and structural design. Manage. Sci. **32**(5), 554–571 (1986)
12. Gallego, G., Van Ryzin, G.: Optimal dynamic pricing of inventories with stochastic demand over finite horizons. Manage. Sci. **40**(8), 999–1020 (1994)
13. Grant, J., Kraus, S., Wooldridge, M.: Intentions in equilibrium. In: Proceedings of the Twenty-Fourth AAAI Conference on Artificial Intelligence (AAAI 2010) (2010)
14. Grosz, B., Kraus, S.: Collaborative plans for group activities. In: Proceedings of the 13th International Joint Conference on Artificial Intelligence (IJCAI 1993), pp. 367–373 (1993)
15. Grosz, B.J., Sidner, C.L.: Attention, intentions, and the structure of discourse. Comput. Linguist. **12**(3), 175–204 (1986)
16. Gutierrez, J., Harrenstein, P., Wooldridge, M.: Reasoning about equilibria in game-like concurrent systems. In: Proceedings of the 14th International Conference on Principles of Knowledge Representation and Reasoning (KR 2014) (2014)
17. Häubl, G., Trifts, V.: Consumer decision making in online shopping environments: the effects of interactive decision aids. Mark. Sci. **19**(1), 4–21 (2000)
18. Hobbs, J.R.: Why is Discourse Coherent?: Technical Note 176. Stanford Research Inst, Menlo Park (1978)
19. Jackendoff, R.: Language, Consciousness, Culture - Essays on Mental Structure. MIT Press, Cambridge (2007)
20. Kauffman, R.J., Walden, E.A.: Economics and electronic commerce: survey and directions for research. Int. J. Electron. Commer. **5**, 5–116 (2001)
21. Mann, W.C.: Discourse structures for text generation. In: Proceedings of the 10th International Conference on Computational Linguistics, pp. 367–375. ACL (1984)
22. Mann, W.C., Thompson, S.A.: Assertions from discourse structure. In: Proceedings of the Workshop on Strategic Computing Natural Language, pp. 257–270 (1986)
23. March, J.G., Simon, H.A.: Organizations. Wiley (1958)
24. McKeown, K.R.: Discourse strategies for generating natural-language text. Artif. Intell. **27**(1), 1–41 (1985)

25. Moore, G.C., Benbasat, I.: Development of an instrument to measure the perceptions of adopting an information technology innovation. Inf. Syst. Res. **2**(3), 192–222 (1991)
26. Moore, J.D., Paris, C.L.: Planning text for advisory dialogues: capturing intentional and rhetorical information. Comput. Linguist. **19**(4), 651–694 (1993)
27. Mudambi, S.M., Schuff, D.: What makes a helpful online review? a study of customer reviews on amazon.com. MIS Q. **34**(1), 185–200 (2010)
28. Myerson, R.B.: Game Theory: Analysis of Conflict. Harvard University Press, Cambridge (1997)
29. Nash, J.: Non-cooperative games. Ann. Math. **54**(2), 286–295 (1951)
30. Osborne, M.J., Rubinstein, A.: A Course in Game Theory. MIT press, Cambridge (1994)
31. Paquette, M.A.: The logic of conversation: from speech acts to the logic of games. In: Proceedings of Agent Communication (2010)
32. Paquette, M.A.: Speech acts, dialogues and the common ground. In: FLAIRS Conference (2012)
33. Parikh, P.: Language and Equilibrium. MIT Press, California (2010)
34. Pollack, M.E.: The uses of plans. Artif. Intell. **57**(1), 43–68 (1992)
35. Searle, J.R.: Speech Acts: An Essay in the Philosophy of Language. University Press, Cambridge (1969)
36. Searle, J.R.: Collective intentions and actions. In: Cohen, P.R., Morgan, J., Pollack, M.E. (eds.) Intentions in Communication. MIT Press, Cambridge (1990)
37. Singh, M.P., Asher, N.M.: Towards a formal theory of intentions. In: Proceedings of the European Workshop on Logics in Artificial Intelligence, pp. 472–486 (1991)
38. Storey, V.C., Burton-Jones, A., Sugumaran, V., Purao, S.: Conquer: a methodology for context-aware query processing on the world wide web. Inf. Syst. Res. **19**(1), 3–25 (2008)
39. Stremersch, S., Tellis, G.J.: Strategic bundling of products and prices: a new synthesis for marketing. J. Mark. **66**(1), 55–72 (2002)
40. Suchman, L.A.: Plans and Situated Actions. Cambridge University Press, New York (1987)
41. Suh, K.S., Milton Jenkins, A.: A comparison of linear keyword and restricted natural language data base interfaces for novice users. Inform. Syst. Res. **3**(3), 252–272 (1992)
42. Tversky, A., Kahneman, D.: The framing of decisions and the psychology of choice. Science **211**(4481), 453–458 (1981)
43. Vassiliou, Y., Jarke, M., Stohr, E.A., Turner, J.A., White, N.H.: Natural language for database queries: a laboratory study. MIS Q. **7**(4), 47–61 (1983)
44. Wooldridge, M.J.: Reasoning About Rational Agents. MIT press, Cambridge (2000)

Sentiment and Opinion Analysis

A Rule-Based Approach to Implicit Emotion Detection in Text

Orizu Udochukwu[✉] and Yulan He

School of Engineering and Applied Science, Aston University, Birmingham, UK
{orizuus,y.he9}@aston.ac.uk

Abstract. Most research in the area of emotion detection in written text focused on detecting explicit expressions of emotions in text. In this paper, we present a rule-based pipeline approach for detecting implicit emotions in written text without emotion-bearing words based on the OCC Model. We have evaluated our approach on three different datasets with five emotion categories. Our results show that the proposed approach outperforms the lexicon matching method consistently across all the three datasets by a large margin of 17–30 % in F-measure and gives competitive performance compared to a supervised classifier. In particular, when dealing with formal text which follows grammatical rules strictly, our approach gives an average F-measure of 82.7 % on "`Happy`", "`Angry-Disgust`" and "`Sad`", even outperforming the supervised baseline by nearly 17 % in F-measure. Our preliminary results show the feasibility of the approach for the task of implicit emotion detection in written text.

Keywords: Implicit emotions · OCC model · Emotion detection · Rule-based approach

1 Introduction

Human emotions are defined as subjective feelings and thoughts, and is a short episode that is coordinated by the brain [4]. Emotions exist in various forms and Ekman [2] made a strong compelling case for the six basic emotion categories. In Natural language Processing (NLP), emotion detection focuses on categorising a piece of text into an emotion category. The expression of emotion in written text is through the use of words and most often emotion-bearing words such as "happy". However, emotions can be adequately expressed without the use of emotion-bearing words. For example, given two sentences "`The outcome of my exam makes me happy.`" and "`I passed my exam.`", both sentences express the emotion of happiness, with the first expressing it explicitly and the second implying it. Most research in the area of emotion detection focuses on explicit emotion detection [6,9]. Implicit emotion detection is a much more difficult task and the approaches which rely on emotion lexicons are inapplicable here. Although it is possible to train supervised classifiers from annotated data, acquiring sufficient annotated data for training requires heavy manual effort.

© Springer International Publishing Switzerland 2015
C. Biemann et al. (Eds.): NLDB 2015, LNCS 9103, pp. 197–203, 2015.
DOI: 10.1007/978-3-319-19581-0_17

We present in this paper a rule-based approach to sentence-level implicit emotion detection based on the OCC model, which was created by Ortony, Clore & Collins in their book "The Cognitive Structure of Emotions" [7] and over the years has become a widely accepted model for emotion. We demonstrate using this resource and with the use of NLP techniques we are able to detect and classify implicit expressions of emotion in text. As opposed to most existing approaches, our approach does not rely on any specific knowledge base or annotated training data and can still offer a reasonably high precision rate given the complex nature of the problem.

2   The OCC Model

The OCC Model provides a clear and convincing structure of the eliciting conditions of emotions and the variables that affect their intensities [7]. It describes a hierarchy that classifies 22 emotion types. The hierarchy contains three branches. The first branch is the *"consequences of events"* branch. The emotions on this branch express pleasure or displeasure with event consequences. The second branch contains emotions in relation to *"actions of agents"*. An agent can be self or other and these are the attribution type of emotions. The third branch contains emotions relating to liking or disliking in regards of *"aspects of objects"*.

The OCC model is dependent on its variables and the rules for implementing/identifying emotion. The variables are grouped into *emotion-inducing variable* and *emotion intensity variables*. For a full list of the emotions and variables, see [7]. The OCC model however is rather complex and full of ambiguity. Steunebrink et al. [10] outlined a number of issues of the original OCC model and proposed changes to remove duplications and ambiguities. In our work, we use the revisited OCC model [10] for emotion detection in text.

3   Our Approach

In order to use the OCC model for emotion detection, we need to first assign values to a list of variables defined in OCC and then use a set of pre-defined rules to identify an emotion for a given text. In this paper, we focus on identifying emotion in relation to events and actions only and leave the detection of emotions associated with objects as future work. The list of rules is shown in Table 1. For example, the first row of Table 1(a) can be read as

If Direction = "Self" **and** Tense = "Future" **and** Overall Polarity = "Positive" **and** Event Polarity = "Positive", **then** Emotion = "Hope".

It is worth noting that emotion-bearing words are different from polarity-bearing words. An emotion-bearing word can be described as words which on their own can convey emotions. For example, the word "passionate" can convey an emotion of *Joy*. Polarity-bearing words, on the contrary, express *positive* or *negative* polarity in a given context. For example, the word "pass" expresses a *positive* polarity as in "I passed my exam.". But the word "pass" does not

Table 1. Rules for emotion detection.

Direction	Tense	*Input Variables* Overall polarity	Event polarity	*Output* Emotion
Self	Future	Positive	Positive	Hope
Self	Future	Negative	Negative	Fear
Self	Present	Positive	Positive	Joy
Self	Present	Negative	Negative	Distress
Self	Past	Positive	Positive	Satisfaction
Self	Past	Negative	Negative	Fears-confirmed
Self	Past	Positive	Negative	Relief
Self	Past	Negative	Positive	Disappointment
Other	All	Positive	Positive	Happy-for
Other	All	Negative	Positive	Resentment
Other	All	Positive	Negative	Gloating
Other	All	Negative	Negative	Sorry-for

(a) **Event-based.**

Input Variables Direction	Polarity	*Output* Emotion
Self	Positive	Pride
Self	Negative	Shame
Other	Positive	Admiration
Other	Negative	Reproach

(b) **Action-based.**

Input Variables Event	Action	*Output* Emotion
Joy	Pride	Gratification
Distress	Shame	Remorse
Joy	Admiration	Gratitude
Distress	Reproach	Anger

(c) **Compound emotions.**

have an explicit prior emotion associated with it. Hence, it is more likely that emotions-bearing words also have a polarity, but not all polarity words convey specific emotions.

We first perform pre-processing on text in order to be able to assign values to the OCC variables which will be discussed subsequently. For pre-processing we carried out sentence splitting and tokenisation, part-of-speech (POS) tagging, word sense disambiguation (WSD), dependency parsing, sentence tense detection based on the POS tags, and polarity detection using majority vote based on the lexicon matching results obtained with three sentiment lexicons, SentiWordNet [3], AFINN [5] and the Subjectivity Lexicon [12].

We now describe how we assign values to each OCC variable.

Direction: The value for this variable can either be "Self" or "Other". The former refers to emotions expressed for oneself while the latter refers to emotions expressed for others. This value is assigned based on the dependency relationship (identified by the dependency parser) of a first person pronoun (such as "I", "we") with an action or event. We identify 3 possible scenarios for assigning a value to this variable: (1) When dealing with a simple sentence, we simply apply the process mentioned above; (2) When dealing with a complex sentence where multiple subject(s) are identified by the parser, we assign values based on respective action/event relations with identified subjects; (3) No subject is identified or no verbs exist in the text, here we just assign the value "Other" to the variable.

Tense: The value for this variable can either be "Present", "Past" or "Future". The value assignment is determined by the POS tags of the verbs in a sentence or by the results obtained from the FrameNet[1] where the tense of the verb is identified as future tense if the frame is associated with "desiring". In the cases where no verbs are used in a sentence, the value of the variable is set to "Present".

Overall Sentence Polarity: The value for this variable can either be "Neutral", "Negative" or "Positive". It is determined by polarity detection through majority vote.

Event Polarity: The event is identified based on the verb-object relations revealed by the dependency parser. The noun phrase which contains an identified object is treated as the event for its relative verb. The polarity of an event is then determined using lexicon-matching.

Action Polarity: The action is identified based on the subject-verb relations revealed by the dependency parser. The verb phrase which contains the identified verb is treated as an action. Similar to event polarity, the action polarity is also determined using lexicon-matching.

We have also implemented a contextual valence shifter as described in [8] to detect polarity change in different context. Once the variable values are identified, the rules defined in Table 1 are then applied to detect the presence of emotions. The compound emotions are results of the output of the event-based and action-based emotions. For the "sorry-for" emotion, we ensure that the subject is of positive valence; otherwise the emotion is identified as "resentment". The same rule is applied to the "admiration" and "reproach" emotion pairs.

4   Experiments

In this section, we present the evaluation results of our rule-based emotion detection approach on three different datasets, which include:

- *The International Survey On Emotion Antecedents And Reactions (ISEAR) Dataset*[2] which was developed by asking nearly 3,000 participants from different cultural background about their emotional experiences.
- *The SemEval-2007 Task 14 Affective Text dataset* [11] consists of news headlines collected from major newspapers.
- *The Alm's Dataset* [1] comprises sentences taken from 176 fairy tale stories. We use only the data extracted from Grimm's and Anderson's tales, which have a total number of 1,040 sentences.

As our goal is to detect emotions in the absence of emotion-bearing words, we filter out sentences which contain emotion words as can be found in the emotion lexicon WordNet-Affect[3]. The total number of sentences before and after filtering of emotion-bearing words in each emotion category for these three datasets are

---

[1] https://framenet.icsi.berkeley.edu/fndrupal/.

[2] http://www.unige.ch/fapse/emotion/databanks/isear.html.

[3] http://wndomains.fbk.eu/wnaffect.html.

Table 2. Statistics of the datasets. "Total" denotes the original number of sentences in each emotion category while "Implicit" denote the number of sentence which do no contain any emotion words according to WordNet-Affect.

Emotion	Total	Implicit
Joy	1095	537
Fear	1095	366
Anger	1096	483
Sadness	1096	488
Disgust	1096	484
Shame	1096	581
Guilt	1093	482
Total	7667	3421

(a) ISEAR

Emotion	Total	Implicit
Joy	362	317
Fear	160	130
Anger	66	60
Sadness	202	182
Disgust	26	24
Surprise	184	160
Total	1000	873

(b) SemEval

Emotion	Total	Implicit
Happy	406	103
Fearful	121	33
Angry-Disgusted	174	84
Sad	247	90
Surprised	92	50
Total	1040	360

(c) Alm's

Table 3. Performance comparison of F-measure results on the three datasets. Bold face values denote the best results obtained in each dataset.

Emotion	ISEAR			SemEval			Alm's		
	Lexicon	NB	Rule	Lexicon	NB	Rule	Lexicon	NB	Rule
Joy/Happy	33.4	61.2	**69.6**	39.7	**71.7**	59.9	58.8	63.5	**81.8**
Fear/Fearful	0	**47.6**	18.3	0	**52.2**	31.8	0	**26.7**	14.0
Anger/Angry-Disgusted	23.0	47.1	**61.3**	55.8	16.2	**61.3**	48.9	58.6	**86.6**
Sadness/Sad	25.6	55.4	**68.0**	47.8	56.0	**71.5**	61.0	56.0	**79.6**
Disgust	25.6	**51.0**	39.2	38.5	34.5	**61.7**	-	-	-
Average	21.5	**52.5**	51.3	36.4	**58.2**	57.3	42.2	56.0	**65.5**
Average (− Fear)	27.0	53.7	**59.5**	45.5	44.6	**63.6**	56.12	65.8	**82.7**

shown in Table 2. We focus specifically on the 5 emotion categories which are shared across these three datasets and map the OCC-output emotions to the five emotion categories in the following ways: (Fear, Fear-confirmed) → Fear, (Joy, Happy-For, Satisfaction, Admiration, Pride) → Joy, (Anger, Reproach) → Anger, (Distress, Sorry-For, Disappointment, Shame) → Sadness, Resentment → Disgust.

We have developed two baseline models. One is a lexicon matching method which uses the NRC emotion Lexicon[4] for sentence-level emotion detection, We also train supervised Naïve Bayes (NB) classifiers using the implementation in Weka[5] on the three datasets with 5-fold cross validation. We report the results in terms of the F-measure scores.

It can be observed from Table 3 that although we have filtered out sentences which contain emotion words from WordNet-Affect, using other emotion lexicons such as the NRC emotion lexicon can still identify emotions of some sentences. Nevertheless, using the NRC lexicon gives quite low F-measure values across

[4] http://www.saifmohammad.com/WebPages/lexicons.html.
[5] http://www.cs.waikato.ac.nz/ml/weka.

the three datasets and it fails to detect any "`Fear`" emotion bearing sentences. Despite using no labelled data, our approach achieves similar performance as supervised NB on the ISEAR and SemEval datasets (with 1 % difference in F-measure on average) and outperforms NB by 9.5 % in F-measure on the Alm's dataset. The results also show that our approach is largely affected by the quality of text. ISEAR contains personal experience expressed by a wide range of participants and hence might contain lots of informal and ill-grammatical text. SemEval contains news headlines which are often incomplete sentences ignoring grammar conventions. The Alm's dataset, on the other hand, contains fairy tales which are formal text following rules of grammar very strictly. As such, the performance obtained from the Alm's dataset by our approach are significantly better than that obtained from the other two datasets.

Our approach relies on results generated from a series of NLP tasks such as POS tagging,word-sense disambiguation, dependency parsing and polarity detection in order to be able to assign values to a set of OCC variables for emotion detection. Thus, any error that occurs will be propagated down the pipeline process. Furthermore, failure in detecting the polarity of text will make it impossible for our approach to identify the underlying emotion. Also, we have not considered ironic and sarcastic sentences in our current work. Nevertheless, we have shown that in the absence of annotated data, the OCC-based approach is able to identify implicit emotion in text with performance competing to supervised classifiers and it even outperforms the supervised approach for formal text (the Alm's dataset). The emotion detection results generated by the OCC-based approach can be used as seed examples to bootstrap more complicated emotion detection methods which require large amount of training data. We will leave it as our future work.

5   Conclusions and Future Work

In this paper, we have proposed a OCC-based approach for implicit emotion detection. Experimental results on three datasets have shown that our approach outperforms the lexicon matching method consistently across all the three datasets by a large margin of 17–30 % in F-measure and gives competitive performance compared to a supervised method. Also, when dealing with formal text which follows grammatical rules strictly, the approach achieves an average F-measure score of 82.7 % in identifying "`Happy`", "`Angry-Disgusted`" and "`Sad`" categories.

In future, we will investigate methods to improve the performance of our approach with informal short text such as tweets and social media posts. We will also improve the identification of emotions involving intensity variables and unexpectedness variables by examining how adverbs and adjectives influence the emotion of sentences (for emotions like "`Surprise`" and "`Shock`") and investigate performance on ironic or sarcastic sentences. We will study possible solutions to deal with the poor performing emotion category such as "`Fear`".

References

1. Alm, C.O., Roth, D., Sproat, R.: Emotions from text: machine learning for text-based emotion prediction. In: Proceedings of the Conference on Human Language Technology and Empirical Methods in Natural Language Processing, pp. 579–586 (2005)
2. Ekman, P.: An argument for basic emotions. Cogn. Emot. **6**(3–4), 169–200 (1992)
3. Esuli, A., Sebastiani, F.: Determining the semantic orientation of terms through gloss classification. In: Proceedings of the 14th ACM International Conference on Information and Knowledge Management, pp. 617–624 (2005)
4. Friedenberg, J., Silverman, G.: Cognitive Science: An Introduction to the Study of Mind. Sage, California (2011)
5. Hansen, L.K., Arvidsson, A., Nielsen, F.Å., Colleoni, E., Etter, M.: Good friends, bad news-affect and virality in twitter. In: Park, James J., Yang, Laurence T., Lee, Changhoon (eds.) FutureTech 2011, Part II. CCIS, vol. 185, pp. 34–43. Springer, Heidelberg (2011)
6. Marsella, S., Gratch, J., Petta, P.: Computational models of emotion. In: Scherer, K.R., Bänziger, T., Roesch, E. (eds.) A Blueprint for Affective Computing: A Sourcebook and Manual, pp. 21–46. Oxford University Press, New York (2010)
7. Ortony, A.: The Cognitive Structure of Emotions. Cambridge University Press, Cambridge (1990)
8. Polanyi, L., Zaenen, A.: Contextual valence shifters. In: Shanahan, J.G., Qu, Y., Wiebe, J. (eds.) Computing Attitude and Affect in Text: Theory and Applications. The Information Retrieval Series, pp. 1–10. Springer, Netherlands (2006)
9. Shivhare, S.N., Khethawat, S.: Emotion detection from text. arXiv preprint arXiv:1205.4944 (2012)
10. Steunebrink, B.R., Dastani, M., Meyer, J.J.C.: The OCC model revisited. In: Proceedings of the 4th Workshop on Emotion and Computing, vol. 65, pp. 2047–2056 (2009)
11. Strapparava, C., Mihalcea, R.: Semeval-2007 task 14: affective text. In: Proceedings of the 4th International Workshop on Semantic Evaluations, SemEval 2007, pp. 70–74 (2007)
12. Wilson, T., Wiebe, J., Hoffmann, P.: Recognizing contextual polarity in phrase-level sentiment analysis. In: Proceedings of the Conference on Human language Technology and Empirical Methods in Natural Language Processing, pp. 347–354. Association for Computational Linguistics (2005)

Deciphering Review Comments: Identifying Suggestions, Appreciations and Complaints

Sachin Pawar[(✉)], Nitin Ramrakhiyani, Girish K. Palshikar,
and Swapnil Hingmire

Systems Research Laboratory, Tata Consultancy Services Ltd., Hadapsar, Pune, India
{sachin7.p,nitin.ramrakhiyani,gk.palshikar,swapnil.hingmire}@tcs.com

Abstract. The problem of classifying sentences into various categories, arises frequently in text mining applications. One of the most important categorization of sentences observed in product reviews, movie reviews, blogs, customer feedbacks is - Suggestions, Appreciations and Complaints. We observed that the document classification techniques do not perform well for these three non-topical sentence classes. We propose to solve this problem using a supervised approach based on Dependency-based Word Subsequence Kernel and its variations. We compare the performance of our approach with the state-of-the-art short text classification techniques on 2 different datasets - Performance Appraisal comments and Product Reviews.

1 Introduction

Performance Appraisal (PA) is a very important process in large organizations. Each year, PA process generates huge amount of text data in terms of self assessments written by the employees and review feedback given by their supervisors. Such large organizations have employee strengths of more than 100,000 and the volume of appraisal comments can be of an order of one million and more. To get insights from this PA data at the organization level, it is necessary to classify the feedback sentences into useful semantic classes such as *suggestions*, *complaints* and *appreciations*. Similar review comments are widely available from sources such as customer feedback during store purchases, online shopping product reviews, expert reviews in media and much more. Classifying sentences in product reviews can be useful to identify customer wishlists, valuable product features and product shortcomings. In addition, sentence level classification of text is also needed in other Natural Language Processing (NLP) applications like Sentiment Analysis for subjective/objective or positive/negative classification.

The task however is challenging due to various reasons. Firstly, it is not solvable by existing document classification algorithms. The sparseness of the sentence vector when represented in a very high-dimensional space with hundreds or thousands of words as dimensions, misses the necessary semantics of the sentence in turn blurring the classification difference. Also, features necessary for sentence classification vary significantly from domain to domain requiring repetitive manual intervention for feature engineering. Problems which arise

© Springer International Publishing Switzerland 2015
C. Biemann et al. (Eds.): NLDB 2015, LNCS 9103, pp. 204–211, 2015.
DOI: 10.1007/978-3-319-19581-0_18

when using document representation techniques for sentences are elaborated by Li et al. [1] in their work on sentence similarity. Also common pre-processing techniques useful for document classification like stop-word removal and lemmatization may have negative effect on sentence classification (Khoo et al. [2]). It is important to consider the nuances of the sentences like their structure, semantics, sentiment, etc. while classification and build effective features around these sentential aspects.

Sentence classification can be carried out based on two aspects - topical and non-topical. Topical classification focuses on the content of the sentences and the classes are derived from the domain for example sports, politics, business, etc. in the news domain. Non-topical classification on the other end focuses on the semantics being conveyed by the sentences and the target classes are valid across various domains. As an example, let us consider the sentence `Lara showed some great foot-work and contributed with the much needed knock`. In the news domain, this sentence, would be classified as from the topic 'Sports', but non-topically this sentence is appreciative in nature. Now let us tweak our example sentence to `Lara should show some great foot-work and contribute with much needed knocks`. As one can observe, with almost same words the semantics of the sentence has changed and non-topically it is more suggestive. Hence, word based topical classification approaches are often insufficient for such non-topical classification.

In this work we focus on non-topical classification of review sentences into classes namely Suggestions, Appreciations and Complaints. Our approach is an extension to an existing dependency-based word subsequence kernel [3] used for sentence similarity. It measures similarity between two sentences by finding number of paths shared by their respective dependency trees. We extend this basic kernel by considering word classes (to deal with data sparsity), dependency relation types (to capture sentence structure) and node heights in the dependency tree. The kernel modifications are generic and can be carried out to make it usable in other applications. We compare our approach with state-of-the-art short text classifiers and bag-of-words based Naive Bayes and SVM classifiers.

In summary, the contributions of the paper are analysing text in a new domain of performance appraisals and use of a novel kernel for sentence classification. The paper is organized as follows. Section 2 presents the related work in the area of sentence classification. Details about the classification approaches are discussed in Sect. 3. Description of the datasets and detailed experimental analysis is presented in Sect. 4. Finally, the conclusions and future work are discussed in Sect. 5.

2   Related Work

Deshpande et al. [4] used the theory of speech acts to build a knowledge based unsupervised classifier to classify sentences into *suggestion* and *complaint* classes. Goldberg et al. [5] described a technique to automatically detect whether any sentence is expressing some *wish*. Although this work is not explicitly concerned with *suggestions* but *wishes* can be thought of as a superset of *suggestions*. Pan [6] aims at classifying sentences in the biomedical domain into five non-topical classes namely *Focus, Polarity, Certainty, Evidence* and *Trend*, using Naive Bayes, SVM and MaxEnt classifiers. Mukherjee and Liu [7] proposed generative models for classifying review comments into non-topical classes such as *Thumbs-up, Thumbs-down, Disagreement, Agreement, Question, Answer-Ack, Answer and None*.

Danescu-Niculescu-Mizil et al. [8] proposed a computational framework for identifying linguistic aspects of politeness and built a classifier to detect whether any sentence is *polite* or not. Wiebe and Riloff [9] developed a classification framework to detect *subjective* sentences using only unannotated texts for training. Wagner et al. [10] propose a sentence classification technique to identify ill-formed sentences from proper ones using classifier voting over decision trees alongwith other basic methods. Kadoya et al. [11] describes a sentence intention classifier which uses deterministic multi-attribute pattern matching on conversation sentences. Other important work in this area are [12–15].

3   Classification Approaches

3.1   Advanced Dependency-Based Word Subsequence Kernel

Kate [3] proposed a kernel (DWSK) which computes similarity between two natural language sentences as the number of paths shared by their dependency trees. This kernel captures the semantic similarity between two sentences in a better way because it only considers linguistically meaningful word sub-sequences which are based on word dependencies. We have designed an advanced version (ADWSK) of this kernel and the revised formulation is as follows:

$$K_{ADWSK}(T_1, T_2) = \sum_{n_1 \in N_1} \sum_{n_2 \in N_2} K_{height}(n_1, n_2) * K_{node}(n_1, n_2) * (1 + CDP(n_1, n_2))$$

$$CDP(n_1, n_2) = 0 \text{ if } K_{node}(n_1, n_2) = 0; \text{ else}$$

$$CDP(n_1, n_2) = \sum_{\substack{c_1, c_2 \text{ s.t.} \\ c_1 \in Ch(n_1) \\ c_2 \in Ch(n_2)}} \alpha \cdot K_{node}(c_1, c_2) \cdot K_{dr}(DR(n_1, c_1), DR(n_2, c_2)) \cdot (1 + CDP(c_1, c_2))$$

Here, $K_{ADWSK}(T_1, T_2)$ computes the semantic similarity between two dependency trees T_1 and T_2. $CDP(n_1, n_2)$ computes weighted common downward paths emanating from n_1 and n_2. $Ch(n)$ represents the set of children of node n and $DR(n, c)$ represents type of dependency relation between a node n and its child c. K_{height}, K_{node} and K_{dr} are similarity functions which would be explained later. To down weight the contribution by the longer paths, the α parameter is used in a similar way as in the original DWSK kernel.

We now describe some of our modifications in the original kernel in detail.

1. We consider only downward paths in the dependency tree instead of considering all paths. This means that we consider all the paths of the form parent-child-grandchild but no paths of the form child1-parent-child2. We observed that this significantly reduces number of complex paths and thus avoids overfitting.
2. We assign more weight to the paths emanating at higher nodes in a dependency tree. The intuition here is that the nodes occurring higher up in the dependency tree contribute more to the overall "meaning" of the sentence. This is achieved by using K_{height} in the above formulation. $K_{height}(n_1, n_2)$ returns minimum of the heights of the nodes n_1 and n_2 in the trees T_1 and T_2 respectively. The node heights are normalized to lie between 0 and 1. This helps in reducing similarity between sentence pairs such as A dog chased a cat. and A dog praised a cat. by assigning lower weight to common path like dog-a and cat-a.
3. We generalize words to various word classes (POS tags, word clusters) to deal with the problem of data sparsity. This is achieved by using K_{node} in the above formulation. $K_{node}(n_1, n_2) = 1$ if $n_1.word = n_2.word$, $K_{node}(n_1, n_2) = 0.5$ if $n_1.word\_cluster = n_2.word\_cluster$ and $K_{node}(n_1, n_2) = 0.2$ if $n_1.POS = n_2.POS$. The word clusters are formed by using distributional similarity between the words obtained by statistics from a large text corpus [16]. Hence, semantically similar words fall in the same cluster. E.g. technology, applications, software.
4. We also take into consideration the dependency relation types[1] when determining similarity between any two nodes in the dependency trees. This is achieved by using K_{dr} in the above formulation. $K_{dr}(DR_1, DR_2) = 1$ if DR_1 and DR_2 are exactly same or semantically similar pairs like *nsubj-agent* or *nsubjpass-dobj* and $K_{dr}(DR_1, DR_2) = 0.5$ otherwise. This helps in reducing similarity between sentence pairs sucwh as A dog chased a cat. and A cat chased a dog. Because in first sentence dog is connected to chased through *nsubj* (Nominal subject) whereas in the second sentence, the relationship is *dobj* (Direct object).
5. We added one more constraint in computing $K_{node}(n_1, n_2)$ by checking *negation compatibility* of two nodes. A node in a dependency tree is said to be negated when it has a child (usually words like not, no, nt) related with dependency relation *neg*. When any one node of n_1 and n_2 is negated but other is not, we make $K_{node}(n_1, n_2) = 0$. This helps in reducing similarity between sentence pairs such as You achieved a certification. and You did not achieve any certification. Because in the first sentence, the root node achieved is not negated but in the second sentence, it is negated.

3.2   A Note on Using Rules for Classification

One of the authors also spent some time (2 days) in coming up with some simple rules for identifying *suggestions*. E.g. one of the rule (regular expression over

---

[1] We use the Stanford Dependency Parser for the typed dependencies [17].

Table 1. Class distributions and examples

Label	Performance appraisals	Product reviews	Example sentences in performance appraisal dataset
Suggestion	432	79	Pls focus more on Technical & Domain certifications
Complaint	43	225	No certification done in last 6 months
Appreciation	344	359	The email quality scores have always been excellent
NONE	181	337	Attitude towards learning is ok

POS-tagged text) tries to identify imperative sentences in order to label them as suggestions:

```
(^|, )([A-Za-z]+/RB(S|R)? )?((?!agree|appreciate)[A-Za-z]+)/VB[ ].*\$.
```

It achieves a precision of 87.8 % on performance appraisals dataset and 61.5 % on product reviews. Similar rules were developed for identifying *Appreciation* and *Complaints*. Table 2 shows the accuracy of this rule-based classifier.

4   Experimental Analysis

4.1   Datasets

As a part of PA process, for each employee, her supervisor sets some goals. After the end of the evaluation period, each employee writes her self assessment comments as plain English text and each supervisor then evaluates her subordinates by writing her assessment for each goal as plain English text.

We randomly selected a set of 1000 sentences from supervisor comments and another set of 1000 sentences from Amazon product reviews [18] for Watches and Cellphones. We tagged both of these datasets[2] manually as *suggestions, appreciations* and *complaints*. The sentences which didn't fall in any of these categories were labelled as NONE. The details about the datasets and distribution of the comments across various classes are presented in Table 1.

4.2   Evaluation and Result Analysis

We used LibSVM [19] package for classification with our ADWSK kernel. We compared our results with various other approaches (Naive Bayes, Decision Tree and SVM with bag-of-words features) and the most recent of them being Lib-ShortText [20] which is a classifier specifically designed for classifying short texts. 5-fold cross-validation was performed for all approaches. Accuracy of classification is nothing but the ratio of number of correctly classified sentences to the total number of sentences.

---
[2] To obtain the datasets, please contact the authors.

Table 2. Classification Accuracies on both the datasets

Algorithm	Performance appraisals (%)	Product reviews (%)
Naive Bayes	72.6	55.8
Decision Tree	71.1	46
SVM (linear)	76.1	58.7
Rule-based classifier	71.2	54.4
DWSK [3] (using LibSVM)	79.9 ($\alpha = 0.25, c = 10$)	60.8 ($\alpha = 0.5, c = 5$)
LibShortText [20]	80.7 ($P = 3$)	64.5 ($P = 1, N = 0, F = 3$)
ADWSK (using LibSVM)	**81.9** ($\alpha = 0.25, c = 10$)	**65.0** ($\alpha = 0.5, c = 5$)
ADWSK not considering height	81.0	61.1
ADWSK not considering dep. relation types	81.8	64.8
ADWSK without word classes	79.7	63.2
ADWSK without negation compatibility	81.6	64.0
ADWSK with all paths	81.5	64.9

The summary of results is presented in Table 2. Our method using ADWSK kernel outperforms all other methods including basic DWSK and LibShortText. We also analyzed the importance of each our advancements to the basic DWSK algorithm by removing these one by one from the ADWSK kernel and then measuring the fall in accuracies. It is observed that on the PA dataset, addition of word classes (POS and word clusters) proves to be most useful whereas in Product Reviews dataset, considering height of nodes in dependency trees is the most useful advancement. We also analyzed whether high precision rules can be used to correct the errors by supervised classifiers. We applied following simple but high precision rules after the ADWSK output is obtained. (i) If root of the tree is any of these words- suggest, advice, recommend, then class = *suggestion*, (ii) If root of the tree is any of these words- love, like, good, awesome, excellent, then class = *Appreciation* and (iii) If root of the tree is any of these words- hate, bad, worst, horrible, mistake, then class = *Complaint*. These 3 rules boosted the performance of ADWSK on the Product Reviews dataset by 1 % to 66 %.

5   Conclusions and Future Work

To the best of our knowledge, this is the first attempt to identify *suggestions, appreciations* and *complaints* in an unexplored domain of performance appraisal comments. We proposed a new kernel ADWSK as an advanced version of the basic dependencies-based word subsequence kernel (DWSK). As per our knowledge, dependency tree based kernels have not been tried before for non-topical sentence classification problems. Our approach outperformed the basic DWSK and the state-of-the-art LibShortText classifier on both the datasets of performance appraisals and product reviews.

In future, we plan to develop linguistic rules based on Speech Act Theory for identifying *suggestions, appreciations* and *complaints*. We believe that using such high precision rules, the errors made by the statistical classifiers can be reduced. We also plan to incorporate this sentence classifier into a large information extraction system to mine insights from PA data.

References

1. Li, Y., McLean, D., Bandar, Z.A., O'shea, J.D., Crockett, K.: Sentence similarity based on semantic nets and corpus statistics. IEEE Trans. Knowl. Data Eng. **18**, 1138–1150 (2006)
2. Khoo, A., Marom, Y., Albrecht, D.: Experiments with sentence classification. In: Proceedings of the 2006 Australasian Language Technology Workshop, pp. 18–25 (2006)
3. Kate, R.J.: A dependency-based word subsequence kernel. In: Proceedings of the Conference on Empirical Methods in Natural Language Processing, pp. 400–409. Association for Computational Linguistics (2008)
4. Deshpande, S., Palshikar, G.K., Athiappan, G.: An unsupervised approach to sentence classification. In: COMAD, p. 88 (2010)
5. Goldberg, A.B., Fillmore, N., Andrzejewski, D., Xu, Z., Gibson, B., Zhu, X.: May all your wishes come true: a study of wishes and how to recognize them. In: Proceedings of Human Language Technologies: The 2009 Annual Conference of the North American Chapter of the Association for Computational Linguistics, pp. 263–271. Association for Computational Linguistics (2009)
6. Pan, F.: Multi-dimensional Fragment Classification in Biomedical Text. Queen's University, Kingston (2006)
7. Mukherjee, A., Liu, B.: Modeling review comments. In: Proceedings of the 50th Annual Meeting of the Association for Computational Linguistics: Long Papers, vol. 1, pp. 320–329. Association for Computational Linguistics (2012)
8. Danescu-Niculescu-Mizil, C., Sudhof, M., Jurafsky, D., Leskovec, J., Potts, C.: A computational approach to politeness with application to social factors. arXiv preprint arXiv:1306.6078 (2013)
9. Wiebe, J., Riloff, E.: Creating subjective and objective sentence classifiers from unannotated texts. In: Gelbukh, A. (ed.) CICLing 2005. LNCS, vol. 3406, pp. 486–497. Springer, Heidelberg (2005)
10. Wagner, J., Foster, J., van Genabith, J.: Judging grammaticality: experiments in sentence classification. CALICO J. **26**, 474–490 (2013)
11. Kadoya, Y., Morita, K., Fuketa, M., Oono, M., Atlam, E.S., Sumitomo, T., Aoe, J.I.: A sentence classification technique using intention association expressions. Int. J. Comput. Math. **82**, 777–792 (2005)
12. Yamamoto, Y., Takagi, T.: A sentence classification system for multi biomedical literature summarization. In: 21st International Conference on Data Engineering Workshops, pp. 1163–1163. IEEE (2005)
13. He, Q., Chang, K., Lim, E.P.: Anticipatory event detection via sentence classification. In: IEEE International Conference on Systems, Man and Cybernetics, SMC 2006, vol. 2, pp. 1143–1148. IEEE (2006)
14. Hachey, B., Grover, C.: Sequence modelling for sentence classification in a legal summarisation system. In: Proceedings of the 2005 ACM symposium on Applied Computing, pp. 292–296. ACM (2005)

15. Kim, Y.: Convolutional neural networks for sentence classification, pp. 1746–1751 (2014)
16. Baroni, M., Bernardini, S., Ferraresi, A., Zanchetta, E.: The wacky wide web: a collection of very large linguistically processed web-crawled corpora. Lang. Resour. Eval. **43**, 209–226 (2009)
17. De Marneffe, M.C., MacCartney, B., Manning, C.D., et al.: Generating typed dependency parses from phrase structure parses. In: Proceedings of LREC, vol. 6, pp. 449–454 (2006)
18. McAuley, J., Leskovec, J.: Hidden factors and hidden topics: understanding rating dimensions with review text. In: Proceedings of the 7th ACM Conference on Recommender systems, pp. 165–172. ACM (2013)
19. Chang, C.C., Lin, C.J.: Libsvm: a library for support vector machines. ACM Trans. Intell. Syst. Technol. (TIST) **2**, 27 (2011)
20. Yu, H., Ho, C., Juan, Y., Lin, C.: Libshorttext: a library for short-text classification and analysis. Technical report (2013). http://www.csie.ntu.edu.tw/~cjlin/papers/libshorttext.pdf

Associating Intent with Sentiment in Weblogs

Mark Kröll[1](✉) and Markus Strohmaier[2]

[1] Know-Center GmbH, Inffeldgasse 13, 8010 Graz, Austria
mkroell@know-center.at
[2] GESIS Leibniz Institute for the Social Sciences, Cologne, Germany
markus.strohmaier@gesis.org

Abstract. People willingly provide more and more information about themselves on social media platforms. This personal information about users' emotions (sentiment) or goals (intent) is particularly valuable, for instance, for monitoring tools. So far, sentiment and intent analysis were conducted separately. Yet, both aspects can complement each other thereby informing processes such as explanation and reasoning. In this paper, we investigate the relation between intent and sentiment in weblogs. We therefore extract ∼ 90,000 human goal instances from the ICWSM 2009 Spinn3r dataset and assign respective sentiments. Our results indicate that associating intent with sentiment represents a valuable addition to research areas such as text analytics and text understanding.

Keywords: Human goals · Intent analysis · Sentiment · Weblogs

1 Introduction

Social media monitoring tools have been gaining attention as people willingly provide more and more information about themselves on the web. Large amounts of personal information are spread amongst social media applications such as weblogs. This information about users' opinions, emotions or goals appears to be valuable, for example, in areas such as economics (cf. [6]), finance (cf. [3]) or politics (cf. [11]).

In this paper, we focus on two aspects of personal information in text - people's goals (Intent) and people's emotions (Sentiment). While a lot of research has been dedicated to sentiment analysis (cf. [15]), the analysis of intent (cf. [7]) with respect to natural language text is still in an early phase. Understanding peoples' goals can help to answer "why questions" about user behavior and user interactions (cf. [4, 19]). As sentiment, intent is capable of providing an orthogonal view on textual content. So far, these views were considered separately. We believe that intent and sentiment have the potential to complement and inform each other. (i) Sentiment can characterize a person's goal and thus provides additional information about the person itself. For instance, assigning a negative sentiment to a person's goal mourn the dead allows us to make assumptions about her emotional state. (ii) Vice versa, intent can serve as a rationale for emotional expressions, providing an explanation for a person's emotional state. A positive emotional state can be explained by a corresponding goal instance such as take responsibility.

© Springer International Publishing Switzerland 2015
C. Biemann et al. (Eds.): NLDB 2015, LNCS 9103, pp. 212–219, 2015.
DOI: 10.1007/978-3-319-19581-0_19

In this paper, we take a first step towards learning more about this relation and its potentials. We therefore develop a pattern-based algorithm to automatically extract ~90.000 human goal instances and assign respective sentiments. In our experiments, we choose the ICWSM 2009 Spinn3r dataset [1] which contains ~44 million blog posts. By conducting quantitative and qualitative analyses, we provide insights into the nature and diversity of human goals expressed in weblogs as well as their relation to sentiment values. Our findings indicate that associating intent and sentiment can be valuable for research areas such as text analytics and text understanding as well as for applications monitoring social media.

2 Related Work

The task of Sentiment Analysis, also known as Opinion Mining [15], is to classify textual content according to expressed emotions and opinions (cf. [23]). It is commonly defined as binary classification task, i.e. to assign a sentence either positive or negative polarity. Turney's work was among the first ones to tackle automated sentiment classification [22]. He used mutual information between a text phrase and the words "excellent" and "poor" as a decision metric. Inspired by supervised machine learning algorithms from text mining, Pang et al. [14] successfully applied these techniques to movie reviews thereby improving previous approaches.

As sentiment analysis, intent analysis provides an orthogonal view on traditional topic categorization. Kröll et al. [7] introduced *Intent Analysis* and examined political speeches from an intentional perspective which allowed further insights into parties' election campaigns. In general, knowledge about human goals has been found to be an important kind of knowledge for a range of challenging research problems, such as goal recognition from people's actions, reasoning about people's goals or the generation of action sequences that implement goals (planning) (cf. [17, 2]). In commonsense enabled applications (cf. [10]), explicit representations of goal knowledge are crucial for plan recognition and planning. Explicit goal knowledge representations are also an enabler for intelligent user interfaces which exhibit traits of commonsense understanding such as goal-oriented search [13] or goal-oriented event planning [18].

3 Extracting Human Goals from Weblogs

In this paper, we choose the ICWSM 2009 Spinn3r Dataset [1] which contains ~44 million blog posts made between August 1st and October 1st, 2008. To prepare the data set for analyzing intent and sentiment, we conducted a number of pre-processing and sanitization steps. For part-of-speech tagging and pattern matching we used python's Natural Language Processing Toolkit (NLTK[1]).

We regard a textual phrase to contain a goal whenever the phrase *(1) contains at least one verb and (2) describes a plausible state of affairs that the user may want to achieve or avoid (cf. [16]) (3) in a recognizable way (cf. [21]).*

---

[1] http://www.nltk.org/.

To extract human goal instances from blog posts, we devised a set of lexico-syntactic patterns and examined them with respect to frequency and quality. Similar to the work of Hearst [5], we manually annotated a small set of blog posts, identified goal instances and examined the textual environment to devise appropriate extraction patterns. In addition, we categorized goal instances either as "Achieve Goal", e.g., getting rich or change my job, or as "Avoid Goal", e.g., being stupid or develop an addiction, and crafted patterns accordingly. For that task, part-of-speech[2] patterns alone did not perform well, since goal instances are in essence verb phrases but not necessarily vice versa.

We thus crafted patterns consisting of a combination of an indicator pattern and a verb phrase pattern. Indicator patterns such as "<WANT > <TO>" signal that the subsequent verb phrase represented a potential goal instance. Verb phrase patterns implement the requirements for the presence of a human goal, i.e. a verb phrase preferably ending in a noun (< NN.* >) or an adjective (< JJ >). Initially, we hand-crafted a set of nine verb phrase patterns to cover also instances such as make her smile. Table 1 shows the resulting three patterns; the other ones produced a lot of false positives and were discarded.

Table 1. Shows three verb phrase patterns with matching goal instance examples. (*) denotes no, one or several occurrences, (+) denotes at least one occurrence and (?) denotes one optionaloccurrence. These patterns were in turn combined with indicator patterns.

Verb phrase pattern	Matching goal instances
<VB > < CD > ? < TO > ? < PRP$ > ? *< DT > ? < IN > ? < JJ > ? < NN.* >+*	dampen our spirits, get basic groceries
<VB > (< CD > ? < TO > ? < PRP$ > ? *< DT > ? < IN > ? < JJ > ? < NN.* >+)+*	go to the park on Sunday
<V. > < JJ >+*	eat healthy, be happy

Table 2. Illustrates the resulting set of indicator patterns to extract human goal instances from weblogs. Relative frequency and precision were manually calculated for every pattern. Thecalculation was based on a random sample containing 1000 goal instances.

Indicator pattern	Goal class	Rel. Freq. (%)	Precision (%)
<WANT > < TO>	Achieve	68.45 %	54.8 %
<INTEND > < TO>	Achieve	2.64 %	70.0 %
<INTENT|PURPOSE|GOAL|OBJECTIVE| AIM > <VBZ > <TO>	Achieve	2.24 %	68.8 %
<NOT > <WANT| INTEND > <TO>	Avoid	25.46 %	68.3 %
<NEVER > <WANT > <TO>	Avoid	1.21 %	50.0 %

---

[2] The part-of-speech tags used in this paper are consistent with the Penn Treebank Tag Set.

We applied our patterns to ~8,9 million blog posts and extracted ~90,000 goal instances. Table 2 shows the relative frequency and precision for every indicator pattern. For our experiments, we required the precision to be at least 50 % for every indicator pattern; otherwise it was discarded.

Having extracted a set of ~90,000 goals, we were interested whether the extraction quality depends on certain aspects of the goal instance, for instance, frequency or goal class (achieve vs. avoid). Table 3 illustrates precision results for various goal subsets each containing 100 extracted instances. The annotation process itself was conducted in form of a question-answer game. Given a goal instance X, the annotator should ask herself: "Is it possible that a person deems X as a potential goal for herself?"

Table 3. Shows precision values for frequent goals (>1) and infrequent goals (=1) according to their goal class. Precision values are based on random samples of 100 goal instances.

	Frequency > 1	Frequency = 1
Achieve goals	63 %	36 %
Avoid goals	68 %	49 %

Precision values in Table 3 indicate that frequently occurring goal instances are of a better quality than infrequent goals. Incorrect instances are mostly due to incorrect part-of-speech information and informal /colloquial language used in weblogs.

To provide insights into the diversity of human goals in weblogs, we decided on (i) generating the frequency distribution and on (ii) categorizing goal instances into a subset of Levin's verb class taxonomy [8]. Pre-processing steps included applying stop word removal, stemming and merging of similar goal instances. To give an example, two goal instances "buying a car" and "buy many cars" were merged into "buy car".

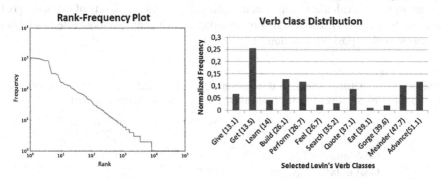

Fig. 1. The left part shows the rank-frequency plot of all goal instances. Popular achieve goals include be a part of something or make money; popular avoid goals include get sick or lose someone. The right part shows the verb class histogram. Levin's verb classes are stated in brackets.

The distribution in Fig. 1 resembles a power-law distribution implying that few goals in weblogs are popular while the majority of goals occur less frequent. To elaborate on the diversity of goals in weblogs, we decided on categorizing goal instances into a subset of Levin's verb class taxonomy. At this point, we remark that a diversity of goals directly transfers to a diversity of Intent/Sentiment pairs. In the resulting verb class histogram (see right part of Fig. 1) we can identify dominant verb classes including "get", "build", "perform" and "advance". Their dominance can be explained by frequent occurrences of verbs such as "make" (class "build"), "get, find, and buy" (class "get"), "take and play" (class "perform") and "come and go" (class "advance").

4 Associating Intent with Sentiment

In this section, we elaborate on the relation between sentiment and intent. For this purpose, we assign sentiment to the extracted goal instances based on simple keyword matching. We merged two available sentiment gazetteer lists (cf. [20, 24]) and used the intersection set in order to increase the quality of the sentiment keywords. Input text is then scanned and matched with our intersection set. Overall sentiment is assigned based on a majority vote principle; if it is even, a neutral sentiment is assigned. In the paper at hand we are interested in studying the principle relation between intent and sentiment postponing optimization steps. We therefore leave the task of applying more sophisticated sentiment analysis approaches to future work.

We focused on goal instances that were either classified as negative or positive sentiments which led to ~6,000 goal instances. To learn more about the nature of Intent/Sentiment pairs and their potential to complement and inform each other, we conduct qualitative and quantitative evaluations.

Quantitative Evaluation: To get an impression of the relationship between intent and sentiment, we first calculate sentiment distributions of various subsets of extracted goal instances (see Table 4). The first row result indicates that goals are more often (~61 %) associated with a positive emotion.

Table 4. Shows relative sentiment distributions for all extracted goal instances (classified as either positive or negative) and for goals filtered by goal class (Achieve/Avoid).

	Positive (%)	Negative (%)
All goal instances	60.9 %	39.1 %
Goal instances | Achieve	64.2 %	35.8 %
Goal instances | Avoid	27.9 %	72.1 %

This could imply that simply having goals might put a person more often in a good mood than in a bad one. "Achieve Goals" are more often assigned positive (~64 %) than negative (~36 %) sentiment and vice versa with "Avoid Goals": negative (~72 %), positive (28 %).

Table 5. Shows selected Intent/Sentiment pairs grouped by goal class. Stems are manually extended to their base form. If necessary, stop words are reinserted to restore original meaning.

Intent/sentiment pair examples			
Achieve Goals		Avoid Goals	
have some vitality, (+)	share my joy, (+)	develop an addiction, (-)	be lonely tonight, (-)
stay healthy, (+)	save our friendship, (+)	endure the humiliation, (-)	spread a disease, (-)
buy fuel efficient car, (+)	have success, (+)	relive the sadness, (-)	hear bad news, (-)
make horror film, (-)	leave a bad impression, (-)	take his pain medication, (-)	be angry, (-)
mourn the dead, (-)	quit my damn job, (-)	be your good luck charm, (+)	admit my prowess, (+)
have revenge, (-)	become a serial killer, (-)	be extraordinary, (+)	hear the sound, (+)

Qualitative Evaluation: We perform a limited qualitative analysis to infer information about the quality of selected examples of Intent/Sentiment pairs (see Table 5). "Achieve Goals" more often describe positive concepts such as success (have success), health (stay healthy, emphasize my natural beauty), friendship (be your friend, save our friendship) or being economic (buy fuel efficient car). "Avoid Goals", on the other hand, more often describe negative circumstances such as develop an addiction, endure the humiliation or spread a disease. Examples in Table 5 suggest that certain Intent/ Sentiment combinations appear more often together and can be used as prior information to approximate emotional states. Combinations such as Achieve Goal/Positive Sentiment (stay healthy, (+)) and Avoid Goal/Negative Sentiment (spread a disease, (-)) potentially have a positive effect on a person's emotional state. To exemplify, the fact, that a person wants to achieve a goal assigned positive sentiment such as stay healthy, might have a positive effect on her emotional state. The same applies to a person wanting to avoid a future state of affairs (described by a goal) which is assigned a negative sentiment.

5 Conclusion

Our findings suggest that associating intent with sentiment represents a valuable addition to research areas such as text analytics and text understanding. We illustrated that intent and sentiment can complement each other thereby potentially informing processes such as explanation and reasoning. Knowing more about a person's goals can help approximating her emotional state. Vice versa, a person's emotional state can be explained by the presence of Intent/Sentiment pairs such as "be confident, (+)" or "mourn the dead, (-)". The combined information can contribute to analyze and thus better understand textual content. In addition, this combination is potentially useful in many tasks and applications. By statistically analyzing Intent/Sentiment pairs, social media monitoring applications can benefit to generate more accurate user profiles. Search applications could use intent and/or sentiment as additional search facet (i) to rank their results based on shared goals and/or emotions or (ii) to identify communities of interest. In addition, the emotional characterization of human goals could be valuable to complement existing commonsense knowledge bases such as CyC [9] or ConceptNet [12] which contain human goals.

Acknowledgments. Thanks to Daniel Lamprecht and Johannes Liegl for participating in this work. This work is funded by the KIRAS program of the Austrian Research Promotion Agency (FFG) (project number 840824). The Know-Center is funded within the Austrian COMET Program under the auspices of the Austrian Ministry of Transport, Innovation and Technology, the Austrian Ministry of Economics and Labor and by the State of Styria. COMET is managed by the Austrian Research Promotion Agency FFG.

References

1. Burton, K., Java, A., Soboroff, I.: The ICWSM 2009 spinn3r dataset. In: Proceedings of the 3rd Annual Conference on Weblogs and Social Media (2009)
2. Carberry, S.: Techniques for plan recognition. J. User Model. User-Adap. Inter. **11**(1–2), 31–48 (2001)
3. De Choudhury, M., Sundaram, H., John, A., Seligmann, D.: Can blog communication dynamics be correlated with stock market activity?. In: Proceedings of the 19th ACM Conference on Hypertext and Hypermedia (2008)
4. Faaborg, A., Lieberman, H.: A goal-oriented web browser. In: Proceedings of the Conference on Human Factors in Computing Systems (2006)
5. Hearst, M.: Automatic acquisition of hyponyms from large text corpora. In: Proceedings of the 14th Conference on Computational Linguistics (1992)
6. Kroha, P., Baeza-Yates, R., Krellner, B.: Text mining of business news for forecasting. In: The International Workshop on Database and Expert Systems Applications (2006)
7. Kröll, M., Strohmaier, M.: Analyzing human intentions in natural language text. In: Proceedings of the 5th International Conference on Knowledge Capture (2009)
8. Levin, B.: English Verb Classes and Alternations: A Preliminary Investigation. University of Chicago Press, Chicago (1993)
9. Lenat, D.: CYC: a large-scale investment in knowledge infrastructure. Commun. ACM **38** (11), 33–38 (1995)
10. Lieberman, H.: Usable AI requires common sense knowledge. In: Workshops and Courses: Usable Artificial Intelligence held in Conjunction with CHI 2008 (2008)
11. Lietz, H., Wagner, C., Bleier, A., Strohmaier, M.: When politicians talk: assessing online conversational practices of political parties on twitter. In: International AAAI Conference on Weblogs and Social Media (2014)
12. Liu, H., Singh, P.: ConceptNet - a practical commonsense reasoning tool-kit. BT Technol. J. **22**(4), 211–226 (2004)
13. Liu, H., Lieberman, H., Selker, T.: GOOSE: a goal-oriented search engine with commonsense. In: Proceedings of the 2nd International Conference on Adaptive Hypermedia and Adaptive Web-Based Systems (2002)
14. Pang, B., Lee, L., Vaithyanathan, S.: Thumbs up? sentiment classification using machine learning techniques. In: Proceedings of the Conference on Empirical Methods in Natural Language Processing (2002)
15. Pang, B., Lee, L.: Opinion mining and sentiment analysis. Found. Trends Inf. Retrieval **2**(1–2), 1–135 (2008)
16. Regev, G., Wegmann, A.: Where do goals come from: the underlying principles of goal-oriented requirements engineering. In: Proceedings of the 13th International Conference on Requirements Engineering (2005)
17. Schank, R., Abelson, R.: Scripts, Plans, Goals, and Understanding: An Inquiry into Human Knowledge Structures. Lawrence Erlbaum Associates, Hillsdale (1977)

18. Smith, D.: EventMinder: a personal calendar assistant that understands events. Master's Thesis at the Massachusetts Institute of Technology (2007)
19. Smith, D., Lieberman, H.: The why UI: using goal networks to improve user interfaces. In: Proceedings of the 14th International Conference on Intelligent User Interfaces (2010)
20. Stone, P., Dunphy, D., Smith, M., Ogilvie, D.: The General Inquirer: A Computer Approach to Content Analysis. M.I.T. Press, Cambridge, Mass (1966)
21. Strohmaier, M., Prettenhofer, P., Kröll, M.: Acquiring explicit user goals from search query logs. In: International Workshop on Agents and Data Mining Interaction (2008)
22. Turney, P.D.: Thumbs up or thumbs down?: semantic orientation applied to unsupervised classification of reviews. In: Proceedings of the 40th Annual Meeting on Association for Computational Linguistics (2002)
23. Wiebe, J., Wilson, T., Cardie, C.: Annotating expressions of opinions and emotions in language. Lang. Resour. Eval. 1(2), 165–210 (2005)
24. Wilson, T., Wiebe, J., Hoffmann, P.: Recognizing contextual polarity in phrase-level sentiment analysis. In: Proceedings of Human Language Technologies Conference/ Conference on Empirical Methods in Natural Language Processing (2005)

PSO-ASent: Feature Selection Using Particle Swarm Optimization for Aspect Based Sentiment Analysis

Deepak Kumar Gupta, Kandula Srikanth Reddy, Shweta, and Asif Ekbal[✉]

Computer Science and Engineering, Indian Institute of Technology Patna,
Patna, India
{deepak.mtmc13,kandula.cs11,shweta.pcs14,asif}@iitp.ac.in

Abstract. The amount of user generated online contents has increased dramatically in the recent past. The phenomenal growth of e-commerce has led to a significantly large number of reviews for a product or service. This provides useful information to the users to take a fully informed decision on whether to acquire the service and/or product or not. In this paper we present a method for automatic feature selection for aspect term extraction and sentiment classification. The proposed approach is based on the principle of Particle Swarm Optimization (PSO) and performs feature selection within the learning framework of Conditional Random Field (CRF). Experiments on the benchmark set up of SemEval-2014 Aspect based Sentiment Analysis Shared Task show the F-measure values of 81.91 % and 72.42 % for aspect term extraction in the laptop and restaurant domains, respectively. The method yields the classification accuracies of 78.48 % for the restaurant and 71.25 % for the laptop domain. Comparisons with the baselines and other existing systems show that our proposed approach attains the promising accuracies with much reduced feature sets in all the settings.

Keywords: Aspect extraction · Sentiment analysis · Feature selection · Conditional random field · Particle Swarm Optimization

1 Introduction

The use of the Internet in our daily life has grown tremendously in the recent past, and this is also true for the activities related to business and commerce. The increasing number of e-commerce portals has made the lives of people more easier in terms of the efforts spent for buying a product or acquiring any service. Blogs, forums, review sites etc. are growing in number day by day, as more people seek the advices of fellow users for making any informed decision on whether to buy the product or acquiring the service or not. These reviews can be useful to both the manufactures as well as the customers. Based on the feedback of the customers, manufacturers can improve or upgrade the quality of the product by focusing on those particular aspects that need to be given more attention.

© Springer International Publishing Switzerland 2015
C. Biemann et al. (Eds.): NLDB 2015, LNCS 9103, pp. 220–233, 2015.
DOI: 10.1007/978-3-319-19581-0_20

Different customers or users may express their views on different aspects of the products, or services and, hence, the amount of information available is significantly large in size. In order to extract the most useful information, any customer or manufacturer has to read all the reviews that have been written. Otherwise she/he may get a biased view of the product or service. Unfortunately, this task is not simple because of the following facts: searching through the entire collection of reviews for any particular aspect is a hugely involved task that consumes lots of time and efforts; communication in these mediums are informal, and the texts are not very well-formed. Therefore, there is an urgent need to develop applications that aid in mining the desired information from this huge collection of online contents.

Sentiment analysis is the task of identifying the sentiments (*positive, negative* or *neutral*) of the users based on the opinions and emotions expressed in the reviews written either for a particular product or service or any of its aspect (or feature/attribute). Classifying sentiment at document and sentence level no longer suffice the user's requirements of getting specific and precise information related to particular *aspects* (or, *features*) of the product. An *aspect* refers to an attribute or a component of the product that has been commented on in a review. As an illustration, let us consider the following review: *"The price is reasonable although the service is poor"*. Here we have two aspect terms, namely *price* and *service*. The sentiments expressed for these two aspects are contrasting in nature, positive for the first aspect but *negative* for the second one. The research of aspect based sentiment analysis aims to extract the relevant aspects of an entity for which opinions have been expressed, and then classifying these opinions into some categories such as positive, negative or neutral [1,2]. Aspect terms can influence sentiment polarity within a single domain. As an example, for the restaurant domain, *cheap* is usually positive with respect to *food*, but it denotes a negative polarity when discussing the decor or ambiance [3]. By performing sentiment analysis at aspect level, we can help users gain more insights on the sentiments of various aspects of the target entity. Hence the decision taken thereafter will be more informative and practical.

Literature shows that aspect based sentiment analysis has attracted attention to the researchers only in the recent past. Earlier studies on sentiment analysis mainly focused on either at the sentence [4], document [5] or phrase level [6]. However, this information is not sufficient for customers or users who are seeking opinions on specific product features (aspects) such as design, battery life, or screen of camera. This fine-grained classification is a topic of aspect-based sentiment analysis [7]. Traditional approaches to aspect extraction are based on frequently used nouns and noun phrases [1,8,9]. Such approaches work well when many aspects are strongly associated with certain categories of words (such as nouns), but often fail when many low frequency terms are used as the aspect. Nowadays with the emergence of various labeled datasets, supervised learning approaches [10,11] are predominantly being used. Some other approaches include the techniques, such as those that define aspect terms using a manually specified subset of the Wikipedia category [12] hierarchy, unsupervised clustering technique [8] and semantically motivated technique [5]. In SemEval-2014, many teams reported their research as part of the evaluation challenge, Aspect based Sentiment Analysis [13].

Most of the features used for aspect term extraction or sentiment classification exploit lexical, syntactic or semantic level features. The features used for a domain often fail to perform well for the other domains. The systematic approach for feature selection for these two tasks have not been attempted so far. In general, heuristics based techniques were used to determine the best fitting feature sets. In this paper we propose a technique for feature selection that can automatically determine the most relevant set of features for aspect term extraction and sentiment classification. This is based on the concept of Particle Swarm Optimization (PSO) [14], an efficient evolutionary algorithm. Feature selection, also known as variable subset selection or dimensionality reduction, is a technique that selects the most relevant features for the target problem. By removing the most irrelevant and redundant features from the data, feature selection helps to improve the performance of a classifier. We use Conditional Random Field (CRF) [15] as a learning algorithm. We implement a robust set of features for aspect term extraction and sentiment classification. A number of experiments were carried out with the various combination of features. The best configuration, thus obtained, is used as the baseline model on which PSO based feature selection technique is applied. The proposed technique is evaluated on the benchmark datasets of SemEval-2014 Aspect based Sentiment Analysis datasets [13]. Experiments show that our system can achieve better performance with much reduced sets of features for the problems. The contribution of the present work is two-fold, *viz.* use of very diverse and rich feature set for the task at hand, and PSO based feature selection for aspect based sentiment analysis.

2 Methods

In this section we present our method that we develop for solving the problems of aspect term extraction and classification. At first supervised machine learning algorithm, CRF [15] is used to construct the models for solving these problems. The individual system is optimized following a heuristics based feature selection technique, where we tried different feature combinations to find the best ones. Thereafter each feature configuration (best) is subjected to the PSO [14] to determine the best fitting feature set for each domain.

2.1 Introduction to Particle Swarm Optimization

Particle Swarm Optimization (PSO) [14] is a population based stochastic optimization method which is founded on the behaviour of bird flocking. Like genetic algorithm (GA), PSO also starts with a set of random solutions and searches for the global optima by updating the generations. In PSO, the potential solutions of the given problem are called as particles and denoted as $\overrightarrow{X}(k) = (x_{(k,1)}, x_{(k,2)}, \ldots\ldots\ldots, x_{(k,n)})$ in an n-dimensional search space. Each co-ordinate $x_{(k,d)}$ of these particles can change with some rate, known as the velocity $v_{(k,d)}$ d = 1,2,...,n. Every particle keeps a record of the best position that it has ever visited. Such a record is called the particle's previous best position and denoted by $\overrightarrow{B}(k)$. The global best position attained by any particle so far is also recorded and stored in a particle denoted by $\overrightarrow{G}$. An iteration comprises evaluation of each particle, then stochastic adjustment of $v_{(k,d)}$ in the direction of particle $\overrightarrow{X}(k)$'s previous best position and the previous best position of any particle in the neighbourhood [14]. Entire process of PSO is governed by three operations, namely *evaluate*, *compare* and *imitate*. The *evaluation* phase measures how well each particle, i.e. the candidate solution solves the problem at hand. The *comparison* process attempts to identify the best particle by comparing different solutions. The *imitation* process produces new particles based on some of the best particles found so far. These three processes are repeated until a given stopping criterion is met. The objective is to find the particle that best solves the target problem. Velocity and neighbourhood are the two important concepts in PSO. Every particle $\overrightarrow{X}(k)$ is associated with a velocity vector. The velocity vector is updated at every generation, and used to generate a new particle $\overrightarrow{X}(k)$. The neighbourhood defines how other particles in the swarm, such as $\overrightarrow{B}(k)$ and $\overrightarrow{G}$, interact with $\overrightarrow{X}(k)$ to modify its respective velocity vector and position.

There are other popular heuristic approaches like the well-known genetic algorithm (GA) [16] and the simulated annealing (SA) [17]. In order to solve the global optimization problems these techniques are widely used to find the good set of solutions in the search space. While SA is a probabilistic metaheuristic approach, GA relies on the concept of survival of fittest. Unlike GA, PSO does not retain only the good solutions. PSO allows the particles to move in the search space on the basis of number of cases, and it generates good set of possible solutions without eliminating any weak particle.

2.2   Feature Selection Using PSO

The proposed feature selection algorithm is based on the binary version of PSO algorithm [18]. Below, we describe the basic steps of our proposed approach:

2.2.1   Encoding of the Particles and Initial Population

Potential solutions to the target problem are encoded as fixed length binary-valued strings $\overrightarrow{X}(i) = (x_{(i,1)}, x_{(i,2)}, ..., x_{(i,n)})$ where $x_{(i,j)} \in \{0,1\}$, $i = 1, 2, ..., N$, N is the number of particles, $j = 1, 2, ...n$ (n denotes the number of features). The length of the particle depends on the number of features present. If we have a set of features $F = (f_1, f_2, ..., f_n)$, then a string of length n is defined to denote the particle. If any bit of $\overrightarrow{X}(i)$ contains a value of "1" then the respective feature is used for classifier's training, and a value of 0 designates that the feature is not used. As an example, for a given list of features, $F = (f_1, f_2, f_3, f_4, f_5, f_6)$ and

$N=3$, the swarm can be represented as follows:

$$\vec{X}(1) = (1, 0, 1, 0, 1, 1)$$
$$\vec{X}(2) = (1, 0, 1, 1, 0, 1)$$
$$\vec{X}(3) = (1, 1, 1, 0, 0, 1)$$

Here for the first particle, the features f_1, f_3, f_5, and f_6 are used for learning as only these positions have the values of 1.

For the initial population we randomly generate such N solutions. The bit positions are randomly initialized to either 0 or 1. We generate a uniform random number μ on the interval (0,1) for every position $X_{(i,d)}$ of $\vec{X}(i)$. Every particle $\vec{X}(i)$ is generated as follows.

$$X_{(i,d)} = \begin{cases} 1 & \text{if } \mu \text{ is } < 0.5 \\ 0 & \text{otherwise} \end{cases}$$

2.2.2   Updating the Global and Best Position Value

Initially, the previous best position of the particle $\vec{X}(i)$, represented by $\vec{B}(i)$, is set to null. As the initial particle is generated, we set the value of $\vec{B}(i)$ to the position vector of the particle $\vec{X}(i)$. The best position vector $\vec{B}(i)$ is updated based on the fitness function (or, objective function). In our case the fitness function is the F-score value of the classifier trained with CRF classifier using the feature combination as represented in the particle $\vec{X}(i)$. If the objective function value of new position vector is better than its previous best position vector, we update the position of the best vector, otherwise it remains as it is. This means that if f($\vec{P}(i)$) > f($\vec{B}(i)$) update the value of $\vec{P}(i)$. For the global best position, i.e. for $\vec{G}$ we also follow the same process for updating. Initially the global best position is also set to empty. The update only happens after all the values of $\vec{B}(i)$ are determined. The value of $\vec{G}$ is set to the fittest $\vec{B}(i)$ found so far. It is updated only when the fittest solution represented by $f(\vec{B}(i))$ in the swarm is superior than $f(\vec{G})$.

2.2.3   Updating the Velocities

Every particle i is associated with an unique velocity vector denoted by $V(i) = (v_{(i,1)}, v_{(i,2)}, ..., v_{(i,n)})$. The components $v_{(i,d)}$ in $V(i)$ determine the rate of change of $x_{(i,d)}$ in $\vec{X}(i), d = 1, 2, ..., n$. Each element $v_{(i,d)}$ in $V(i)$ is updated following the process as mentioned below [18]:

$$v_{(i,d)} = w * v_{(i,d)} + \mu_1(b_{i,d} - x_{(i,d)}) + \mu_2(g_d - x_{(i,d)})$$

where $w(0 < w < 1)$, known as inertia weight, The $b_{(i,d)}$, $x_{(i,d)}$, $g_{(d)}$ denote the dth components of $\vec{B}(i)$, $\vec{X}(i)$ and $\vec{G}$, respectively. The concept of inertia weight is used by Shi and Eberhart [19] to better control the exploration and

Table 1. Parameter settings for aspect term extraction and polarity classification

Parameters	Restaurants		Laptops	
	Aspect extraction	Polarity	Aspect extraction	Polarity
Inertia weight	0.89	0.95	0.3925	0.3925
No. of particle	10	10	25	25
Iteration	100	50	200	100
μ_1	2.1	1.49618	2.5586	2.5586
μ_2	1.9	1.49618	1.3358	1.3358

exploitation. It is a force that pulls the particle to continue its current direction. The factors μ_1 and μ_2 are known as the cognitive and social scaling parameters, respectively. The values of these factors are randomly generated from a number drawn from a uniform distribution. We perform 3-fold cross validation on the training data to set the values of these parameters. We present in the following table the values of different parameters used in the experiments (Table 1):

2.2.4 Sampling New Particles

For each particle i and each dimension d the new particle $x_{(i,d)}$ in $\vec{X}(i)$ can be either 0 or 1. The decision is based on the following equation:

$$X_{(i,d)} = \begin{cases} 1 & \text{if r } < S(v_{(i,d)}) \\ 0 & \text{otherwise} \end{cases}$$

where $0 \leq r \leq 1$ is a uniform random number and

$$S(v_{(i,d)}) = \frac{1}{1 + exp(\overrightarrow{v_{(i,d)}})}$$

2.3 Pre-processing

We use the datasets as provided in the SemEval-2014 shared task. We pre-process this data to remove all the XML tags, and then run Stanford CoreNLP[1] suite in order to extract the information such as lemma, Part-of-Speech (PoS) and named entity (NE). In order to cast the aspect term extraction as a sequence labelling problem, it is essential to identify the boundary of any aspect term properly. We follow the standard BIO notation where **"B-ASP"**, **"I-ASP"** and **"O"** denote the beginning, intermediate and outside tokens of a multi-word aspect term.

2.4 Features for Aspect Term Extraction

We identify and implement the following set of features for aspect term extraction. We use the same set of features for both the domains, namely restaurant

---

[1] http://nlp.stanford.edu/software/corenlp.shtml.

and laptop. We restricted ourselves in not using much domain-dependent external resources.

1. **Word and local context:** The words and its contexts are used as the features of the model. We convert the words into the lower case forms, and use it as a feature along with the surface forms. Local contexts provide useful evidence for the detection of aspect terms. In general the words that span the preceding and following few tokens of the target term are used as the features. Here we use the context of preceding two and following two tokens as the features.

2. **Part-of-Speech (PoS) information:** An aspect can be expressed by a noun, adjective, verb or adverb. We use the PoS information of the previous two and next tokens along with the current one as features.

3. **Head word:** Aspect terms generally belong to the noun phrase. We use the head word of the noun phrase as the feature. The words that do not belong to a noun phrase are assigned the value "NULL".

4. **Head word PoS:** PoS of the head word is used as a feature of the model.

5. **Chunk information:** A review can often have multi-word aspect terms such as *"battery life"*, *"spicy tuna rolls"* etc. Chunk information helps in identifying the boundaries of these multi-word aspect terms. We use the chunk information of the current token as the feature.

6. **Lemma:** The goal of lemmatization is to reduce inflectional forms and sometimes derivationally related forms of a word to a common base form. For example the words *serve, serves, served* and *serving* in restaurant domain can be identified as different inflectional forms of the word *serve*. We use lemma as feature of CRF.

7. **Stop word:** Stop words are those words that don't contribute for aspect term extraction (for example, *the, is, at*). A feature is defined that takes the value equal to 1 or 0 depending upon whether it appears in the list of stopwords or not.

8. **Length:** Length of token sometimes be effective in identifying the aspect terms. More is the length of a token, higher is the chance of it being an aspect term. We assume an entity as the candidate aspect term if its length exceeds a predefined threshold value, which is equal to five in our case.

9. **Prefix and suffix:** Prefix and suffix of fixed length character sequences are stripped from each token and used as the features of classifier. Here we use the prefixes and suffixes of length upto three characters as the features.

10. **Frequent aspect term:** We extract the aspect terms from the training data, and prepare a list by considering the most frequently occurring terms. We consider an aspect term to be frequent if it appears at least five times in the training data. A feature is then defined that fires if and only if the current token appears in this list. This feature is useful for correlating already seen aspect terms in the train dataset.

11. **Dependency relation:** Dependency relations provide representation of grammatical relations between the words in a sentence. We use Stanford dependency parser[2] to extract the dependency relation. Two different

---

[2] http://nlp.stanford.edu/software/lex-parser.shtml.

features were defined in line with [20]. The first feature denotes the dependency relation in case the current token is the governor (i.e. head of the relation). The second feature represents the relation where current token is the dependent.

12. **WordNet:** In WordNet [21] words from the same lexical category that are roughly synonymous are grouped into synsets. We define a feature which represents synsets of each token. This feature can be helpful in grouping tokens with identical senses. For example, the tokens *lunch* and *dinner* are related as the hyponyms of *meal* in the WordNet hierarchy. We consider only the noun synsets. We define this feature similar to the line as defined in [20].

13. **Named entity information:** We use the named entity (NE) tag of the current token as the feature. Ambiguity arises in distinguishing NE from aspect terms. Let us consider the following two examples sentences: (i). Ex-1: "Certainly not the best sushi in New York." and (ii). Ex-2: "I trust the people at Go Sushi, it never disappoints." In the first sentence, *sushi* is a aspect term but in the second sentence *Sushi* is part of a NE.

14. **Character n-grams:** Character n-gram is a contiguous sequence of n character extracted from a given word. We extract character n-grams of length one (unigram), two (bigram), three (trigram), four(quad), five(five) and use these as features of the classifier.

15. **Aspect term list:** For each domain we create two different lists of aspect terms from the respective training set. The first list contains the aspect terms that appear in the training set with a frequency of more than f_1. The second list was created in order to handle the multi-word aspect terms. This list is used to record those single-word aspect terms whose counts are above a predefined threshold frequency, f_2. A probability p is computed in lines with [20] in order to take into account those words that appear as aspect terms independently.

16. **Word cluster:** From the respective domain of dataset, we generate the clusters using Brown clustering algorithm presented in [22]. This is a hierarchical clustering algorithm which clusters words based on the contexts in which they occur. We induce 1000 Brown clusters for each user review dataset. For each token the default prefix string as obtained in the output of this clustering process is used as the value of this feature. Ideally clusters should be induced from a larger sized datasets. But in order to preserve the spirit of domain- independent behaviour we did not make use of any external product review.

17. **Semantic orientation (SO) score:** Sentiment orientation (SO) score [23] is a measure of positive or negative sentiment expressed in a phrase. The sentiment score of each token is determined with Pointwise Mutual Information (PMI) as follows

$$SO(t) = PMI(t, prev) - PMI(t, nrev)$$

where PMI is a measure of association of token t with respect to positive review (i.e. *prev*) or negative review (i.e. *nrev*). We compute the SO score from the training data of the respective domain. A positive SO score implies that the token is more related to positive than negative reviews.

18. **Orthographic features:** We define two features based on the constructions of words. These check whether the token starts with a capitalized letter or starts with a digit.

2.5   Features for Sentiment Classification

Here the opinions are classified into the semantic classes such as *positive*, *negative*, *neutral* and *conflict*. The class *conflict* is used where the reviewer has made both *positive* and *negative* comments on an aspect term in a single review. For example the review *"The food was delicious but do not come here on a empty stomach."* in the restaurant domain is both positive and negative with respect to the *food* aspect. For classification we make use of some of the features such as local context, PoS, Chunk, prefix, suffix etc., as defined in the previous subsection. Some other problem-specific features that we implement for the sentiment classification are defined as below:

1. **Aspect term and context:** Actual surface forms of aspect terms are converted into the lowercase forms and used as the feature along with the actual aspect terms. The sentiment of a review also heavily depends on the context where the aspect term appears. We include previous five and next five tokens as the local contextual information.

2. **Lexicon:** Lexicons are the useful sources and provide useful information for analysing the sentiment. We use the following lexicons and extract few features:

   1. **MPQA lexicon:** We make use of MPQA subjectivity lexicon [24] that contains the words along the scores denoting the positive, negative and neutral sentiments. A feature is defined that takes the values as follows: 1 for positive, −1 for negative, 0 for neutral and 2 for those that do not appear in the lexicon. For each aspect term, we compute the sum of the polarity scores of all the terms that appear in the surrounding context of previous five and next five words.

   2. **Bing Liu lexicon:** For each term in the training and test set we set the values in the following way depending upon whether it appears in the Bing Liu Lexicon [25] or not: 1 for positive; −1 for negative and 2 for those that appear in neither. Two features were then defined as:

      i. For each aspect term, we compute the sum of the polarity scores of all the words that appear in the context of previous five and next five words, and use it as a value of the feature.

      ii. In this feature we calculate the sum of the polarity scores of only those words which have a *direct dependency relation* with the target aspect term.

   3. **SentiWordNet lexicon**[3]: This is one of the most widely used source for sentiment analysis. The sentiment scores of all the words that appear in

---

[3] http://sentiwordnet.isti.cnr.it/.

the surrounding context (five left and five to the right) of the target aspect term are retrieved from the SentiWordNet. The feature value is set equal to the sum of the scores of these words.

3. **Domain-Specific words:** The general lexicons cannot cover words which are specifically used in a domain which express sentiment. Some of these examples are *mouth watering, yummy* and *over cooked* for the restaurant domain. We hand-made a lexicon by adding a list of words from general intuition and an online site[4] that describes food. We define the feature value in such a way that it takes a value of 1 for the positive, −1 for the negative and 2 for those that don't appear in the list. For each target aspect term, we compute the sum of the polarity scores of all the words that appear in an window of context of size 10 (i.e. the previous five and next five words of the target token).

3   Experiments and Analysis

We perform experiments with the benchmark set up of SemEval-2014 Task-4 on Aspect based Sentiment Analysis evaluation challenge.

3.1   Datasets

The datasets comprise of the domains of restaurants and laptop reviews. The training sets consist of 3,044 and 3,045 reviews for these two domains, respectively. There are 3,699 and 2,358 aspect terms, respectively. The test sets contain 800 reviews for each domain with 1,134 and 654 test instances, respectively.

3.2   Results of Heuristics Based Feature Selection

For each domain we build several models of CRF by varying the different combinations of feature templates, and select the best ones depending upon the performance achieved on the development set[5]. These heuristic based feature selection models are treated as the baselines for comparisons. For our implementation we make use of the C^{++} based CRF^{++} package[6] for our implementation. Results of these models for the aspect term extraction are reported in Table 2. For the restaurant domain we obtain the recall, precision and F-score values of 77.95 %, 85.16 % and 81.39 %, respectively with 67 features. Though we make use of more features for the laptop domain, we don't achieve higher performance. The official baseline model as defined in [13] showed the recall, precision and F-score values of 42.77 %, 52.55 % and 47.16 % for the restaurant domain. This baseline was constructed by making a list of aspect terms from the training data, and then matching each test instance. This official baseline model for the laptop domain had the recall, precision and F-score values of 29.82 %, 44.32 % and

---

[4] http://world-food-and-wine.com/describing-food.

[5] A part of each training set is used as the development set.

[6] http://crfpp.sourceforge.net.

Table 2. Results of aspect term extraction

Data Set	Restaurant		Laptop	
	CRF [Heuristic]	CRF [PSO]	CRF[Heuristics]	CRF [PSO]
No. of features	67	41	83	35
recall	77.95	78.48	63.14	64.06
precision	85.16	85.65	82.10	83.30
F-score	81.39	81.91	71.39	72.42

Table 3. Results of polarity classification

Data Set	Restaurant		Laptop	
	CRF [Heuristics]	CRF [PSO]	CRF [Heuristics]	CRF [PSO]
No. of features	38	20	26	14
Correct identification	878	890	457	466
Accuracy	77.42	78.48	69.87	71.25

35.65 %, respectively. Results of polarity classification are reported in Table 3. We obtain the accuracies of 77.42 % and 69.87 % for the restaurant and laptop domains, respectively for the CRF based classifier that is trained with the feature set selected through heuristic based method. This is much above the official baseline model [13] which was constructed by assigning the most frequent sentiment class that appears in the training set for the target aspect term. The baseline accuracies were 63.89 % for the restaurant domain and 65.65 % for the laptop domain.

3.3 Results Using PSO

In order to determine the most relevant set of features for each task, we apply our proposed PSO based feature selection on the baseline models. Results are reported in Table 2 for aspect term extraction and in Table 3 for sentiment classification. Results show that only a smaller subset can actually achieve better performance for each of the domains. This shows the efficacy of our proposed approach for both aspect term extraction and polarity classification. For aspect extraction in the restaurant domain, only 41 out of 67 were actually proved to be effective. In case of laptop domain we observe that out of 83 features, only 35 are actually effective. For sentiment classification we also observe that for both the domains the proposed approach actually attains better classification accuracies with the reduced feature sets. For the restaurant and laptop domains, we see that only 20 (out of 38) and 18 (out of 26) features actually help to gain better classification accuracies.

3.4   Comparisons

We compare the performance of our proposed systems with those, developed on the same benchmark datasets. Comparisons show that we obtain the performance which are at par the state-of-the-art methods. For aspect term extraction in the restaurant domain, the CRF-based system reported in [26] achieved the recall, precision and F-score values of 82.72 %, 85.35 % and 84.01 %, respectively. This was unconstrained in nature in the sense that they made use of external lexicons and rule-based system for improving the performance. But we obtain the F-score of 81.91 % without using any such external resources. Another notable advantage of our proposed method is that we obtain this accuracy with much reduced set of features. For the laptop domain we are also at par with the best system [27] that reported the F-score value of 74.55 %. This system was developed with additional resources and a rule-based sentiment analysis tool. For polarity identification the highest accuracy reported was 70.48 % for the laptop domain [28,29] and 81.00 % for the restaurant domain [28]. Hence our system attains better accuracy for the laptop data and quite comparable accuracy for the restaurant data. The method proposed in [28] incorporated several other lexicon based features (combining several lexicons etc.), rules as well as the bag-of-N-gram features, which we did not use. Unlike the system proposed in [29] we did not use various external resources such as unlabelled product reviews of Amazon or Yelp Restaurant Word–Aspect Association Lexicon. The most appealing attribute of our proposed method is that we achieve quite comparable performance with much smaller set of features.

4   Conclusion

In this paper we propose a feature selection method for aspect based sentiment analysis. The method is grounded on the principle of PSO. As a learning framework we used CRF. We implemented a diverse set of feature that incorporates lexical, syntactic and semantic level features. We performed our experiments on the benchmark datasets of SemEval-14 shared task datasets. Evaluation shows that our proposed method achieves encouraging performance for both the tasks, *viz.* aspect term extraction and sentiment classification.

In future we target to implement some more features for the problems. We would also like to bring into consideration other classifiers for solving the problem. Future study includes developing method for feature selection by multi-objective PSO algorithm. For multi-objective optimization we can optimize more than one fucntions simultaneously (e.g., recall and precision).

References

1. Hu, M., Liu, B.: Mining and summarizing customer reviews. In: Proceedings of the 10th KDD, Seattle, WAs, pp. 168–177 (2004)
2. Liu, B.: Sentiment Analysis and Opinion Mining. Synthesis Lectures on Human Language Technologies. Morgan & Claypool Publishers (2012)

3. Brody, S., Elhadad, N.: An unsupervised aspect-sentiment model for online reviews. In: Proceedings of NAACL, Los Angeles, CA, pp. 804–812 (2010)
4. Kim, S.M., Hovy, E.: Determining the sentiment of opinions. In: Proceedings of the 20th International Conference on Computational Linguistics, p. 1367. Association for Computational Linguistics (2004)
5. Turney, P.D.: Thumbs up or thumbs down?: semantic orientation applied to unsupervised classification of reviews. In: Proceedings of the 40th ACL, pp. 417–424 (2002)
6. Jagtap, V., Pawar, K.: Analysis of different approaches to sentence-level sentiment classification. Int. J. Sci. Eng. Technol. **2**, 164–170 (2013). ISSN: 2277–1581
7. Moghaddam, S., Ester, M.: Aspect-based opinion mining from online reviews. In: Tutorial at SIGIR Conference (2012)
8. Popescu, A.M., Etzionir, O.: Extracting product features and opinions from reviews. In: Proceedings of the Conference on HLT/EMNLP, pp. 339–346 (2005)
9. Blair-Goldensohn, S., Hannan, K., McDonald, R., Neylon, T., Reis, G.A., Reynar, J.: Building a sentiment summarizer for local service reviews. In: WWW Workshop on NLP in the Information Explosion Era, vol. 14 (2008)
10. Zhuang, L., Jing, F., Zhu, X.Y.: Movie review mining and summarization. In: Proceedings of the 15th ACM International Conference on Information and Knowledge Management, CIKM 2006 (2006)
11. Mukherjee, A., Liu, B.: Aspect extraction through semi-supervised modeling. In: Proceedings of the 50th Annual Meeting of the Association for Computational Linguistics: Long Papers, ACL 2012, vol. 1, pp. 339–348 (2012)
12. Fahrni, A., Klenner, M.: Old wine or warm beer: target-specic sentiment analysis of adjectives. In: Symsposium on Affective Language in Human and Machine, The Society for the Study of Artificial Intelligence and Simulation of Behavior (AISB), pp. 60–63 (2008)
13. Pontiki, M., Galanis, D., Pavlopoulos, J., Papageorgiou, H., Androutsopoulos, I., Manandhar, S.: Semeval-2014 task 4: aspect based sentiment analysis. In: Proceedings of the 8th International Workshop on Semantic Evaluation (SemEval 2014), Dublin, Ireland, August 2014
14. Kennedy, J., Eberhart, R.C.: Swarm Intelligence. Morgan Kaufmann Publishers Inc., San Francisco (2001)
15. Lafferty, J.D., McCallum, A., Pereira, F.C.N.: Conditional random fields: probabilistic models for segmenting and labeling sequence data. In: ICML, pp. 282–289 (2001)
16. Goldberg, D.E.: Genetic Algorithms in Search, Optimization and Machine Learning, 1st edn. Addison-Wesley Longman Publishing Co. Inc., Boston (1989)
17. Kirkpatrick, S., Gelatt, C.D., Vecchi, M.P.: Optimization by simulated annealing. SCIENCE **220**(4598), 671–680 (1983)
18. Kennedy, J., Kennedy, J.F., Eberhart, R.C.: Swarm Intelligence. Morgan Kaufmann, San Francisco (2001)
19. Shi, Y., Eberhart, R.: A modified particle swarm optimizer. In: IEEE World Congress on Computational Intelligence, pp. 69–73. IEEE (1998)
20. Toh, Z., Wang, W.: DLIREC: aspect term extraction and term polarity classification system. In: Proceedings of the 8th International Workshop on Semantic Evaluation (SemEval 2014), pp. 235–240 (2014)
21. Miller, G.A.: WordNet: a lexical database for English. Commun. ACM **38**(11), 39–41 (1995)
22. Brown, P.F., Desouza, P.V., Mercer, R.L., Pietra, V.J.D., Lai, J.C.: Class-based n-gram models of natural language. Comput. Linguist. **18**(4), 467–479 (1992)

23. Hatzivassiloglou, V., McKeown, K.R.: Predicting the semantic orientation of adjectives. In: Proceedings of the ACL/EACL, pp. 174–181 (1997)
24. Wiebe, J., Mihalcea, R.: Word sense and subjectivity. In: Proceedings of the COLING/ACL, pp. 1065–1072 (2006)
25. Ding, X., Liu, B., Yu, P.S.: A holistic lexicon-based approach to opinion mining. In: Proceedings of the 2008 International Conference on Web Search and Data Mining, WSDM 2008 (2008)
26. Brun, C., Popa, D.N., Roux, C.: XRCE: hybrid classification for aspect-based sentiment analysis. In: SemEval 2014, pp. 838–842 (2014)
27. Chernyshevich, M.: IHS R&D belarus: cross-domain extraction of product features using conditional random fields, pp. 309–313 (2014)
28. Wagner, J., Arora, P., Cortes, S., Barman, U., Bogdanova, D., Foster, J., Tounsi, L.: DCU: aspect-based polarity classification for semeval task 4. In: Proceedings of the 8th International Workshop on Semantic Evaluation (SemEval 2014), pp. 223–229 (2014)
29. Kiritchenko, S., Zhu, X., Cherry, C., Mohammad, S.: NRC-Canada-2014: detecting aspects and sentiment in customer reviews. In: Proceedings of the 8th International Workshop on Semantic Evaluation (SemEval 2014) (2014)

Improving Spanish Polarity Classification Combining Different Linguistic Resources

Eugenio Martínez-Cámara[1]([✉]), Fermín L. Cruz[2],
M. Dolores Molina-González[1], M. Teresa Martín-Valdivia[1],
F. Javier Ortega[2], and L. Alfonso Ureña-López[1]

[1] SINAI Research Group, University of Jaén,
Campus Las Lagunillas, 23071 Jaén, Spain
emcamara@ujaen.es
[2] Department of Languages and Computer Systems, University of Seville,
Av. Reina Mercedes s/n, 41012 Sevilla, Spain

Abstract. Sentiment analysis is a challenging task which is attracting the attention of researchers. However, most of work is only focused on English documents, perhaps due to the lack of linguistic resources for other languages. In this paper, we present several Spanish opinion mining resources in order to develop a polarity classification system. In addition, we propose the combination of different features extracted from each resource in order to train a classifier over two different opinion corpora. We prove that the integration of knowledge from several resources can improve the final Spanish polarity classification system. The good results encourage us to continue developing sentiment resources for Spanish, and studying the combination of features extracted from different resources.

Keywords: Sentiment analysis · Polarity classification · Lexicon-based approach · Sentiment feature generation

1 Introduction

Sentiment classification or polarity detection is an opinion mining task oriented to determine the overall sentiment-orientation of the opinions contained within of a given document. The document is supposed to contain subjective information such as product reviews or opinionated posts in blogs. This task has been widely studied, but most of the research is focused on dealing with English documents, perhaps due to the lack of resources in other languages. However, opinions and comments in the Internet are expressed using other languages different from English such as Chinese, Spanish or Arabic. The development of

This work has been partially supported by a grant from the Fondo Europeo de Desarrollo Regional (FEDER), ATTOS project (TIN2012-38536-C03-0) from the Spanish Government. The project AORESCU (P11-TIC-7684 MO) from the regional government of Junta de Andalucía partially supports this manuscript, and the project CEATIC-2013-01 from the University of Jaén.

C. Biemann et al. (Eds.): NLDB 2015, LNCS 9103, pp. 234–245, 2015.
DOI: 10.1007/978-3-319-19581-0_21

new linguistic resources is essential to make progress in solving the problem of the sentiment analysis, being even higher in languages other than English, like Spanish. Therefore, it is important to develop resources to help researchers to work with these languages.

In this paper we combine several Spanish resources for opinion mining in order to improve the final system for polarity classification. Specifically, we are going to combine two Spanish lexicons that have been developed in very different ways. On the one hand, iSOL (improved Spanish Opinion Lexicon) resource [21] has been derived from the English Lexicon of Bing Liu [14] and then it has been manually revised and improved. On the other hand, the ML-SentiCon [6,8] (Multi-Layered, Multilingual Sentiment Lexicon) resource is based on an improved version of SentiWordNet [1]. Several features were calculated using iSOL and ML-SentiCon to represent each document. Also, the combination of the features calculated using the two lexicons has been study. Support Vector Machine was the algorithm selected to analyse the goodness of the different sets of features. Our main goal is to demonstrate that the combination of both different resources can improve the final result.

In order to prove the robustness of the method, we have carried out experiments over two different corpora of reviews: a corpus of movie reviews called MuchoCine [5] and a corpus of opinions about hotels called COAH (Corpus of Opinions from Andalusian Hotels) [22]. Both corpora include comments and reviews written in Spanish. However, they have meaningful differences not only related to the domain tackled but also according to the number of reviews, comment size, balance between positive and negative samples, and even the nature of the information contained, more descriptive in movies and more subjective in hotels. In addition, the experiments also reinforce the idea already depicted in other works that the movie domain is more difficult to learn than hotel domain [3]. Therefore, the polarity in sentiment analysis not only remains through different languages [20], but also we have proved that the difficulty of a domain is also preserved across the languages.

The paper is organized as follows: Next section includes a review of some works related to polarity classification using other languages than English and specifically dealing with Spanish. Section 3 presents the resources used in our experiments paying more attention to the corpus COAH and the lexicon ML-SentiCon. Experimental framework is described in Sect. 4 and the analysis of results is presented in Sect. 5. Finally, main conclusions and further work are expounded in Sect. 6.

2   Related Works

Polarity classification has been mainly faced from two points of view: machine learning techniques based approaches and approaches based on the semantic orientation of words. The first group is wider used for the classification of reviews. In this type of approaches, the document is represented by different features that may include the use of n-grams or defined grammatical roles like adjectives or other linguistic feature combinations. Then a machine learning algorithm is applied, usually Support Vector Machine (SVM), Maximum Entropy (ME) and Nave Bayes (NB). A survey of studies using machine learning on this task can be found in [17, 23].

On the other hand, there is a lot of work based on the semantic orientation approach, which represent the document as a collection of words, computing the polarity of the document as an aggregation of the polarity of its words. The sentiment of each word can be determined by different methods, for example using a web search [12] or consulting a lexical database like WordNet [15]. Among the methods that consider some linguistic features in order to determine the sentiment at a word-level we can highlight many studies in the literature [9, 11, 27, 30]. Part of our proposal is based on the work [14].

Regarding polarity classification on non-English languages, there are a number of interesting studies that apply a semantic orientation approach based on sentiment words. Many of them use resources on English and then apply some translation process in order to apply the English-based resources to the target language. Kim and Hovy [16] compare opinion expressions within an aligned corpus of emails in German and English. They translate English opinion-bearing words into German and then analyze German emails using those translated words. Zhang et al. [31] apply sentiment analysis to two Chinese datasets: the first one contains opinions on euthanasia from different web sites, while the second dataset included Chinese reviews about products in six categories from Amazon. They propose a rule-based approach including two phases: firstly, determining each sentences sentiment based on word dependency, and secondly, aggregating those values of sentiment at a sentence level in order to predict the document sentiment. Wan [29] studied how to reduce the need of using linguistic resources for sentiment analysis for texts in Chinese. The author followed a supervised approach and proposed a co-training system based on the use of an English corpus for polarity classification of Chinese products reviews applying a machine translation system. There are also some remarkable studies using SO based on bearing-words lists. For example, Banea et al. [2] proposed several approaches to cross-lingual subjectivity analysis by directly applying the translations of opinion corpus in English to the training of an opinion classifier in Romanian and Spanish. This work showed that automatic translation is a viable alternative for the construction of resources and tools for subjectivity analysis in a new target language. Cruz et al. [5] gathered a corpus of Spanish movie reviews from the MuchoCine website. MuchoCine (MC) corpus was manually annotated and used for the development of a polarity classifier based on the semantic orientation of the words.

On the other hand, Brooke et al. [3] presented several experiments dealing with Spanish and English resources. They conclude that, although the machine learning techniques can provide a good baseline performance, it is necessary to integrate language-specific knowledge and resources in order to achieve a noticeable improvement. They proposed three approaches: the first one uses manually and automatically generated resources for Spanish. The second one applies machine learning to a Spanish corpus. The last one translates the Spanish corpus into English and then applies the SO-CAL, (Semantic Orientation CALculator), a tool developed by themselves [26]. Finally, it is worth to mention the work in [24], where a framework for the generation of sentiment lexicons in a target language is presented. They use manually and automatically annotated English resources and then map these annotations to other languages by using the multilingual aligned WordNet family. Using this method, they build two Spanish lexicons with 1,347 and 2,496 terms, respectively. It is interesting to highlight the evaluations performed through the manual annotation of a test set of 100 terms from each lexicon, achieving an accuracy of 90 and 74 %, respectively.

3   Resources

In this paper we will use four Spanish resources for opinion mining. Firstly, a Spanish corpus of hotel reviews has been compiled. This corpus, called COAH (Corpus of Opinion about Andalusian Hotels) [22], is a new resource for Spanish community in opinion mining. Secondly ML-SentiCon [6,8], a new resource composed by lemma-level sentiment lexicons for English, Spanish and other three official languages in Spain (Catalan, Basque and Galician). Finally, the other corpus and lexicon used for our experiments are the Spanish corpus of movie reviews called MuchoCine Corpus [5] and the domain-independent lexicon iSOL [21] already presented in other works. We briefly introduce these resources in next subsections.

3.1   Lexicons

The iSOL resource was generated from the Bing Liu English Lexicon [14] by automatically translating it into Spanish and obtaining the SOL (Spanish Opinion Lexicon) resource. Then this resource was manually reviewed in order to improve the final list of words obtaining iSOL (improved SOL). The iSOL is composed of 2,509 positive and 5,626 negative words, thus the Spanish lexicon has 8,135 opinion words in total. This resource has been successfully evaluated in [21] using the MuchoCine corpus. The results showed that the use of an improved list of sentiment words from the same language could be considered a good strategy for unsupervised polarity classification.

On the other hand, ML-SentiCon is a set of lemma-level sentiment lexicons for English, Spanish and other three official languages in Spain (Catalan, Basque and Galician). The lexicons are induced using an automatic, semi-supervised method and are formed by 8 layers, allowing applications to choose different compromises between the amount of available words and the accuracy of the estimations of their prior polarities. For each POS tagged lemma in the resource, they are provided two scores: a real value representing the prior polarity, between -1 and 1, and the standard deviation reflecting the ambiguity of that value. According to manual verification of a significant sample, the lexicons for English and Spanish have both high accuracies, over 90 % for layers 1–6 and 1–5, respectively

Table 1. Sizes and accuracies of English and Spanish ML-SentiCon lemma-level lexicons

Layer	English		Spanish	
	#Lemmas	Accuracy	#Lemmas	Accuracy
1	157	99.36 %	353	97.73 %
2	982	98.88 %	642	97.20 %
3	1,600	97.75 %	891	94.95 %
4	2,258	96.24 %	1,138	93.06 %
5	3,595	93.95 %	1,779	91.75 %
6	6,177	91.99 %	2,849	86.09 %
7	13,517	85.29 %	6,625	77.69 %
8	25,690	74.06 %	11,918	61.29 %

(Table 1). In the case of the Spanish lexicon, the accuracy is sensibly better than the accuracy reported in other recent work [24].

The lemma-level lexicons were automatically created from a synset-level lexicon for English, which in turn were built with an enhanced version of the method used by Baccianella et al. [1] to build SentiWordNet 3.0, one of the most used sentiment lexicons nowadays. This method comprises two steps, one involving the classification of individual synsets from WordNet as positive, negative or neutral, and another one involving a global, graph-based quality improvement of the positivity and negativity scores of the synsets. Several improvements were added in both steps. In the first one, a new source of information was used for training the classifiers, WordNet-Affect 1.1 [25] and a meta-learning scheme for combining multiple classifiers was applied. In the second step of the method, they were proposed new kinds of WordNet-based graphs, including positive and negative arcs, and a different random-walk algorithm called PolarityRank [7]. Evaluations of the positivity and negativity scores obtained in each step show significant improvements with respect to the original method.

The lemma-level English lexicon was built by computing the means of positivity and negativity scores from those synsets corresponding to each POS tagged lemma. The lemmas were distributed over layers by gradually relaxing a set of restrictions (Table 1). In this way, the number of lemmas that satisfy the restrictions increases in each layer, at the same time as the reliability of those lemmas as indicators of positivity and negativity decreases. The Multilingual Central Repository 3.0 [10] and some resources from the EuroWordNet project [28] up to November 2009 were used in order to link synsets to lemmas for other languages in ML-SentiCon: Spanish, Catalan, Basque and Galician.

3.2   Corpora

In this paper we have used two different corpora. MuchoCine has been already described in [5] and it has widely used in several works [13,18,19,21]. The corpus consists of 3,878 movie reviews collected from the MuchoCine website. The reviews are written by web users instead of professional film critics. This increases the difficulty of the task because the sentences found in the documents may not always be grammatically correct, or they may include spelling mistakes or informal expressions. The corpus contains about 2 million words and an average of 546 words per review.

On the other hand, we have collected the Corpus of Opinion about Andalusian Hotels COAH, from the TripAdvisor site. The collection contains 1,816 reviews which were written by non-professional reviewers, rather than web users. Similarly to what happens with movie reviews, the texts in hotel reviews may not be grammatically correct, or they can include spelling mistakes or informal expressions. We have selected only Andalusian Hotels: per each province of Andalusia (Almera, Cdiz, Crdoba, Granada, Jan, Huelva, Mlaga and Sevilla) we have selected ten hotels, five of them with higher ratings and the other five with worse ratings. All the hotels must have at least twenty opinions in the latter years written in Spanish. As a result of these constraints, we have obtained 1,816 reviews. Finally, the corpus contains reviews for 80 hotels with an average of 23 reviews per hotel. We want to highlight that the hotel reviews are composed by about 145 words with a mean close to ten adjectives.

For both corpora the opinions are rated on a scale from 1 to 5. A rank of 1 means that the opinion is very bad, and 5 means very good. Reviews with a rating of 3 can be categorized as neutral which means the user consider the hotel/movie is neither bad nor good. Table 2 shows the number of reviews per rating for each corpus.

Finally, in our experiments the neutral reviews were discarded. In this way, opinions rated with 3 were not considered, the opinions with ratings of 1 or 2 were considered as positive and those with ratings of 4 or 5 were considered as negative. A total of 2,625 reviews were processed for MuchoCine Corpus (1,274 positives and 1,351 negatives) and 1,816 comments about hotels were considered for COAH corpus (1,020 positives and 511 negatives).

Table 2. Rating distribution for MuchoCine and COAH corpora

Rating	#Reviews in MC	Reviews in COAH
1	351	312
2	923	199
3	1,253	285
4	890	489
5	461	531
Total	3,875	1,816

4   Experimental Frameworks

As it has been mentioned above, two are the main contributions of this research. The first one is focused on the comparison of two linguistic resources for opinion mining in Spanish: iSOL and ML-SentiCon. Thus, two polarity classification systems were developed and they were assessed over two opinion mining corpora in Spanish. The fact that the two corpora are centred in two different domains must also be highlighted.

One of the main issues of a text classification system is the selection of a good set of features. The selection of the features in a sentiment classification system is critical, because those features have to represent the polarity or the intention of the author. The two lexicons herein compared offer different kind of sentiment information, so the second bunch of experiments are focused in the combination of the features calculated with iSOL and with ML-SentiCon with the aim of improving the polarity classification in Spanish. The combination of features from two different linguistic sources constitutes the second main contribution of this article.

4.1   Individual Experiments

Due to the different nature of the linguistic resources, two polarity classification systems have been developed. Concerning iSOL, two polarity classification systems have been evaluated. Each review is represented as a vector of features that are computed using the lexicon iSOL. The two corpora have two sections where the authors briefly express their overall opinions (a title or summary of the review), and the complete opinion about the movie or the hotel (body of the review). The summary is not necessarily an excerpt of the body. The main characteristic of the summary is that it is more concise and the author expresses the opinion more clearly. This is the main reason of treating independently the two sections of the documents. So, the system extracts features from the two sections separately. Thus, per each document four features are calculated:

1. Number of positive words in each part of the document (two features).
2. Number of negative words in each part of the document (two features).

After calculating the features of the documents, a 10-fold cross-validation evaluation is carried out with the goal of assessing the goodness of the bunch of features. The machine learning algorithm selected was SVM[1], using a linear kernel and normalizing the feature vectors. The results obtained are shown in Table 3.

Regarding ML-SentiCon, a supervised polarity classification system has been developed. Each review, taken from each of the two corpora, is represented as a vector of 48 features. The features are calculated taking the information of each of the layers of the ML-SentiCon. As this resource is composed of POS-tagged

---

[1] The SVM implementation was LibSVM [4].

Table 3. Results obtained with iSOL lexicon

Corpus	Accuracy	Precision	Recall	F1
MC	66.01 %	66.12 %	66.01 %	66.02 %
COAH	92.09 %	92.21 %	92.09 %	91.94 %

Table 4. Results obtained with ML-SentiCon lexicon

Corpus	Accuracy	Precision	Recall	F1
MC	65.37 %	65.47 %	65.37 %	65.37 %
COAH	89.09 %	89.07 %	89.09 %	88.88 %

lemmas, each review needs to be lemmatized and POS-tagged. MuchoCine corpus already contains this information. In the case of COAH corpus, we applied the same analysis tool used in the MuchoCine corpus, FreeLing (Padr and Stanilovsky, 2012). Per each layer and section of the documents three features are calculated:

1. Sum of polarities of all lemmas from that layer appearing in that section.
2. Sum of polarities of positive lemmas from that layer appearing in that section.
3. Sum of polarities of negative lemmas from that layer appearing in that section (Table 4).

Once the documents are represented as vectors of features, a 10-fold cross-validation evaluation is applied with the aim of assessing the goodness of the set of features. The machine learning algorithm selected was the same as the experiments realised with iSOL, i.e. SVM, using the same exact parameters. The results obtained over the two corpora are shown in Table 5.

4.2   Combining Lexicons

The similar results obtained with the two sets of features encourage us to combine the features generated with each lexicon. Both resources have been developed with quite different approaches, which suggest they provide different information to a certain extent. Since each document in ML-SentiCon is represented with 48 features and documents in iSOL have 4 features, in the union of the two sets of features each document is represented with 52 features. The same evaluation process has been repeated, i.e., 10-fold cross-validation evaluation and SVM as machine learning algorithm. The results reached are shown in Table 5.

5   Analysis of Results

In the analysis of the results three points could be highlighted: the differences between the two domains (movies, hotels), the differences between the two linguistic resources, and the improvement reached with the combination of the two sets of features.

Table 5. Results of the lexicon features union

Corpus	Accuracy	Precision	Recall	F1
MC	68.38 %	68.41 %	68.38 %	68.38 %
COAH	93.79 %	93.78 %	93.79 %	93.74 %

It is evident that there is an important difference between the results obtained with the two domains. The first conclusion is that it is easier to classify opinions in reviews of hotels than in reviews of movies. This assert is confirmed when the documents of each corpus are analysed. The main difference is the length of the reviews: meanwhile the movie reviews are long, hotel reviews are concise. The greater length of the movie reviews indicates that the authors not only express their feelings about the movie, but also wrote a synopsis. The authors use in the description of the movies polar words that are in the lexicons, but they are not been used as a part of an opinion of the author. Furthermore, the systems developed to evaluate the lexicons do not distinguish whether a sentence is subjective or objective, the systems are only polarity classifiers. Thus, the systems developed tend to misclassify the movie reviews where the author also summarizes the plot of the movie. On the other hand, the hotel reviews are more succinct, and the authors express their opinion on the hotel and not a description of the accommodation. Thus, when the authors use a polar word is more likely to be part of an opinion. This conclusion about the different difficulties learning a specific domain has been already depicted in other works, however this is the first time that has been proved for Spanish.

It is very clear that the two linguistic resources are totally different, while iSOL is a list of positive and negative words, ML-SentiCon is a ranking of words layered. The results reached with the two sentiment lexicons are good, but we have to say that iSOL reached slightly better results. This is because iSOL is a sentiment lexicon compiled semi-automatically and revised manually, while ML-SentiCon has been compiled fully automatically and has not been manually corrected.

An interesting result is reached when the features calculated with the two sets of resources are joined. In both domains the results are improved, so it proves that the information obtained from the two lexicons is complementary. Thus, it is not fair to say that a resource is better than the other because each of them obtains different information that allows performing a good polarity classification in Spanish, but when they are combined the classification is even better than when the features are classified separately.

6   Conclusions

The main contributions of this paper are the use of different Spanish resources for opinion mining. Specifically, the new sentiment lexicon for Spanish, ML-SentiCon and the new corpus called COAH, have been firstly used for Spanish opinion mining. In addition, our proposal uses, as well, the iSOL Spanish lexi-con and the MC corpus, a well-known resource for the sentiment analysis Span-ish research community. The results show that the two lexicons achieve similar results for this task. The most relevant observation is that the combination of the information extracted from both lexicons improves the performance of the polarity classification system.

As further work, we plan to study the application of ensemble methods that could improve the performance of a set of classifiers that use the information obtained from a set of lexicons. The treatment of the negation in Spanish texts following a linguistic strategy can be also an interesting line of research.

References

1. Baccianella, S., Esuli, A., Sebastiani, F.: Sentiwordnet 3.0: an enhanced lexical resource for sentiment analysis and opinion mining. In: Proceedings of the Seventh International Conference on Language Resources and Evaluation (LREC 2010), Valletta, Malta, pp. 2200–2204, May 2010
2. Banea, C., Mihalcea, R., Wiebe, J., Hassan, S.: Multilingual subjectivity analysis using machine translation. In: Proceedings of the Conference on Empirical Methods in Natural Language Processing, EMNLP 2008, Association for Computational Linguistics, USA, pp. 127–135 (2008). http://dl.acm.org/citation.cfm?id=1613715. 1613734
3. Brooke, J., Tofiloski, M., Taboada, M.: Cross-linguistic sentiment analysis: from english to spanish. In: Proceedings of the International Conference RANLP-2009, ACL, Bulgaria, pp. 50–54, September 2009. http://www.aclweb.org/anthology/ R09-1010
4. Chang, C.C., Lin, C.J.: Libsvm: a library for support vector machines. ACM Trans. Intell. Syst. Technol. **2**(3), 27:1–27:27 (2011). http://doi.acm.org/10.1145/ 1961189.1961199
5. Cruz, F., Troyano, J.A., Enriquez, F., Ortega, J.: Clasificación de documentos basada en la opinión: experimentos con un corpus de críticas de cine en español. Procesamiento del Lenguaje Natural 41, pp. 73–80, September 2008. http:// journal.sepln.org/sepln/ojs/ojs/index.php/pln/article/view/2551
6. Cruz, F.L., Troyano, J.A., Pontes, B., Ortega, F.J.: Ml-senticon: a multilingual, lemma-level sentiment lexicon. Procesamiento del Lenguaje Natural **53**, 113–120 (2014). http://journal.sepln.org/sepln/ojs/ojs/index.php/pln/article/view/5041
7. Cruz, F.L., Vallejo, C.G., Enríquez, F., Troyano, J.A.: Polarityrank: Find-ing an equilibrium between followers and contraries in a network. Inf. Process. Manage. **48**(2), 271–282 (2012). http://www.sciencedirect.com/science/ article/pii/S0306457311000823
8. Cruz, F.L., Troyano, J.A., Pontes, B., Ortega, F.J.: Building layered, multilingual sentiment lexicons at synset and lemma levels. Expert Syst. Appl. **41**(13), 5984–5994 (2014). http://www.sciencedirect.com/science/article/pii/S095741741400 1997
9. Ding, X., Liu, B.: The utility of linguistic rules in opinion mining. In: Proceed-ings of the 30th Annual International ACM SIGIR Conference on Research and Development in Information Retrieval, SIGIR 2007, pp. 811–812. ACM, New York (2007). http://doi.acm.org/10.1145/1277741.1277921

10. Gonzalez-Agirre, A., Laparra, E., Rigau, G.: Multilingual central repository version 3.0. In: Proceedings of the Eight International Conference on Language Resources and Evaluation (LREC 2012). European Language Resources Association (ELRA), Istanbul, Turkey, May 2012

11. Hatzivassiloglou, V., McKeown, K.R.: Predicting the semantic orientation of adjectives. In: Proceedings of the Eighth Conference on European Chapter of the Association for Computational Linguistics, EACL 1997, pp. 174–181. Association for Computational Linguistics, Stroudsburg (1997). http://dx.doi.org/10.3115/979617.979640

12. Hatzivassiloglou, V., Wiebe, J.M.: Effects of adjective orientation and gradability on sentence subjectivity. In: Proceedings of the 18th Conference on Computational Linguistics, COLING 2000, vol. 1, pp. 299–305. Association for Computational Linguistics, Stroudsburg (2000). http://dx.doi.org/10.3115/990820.990864

13. del Hoyo, R., Hupont, I., Lacueva, F.J., Abadía, D.: Hybrid text affect sensing system for emotional language analysis. In: Proceedings of the International Workshop on Affective-Aware Virtual Agents and Social Robots, AFFINE 2009, pp. 3:1–3:4. ACM, New York (2009). http://doi.acm.org/10.1145/1655260.1655263

14. Hu, M., Liu, B.: Mining and summarizing customer reviews. In: Proceedings of the Tenth ACM SIGKDD International Conference on Knowledge Discovery and Data Mining, KDD 2004, pp. 168–177. ACM, New York (2004). http://doi.acm.org/10.1145/1014052.1014073

15. Kamps, J., Marx, M., Mokken, R.J., de Rijke, M.: Using wordnet to measure semantic orientations of adjectives. In: Proceedings of the Fourth International Conference on Language Resources and Evaluation, LREC 2004, Lisbon, Portugal, 26–28 May (2004). http://www.lrec-conf.org/proceedings/lrec2004/pdf/734.pdf

16. Kim, S.M., Hovy, E.: Identifying and analyzing judgment opinions. In: Proceedings of the Main Conference on Human Language Technology Conference of the North American Chapter of the Association of Computational Linguistics, HLT-NAACL 2006, pp. 200–207. Association for Computational Linguistics, Stroudsburg (2006). http://dx.doi.org/10.3115/1220835.1220861

17. Liu, B.: Sentiment analysis and opinion mining. Synth. Lect. Hum. Lang. Technol. 5(1), 1–167 (2012). http://dx.doi.org/10.2200/S00416ED1V01Y201204HLT016

18. Martín-Valdivia, M.T., Martínez-Cámara, E., Perea-Ortega, J.M., na López, L.A.U.: Sentiment polarity detection in spanish reviews combining supervised and unsupervised approaches. Expert Syst. Appl. 40(10), 3934–3942 (2013). http://www.sciencedirect.com/science/article/pii/S0957417412013267

19. Martínez-Cámara, E., Martín-Valdivia, M.T., Perea-Ortega, J.M., Ureña-López, L.A.: Opinion classification techniques applied to a spanish corpus. Procesamiento del Lenguaje Natural 47, 163–170 (2011). http://journal.sepln.org/sepln/ojs/ojs/index.php/pln/article/view/958

20. Mihalcea, R., Banea, C., Wiebe, J.: Learning multilingual subjective language via cross-lingual projections. In: Proceedings of the 45th Annual Meeting of the Association of Computational Linguistics, pp. 976–983. Association for Computational Linguistics (2007). http://aclweb.org/anthology/P07-1123

21. Molina-González, M.D., Martínez-Cámara, E., Martín-Valdivia, M.T., Perea-Ortega, J.M.: Semantic orientation for polarity classification in spanish reviews. Expert Syst. Appl. 40(18), 7250–7257 (2013). http://www.sciencedirect.com/science/article/pii/S0957417413004752

22. Molina-González, M.D., Martínez-Cámara, E., Martín-Valdivia, M.T., Ureña-López, L.A.: Cross-domain sentiment analysis using spanish opinionated words. In: Métais, E., Roche, M., Teisseire, M. (eds.) NLDB 2014. LNCS, vol. 8455, pp. 214–219. Springer, Heidelberg (2014)
23. Pang, B., Lee, L.: Opinion mining and sentiment analysis. Found. Trends Inf. Retr. **2**(1–2), 1–135 (2008). http://dx.doi.org/10.1561/1500000011
24. Perez-Rosas, V., Banea, C., Mihalcea, R.: Learning sentiment lexicons in spanish. In: Proceedings of the Eight International Conference on Language Resources and Evaluation (LREC 2012). European Language Resources Association (ELRA), Istanbul, Turkey, May 2012
25. Strapparava, C., Valitutti, A., Stock, O.: The affective weight of lexicon. In: Proceedings of the Fifth International Conference on Language Resources and Evaluation, pp. 423–426 (2006)
26. Taboada, M., Brooke, J., Tofiloski, M., Voll, K., Stede, M.: Lexicon-based methods for sentiment analysis. Comput. Linguist. **37**(2), 267–307 (2011). http://dx.doi.org/10.1162/COLI_a_00049
27. Turney, P.D.: Thumbs up or thumbs down?: Semantic orientation applied to unsupervised classification of reviews. In: Proceedings of the 40th Annual Meeting on Association for Computational Linguistics, ACL 2002, pp. 417–424. ACL, Stroudsburg (2002). http://dx.doi.org/10.3115/1073083.1073153
28. Vossen, P. (ed.): EuroWordNet: A Multilingual Database with Lexical Semantic Networks. Kluwer Academic Publishers, Norwell (1998)
29. Wan, X.: Co-training for cross-lingual sentiment classification. In: Proceedings of the Joint Conference of the 47th Annual Meeting of the ACL and the 4th International Joint Conference on Natural Language Processing of the AFNLP, ACL 2009, vol. 1, pp. 235–243. Association for Computational Linguistics, Stroudsburg (2009). http://dl.acm.org/citation.cfm?id=1687878.1687913
30. Wiebe, J.: Learning subjective adjectives from corpora. In: Proceedings of the Seventeenth National Conference on Artificial Intelligence and Twelfth Conference on Innovative Applications of Artificial Intelligence, pp. 735–740. AAAI Press (2000). http://dl.acm.org/citation.cfm?id=647288.721121
31. Zhang, C., Zeng, D., Li, J., Wang, F.Y., Zuo, W.: Sentiment analysis of chinese documents: From sentence to document level. J. Am. Soc. Inf. Sci. Technol. **60**(12), 2474–2487 (2009). http://dx.doi.org/10.1002/asi.21206

Information Extraction
and Social Media

Tree-Structured Named Entities Extraction from Competing Speech Transcriptions

Davy Weissenbacher[✉] and Christian Raymond

INSA - IRISA, INRIA de Rennes, 20 Avenue des Buttes de Coësmes, Rennes, France
davy.weissenbacher@inria.com, christian.raymond@irisa.fr

Abstract. When real applications are working with automatic speech transcription, the first source of error does not originate from the incoherence in the analysis of the application but from the noise in the automatic transcriptions. This study presents a simple but effective method to generate a new transcription of better quality by combining utterances from competing transcriptions. We have extended a structured Named Entity (NE) recognizer submitted during the ETAPE Challenge. Working on French TV and Radio programs, our system revises the transcriptions provided by making use of the NEs it has detected. Our results suggest that combining the transcribed utterances which optimize the F-measures, rather than minimizing the WER scores, allows the generation of a better transcription for NE extraction. The results show a small but significant improvement of 0.9 % SER against the baseline system on the ROVER transcription. These are the best performances reported to date on this corpus.

Index Terms: Speech transcription · Structured named entities · Multi-pass decoding

When real applications are working with automatic speech transcription, the first error does not originate from the incoherence in the analysis of the application, but from the noise of the automatic transcription outputs. With a rate often close to one in three words incorrect in the transcription, the quality of the preprocessing is low and, as a result, the output analysis of the application is often unexploitable. An explanation for this low performance of speech recognizers can be found in [8]. Little lexical and syntactic information is effectively used to enable the computation of the decoding of the acoustic output. More complex information are reintegrated in a second decoding pass where only the best sequences of words produced during the first pass are considered.

The main contribution of this study is to present a simple but effective method to generate a new transcription of better quality by combining several competing transcriptions. Current Automatic Speech Recognition (ASR) systems rely on various strategies and/or resources to discover the original utterances pronounced. As a consequence, errors made by competing ASRs are different, which make the transcriptions complementary. The Rover method exploits such complementarity to recombine several transcriptions and output a new

© Springer International Publishing Switzerland 2015
C. Biemann et al. (Eds.): NLDB 2015, LNCS 9103, pp. 249–260, 2015.
DOI: 10.1007/978-3-319-19581-0_22

transcription [3]. Previous studies to recombine the transcriptions focus on minimizing the Word Error Rate (WER) measure[1]. We claim that the WER measure is not the measure of importance and should be ignored [4]. The measure that is more important is the measure of the performance of the system on the final application, and that is the one to be optimized.

To test this hypothesis we have run experiments on structured Named Entity (NE) extraction using the corpus released during the recent ETAPE Challenge. This challenge aimed to evaluate the state of the art in NE extraction on automatic speech transcription of TV and radio French programs. We found promising results, with the best performances achieved to date on this corpus.

In Sect. 1, we first describe the task and the corpus of the evaluation campaign ETAPE, and then provide an overview of the system submitted to extract structured NEs and which ranked first during the campaign. Section 2 details our first investigation to use the NEs extracted to recombine the complementary transcriptions. We report the gain observed during our experiments and the perspectives of this work in Sect. 3.

1 The ETAPE Challenge

The goal of the ETAPE challenge in 2012 was to extract named entities (NEs) from automatic transcription output[2]. The ETAPE corpus [5] consists of 13.50 h of radio news broadcast and 28.40 h of TV shows. The corpus was chosen to be difficult to process, with the programs in French language chosen not only from French channels, but also from Moroccan and African radio stations. The programs were selected to include mostly non planned speech and reasonable proportions of conversations with multiple speakers. The data was split into 8.20, 25.50 and 8.20 h for development, training and testing respectively. Five speech recognizers have been applied on the corpus. Their performances on our test data range from 23 % to 35 % WER.

The originality of the ETAPE challenge was in its definition of the NEs [18]. A NE is a rigid designator [9], like a proper name or a company name, and is commonly viewed as a simple object, that is a sequence of words. However, a NE can also be seen as a structured object. According to the definition of the ETAPE challenge, NEs have a tree structure and are both hierarchical and compositional. For instance, type *pers* (person) is split into two subtypes, *Pers.ind* (individual person) and *Pers.coll* (collective person). *Pers* entities are composed like in the individual person *Nicolas Sarkozy* where *Nicolas* is the first name and *Sarkozy* the last name. Figure 1 enumerates the 7 main types and the 32 subtypes of the taxonomy. Figure 2 shows all the components.

Learning trees from data is known to be a difficult task. Since complexity issues rise quickly, learning the full tree in one step is often impossible [13].

---

[1] $WER = \frac{S+D+I}{N}$, where D, I, S stand for the number of deletions, insertions, substitutions of words and N for the total number of words in the reference.

[2] More information about the challenge can be found at www.afcp-parole.org/etape/workshop.html.

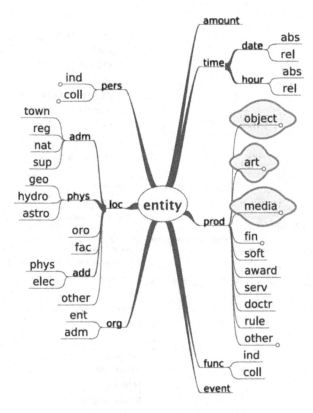

Fig. 1. Named entity hierarchy

Standard approaches to build trees, as grammar and formal based approaches, fail to operate on noisy inputs like automatic speech transcriptions. In contrast, statistical approaches have been proven to be very efficient on both clean and noisy texts. To the best of our knowledge, most statistical NE recognition systems deal with the structure thanks to cascade approaches [1,15]. But the cascade methodology has an important limitation, errors made in the early stages are propagated through the whole process. The propagation of errors particularly problematic when the inputs are very noisy.

The winning system of the ETAPE challenge avoids the cascade approach by building the trees in two steps. In the first step, it extracts the nodes of all possible trees which may be contained in an utterance. The detection of all the nodes is performed independently in order to avoid complexity issues. For detecting the nodes Conditional Random Fields (CRFs) are used. This sequence labeler is currently one of the best statistical frameworks for NE recognition [12]. Each CRF is trained to recognize a unique type of node, resulting in a total of 68 binary CRFs. Segmentation and labelling are performed in the same time using the *BIO* annotation format. Each CRF uses a common set of features: the words themselves, their associated Part-Of-Speech tags and the mentions

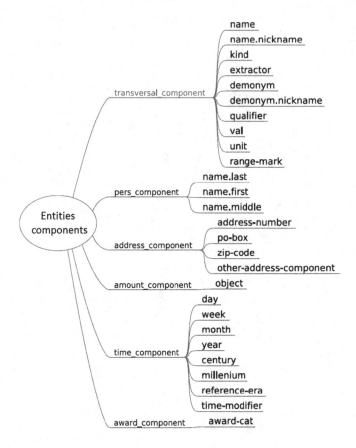

Fig. 2. Entities components

of predefined classes like cities, countries etc. These mentions are extracted by utilizing dictionaries [17].

In the second step the trees are rebuilt from the nodes extracted in the previous step. The simplest and most effective way to rebuild the tree is to choose the nodes of the best analysis for all binary CRFs. To reconstruct a coherent tree one needs to know the subsumption relations between nodes, such as nodes *Pers.ind* always dominating nodes *First.Name*. These relations are learned from the training data. Since nodes are extracted independently by the CRFs, incoherences between their segmentations may occurred. Simple heuristics are employed to recover coherence between erroneous nodes annotations. Despite of its simplicity this algorithm distinctly ranked first among eight participants during the ETAPE challenge with a score of 55.51 % WER on the ROVER transcription [16].

Having presented the NE recognizer, in the next section we turn to its use for revising existing transcriptions with the goal of improving the NE extraction scores on the ETAPE corpus.

2   Automatic Transcription Revision Driven by NE Recognition

In this study, we claim that when revising existing transcriptions, the measure of the final task should be optimized rather than minimizing the WER of the new transcription. To improve the quality of the transcription, we select the transcription which maximizes the F-measure of our NE recognizer from the competing transcriptions of an utterance. Considering the same utterance transcribed by two ASRs, the underlying idea is that if a structured NE is recognized in the first transcription and not in the second, it is more likely that the first transcription is correct. Since the final application makes only use of the NEs, the overall quality of the transcription in terms of WER doesn't have to be perfect as long as the NEs can be discovered by the NE recognizer.

We now explain the algorithm followed to generate a new transcription from transcriptions of competing ASRs. Our algorithm takes as input a set of transcriptions output by several ASRs. We have segmented the utterances of all transcriptions to avoid complexity problems[3]. Considering an utterance U_i in the gold transcription, we call the transcriptions output by all competing ASRs for this utterance the *set of competing utterances for U_i*. Each competing utterance in a given set are aligned with the longest utterance in the set using the SCLite algorithm[4]. Each set of competing utterances is then processed sequentially to find the best utterance for each set of competing utterances. In order to select the best utterances for a given set, we apply the NER approach described in the previous section on all competing utterances of the set. The annotated competing utterances are then passed to a Machine Learning (ML) system. The ML, described in the section below, was trained to recognize the best utterance based on the presence or absence of NEs in the utterances. When all best utterances are selected, they are merged to generate a new transcription of the document. The quality of this new transcription is finally evaluated using the official tools provided during the ETAPE challenge for evaluating the original competing transcriptions.

Our algorithm can be illustrated on two competing transcriptions of the following utterances: *U1*, *U1'* for the first set and *U2*, *U2'* for the second set:

Reference: nous sommes ensemble pour soixante minutes une heure au coeur de tout ce qui fait l'actualité
[we are together for sixty minutes one hour at the heart of everything which make the news].
U1: nous sommes ensemble pour soixante minutes une heure au coeur de l'actualité
[we are together for sixty minutes one hour at the heart of the news]
U1': nous sommes ensemble pour soixante minutes une trop grande tout ce qui fait l'actualité

---

[3] As a first working hypothesis, we have segmented the transcriptions based on the gold standard utterances.

[4] SCLite: www1.icsi.berkeley.edu/Speech/docs/sctk-1.2/sclite.htm.

[we are together for sixty minutes a too big everything which make the news]
Reference: *c'est ce qu'a dit le ministre Bruno Le Maire ministre de l'agriculture*
[this is what the minister Bruno Le Maire minister of the agriculture said]
U2: c'est ce qu'a dit la ministre de l'agriculture
[this is what the minister of the agriculture said]
U2': c'est ce qu'a dit le ministre Bruno Lemaire ministre de l'agriculture
[this is what the minister Bruno Lemaire minister of the agriculture said].

The algorithm has to select between *U1* and *U1'* in the first set, and between *U2* and *U2'* in the second. Our algorithm retains utterances with the maximum number of correct NEs. In the first set it is straightforward to select *U1*. It is possible to extract a NE from *U1*, *<Amount> <Val> une </Val> <Unit> heure </Unit> </Amount>*, but not from *U1'* by applying our NE Recognizer on each utterance. Therefore, *U1* is selected as best utterance for the first set. The choice for the second set is less obvious since *U2* and *U2'* both contain NEs. The algorithm has to arbitrate based on the quality of each NE. The problem has been formulated as a classification problem to optimize the decision. An ideal ML framework should prefer *U2'* against *U2*, as the NE *<Func.ind> ministre <Pers.ind> <First.name> Bruno </First.name> <Last.name> Lemaire </Last.name> </Pers.ind> </Func.ind>* is longer and perfectly valid. Once all utterances have been selected, a new transcription composed of *U1* and *U2'* is output. This new transcription is ready for being evaluated on the structured NE extraction task. Since the presence of the NEs is optimized in all utterances of the new transcription, rather than the WER, better performance is expected for the NER task when using this new transcription.

The selection of the best utterance is a ML problem which can be expressed in different ways. We describe here the ML frameworks studied in this work.

Transcriptions Classification. In this framework all competing utterances are submitted to a multi-class classifier. Features available allow the classifier to describe and compare the utterances in order to choose one among them. The features employed here are explained in the Table 1. The ML framework which gave the best results on the training corpus was a Bayesian Network. The structure and the conditional probabilities of the Bayesian Network were learned automatically.

Transcriptions Regression. Another framework is to learn directly the F-measure of an utterance using a regression classifier. The selection of the best utterance is done afterwards by picking up the utterance exhibiting the highest F-measure estimation. A bagging-Regression tree for regression obtained the highest performances on our training data and has been chosen for our test. Features used for the regression were similar to those for the *Transcription Classifier*.

Phrases Classification. The features of the previous ML systems provide a global description of the utterances, but this level may appear to be too broad for our purpose. Not only does the system have to find the utterance containing the maximum number of NEs, but it also has to ensure that the NEs obtained have appropriate qualities. For that reason, we redesigned our system to be able to describe and evaluate independently all phrases annotated as NE in an utterance. The utterance containing the highest number of selected NEs is kept as best utterance.

Let us consider our previous example. When a phrase is annotated as NE by at least one CRF, the phrase and all corresponding phrases in the competing utterances become subject to decision. In *U1*, the phrases *soixante minutes* and *une heure* have been found to be NEs. According to the SCLite alignment, the corresponding phrases in *U1'* are *soixante minutes* and *une trop*. Only *soixante minutes* has been annotated as possible NE in *U1'*. The annotation nodes of *soixante minutes* are the same for *U1* and *U1'*. The choice only relies on the decision taken by the algorithm for the quality of the annotations of *une heure* and *une trop*. The algorithm based on a ML inference gives better credit to the phrase *une heure*, which qualifies the transcription *U1* with two NEs selected against one in *U1'*.

The main component in our algorithm is the ML model used to gauge the phrases. We opt for a multi-class classification to select the best phrases among the competing phrases of each set of utterances. We did not change the ML framework and continue to train a Bayesian Network in the same way as in the previous experiences. The features of Table 1 were adapted for describing phrases and completed with the features of the Table 2.

Oracle and Baseline Systems. To reveal the maximum improvement possible with our approach, we have computed the performance of an oracle. For all NEs discovered in an utterance by our NER, the oracle is informed with the true value of the NEs. Therefore, it always outputs the best possible utterances for each set of competing utterances given the NE resolution. As a baseline system we have chosen the ROVER transcription. This baseline is a strong baseline since the winning NER achieved its best performances on the ROVER transcription during the challenge.

3 Results and Discussion

In Table 3, we report the NE recognition scores of our system for each recombined transcription given by the ML framework tested. Standard measures of Precision and Recall are completed by the Slot Error Rate (SER) [10], a measure similar to WER which also considers errors made for the segmentation and the labelling of NEs. Both *Transcriptions* and *Phrases* classifiers output a transcription of better quality than the baseline ROVER transcription for our final task. The improvement of 0.9 % is shown to be statistically significant with a one-tailed t-test with a degree of liberty = 28 and $\alpha = 0.1$.

These results demonstrate the interest of maximizing the F-measure over minimizing the WER measure when recombining competing transcriptions. The improvement of the recombined transcriptions in terms of WER is not important. The ROVER transcription exhibits a WER of 39.0 % whereas the recombined transcriptions produced by the *Transcription classification* shows a WER of 39.2 %. This finding corroborates the findings of the ETAPE challenge. NE

Table 1. Features describing the set of the 6 Transcriptions in competition.

Feature	Description
CRFs score (for transcription i)	A global score computed by summing all binaries CRFs' probabilities of the words
Max Nodes	The name of the transcription containing the highest number of nodes
Impossible bigram (for transcription i)	The number of sequence of two words never co-occurring in the training corpus
Length (for transcription i)	Total number of words in the transcriptions
Min/Max CRFs scores	The CRFs score of the node which is found to be the min/max score in the competing transcriptions
Mean Node scores	The mean of CRFs scores in the competing transcriptions

Table 2. Features describing phrases in competing transcriptions.

Feature	Description
Max Depth	The size of the longest branch of the structured NE covering the phrase
Max score node	The node which has the highest CRFs score
Existing phrase	1 if an occurrence of the phrase has been found in the training data
Phrase coherence	1 if the phrase is covered by a node known to be a subtype in Fig. 1

Recognition performances are not necessarily better on transcriptions with lower WER. When comparing the scores of our system on two transcriptions of the ETAPE data, we found a score of 63.4 % SER on the first transcription with 24 % WER, and a score of 62.53 % SER when the second transcription's WER is of 25 % [16]. That is, the SER diminishes by 0.83 % whereas the WER increases of 1 %.

Analysis of the *Transcription classifier* model informs that this classifier tends to select the ROVER classification of an utterance by default, except when another transcription of the utterance is found with a higher CRF score going along with a higher number of nodes in the utterance. This confirms our intuition:

Table 3. Performances of NE recognition on recombined transcriptions, in term of Slot Error Rate.

	SER	Precision	Recall
Baseline rover	.563	.734	.449
Transcriptions classification	**.554**	.728	.461
Transcriptions regression	.640	.586	**.463**
Phrases classification	**.554**	**.738**	.454
Oracle	.509	.751	.499

the detection of the NEs is possible only when the transcription reaches a certain threshold of quality and this, in turn, reveals the best transcription among the candidate transcriptions.

A surprising result is the counter-performance of the *Regression* system. This system takes more risks by often picking up utterances that are different from the most reliable ones (*i.e.* ROVER or s23). Although rewarded by a higher recall, it is punished by a drop of precision. The opposite phenomenon is noticed for the *Phrase* classification system. The description of phrases allows the system to discriminate the expected ones and increase its precision with a slight drop of its recall.

4   Related Work

A significant number of errors of ASR systems are caused by the Out-Of-Vocabulary words (OOV) since ASR systems rely on a finite lexicon to interpret phonetic inputs [19]. Due to the nature of most of the NEs, that is the open class of Proper Nouns, a large proportion of OOV are unknown NEs. Therefore, a considerable amount of literature has been published on OOV-NEs detection and revision. To date, two complementary approaches have been explored.

The method proposed in this paper is close to the first approach which extends the search space by exploiting multiple sources of information. The simple method is to use multiple ASR system transcriptions as in [3], which results in an important improvement of the WER. More sophisticated methods, with a cost of higher computation complexity, introduce NEs hypotheses directly in the decoding model. In their seminal article, [4] encode the output of a NE recognizer into the loss function of a Minimum Bayes-Risk Classifier to reorder a N-Best list of transcriptions. In a study which worked on a corpus similar to our own [6], [2] make use of the release time of the news to enrich the list of NEs available to the system by adding an external list of NEs known to occur in the documents published during this period of time.

The second type of approach targets specifically the strange grammatical constructions caused by the presence of OOV words with the aim of identifying the underlying NE(s) [14]. At the last resort, when no NE can be found, a phonetic transcription of the OOV is generally suggested. In 2006, [7] investigated the interest of training a classifier to recognize distortions caused by the unknown NEs. More recently, by noticing that not only do the OOVs deteriorate the transcription at their position in the utterance, but also the immediate context where they appear, [11] rebuilt the best parse of the utterance from a word confusion network with the distinct mentions of OOVs. These latter methods are not in contradiction to our approach, but complement it. While we have empirically established that if the expected NEs occurred in one competing translations, our algorithm will more likely find it, its strong limitation lies in the cases when NEs are absent in the translations. In such cases, nothing can be done to recover and the system relies on the ROVER transcriptions. A module implementing the latter algorithm may be able to detect a NE position and pass to our own system an anonymous NE (or an attempted assertion of the unknown NE) in order to help our system to output the best transcription.

5   Conclusion

Our findings emphasize the interest of optimizing the measure of the final task to improve the quality of the transcription when complementary transcriptions are available. Building on our achievements during the ETAPE Challenge, we have used the structured NEs detected by our NE recognizer to drive the revision of the transcriptions provided to the system. Our results suggest that selecting the competing transcription of the utterances by optimizing the F-measure leads to a better global transcription for NE extraction compared to selection based on a lower WER.

Taking into account the difficulty of the corpus, the results obtained are mainly positive, with a small but significant improvement of 0.9 % SER on the recombined transcription against the ROVER baseline. There is, however, still a lot of room for improvement. A promising approach is to recombine the transcriptions by merging all the best transcriptions of phrases, and this even if two distinct phrases of the same utterance belong to different transcriptions. This method will be somehow similar to Word Confusion Network based methods which already have been demonstrated to provide better recombined transcriptions compared to a simple N-Best list recombination [20]. To obviate the unrecoverable limitation when expected NEs do not occur in any transcriptions, as further work, we are considering to integrate a procedure to detect unreliable sequences of words caused by OOV-NEs. This will enable a dedicated algorithm to track down the hidden NEs within external resources before the recombining stage of the transcriptions.

Acknowledgments. We thank Dr. Abeed Sarker and Dr. Graciela Gonzalez for their helpful comments and remarks.

References

1. Dinarelli, M., Rosset, S.: Models cascade for tree-structured named entity detection. In: Proceedings of International Joint Conference on Natural Language Processing (IJCNLP), pp. 1269–1278 (2011)
2. Favre, B., Béchet, F., Nocéra, P.: Robust named entity extraction from large spoken archives. In: Proceedings of Human Language Technology Conference and Conference on Empirical Methods in Natural Language Processing (HLT/EMNLP), pp. 491–498 (2005)
3. Fiscus, J.: A post-processing system to yield reduced word error rates: recognizer output voting error reduction (rover). In: Proceedings IEEE Automatic Speech Recognition and Understanding Workshop, pp. 347–352 (1997)
4. Goel, V., Byrne, W.: Minimum bayes-risk automatic speech recognition. Comput. Speech Lang. **14**(2), 115–135 (2000)
5. Gravier, G., Adda, G., Paulson, N., Carré, M., Giraudel, A., Galibert, O.: The ETAPE corpus for the evaluation of speech-based TV content processing in the french language. In: International Conference on Language Resources, Evaluation and Corpora (2012)
6. Gravier, G., Bonastre, J., Geoffrois, E., Galliano, S., McTait, K., Choukri, K.: Ester, une campagne d'évaluation des systèmes d'indexation automatique d'émissions radiophoniques en franais. In: Proceedings Journées d'Etude sur la Parole (JEP) (2004)
7. Hakkani-Tr, D., Béchet, F., Riccardi, G., Tur, G.: Beyond ASR 1-best: using word confusion networks in spoken language understanding. Comput. Speech Lang. **20**, 495–514 (2006)
8. Jurafsky, D., Martin, J.: Speech and Language Processing. Prentice Hall, Englewood Cliffs (2008)
9. Kripke, S.: Naming and necessity. In: Davidson, D., Harman, G. (eds.) Semantics of Natural Language. Harvard University Press, Cambridge (1972)
10. Makhoul, J., Kubala, F., Schwartz, R., Weischedel, R.: Performance measures for information extraction. In: Proceedings of DARPA Broadcast News Workshop, pp. 249–252 (1999)
11. Marin, A., Kwiatkowski, T., Ostendorf, M., Zettlemoyer, L.: Using syntactic and confusion network structure for out-of-vocabulary word detection. In: Proceedings IEEE Spoken Language Technology Workshop (SLT), pp. 159–164 (2012)
12. McCallum, A., Li, W.: Early results for named entity recognition with conditional random fields, feature induction and web-enhanced lexicons. In: Proceedings of CoNLL-2013, pp. 188–191 (2013)
13. Nowozin, S., Lampert, C.: Structured learning and prediction in computer vision. Found. Trends Comput. Graph. Vis. **6**, 185–365 (2010)
14. Palmer, D., Ostendorf, M.: Improving information extraction by modeling errors in speech recognizer output. In: Proceedings of the First International Conference on Human Language Technology Research (2001)
15. Punyakanok, V., Roth, D., Tau Yih, W., Zimak, D.: Learning and inference over constrained output. In: Proceedings of International Joint Conferences on Artificial Intelligence (2005)
16. Raymond, C.: Robust tree-structured named entities recognition from speech. In: Proceedings of International Conference on Acoustic Speech and Signal Processing, ICASSP 2013 (2013)

17. Raymond, C., Fayolle, J.: Reconnaissance robuste d'entités nommées sur de la parole transcrite automatiquement. In: Proceedings of Traitement Automatique des Langues Naturelles (2010)
18. Rosset, S., Grouin, C., Zweigenbaum, P.: Entités nommées structurées: guide d'annotation quaero. Technical report, LIMSI-Centre national de la recherche scientifique (2011)
19. Subramaniam, L., Roy, S., Faruquie, T., Negi, S.: A survey of types of text noise and techniques to handle noisy text. In: Proceedings of The Third Workshop on Analytics for Noisy Unstructured Text Data, pp. 115–122 (2009)
20. Tur, G., Deoras, A., Hakkani-Tr, D.: Semantic parsing using word confusion networks with conditional random fields. In: Proceedings of Interspeech 2013, pp. 2579–2583 (2013)

Interactive Learning with TREE: Teachable Relation and Event Extraction System

Maya Tydykov(✉), Mingzhi Zeng, Anatole Gershman, and Robert Frederking

Carnegie Mellon University, Pittsburgh, PA 15213, USA
{mtydykov,mingzhiz,anatoleg,ref}@cs.cmu.edu

Abstract. Information extraction, and specifically event and relation extraction from text, is an important problem in the age of big data. Current solutions to these problems require large amounts of training data or extensive feature engineering to find domain-specific events. We introduce a novel Interactive Learning approach that greatly reduces the number of training examples needed and requires no feature engineering. Our method achieves event detection precision in the 80 s and 90 s with only 1 h of human supervision.

1 Introduction

There has recently been considerable progress in the field of event and relation detection and extraction to address the need of acquiring semantic frames from text. This includes the lexical semantic domain (e.g., FrameNet [1]) and the information extraction domain (e.g., MUC [2]). Acquiring frames turns out to be a difficult and unsolved task, and most systems to date have either required manual methods which are expensive and often require expert knowledge [3,4] or are fully automatic but have not been able to achieve high levels of precision [5]. We propose a hybrid system that leverages both automatic techniques and human intervention in order to decrease the amount of human effort needed to introduce a new frame without sacrificing precision.

The goal of our work is to enable an analyst without special linguistic training to teach new events or relations that the system can later extract with high precision. The extracted information is then used to populate a back-end knowledge base. We present a three-stage Interactive Learning approach to teach the system a new event or relation. The pipeline for this process is shown in Fig. 1. In the first step, the teacher introduces the event or relation and the roles she is interested in having extracted. She then annotates a small batch of simple sentences to teach the system this event or relation, using our Assisted Active Teaching module. The next step is active learning, during which the system uses several heuristics to find sentences with a potentially high learning impact (e.g., potentially confusing). The final step is validation, where the system attempts to extract the event in randomly-chosen sentences from a pre-selected corpus and asks the user to correct these. During all three steps, the teacher and the system work together, using the Ontology Builder module, to update a concept ontology which is then used in the extraction process.

© Springer International Publishing Switzerland 2015
C. Biemann et al. (Eds.): NLDB 2015, LNCS 9103, pp. 261–274, 2015.
DOI: 10.1007/978-3-319-19581-0_23

Our main contribution is a simpler, less frustrating way of teaching a system to extract events and relations via an Interactive Learning process. Our process works by progressing from simple to more complex examples, much in the same way that humans are taught new things from childhood to adulthood.

2   Related Work

Common approaches to the problem of event and relation extraction include pattern-matching [6–9], bootstrapping [10], or a combination of the two [11, 12]. A technique common to many of these approaches is to break the event extraction process into two parts, with the first part devoted to detecting event mentions using indicators in the text, and the second part used to extract roles, or arguments [4]. Our system also uses this two-step process for event and relation detection and extraction.

Active learning has previously been shown to be a helpful aid in training information extraction systems. In [13], a user-centric active learning framework called DUALIST was presented, with the goal being to allow for semi-supervised active learning. This framework was applied to text classification tasks with state-of-the-art results. Reference [14] presented an active learning framework for named entity recognition that focused on two techniques - persistent learning and corrective feedback. These two systems had similar goals of reducing user effort while achieving and maintaining high accuracy, and they demonstrate the high potential of applying active learning to various natural language processing tasks. In [15], active learning was applied to train a system to extract noun phrases belonging to specific semantic classes. In [16], it was used to train an event detection system, but focused on finding more positive examples for the user to label rather than negative or potentially confusing ones. Reference [17] used measures of uncertainty and representativeness as their active learning criterion, collecting entity pairs from a corpus of sentences and classifying relations between the pairs, with the training data provided by Amazon Mechanical Turk. Unlike these other works, our system performs both event detection and role extraction for events and relations - a highly difficult task. Our system focuses on the events and relations that the analyst is interested in. Most importantly, in our system, the teacher is an active participant rather than a passive oracle, with the system assisting the teacher in producing good examples.

3   Extraction

Our extraction procedure uses techniques similar to ones used in other systems such as [4], with the event and relation detection process being the first step, and the role extraction process following after that. First, the system annotates a document using the Stanford NLP toolkit to get part of speech tags, dependency parse trees, and co-reference resolution between entities.

The system then filters each sentence for indicators relevant to each type of event or relation. Indicators, often also called triggers in the literature, are groups of items that must be present in the sentence in order for the event

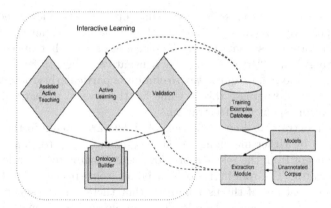

Fig. 1. Current TREE teaching pipeline.

detection process to be initiated, where each item can be the surface form of a word or phrase in the sentence, a named entity type as identified by the named entity recognizer, or a concept from a user-defined set. We build our concept sets interactively using our Concept Builder. Constraints on indicators are event-specific and are added to a teaching database during the teaching process. For example, in an Athlete-Sport relation, a good indicator that the relation may be present in a sentence is if the sentence contains a person and a sport. Thus, in the sentence *"Mary is a hockey player"*, the presence of *"Mary"* and *"hockey"* would satisfy this constraint.

Once an indicator constraint has been satisfied, the system uses a Maximum Entropy classifier and the following features to determine whether or not to trigger an event frame:

1. Features extracted between indicator components:
   (a) Largest word distance
   (b) Largest dependency distance
   (c) Pairwise dependency relations between "types" of indicator components, where types are concepts used to identify the components
2. Total number of indicator components
3. Features extracted with respect to each indicator component:
   (a) Component's number in the sentence paired with POS, NE and text features
   (b) Component's identifying concept paired with POS, NE and text features
   (c) Component's number in the sentence and identifying concept with POS, NE and lexical form features
   (d) POS, NE, and lexical form features are the POS, NE, and text, respectively, of the component, the word to its left, and the word to its right.

If the event classifier labels the sentence as positive, then the sentence becomes a candidate for extraction of event details.

Before proceeding to the role extraction step, the system also consults a negative event trigger classifier. This classifier extracts the same set of features as the first event trigger classifier, but will stop the extraction process if the outcome is positive. For example, in the sentence "John had a heart attack", the teacher can inform the system during the teaching process that the combination of words "heart" and "attack" serve as a negative indicator for an "Attack" event. This step allows for a quick way to eliminate particularly troublesome false positives that may otherwise take a while for the system to learn, minimizing the teacher's time spent and frustration.

In the role extraction step, potential role fillers are filtered in the same way as indicators - by either matching the surface form of a word, a named entity type, or a concept from the relevant ontology sets. Each potential role may have one role indicator. The role indicator can be an item specially selected by the teacher during teaching if the user believes that there is some specific word or concept that can help identify the role. For example, in the sentence "Karen won a gold medal", the word "medal" is a good indicator for the role "Placement", which should be filled by the word "gold". To the best of our knowledge, the use of role indicators is novel in our work.

If no such specific role indicator exists or if it is not found in the sentence, the system uses the closest component in the event indicator by distance in the dependency parse tree. If the role and event indicator items are not connected in the dependency parse tree, the system chooses the closest event indicator component by word distance. We refer to the chosen component as the optimal indicator component.

Once the role's optimal indicator component has been identified, the system extracts the following set of features with respect to the pair of role and optimal indicator component:

1. If any indicator component is the same entity as the role filler
2. If the optimal indicator component is a role indicator
3. Features specific to role indicators:
   (a) Dependency relationships between role indicator and the types of indicator components in the sentence; "type" is the concept used to identify the indicator component
   (b) Dependency distance to each type of indicator component
4. Features between optimal indicator component and role filler:
   (a) Dependency relationship
   (b) Dependency distance
   (c) Word distance
   (d) Number of organizations, people, dates, and locations between them
5. If the optimal indicator component is before or after the role filler
6. Dependency relationship between the optimal indicator component and the closest alternative role filler
7. Features extracted for both the optimal indicator component and role filler:
   (a) POS and lexical features, where POS and lexical features are the POS and lexical, respectively, of the item, the word to its left, and the word to its right
   (b) Type of concept used to identify the item.

The system then uses a Maximum Entropy classifier to determine whether or not the pair of optimal indicator component and potential role filler is a good one. Once a set of roles has been identified for the frame, the system makes final role assignments by using the confidence of the role classifier in the case of multiple entities being assigned the same role or multiple roles being assigned to the same entity.

We are currently working under the simplifying assumption of one event frame per event type per sentence, so if a sentence contains multiple potential event indicators for the same event, it needs to choose one of those (along with the roles extracted with respect to that event indicator). Thus, in the final step of the extraction process, the system groups together each potential event indicator in the sentence with the final set of role fillers and uses a Maximum Entropy classifier to get a final classification for the entire event frame. It uses classifier confidences from previous steps as features, specifically:

1. Role extractor classifier confidence.
2. Event detection classifier confidence.

The system will then select the event frame that received the highest final confidence score from the classifier. In the event of a tie, the system uses several rules to rank candidate event frames and chooses the frame with the highest rank.

After performing extraction, the system populates the knowledge base with the events and relations it detects along with the extracted roles. It performs simple string matching to merge the entities by name in the knowledge base.

4 Interactive Learning Process

4.1 Assisted Active Teaching

The teacher starts the process by introducing an event or relation to the system along with the roles she is interested in having extracted. The teacher annotates 10 simple sentences (i.e., short as compared to longer, more convoluted sentences in news articles) that she can either find in relevant documents or come up with herself. During the annotation process, the teacher marks both role fillers and indicators in the sentence. We hypothesize that the teacher can mark indicators as well as role fillers adequately using just common sense (i.e., without special linguistic training). When marking either an indicator or a role filler, the teacher is presented with three kinds of options in order to provide the system with a rule about how to find this indicator or role filler during extraction. These options are the lexical form of the word, the named entity type of the word as recognized by the Stanford NER pipeline with some simple filtering in place to rule out relative dates, or a concept. The teacher selects one of these options, and a new rule to find an indicator or role filler is then added to the teaching database.

4.2 Concept Builder: Adding Concepts to the Ontology

When the teacher selects a concept as the rule to use in identifying a role filler or an indicator, the system works with the teacher to interactively define

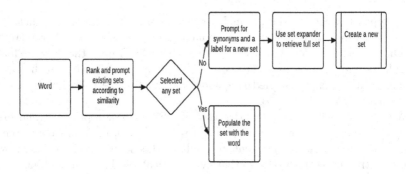

Fig. 2. Current Ontology pipeline.

concepts that best fit the user's intention, while simultaneously making each user-added example worth several examples. An important component of our Concept Builder is SEAL, a set expansion tool that automatically scours the web for lists of items and ranks these items to form similar sets [18]. We also created test sets to tune SEAL to reach better accuracy on our corpus. The current work flow of the ontology part of the system is shown in Fig. 2.

When the user wishes to add an indicator or a role label during any of the teaching phases, she is prompted to give the system more information about what makes the particular selection important. One of the choices is a category type. When the user selects the category option, the system can either create a new set or merge the selected entity with an existing set. A new set is created only when the entity is not in any existing sets and the user chooses not to add it to any existing set.

The system first tries to rank existing sets according to the WuPalmer [19] similarity measure based on the depths of the two synsets and the depth of their LCS in the WordNet taxonomies. Then we prompt the user with the ranked sets. If the user chooses any set to merge with, then we further expand the set. The merging process will be discussed later. If the user chooses none of the given sets, then the system prompts the user for two more, similar seed entities to add to the original selected entity. The resulting three seed entities are then sent to SEAL, which expands the category and returns a list of potential additions to the new category. The top K items with the highest belief in the list are then shown to the user, who selects which of these should actually be added to the new set.

The merging process between a user-selected entity and a set in ontology is as follows. When an entity is added to an existing set, the system attempts to further expand the existing set based on the addition of the new entity. It uses several iterations to choose a subset of entities that are already in the set, along with the new entity, as seed entities for further SEAL expansion and then adds the top K entities from the final list of candidates based on a thresholding of SEAL's belief and frequency values for those candidates. The final list is then shown to the user so that she can select which entities will be added to the category being expanded. The merge work flow is shown in Fig. 3.

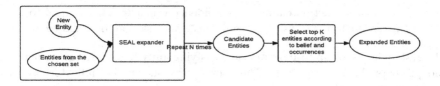

Fig. 3. Ontology Merge pipeline.

Since the ontologies are built on the fly (as the user teaches the system), there can be points during which there are many distinct sets which are actually very similar to one another. The current implementation uses hierarchical clustering to re-organize the sets in the ontology. The dissimilarity of two entities is defined as the inverse WuPalmer similarity between them. The distance of two sets is defined as the average linkage between the two sets. Initially we set all the existing sets to be in one cluster and then perform hierarchical clustering on these sets. In each step we merge the two clusters with the shortest average linkage and stop when the shortest average linkage between any two sets are above a threshold.

4.3   Active Learning

Once the system has at least 10 examples of the event or relation added via the Assisted Active Teaching process, it can perform active learning for the new event or relation. The goal of the active learning stage is to help the teacher by finding more examples that will be particularly helpful for the system to learn. Finding potential negative examples is one way to achieve this goal. We use several active learning heuristics, several of which are novel, and several of which are commonly used in active learning tasks.

1. Novel heuristics are:
   (a) The system looks through its database of old examples that were taught for other events or relations and tries to extract the new event or relation from those. If it succeeds, it presents up to 5 such sentences to the user for correction. The reasoning behind this heuristic is that while possible, it is unlikely for old teaching sentences containing other events or relations to also contain this new event. Thus, these are good candidates for potential false positives.
   (b) The system looks through a corpus of documents for previously unseen sentences where it is able to extract the event, targeting likely confusing sentences. Conditions to be satisfied are:
      i. There are multiple potential role fillers for a given role.
      ii. There are multiple potential roles for one entity.
2. Standard, confidence-based heuristics are:
   (a) When looking through previously unseen sentences:
      i. Event detection classifier's confidence was less than or equal to .6.

ii. Role extraction classifier's confidence was less than or equal to .6.

iii. Event frame classifier's confidence was less than or equal to .6.

The system looks for up to 10 sentences satisfying any one of the above heuristics and presents them to the user. The user is first shown the indicator used to detect the event and can mark whether or not it is correct. If correct, the system proceeds to ask the user about each role it extracted. The user can mark each role as correct or incorrect. The user can then add any missing roles for the event frame. If the user marks an indicator as incorrect, this adds a negative example for the event detection classifier, the role extraction classifier and the event frame classifier. If the user marks the indicators as correct but some of the roles as incorrect, this adds negative examples for the role extraction classifier. Otherwise, positive examples are added for the classifiers. Finally, the user can add any frames that the system missed in the sentence. The confidence threshold of 0.6 was manually set based on pilot tests, but other confidence thresholds or methods may also be appropriate.

4.4 Validation

The final mode in the 3-step process is validation. In this mode, the system randomly selects 10 sentences from a provided corpus, performs extraction on these sentences, and presents the results to the user for correction. Once the system has finished selecting sentences, the user can correct the system in the same way as described for active learning. The goal of this mode is to pick representative sentences from the corpus for validating the quality of the model trained so far.

5   Experiments

We taught the system 6 events and relations based on a corpus of approximately 75,000 news articles about the 2014 Winter Olympics[1]. A definition of the frames is shown in Table 1.

We went through 3 cycles of the 3-step process for each event and relation. One cycle consisted of, for each event/relation, teaching the system 10 basic, previously prepared sentences, performing active learning and, finally, performing validation. We used pre-filtered corpora for active learning and validation so that the event or relation in question was more likely to be found. Although one could easily change the order of teaching modes, resulting in a different teaching configuration, we used this cycle because it allowed each event to make use of a relatively large number of sentences taught previously for the other events.

For our baseline, we trained Conditional Random Field (CRF) models using the MALLET toolkit to detect each role for each event and relation, where the roles were identical to the ones defined above. A CRF is a statistical modeling

---

[1] The corpus is a collection of articles from mainstream English-language press provided by a news aggregator who wished to remain anonymous.

Table 1. Definitions of the event and relation frames taught to TREE.

Event/Relation name	Roles
Athlete-country	Athlete, Country
Athlete-sport	Athlete, Sport
Defeat	Winner, Loser, Date, Location, Sport
Withdrawing from competition	Person, Location, Date, Sport
Placing in competition	Person, Location, Date, Sport, Placement
Injury	Person, Location, Date, Body part injured

Table 2. ER = Event Recall; EP = Event Precision; F1 = F1 Score

Event-relation name	ER TREE	ER CRF	EP TREE	EP CRF	F1 TREE	F1 CRF
Athlete-country	0.54	1.0	**0.90**	0.66	0.68	0.79
Athlete-sport	0.77	0.98	**0.90**	0.64	0.83	0.78
Defeat	0.44	0.49	**0.91**	0.71	0.60	0.58
Withdrawing from competition	0.52	0.80	**1.0**	0.71	0.68	0.75
Placing in competition	0.78	0.98	**0.97**	0.68	0.86	0.80
Injury	0.64	0.88	**0.94**	0.72	0.76	0.79

method that takes context into account when making predictions. CRFs are one of the methods commonly used for information extraction [20–22]. We trained the CRF classifier on the same training data produced by the user's interaction with the system and used the following features for training the CRF model:

1. Word lemma
2. POS of the word
3. Named entity type of the word.

For each event and relation, we annotated 50 positive and 50 negative sentences that were not used in training the tested event. For each of these test sets, we then ran the extraction process on each sentence individually, without performing co-reference. TREE was only scored on its results with respect to the event being tested in each test set. Preliminary results for event detection are shown in Tables 2 and 3, with highest precision scores for each test shown in bold.

Scores were calculated only for events and roles of interest for each test set. Event recall was defined as: $R_e = \frac{S_c}{S_a}$, where S_c is the total number of sentences where the system correctly extracted the event, regardless of the correctness of role assignments for the event, and S_a is the total number of sentences where the event was annotated in the test set. Event precision was defined as: $P_e = \frac{S_c}{S_t}$, where S_t was the total number of sentences where the system had extracted the event. Role recall was defined as: $R_r = \frac{R_c}{R_a}$, where R_c was the total number of role fillers that the system got correct in the test set, and R_a was the total

Table 3. RR = Role Recall, RP = Role Precision, F1 = F1 Score

Event-relation name	RR TREE	RR CRF	RP TREE	RP CRF	F1 TREE	F1 CRF
Athlete-country	0.42	0.70	**0.83**	0.51	0.56	0.59
Athlete-sport	0.45	0.60	**0.70**	0.52	0.55	0.56
Defeat	0.21	0.14	**0.52**	0.51	0.30	0.22
Withdrawing from competition	0.42	0.57	**0.82**	0.55	0.55	0.56
Placing in competition	0.47	0.60	**0.72**	0.61	0.57	0.61
Injury	0.48	0.48	**0.79**	0.56	0.59	0.52

number of role fillers annotated in the test set, where a role filler is defined as a span of text paired with a role for the event type. Role precision was defined as: $P_r = \frac{R_c}{R_t}$, where R_t was the total number of role fillers that the system extracted in the sentence for the event frame. The precision trends of both TREE and the CRF are compared in Fig. 4(a)–(f), plotted over the course of each teaching round. The trendlines also include a soft role precision metric which is defined in the same way as the role precision metric except that any overlap in the text between what the system extracted and the gold standard is considered valid.

The results from our experiments show that the TREE system can achieve high precision for both event and relation detection and role extraction for most events and relations. Role precision for the Defeat relation is lower than the others, most likely because this is the only relation that, as defined, requires two entities of the same kind to fill two distinct roles (Winner and Loser), which presents both a challenge for the system and an opportunity to explore different teaching methods and configurations. Our system outperforms the CRF classifier in all precision metrics when both are trained on the same, small number of examples. The CRF beats TREE in some recall metrics, but our goal is to maintain high precision. It is also important to note that the CRF had the benefit of training on sentences specifically selected by our Interactive Learning process rather than from randomly selected sentences in a large training corpus. While it is possible that the CRF would outperform our system in precision if trained on a significantly larger dataset, our goal is precisely to avoid the use of such a large amount of data, aiming instead at extracting quality information based on a minimal amount of data.

We were able to achieve these results having spent approximately 5 h teaching the system, which presents a significant advantage over the usual requirement of many person-hours needed to label thousands of examples. Figure 5a shows how the amount of time spent varies with each teaching round (averaged for all events) and Fig. 5b how much time was spent in total on each round for all 6 events and relations.

Initially, TREE spends much of its time in the Ontology Builder module, learning how to expand concepts the teacher teaches it. Once it has acquired a sufficient knowledge of these concepts, teaching time decreases. Time spent

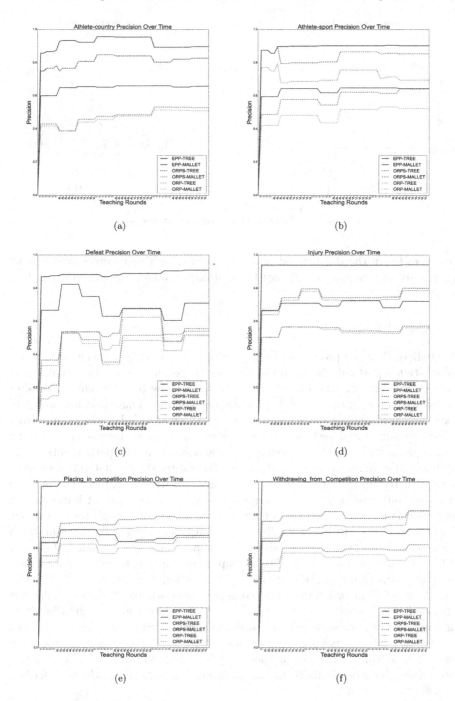

Fig. 4. Plots of precision over time. Time is measured via teaching rounds. T = Assisted Active Teaching; AL = Active Learning; VL = Validation; EPP = Event Presence Precision; ORPS = Overall Role Precision Soft; ORP = Overall Role Precision.

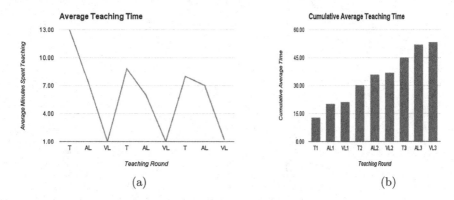

Fig. 5. Graphs of teaching time spent per round.

increases (as in the third round of active learning) if TREE comes across many previously unseen words and uses the Ontology Builder module to expand its concept sets.

6   Conclusion

We believe that the problems of information extraction and, specifically, frame acquisition - particularly when the user wants perform quick data exploration on several events or relations - will not be solvable by fully automated systems or by systems requiring extensive feature engineering. Thus, it is important to explore hybrid methods which can leverage human knowledge while minimizing effort via automated techniques. Our system presents an Interactive Learning technique in which both the system and the user are active participants in the system's learning process. A preliminary evaluation shows that this technique results in reasonable precision. In the future, we wish to explore the optimal ways to configure such hybrid systems as well as what kind of improvement in performance and reduction in effort can be achieved through these systems. Furthermore, we can explore different teaching strategies (i.e., what makes a good or bad teacher). We can also perform more evaluations aimed at testing different parts of the system. For example, we can try to incorporate other, standard corpora used in the Information Extraction domain such as TAC-KBP, MUC, or ACE in order to better compare our work to other work. We can also evaluate the active learning component of our system by comparing it to another system or classifier trained on data via other criteria, e.g. randomly from a relevant corpus. Another direction for future development is to introduce a probabilistic framework into our knowledge base and also into our entity merging procedure, perhaps similarly to the confidence estimation methods described in [23].

Acknowledgments. We would like to thank Jacob Joseph and Eduard Hovy for their valuable advice.

References

1. Baker, C.F., Fillmore, C.J., Lowe, J.B.: The berkeley framenet project. In: Proceedings of the COLING-ACL, pp. 86–90 (1998)
2. Grishman, R., Sundheim, B.: Message understanding conference-6: a brief history. In: Proceedings of the 16th Conference on Computational Linguistics - vol. 1, COLING 1996, pp. 466–471. Association for Computational Linguistics, Stroudsburg (1996)
3. Soderland, S., Fisher, D., Aseltine, J., Lehnert, W.: Crystal inducing a conceptual dictionary. In: Proceedings of the 14th International Joint Conference on Artificial Intelligence, IJCAI 1995, vol. 2, pp. 1314–1319. Morgan Kaufmann Publishers Inc., San Francisco (1995)
4. Ahn, D.: The stages of event extraction. In: Proceedings of the Workshop on Annotating and Reasoning About Time and Events, ARTE 2006, pp. 1–8. Association for Computational Linguistics, Stroudsburg (2006)
5. Vlachos, A., Buttery, P., Séaghdha, D.Ó., Briscoe, T.: Biomedical event extraction without training data. In: Proceedings of the BioNLP 2009 Workshop Companion Volume for Shared Task, pp. 37–40. Association for Computational Linguistics, Boulder, June 2009
6. Grishman, R., Westbrook, D., Meyers, A.: NYU's English ACE 2005 system description. Technical report, Department of Computer Science, New York University (2005)
7. Liao, S., Grishman, R.: Filtered ranking for bootstrapping in event extraction. In: Proceedings of the 23rd International Conference on Computational Linguistics, COLING 2010, pp. 680–688. Association for Computational Linguistics, Stroudsburg (2010)
8. Aone, C., Ramos-Santacruz, M.: Rees: a large-scale relation and event extraction system. In: Proceedings of the Sixth Conference on Applied Natural Language Processing, pp. 76–83. Association for Computational Linguistics, Seattle, April 2000
9. Dzendzik, D., Serebryakov, S.: Semi-automatic generation of linear event extraction patterns for free texts. In: SYRCoDIS 2013, pp. 5–9 (2013)
10. Creswell, C., Beal, M.J., Chen, J., Cornell, T.L., Nilsson, L., Srihari, R.K.: Automatically extracting nominal mentions of events with a bootstrapped probabilistic classifier. In: Proceedings of the COLING/ACL 2006 Main Conference Poster Sessions, pp. 168–175. Association for Computational Linguistics, Sydney, July 2006
11. Xu, F., Uszkoreit, H., Li, H.: Automatic event and relation detection with seeds of varying complexity. In: Proceedings of the 2006 AAAI Workshop on EventExtractionand Synthesis, pp. 12–17 (2006)
12. Huang, R., Riloff, E.: Bootstrapped training of event extraction classifiers. In: Proceedings of the 13th Conference of the European Chapter of the Association for Computational Linguistics, EACL 2012, pp. 286–295. Association for Computational Linguistics, Stroudsburg, (2012)
13. Settles, B.: Closing the loop: fast, interactive semi-supervised annotation with queries on features and instances. In: Proceedings of the Conference on Empirical Methods in Natural Language Processing, EMNLP 2011, pp. 1467–1478. Association for Computational Linguistics, Stroudsburg (2011)
14. Culotta, A., Kristjansson, T., McCallum, A., Viola, P.: Corrective feedback and persistent learning for information extraction. Artif. Intell. **170**(14–15), 1101–1122 (2006)

15. Jones, R., Ghani, R., Mitchell, T., Rilo, E.: Active learning for information extraction with multiple view feature sets. In: ATEM-2003 (2003)
16. Altmeyer, R., Grishman, R.: Active Learning of Event Detection Patterns. New York University, New York (2009)
17. Angeli, G., Tibshirani, J., Wu, J.Y., Manning, C.D.: Combining distant and partial supervision for relation extraction. In: EMNLP (2014)
18. Wang, R.C., Cohen, W.W.: Language-independent set expansion of named entities using the web. In: Proceedings of the 2007 Seventh IEEE International Conference on Data Mining, ICDM 2007, pp. 342–350. IEEE Computer Society, Washington, DC, USA (2007)
19. Wu, Z., Palmer, M.: Verbs semantics and lexical selection. In: Proceedings of the 32nd Annual Meeting on Association for Computational Linguistics, ACL 1994, pp. 133–138. Association for Computational Linguistics, Stroudsburg (1994)
20. Sarafraz, F., Eales, J., Mohammadi, R., Dickerson, J., Robertson, D., Nenadic, G.: Biomedical event detection using rules, conditional random fields and parse tree distances. In: Proceedings of the BioNLP 2009 Workshop Companion Volume for Shared Task, pp. 115–118. Association for Computational Linguistics, Boulder, June 2009
21. Wang, D.Z., Michelakis, E., Franklin, M.J., Garofalakis, M.N., Hellerstein, J.M.: Probabilistic declarative information extraction. In: Li, F. (ed.) ICDE, pp. 173–176. IEEE, Long Beach, CA (2010)
22. Peng, F., McCallum, A.: Accurate information extraction from research papers using conditional random fields. In: HLT-NAACL 2004, pp. 329–336 (2004)
23. Wick, M., Singh, S., Kobren, A., McCallum, A.: Assessing confidence of knowledge base content with an experimental study in entity resolution. In: Proceedings of the 2013 Workshop on Automated Knowledge Base Construction, AKBC 2013, pp. 13–18. ACM, New York (2013)

Identification and Ranking of Event-Specific Entity-Centric Informative Content from Twitter

Debanjan Mahata[1]([⊠]), John R. Talburt[1], and Vivek Kumar Singh[2]

[1] Department of Information Science,
University of Arkansas at Little Rock, Little Rock, USA
{dxmahata,jrtalburt}@ualr.edu
[2] Department of Computer Science, South Asian University, New Delhi, India
vivek@cs.sau.ac.in

Abstract. Twitter has become the leading platform for mining information related to real-life events. A large amount of the shared content in Twitter are non-informative spams and informal personal updates. Thus, it is necessary to identify and rank informative event-specific content from Twitter. Moreover, tweets containing information about named entities (like person, place, organization, etc.) occurring in the context of an event, generates interest and aids in gaining useful insights. In this paper, we develop a novel generic model based on the principle of mutual reinforcement, for representing and identifying event-specific, as well as entity-centric informative content from Twitter. An algorithm is proposed that ranks tweets in terms of event-specific, entity-centric information content by leveraging the semantics of relationships between different units of the model.

1 Introduction

Twitter is a social media platform that has become an indispensable source for disseminating news and real-time information about current events. It is a microblogging application that allows its users to post short messages of 140 characters known as tweets. Twitter is widely accepted as a source for first-hand citizen journalistic content and has been harnessed in detection, extraction and analysis of real-life events [3,4].

Motivation: A significant amount of tweets are related to real-life events (e.g., football matches, music shows, etc.). Majority of these event related tweets are pointless babbles, personal updates and spams providing no information to the general audience interested to know about an event. On the other hand there are tweets that presents newsworthy content, recent updates and real-time coverage of on-going events. These tweets are informative and are very useful for users who follow an event, and search for related information in Twitter.

Occurrence of a real-life event in general is characterized by participation of entities like people, organizations, or things at a certain place over a period

© Springer International Publishing Switzerland 2015
C. Biemann et al. (Eds.): NLDB 2015, LNCS 9103, pp. 275–281, 2015.
DOI: 10.1007/978-3-319-19581-0_24

of time [5]. While sharing information about an event in Twitter, users often mention these entities (e.g. *Update: Statement from Australian Prime Minister Tony Abbott on the Hostage incident #SydneySiege http://t.co/b4tO4A8CQj*). We consider such user updates as entity-centric messages related to the event. The consumers of event related information are most often interested in such entity-centric messages in the context of the event. Also, informative content shared about the entities during an event helps in gaining useful insights about the event as well as the related entities.

Objective and Contribution: The main objective of the work presented in this paper is to automatically identify and rank event-specific informative tweets mentioning relevant entities in their content. Towards this objective, we propose a novel generic model based on principle of mutual reinforcement for representing relationships between event-specific information cues and relevant named entities extracted from the tweet content. We implement an algorithm that leverages the mutually reinforcing relationships represented by the model for ranking tweets in terms of event-specific informative content sharing information about entities related to the event.

Problem: Events have been defined from various perspectives and in different contexts. In the context of our work we adopt a definition similar to [1]. An **event** is defined as a real-world occurrence (E_i) with an associated time period T_{E_i} ($t_{E_i}^{start}$-$t_{E_i}^{end}$), and a time ordered stream of tweets M_{E_i}, of substantial volume, discussing the occurrence of the event and posted in time T_{E_i}. We formally state the problem as,

Given an event E_i, a time ordered stream of n tweets $M_{E_i} = \{m_1, m_2, ..., m_n\}$ related to the event posted in time period T_{E_i}, the problem is to find a ranked set of tweets $\hat{M}_{E_i} = \{m_1 \geq ... \geq m_i \geq m_j \geq ... \geq m_n \mid i < j\}$, ordered in decreasing order of its event-specific informative content sharing information about event related entities.

2   Methodology

Content of a tweet is primarily composed of hashtags, words for expressing and conveying information, and URLs that lead to additional information about the content. While conveying information about an event the users also mention named entities in the textual content of the tweets. The example tweet in the previous section, not only provides information about the Sydney Siege crisis event, but also informs about "Tony Abbott" in the context of the event. The tweets are posted by users. It is also intuitive that users having high follower count tends to post informative posts, as tweets posted by such users are read by a larger audience and vice versa. Therefore, for an event E_i, in order to identify and rank event-specific informative tweets discussing about entities relevant to the event, we consider the following as event-specific information units:

– set of *tweets* ($M_{E_i} = \{m_1, m_2, ..., m_n\}$) related to the event.
– set of *hashtags* ($H_{E_i} = \{h_1, h_2, ..., h_p\}$) used for annotating the event related tweets.

- set of *entities* ($W_{E_i} = \{w_1, w_2, ..., w_r\}$) mentioned in the event related tweets.
- set of *users* ($U_{E_i} = \{u_1, u_2, ..., u_s\}$) posting tweets ($\in M_{E_i}$) about the event.
- set of *URLs* ($L_{E_i} = \{l_1, l_2, ..., l_t\}$) linking to external sources related to the event.

The informativeness w.r.t an event for any information unit depends upon its occurrence with other information units. We define event-specific informativeness of each unit based on the assumption below. For an event E_i.

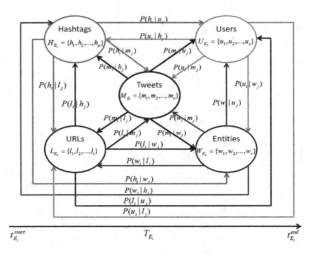

Fig. 1. Mutual reinforcement chains in twitter.

- a *tweet is considered to be event-specific informative* if it is strongly associated with: **(a)** *event-specific informative hashtags,* **(b)** *event-specific informative entities,* **(c)** *event-specific informative users,* **(d)** *event-specific informative URLs.*

We similarly define event-specific hashtags, entities, users and URLs forming a circular mutually reinforcing relationships between each other as shown in Fig. 1.

The relationships between the event-specific *information units* for an E_i forms a *Mutual Reinforcement Chain* [6], as shown in Fig. 1. We represent this relationship in a graph $G = (V, D)$, where $V = M_{E_i} \cup H_{E_i} \cup W_{E_i} \cup U_{E_i} \cup L_{E_i}$, is the set of vertices and D is the set of directed edges between different vertices. Whenever two vertices are associated, there are two edges between them that are oppositely directed. Each directed edge is assigned a weight, which determines the degree of association of one vertex with the other. The weights for each edge is calculated according to the conditional probabilities given in Table 1. We do not consider an edge between two vertices of same type.

We assign an initial event-specific score to all the vertices of the graph. The formulations of the scores assigned to the vertices $\in H_{E_i}, W_{E_i}, U_{E_i}, L_{E_i}$ can be found in Table 1. For initializing the tweets ($\in M_{E_i}$) with a generic informativeness score we develop a logistic regression model with an accuracy of 76.32 % after 10-fold cross validation. For training the model we used an annotated dataset provided by [2]. The tweets labeled as *related and informative* were assigned a score of 1 and all the other tweets labeled as *related - but not informative* and *not related* were assigned a score of 0. The model was then used for assigning informativeness score between 0 and 1 to all the tweets in the dataset, with 0 being least informative and 1 being most informative. The assigned initial scores gives an initial ranking of the vertices. We aim to refine the initial scores

Table 1. Equations for mutual reinforcement chains, affinity scores and event-specific initialization scores of nodes $\in G$.

Affinity scores between different nodes $\in M_{E_i}, H_{E_i}, W_{E_i}, U_{E_i}, L_{E_i}$:
$P(h_i|w_j) = \dfrac{No.\,of\,tweets\,h_i\,and\,w_j\,occur\,together}{No.\,of\,tweets\,w_j\,occurs}$, $P(w_i|h_j) = \dfrac{No.\,of\,tweets\,w_i\,and\,h_j\,occur\,together}{No.\,of\,tweets\,h_j\,occurs}$,
$P(h_i|l_j) = \dfrac{No.\,of\,tweets\,h_i\,and\,l_j\,occur\,together}{No.\,of\,tweets\,l_j\,occurs}$, $P(l_i|h_j) = \dfrac{No.\,of\,tweets\,l_i\,and\,h_j\,occur\,together}{No.\,of\,tweets\,h_j\,occurs}$,
$P(h_i|u_j) = \dfrac{No.\,of\,tweets\,h_i\,and\,u_j\,occur\,together}{No.\,of\,tweets\,u_j\,occurs}$, $P(u_i|h_j) = \dfrac{No.\,of\,tweets\,u_i\,and\,h_j\,occur\,together}{No.\,of\,tweets\,h_j\,occurs}$,
$P(w_i|l_j) = \dfrac{No.\,of\,tweets\,w_i\,and\,l_j\,occur\,together}{No.\,of\,tweets\,l_j\,occurs}$, $P(l_i|w_j) = \dfrac{No.\,of\,tweets\,l_i\,and\,w_j\,occur\,together}{No.\,of\,tweets\,w_j\,occurs}$,
$P(w_i|u_j) = \dfrac{No.\,of\,tweets\,w_i\,and\,u_j\,occur\,together}{No.\,of\,tweets\,u_j\,occurs}$, $P(u_i|w_j) = \dfrac{No.\,of\,tweets\,u_i\,and\,w_j\,occur\,together}{No.\,of\,tweets\,w_j\,occurs}$,
$P(u_i|l_j) = \dfrac{No.\,of\,tweets\,u_i\,and\,l_j\,occur\,together}{No.\,of\,tweets\,l_j\,occurs}$, $P(l_i|u_j) = \dfrac{No.\,of\,tweets\,l_i\,and\,u_j\,occur\,together}{No.\,of\,tweets\,u_j\,occurs}$,
$P(h_i|m_j) = P(m_i|h_j) = P(w_i|m_j) = P(m_i|w_j) = P(u_i|m_j) = P(m_i|u_j) = P(l_i|m_j) = P(m_i|l_j) = 1.0$
Note: $P(h_i \mid w_j)$ should be read as the probability of occurrence of hashtag h_i given the occurrence of the entity w_j in the stream of tweets M_{E_i} related to event E_i collected over the time period T_{E_i}.
Event-specific initialization scores of nodes $\in H_{E_i}, W_{E_i}, U_{E_i}, L_{E_i}$:
$Score(h_i) = \dfrac{freq(h_i)}{max\{freq(h_1), freq(h_2), \dots, freq(h_p)\}}$ **(1)** $Score(w_i) = \dfrac{freq(w_i)}{max\{freq(w_1), freq(w_2), \dots, freq(w_r)\}}$ **(2)**
$Score(u_i) = \dfrac{followers(u_i)}{max\{followers(u_1), \dots, followers(u_r)\}}$ **(3)**, $Score(l_i) = \dfrac{freq(l_i)}{max\{freq(l_1), freq(l_2), \dots, freq(l_r)\}}$ **(4)**
where, $freq(h_i)$ is the frequency of occurrence of the ith hashtag ($\in H_{E_i}$) in the stream of tweets M_{E_i}. Similarly, $freq(w_i)$ denotes the frequency of occurrence of the ith entity ($\in W_{E_i}$) and, $freq(l_i)$ denotes the frequency of occurrence of the ith url ($\in L_{E_i}$). $followers(u_i)$ denotes the number of followers of user $u_i \in (U_{E_i})$.
Equations representing mutual reinforcement chains between $M_{E_i}, H_{E_i}, W_{E_i}, U_{E_i}, L_{E_i}$:
$R_{E_i}^{M(k+1)} = A_{E_i}^{MM(k)} R_{E_i}^{M(k)} + A_{E_i}^{MH(k)} R_{E_i}^{H(k)} + A_{E_i}^{MW(k)} R_{E_i}^{W(k)} + A_{E_i}^{MU(k)} R_{E_i}^{U(k)} + A_{E_i}^{ML(k)} R_{E_i}^{L(k)}$ **(5)**
$R_{E_i}^{H(k+1)} = A_{E_i}^{HM(k)} R_{E_i}^{M(k)} + A_{E_i}^{HH(k)} R_{E_i}^{H(k)} + A_{E_i}^{HW(k)} R_{E_i}^{W(k)} + A_{E_i}^{HU(k)} R_{E_i}^{U(k)} + A_{E_i}^{HL(k)} R_{E_i}^{L(k)}$ **(6)**
$R_{E_i}^{W(k+1)} = A_{E_i}^{WM(k)} R_{E_i}^{M(k)} + A_{E_i}^{WH(k)} R_{E_i}^{H(k)} + A_{E_i}^{WW(k)} R_{E_i}^{W(k)} + A_{E_i}^{WU(k)} R_{E_i}^{U(k)} + A_{E_i}^{WL(k)} R_{E_i}^{L(k)}$ **(7)**
$R_{E_i}^{U(k+1)} = A_{E_i}^{UM(k)} R_{E_i}^{M(k)} + A_{E_i}^{UH(k)} R_{E_i}^{H(k)} + A_{E_i}^{UW(k)} R_{E_i}^{W(k)} + A_{E_i}^{UU(k)} R_{E_i}^{U(k)} + A_{E_i}^{UL(k)} R_{E_i}^{L(k)}$ **(8)**
$R_{E_i}^{L(k+1)} = A_{E_i}^{LM(k)} R_{E_i}^{M(k)} + A_{E_i}^{LH(k)} R_{E_i}^{H(k)} + A_{E_i}^{LW(k)} R_{E_i}^{W(k)} + A_{E_i}^{LU(k)} R_{E_i}^{U(k)} + A_{E_i}^{LL(k)} R_{E_i}^{L(k)}$ **(9)**
Other equations:
$\Delta_{E_i} \cdot R_{E_i} = \lambda . R_{E_i}$ **(10)** $\overline{\Delta}_{E_i} = \alpha \hat{\Delta}_{E_i} + (1 - \alpha)E$ **(11)** $E = p \times [1]_{1 \times k}$ **(12)**

and assign a final score for ranking the vertices by leveraging the relationships between them and propagating the initial scores accordingly, from one vertex to another. Next, we formalize our ranking methodology and present our proposed algorithm step-by-step.

The relationships between two sets of vertices in the graph G is denoted by an affinity matrix. For example, $A_{E_i}^{MH}$ denotes the $M_{E_i} - H_{E_i}$ affinity matrix for event E_i, where $(i, j)^{th}$ entry is the edge weight quantifying the association between i^{th} tweet ($\in M_{E_i}$) and j^{th} hashtag ($\in H_{E_i}$), calculated using Table 1, and so on. The rankings of tweets, hashtags, entities, users and URLs in terms of event-specific informativeness, can be iteratively derived from the Mutual Reinforcement Chain for the event. Let $R_{E_i}^M$, $R_{E_i}^H$, $R_{E_i}^W$, $R_{E_i}^U$ and $R_{E_i}^L$ denote the ranking scores for M_E, H_{E_i}, W_{E_i}, U_{E_i}, and L_{E_i}, respectively. Therefore, the Mutual Reinforcement Chain ranking for the kth iteration can be formulated using Eqs. (5–9) in Table 1. The Eqs. 5–9 can be represented in the form of a block matrix Δ_{E_i}, where,

$$\Delta_{E_i} = \begin{pmatrix} A_{E_i}^{MM} & A_{E_i}^{MH} & A_{E_i}^{MW} & A_{E_i}^{MU} & A_{E_i}^{ML} \\ A_{E_i}^{HM} & A_{E_i}^{HH} & A_{E_i}^{HW} & A_{E_i}^{HU} & A_{E_i}^{HL} \\ A_{E_i}^{WM} & A_{E_i}^{WH} & A_{E_i}^{WW} & A_{E_i}^{WU} & A_{E_i}^{WL} \\ A_{E_i}^{UM} & A_{E_i}^{UH} & A_{E_i}^{UW} & A_{E_i}^{UU} & A_{E_i}^{UL} \\ A_{E_i}^{LM} & A_{E_i}^{LH} & A_{E_i}^{LW} & A_{E_i}^{LU} & A_{E_i}^{LL} \end{pmatrix}$$

Let

$$R_{E_i} = \begin{pmatrix} R_{E_i}^M & R_{E_i}^H & R_{E_i}^W & R_{E_i}^U & R_{E_i}^L \end{pmatrix}^T$$

Input : Sets of vertices $M_{E_i}, H_{E_i}, W_{E_i}, U_{E_i}, L_{E_i}$ of graph G, $\alpha = 0.85$,
$\varepsilon = 1e - 08$.
Output: Ordered set of vertices $\hat{M}_{E_i}$, containing tweets ranked in order
of event-specific informative content sharing information about
event related entities.
Initialize rank vectors $[R_{E_i}^{M(0)}, R_{E_i}^{H(0)}, R_{E_i}^{W(0)}, R_{E_i}^{U(0)}, R_{E_i}^{L(0)}]$;
Assign $R_{E_i}^0 = [R_{E_i}^{M(0)}, R_{E_i}^{H(0)}, R_{E_i}^{W(0)}, R_{E_i}^{U(0)}, R_{E_i}^{L(0)}]^T$;
Normalize $R_{E_i}^0$ such that $||R_{E_i}^0||_1 = 1$;
Construct matrix Δ_{E_i};
Make matrix Δ_{E_i} stochastic and irreducible converting it to $\overline{\Delta}_{E_i}$;
$k \leftarrow 1$
repeat
$\quad | \quad R_{E_i}^k \leftarrow \overline{\Delta}_{E_i} R_{E_i}^{k-1}$;
$\quad | \quad k \leftarrow k + 1$;
until $|| R_{E_i}^k - R_{E_i}^{k-1} ||_1 < \varepsilon$ OR $k \geq 100$;
$R_{E_i}^M \leftarrow R_{E_i}^{M(k)}, R_{E_i}^H \leftarrow R_{E_i}^{H(k)}, R_{E_i}^W \leftarrow R_{E_i}^{W(k)}, R_{E_i}^U \leftarrow R_{E_i}^{U(k)}, R_{E_i}^L \leftarrow R_{E_i}^{L(k)}$;
$\hat{M}_{E_i} \leftarrow R_{E_i}^M$;
return $\hat{M}_{E_i}$

Fig. 2. Algorithm for ranking nodes of graph G.

then, R_{E_i} can be computed as the dominant eigenvector of Δ_{E_i}, as shown in Eq. 10 in Table 1. In order to guarantee a unique R_{E_i}, Δ_{E_i} must be forced to be stochastic and irreducible. To make Δ_{E_i} stochastic we divide the value of each element in a column of Δ_{E_i} by the sum of the values of all the elements in that column. This finally makes Δ_{E_i} column stochastic. We now denote it by $\hat{\Delta}_{E_i}$. Next, we make $\hat{\Delta}_{E_i}$ irreducible. This is done by making the graph G strongly connected by adding links from one node to any other node with a probability vector p. Now, $\hat{\Delta}_{E_i}$ is transformed to $\overline{\Delta}_{E_i}$ using Eqs. 11 and 12 in Table 1, where $0 \leq \alpha \leq 1$ is set to 0.85, and k is the order of $\hat{\Delta}_{E_i}$. We set $p = [1/k]_{k \times 1}$ by assuming a uniform distribution over all elements. Now, $\overline{\Delta}_{E_i}$ is stochastic and irreducible and it can be shown that it is also primitive by checking $\overline{\Delta}_{E_i}^2$ is greater than 0. The final ordered set of tweets $\hat{M}_{E_i}$ ranked in terms of their event-specific informative content sharing information about entities related to the event is obtained using the algorithm in Fig. 2.

3   Results and Future Work

Table 2. Details of data collected for the experiment.

Event Name and Query Hashtag	No. of Tweets	Time Period (UTC)
Millions March NYC (#millionsmarchnyc) ($http://goo.gl/I8WR4B$)	56927	13th Dec, 2014; 20:25:43 to 14th Dec, 2014; 03:30:41
Sydney Siege (#sydneysiege) ($http://goo.gl/qLguvG$)	398204	15th Dec, 2014, 07:21:16 to 15th Dec, 2014; 22:46:45

For implementing our proposed framework we collected 455, 131 tweets from two real-life events, 'Millions March NYC' and 'Sydney Siege', using Twitter Streaming API. Details of the dataset is presented in Table 2. Tweets for each event was collected over the given period of time, by providing a popular hashtag corresponding to each event to the API. We only considered English language tweets. We performed a series of data preparation steps before implementing the logistic regression model and our algorithm. Some of the steps that we took are deduplication of tweets, tokenization, POS tagging, detection of slang words, English stop

Table 3. Top 5 informative hashtags, entities, URLs and tweets for Sydney Siege

Event	Sydney Siege
Top 5 Informative Hashtags	1. #sydneysiege, 2. #SydneySiege, 3. #Sydneysiege, 4. #MartinPlace, 5. #9News
Top 5 Informative Entities	1. police, 2. sydney, 3. reporter, 4. lindt, 5. isis
Top 5 Informative URLs	1. http://www.cnn.com/2014/12/15/world/asia/australia-sydney-hostage-situation/index.html 2. http://www.bbc.co.uk/news/world-australia-30474089, 3. http://edition.cnn.com/2014/12/15/world/asia/australia-sydney-siege-scene/index.html, 4. http://rt.com/news/214399-sydney-hostages-islamists-updates/, 5. http://www.newsroompost.com/138766/sydney-cafe-siege-ends-gunman-among-two-killed
Top 5 Informative Tweets	1. RT faithcnn: Hostage taker in Sydney cafe has demanded 2 things: ISIS flag and; phone call with Australia PM Tony Abbott #SydneySiege http://t.co/a2vgrn30Xh, 2. Aussie grand mufti and; Imam Council condemn #Sydneysiege hostage capture http://t.co/ED98YKMxqM - LIVE UPDATES http://t.c..., 3. RT PatDollard: #SydneySiege: Hostages Held By Jihadis In Australian Cafe - WATCH LIVE VIDEO COVERAGE http://t.co/uGxmd7zLpc #tcot #pjnet, 4. RT FoxNews: MORE: Police confirm 3 hostages escape Sydney cafe, unknown number remain inside http://t.co/pcAt91LIdS #Sydneysiege, 5. Watch #sydneysiege police conference live as hostages are still being held inside a central Sydney cafe http://t.co/OjulBqM7w2 #c4news

words, feeling words, and special characters[1]. We extracted named entities from the tweets using AlchemyAPI (http://alchemyapi.com). The entities containing slang words were removed. Removal of slang hashtags and entities was done in order to obtain high quality results as intuitively high quality informative tweets should not contain a lot of slangs.

Given the space constraint, we show the top 5 hashtags, entities URLs and tweets for the Sydney Siege event in Table 3. We do not report the users for privacy concerns. Apart from identifying event-specific informative tweets containing information about event related entities, our proposed model has an additional advantage of identifying and ranking top event-specific informative hashtags, entities, URLs and users for an event. In this paper we presented our methodology and some preliminary results. Our next step would be to evaluate our results rigorously and extend the developed framework in a distributed computing environment.

References

1. Becker, H., Naaman, M., Gravano, L.: Beyond trending topics: real-world event identification on twitter. In: ICWSM, vol. 11, pp. 438–441 (2011)
2. Olteanu, A., Castillo, C., Diaz, F., Vieweg, S.: Crisislex: a lexicon for collecting and filtering microblogged communications in crises. In: Proceedings of the 8th International AAAI Conference on Weblogs and Social Media (ICWSM 2014), number EPFL-CONF-203561 (2014)
3. Popescu, A.-M., Pennacchiotti, M., Paranjpe, D.: Extracting events and event descriptions from twitter. In: Proceedings of the 20th International Conference Companion on World Wide Web, pp. 105–106. ACM (2011)
4. Sakaki, T., Okazaki, M., Matsuo, Y.: Tweet analysis for real-time event detection and earthquake reporting system development. IEEE Trans. Knowl. Data Eng. **25**(4), 919–931 (2013)

---

[1] List of resources like slang words, stopwords and feeling words used can be obtained from https://github.com/dxmahata/EIIMFramework/tree/master/CodeBase/EventIdentityInformationManagement/Resources.

5. Shaw, R., Troncy, R., Hardman, L.: LODE: linking open descriptions of events. In: Gómez-Pérez, A., Yu, Y., Ding, Y. (eds.) ASWC 2009. LNCS, vol. 5926, pp. 153–167. Springer, Heidelberg (2009)
6. Wei, F., Li, W., Lu, Q., He, Y.: Query-sensitive mutual reinforcement chain and its application in query-oriented multi-document summarization. In: Proceedings of the 31st Annual International ACM SIGIR Conference on Research and Development in Information Retrieval, pp. 283–290. ACM (2008)

Automatic Classification and PLS-PM Modeling for Profiling Reputation of Corporate Entities on Twitter

Jean-Valère Cossu[(⊠)], Eric Sanjuan, Juan-Manuel Torres-Moreno, and Marc El-Bèze

LIA/Université d'Avignon et des Pays de Vaucluse, 39 chemin des Meinajaries, Agroparc BP 91228, 84911 Avignon cedex 9, France
jean-valere.cossu@univ-avignon.fr

Abstract. In this paper, we address the task of detecting the reputation alert in social media updates, that is, deciding whether a new-coming content has strong and immediate implications for the reputation of a given entity. This content is also submitted to a standard typology of reputation dimensions that consists in a broad classification of the aspects of an under public audience company. Reputation manager needs a real-time database and method to report what is happening right now to his brand. However, typical Natural Language Processing (NLP) approaches to these tasks require external resources and show non-relational modeling. We propose a fast supervised approach for extracting textual features, which we use to train simple statistical reputation classifiers. These classifiers outputs are used in a Partial Least Squares Path Modeling (PLS-PM) system to model the reputation. Experiments on the RepLab 2013 and 2014 collections show that our approaches perform as well as the state-of-the-art more complex methods.

1 Introduction

Recently, with the emerging trend of the online networked information, control has moved to users. Each act of a public entity become scrutinized by a powerful global audience. We can understand reputation as the general recognition by other people of some characteristics for a given entity. Specifically, in business or politics, reputation comprises the decisions taken and how it is perceived by the population. This requires new reputation management tools and strategies able to consider the variability of interpretations for a given document. The rise of online social media has become an interesting way to process large amount of opinions about entities even though in the case of tweets there are no explicit ratings to be directly used in an opinion processing. Although significant advances have been made in RepLab[1] [1,2]. Analyzing reputation about companies and individuals is a hard problem requiring a complex modeling of these entities and

---

[1] Replab provides a framework to evaluate Online Reputation Management systems on Twitter http://www.limosine-project.eu/events/replab2013.

© Springer International Publishing Switzerland 2015
C. Biemann et al. (Eds.): NLDB 2015, LNCS 9103, pp. 282–289, 2015.
DOI: 10.1007/978-3-319-19581-0_25

it is still a significant research challenge because unlike products, opinions about entities may vary from the point-of-view (the person who speaks, who reads) and the context. This modeling can be considered similar as tweets taxonomy or and handmade classification.

In this paper, we are interested in reputation alert detection and reputation dimension assignment. It consists in identifying relevant interests for stakeholders in the companies that are also considered as key issues for the entities reputation, in order to contribute to a better understanding of a group of tweets and their topic. These aspects are subjective and may depend on each expert. We aim at guiding this expert to consider that a given tweet remains more important than another one. We use NLP based classifiers to project each tweet in the classes space in order to generalize the experts point-of-view. Then by observing interactions between classes using PLS-PM [3] we can provide a hierarchy to visualize how tweets and classes are perceived in the tweets stream.

2 Related Work

Previous research has exploited supervised methods for topic categorization of short social chat messages as [4]. But these kind of required a costly human annotations which is usually not available in real time for large-scale micro-blogging messages. Due to a lack of applicable performance metrics and exploitable gold-standard labels, it is hard to report the systems performance, or comment on the generalization of approaches such as TweetMotif [5]. Most of the contributions on reputation monitoring to extract sets of tweets requiring a particular attention from a reputation manager have been proposed in the last editions of RepLab [1,2]. Issues were tackled with Social Network Analysis [6] and both supervised and unsupervised algorithms [7] combined with terms selection strategies. Joint work between tweet clustering and priority detection over a NLP-based classification was also proposed by [8]. RepLab'2014 [2] focused on the reputation dimension classification. Which can be viewed as a complement to topic detection it is nearer a stress classification of the company's aspects. The stresses are introduced by the experts and only reflect their interest. Some approaches considered information beyond the tweet textual content such as pseudo-relevant document [9] or semantic expansion [10] and Wikipedia categories [11] but also psychometric and linguistic information [12].

In spite of the great significance of extracting information to obtain high performances, the amount of research dedicated to understand the experts' stress effect is really limited. This leads us to investigate not the best possible performance but to propose a toolbox that allows a better understanding of scores given by classifiers for each tweet-content.

3 Outer Model Learning

We understand the problem of detecting reputation using a supervised classification method based on a threshold intersection graph computed over the discriminant bag-of-words representation of each tweet. Vertices are tweets, edges are between tweets sharing at least one word (lexical relation) and are weighted using Cosine distance and Jaccard similarity index which we add Multi-Class Support Vector Machin[2]. We start with annotated documents based on their following reputation tags: priority level and dimension. We use Term Frequency-Inverse Document Frequency (TF-IDF) [14,15] combined with the Gini purity criterion [16]. Then we estimate the similarity of a given tweet by comparing it to each class bag-of-words and rank tweets according to the score provided by the classifiers.

4   Inner Model Learning

4.1   PLS-PM Models

The statistical method PLS-PM allows to estimate parameters of linear regression by calculating the solution of the general underlying model of multi-variate PLS [17]. Dealing with ranked classes and several ranking led to meaningfully combine these sets and PLS-PM is an interesting way to combine systems outputs. Each tweet is analyzed like a structure made of blocks of manifest (observed) variables (scores in each class). Each block is summarized by a latent variable. PLS-PM is used to find the best weight (for system-class pairs) to predict a conditional "*ALERT*" priority level using dimension and priority probabilities given by each classifier.

4.2   Model for Profiling Reputation

We propose a model combining pre-defined dimensions suggested by e-watcher specialists[3] like Products and services, Citizenship, Governance, Innovation, Leadership, Performance, Workplace with the concepts of Alert and Importance for Profiling reputation. The objective is to be able to explain why some action should be taken after some tweets based on these dimensions. The model also allows to follow the impact of these dimensions over the time and improves the robustness of alerts.

Each dimension is modeled as a latent variable combining several textual classifiers and entailing one and only one of the two concepts Alert or Importance. Alert is also supposed to entail Importance. Therefore dimensions are separated into 2 groups: those that will directly induce Alert and those that seem to be less strategic (thus considered as unimportant). Since all classifier outputs have been normalized between 0 and 1, the complement of the dimension $1 - x$ can be considered and supposed to entail Importance: the underlying 3 valued logic (Alert, Important, NonImportant) is that the complement of a *Non Alert* entails *Important*.

---

[2] See [13] http://www.cs.cornell.edu/people/tj/svm_light/svm_multiclass.html.

[3] Reputation Institute's Reptrak framework http://www.reputationinstitute.com/about-reputation-institute/the-reptrak-framework.

Dimensions are split according to the learning corpus keeping in mine that a tweet can have multiple dimensions and that annotators agreement cannot be expected to be high since most of these dimensions like Innovation or Leadership are vague. Moreover tweets can be misleading and ambiguous. Alert is also estimated based on these classifiers meanwhile Importance is estimated as non-unimportant. The PLS-PM algorithm is then used to estimate each inner variable as a vector minimizing square distance to both classifiers normalized output scores and related inner latent variables. We use the R index to estimate the model quality (maximizes the square sum of correlations inside latent variables and between related variables).

5 Experimental Evaluation

5.1 Evaluation Protocol

We perform a supervised classification over Replab'2013-14 dataset [1,2]. The dataset includes a training set and a test set both annotated in either of several reputation monitoring axis: filtering, polarity, priority-level (*Alert, Important, Unimportant*), clustering and dimension (only for Automotive and Banking domains). This paper only addresses priority-level and dimension classification issues. We consider lower-cased and cleaned text[4]. We choose to be entity and language independent, since we want to detect a tweet reputation regardless of its association with an entity. We compare our system to RepLab'2013-4 baselines and best systems according to official metrics: Reliability/Sensitivity [18] and the overall accuracy which we added typical F-Score (based on Precision and Recall).

5.2 Outer Model Evaluation

Priority Detection. The best F-measure (Best F in the tables) reported in this task was obtained with a kNN based classification method [19]. Other performances ranked with regards to F-measure are summarized in Table 1. Both SVM and Cosinus approaches are competitive according to Accuracy. Although SVM

Table 1. Priority detection performances ordered by F-Measure (R,S).

System	F-Score	Accuracy	Reliability	Sensitivity	F-Measure (R,S)
Best F	**.571**	.636	.387	.315	**.335**
SVM	.553	**.643**	.344	.294	.304
Baseline	.512	.570	**.403**	.248	.274
Cosinus	.566	.637	.344	.236	.260
Jaccard	.492	.561	.342	.212	.233

---

[4] We remove links, stop-words and punctuation marks.

Table 2. Dimensions classification task Performances ordered according to Accuracy.

System	F-Score	Accuracy
Cosinus	**.505**	**.741**
SVM	.467	.733
Best_Acc	.473	.731
Baseline	.380	.622
Jaccard	.378	.476

shows significant improvement with Accuracy it falls with F-Score in contrast with the Cosinus that completely collapses regarding F(R,S).

Dimensions Classification. The best RepLab'2014 participating systems (noted Best_Acc in Table 2) system used tweet enrichment via pseudo-relevant document [9]. As interesting results, we can see that the Cosinus classifier outperforms all performances reported up to now in this task.

5.3   Inner Model Evaluation

Based on the Replab Learning data it appeared that there is a common model for all four domains Bank, Automotive, Music and University in this reputation task: three dimensions entail Alert: Governance, Innovation, Leadership, Performance and WorkPlace. The complement of the two remaining dimensions entail Importance (i.e.: NonUnimportant): NonCitizenship and NonProduct & Services. Figure 1 shows the inner model with its estimated coefficients using the R PLS-PM library[5] where all variables are estimated using Cosine, Jaccard and SVM classifier scores. Figure 2 shows the same model but where inner variable Alert has been replaced by the reference from CLEF Replab. The standard deviation between the predicted model and the reference is significant (t-test p-value < 0.05) but the dimensions are ranked in the exact same order. The Pearson's product-moment correlation between the predicted inner latent variable Alert and the reference is 0.49 which is significantly high (p-value $< 10^{-3}$) but not significantly higher than the single Cosine estimate. Therefore, for the bank domain, the PLS-PM model helps in prioritizing dimensions but not in improving alert prediction. For the automotive domain the results are slightly different, the ranking of dimensions is not exactly the same since the predictive model ranks Governance (0.53 path coefficient towards Alert) before Innovation (0.15) meanwhile in the reference Innovation has an higher impact on Alert (0.23 vs 0.14). However both rankings are highly correlated (Kendall test: $80 < \tau < 1$, p-value < 0.05) and the inner variable Alert is significantly highly correlated to the reference (Pearson 95 percent confidence interval between 0.24 and 0.25). For

---

[5] http://cran.r-project.org/web/packages/plspm/.

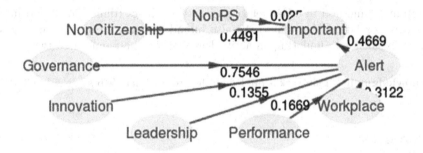

Fig. 1. Inner Model for Bank domain with inferred Alert

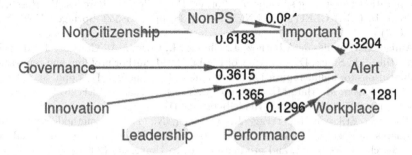

Fig. 2. Inner Model for Bank domain with reference Alert

the Music and the University domain, correlations are lower (Pearson's product-moment correlation of 0.15 and 0.11) because classifiers were less efficient on these domains for Alert prediction, but PLS-PM models significantly improves them by 10 %. Even though we do not have a reference annotation for dimensions in these domains. Classifiers have been trained without domain distinction, since Bank and Automotive are the largest domains, they tend to expand rules from Banks to non related domains but do not infer inverted relations.

6   Conclusions

The experimental evaluations on RepLab establish that tweets lexical content is sufficient to tackle the tasks of identifying the reputation alerts and dimensions of micro-blog posts using simple machine learning approaches. We experimented simple and more complex statistical lexical corpus-based NLP methods that use discriminating textual features inferred from labeled data. Our approaches turn out to be very effective in addressing the priority and reputation dimensions detection task. It then appeared that PLS-PM modeling based on a three valued scale: *Alert, Important, Unimportant* could compensate this lack by detecting uncorrelated tweets with existing topics. In future work, we plan to examine an automatic method to order relations between classes (dimensions) and infer more complex latent hierarchies. We also intend to study an interesting lexical context expansions simulating an active learning over non-annotated tweets.

Acknowledgment. This work is funded by the project ImagiWeb ANR-2012-CORD-002-01.

References

1. Amigó, E., Carrillo de Albornoz, J., Chugur, I., Corujo, A., Gonzalo, J., Martín, T., Meij, E., de Rijke, M., Spina, D.: Overview of RepLab 2013: evaluating online reputation monitoring systems. In: Forner, P., Müller, H., Paredes, R., Rosso, P., Stein, B. (eds.) CLEF 2013. LNCS, vol. 8138, pp. 333–352. Springer, Heidelberg (2013)
2. Amigó, E., Carrillo-de-Albornoz, J., Chugur, I., Corujo, A., Gonzalo, J., Meij, E., de Rijke, M., Spina, D.: Overview of RepLab 2014: author profiling and reputation dimensions for online reputation management. In: Kanoulas, E., Lupu, M., Clough, P., Sanderson, M., Hall, M., Hanbury, A., Toms, E. (eds.) CLEF 2014. LNCS, vol. 8685, pp. 307–322. Springer, Heidelberg (2014)
3. Wold, S., Eriksson, L., Trygg, J., Kettaneh, N.: The pls method-partial least squares projections to latent structures-and its applications in industrial rdp (research, development, and production). Unea University (2004)
4. Ranganath, R., Jurafsky, D., McFarland, D.: It's not you, it's me: detecting flirting and its misperception in speed-dates. In: Proceedings of the 2009 Conference on Empirical Methods in Natural Language Processing, vol. 1, pp. 334–342. Association for Computational Linguistics (2009)
5. O'Connor, B., Krieger, M., Ahn, D.: Tweetmotif: exploratory search and topic summarization for twitter. In: ICWSM (2010)
6. Berrocal, J.L.A., Figuerola, C.G., Rodríguez, Á.Z.: Reina at replab2013 topic detection task: community detection. In: CLEF 2013 Eval. Labs and Workshop Online Working Notes (2013)
7. Sánchez-Sánchez, C., Jiménez-Salazar, H., Luna-Ramirez, W.: Uamclyr at replab2013: monitoring task. In: CLEF 2013 Eval. Labs and Workshop Online Working Notes (2013)
8. Cossu, J.V., Bigot, B., Bonnefoy, L., Senay, G.: Towards the improvement of topic priority assignment using various topic detection methods for e-reputation monitoring on twitter. In: Métais, E., Roche, M., Teisseire, M. (eds.) NLDB 2014. LNCS, vol. 8455, pp. 154–159. Springer, Heidelberg (2014)
9. McDonald, G., Deveaud, R., McCreadie, R., Gollins, T., Macdonald, C., Ounis, I.: University of glasgow terrier team/project abacá at replab 2014: reputation dimensions task (2014)
10. Rahimi, A., Sahlgren, M., Kerren, A., Paradis, C.: The stavicta group report for replab 2014 reputation dimensions task. In: CLEF 2014 Evaluation Labs and Workshop-Working Notes Papers (2014)
11. Qureshi, M.A., ORiordan, C., Pasi, G.: Exploiting wikipedia for entity name disambiguation in tweets. In: Métais, E., Roche, M., Teisseire, M. (eds.) NLDB 2014. LNCS, vol. 8455, pp. 184–195. Springer, Heidelberg (2014)

12. Vilares, D., Hermo, M., Alonso, M.A., Gómez-Rodríguez, C., Vilares, J.: Lys at clef replab 2014: Creating the state of the art in author influence ranking and reputation classification on twitter. In: CLEF, pp. 1468–1478 (2014)
13. Crammer, K., Singer, Y.: On the algorithmic implementation of multiclass kernel-based vector machines. J. Mach. Learn. Res. **2**, 265–292 (2002)
14. Salton, G., Buckley, C.: Term-weighting approaches in automatic text retrieval. Inf. Process. Manag. **24**(5), 513–523 (1988)
15. Robertson, S.: Understanding inverse document frequency: on theoretical arguments for idf. J. Documentation **60**(5), 503–520 (2004)
16. Torres-Moreno, J., El-Beze, M., Bellot, P.: Bechet, opinion detection as a topic classification problem in in textual information access. chap. 9 (2013)
17. Henseler, J.: On the convergence of the partial least squares path modeling algorithm. Comput. Statistics **25**(1), 107–120 (2010)
18. Amigó, E., Gonzalo, J., Verdejo, F.: A general evaluation measure for document organization tasks. In: Proceedings of the 36th International ACM SIGIR Conference on Research and Development in Information Retrieval, pp. 643–652. ACM (2013)
19. Cossu, J., Bigot, B., Bonnefoy, L., Morchid, M., Bost, X., Senay, G., Dufour, R., Bouvier, V., Torres-Moreno, J., El-Beze, M.: Lia@ replab 2013. In: CLEF 2013 Eval. Labs and Workshop Online Working Notes (2013)

NLP and Usability

An Adaptable and Personalised E-Learning System Based on Free Web Resources

Eiman Aeiad[(✉)] and Farid Meziane

School of Computing, Science and Engineering,
University of Salford, Salford M5 4WT, UK
E.Aeiad@edu.salford.ac.uk, f.meziane@salford.ac.uk

Abstract. A personalised and adaptive E-Learning system architecture is developed to provide a comprehensive learning environment for learners who cannot follow a conventional programme of study. The system extract information from freely available resources on the Web, and taking into consideration the learners' background and requirements to design modules and a planner system to facilitate the learning process. The process is supported by the development of an ontology to optimise the information extraction process. An application in the computer science field is used to evaluate the proposed system based on the IEEE/ACM Computing curriculum.

Keywords: E-Learning · Learning styles · Personalized learning

1 Introduction and Motivation

Learning is greatly influenced by the development of Information and Communication Technologies and advanced digital media. Learning using these new media is referred to as E-Learning [6]. It allows access to education to those who find it difficult to be physically present in the traditional classroom based learning. Together with the development of the Web, a plethora of freely available resources is accessible and used as part of E-Learning. Although, E-Learning may have addressed the needs of the masses, it failed to satisfy the needs of individual learners by providing them with personalised learning plans and resources.

In this paper, we present the Adaptive and Personalised E-Learning System APELS that extends the current understanding and use of conventional E-Learning resources, by using freely available resources on the Web to design and deliver content for individual learners. The system would be used by individual learners, universities that may not have the resources and the expertise to develop learning resources and anyone who wishes to learn a specific filed. At this stage, by learning we mean academic learning and by resources those available for academic learning. The APELS system will enable users to design their own learning material based on internationally recognised curricula and contents. Using standard search engines to find learning material that is suitable for individual learners is time consuming and may not lead to a suitable

© Springer International Publishing Switzerland 2015
C. Biemann et al. (Eds.): NLDB 2015, LNCS 9103, pp. 293–299, 2015.
DOI: 10.1007/978-3-319-19581-0_26

outcome. The major contribution of this paper is therefore the development of an intelligent system to support online course design based solely on freely available resources on the Web.

The APELS system will address three main issues. The first is the identification of the learner's requirements, the second is the identification of the learner's learning style and the third is the ability of the APELS system to adapt and modify the content and learning style based on the interactions of the users with the system over a period of time. In addition, the information extracted by the system will be passed to a Planner module that will structure it into lectures/tutorials and workshops based on some predefined learning times. The evaluation of the learning process will be against a set of learning outcomes as defined by standard curricula. In this paper, a case study using the IEEE/ACM Computing Curriculum [11] will be used to illustrate the functionality of the APELS system.

2   Background and Related Work

Over the last few years, researchers have started taking a keen interest in developing personalised learning systems, whose main purpose is to provide content, learning style and environments for the needs of specific learners. First, it is crucial to determine the learners' profile such as their background, level and needs, which can be determined by using a questioner. An example of such system is SPERO [1], a personalised E-learning system based on the IEEE Learning Technology Systems Architecture. The purpose of SPERO is to help teacher training in special education, which could provide different contents for foreign languages learners according to their interests and levels. However, SPERO is largely using questionnaires and e-surveys to build user profiles, which adds extra work to the users and is time consuming. Moreover, adaptive systems can be made more flexible by examining the learners' progress to make changes according to their needs. For instance, Baylari et al. [2] presented a personalized multi-agents E-Learning system based on Item Response Theory, which presents adaptive tests to estimate learners' knowledge and enables it to make changes according to their needs and background. Zhuge and Yanyan [4] proposed the KGTutor; a Knowledge grid based intelligent system personalised E-Learning system that examines each learner requirements such as previous knowledge and the learners' targets and then provides them with a personalised course. The system can also provide the learner with progress reports, evaluations and suggestions based on their performance. Ontologies are also becoming a great tool commonly used by researchers to extract educational information with the aim of assisting students to learn about specific topics. Cassin et al. [8] used the notion of ontology extraction for educational knowledge, aiming at helping students who need to learn about some specific topics.

In our work, learners' levels, background and needs will be addressed by questioners, before starting any module and their progress will be tested after each level and they will not be able to proceed until they have passed an assessment. Furthermore, this system will have an ontology structure to easily extract the knowledge in domain area on the WWW in order to improve information retrieval, organize and update learning resources specific to the user. In E-learning, ontologies can endorse adaptive learning services to help in the online learning process. Yarandi et al. [7] proposed a novel approach for developing personalised E-learning systems for individual learners using an ontology to create a user model to recognize learners and describe their behaviours. Moreover, they developed the learning content using an ontology to build the hierarchical and navigational relations between different parts of learning materials and how these can be determined based on users profiles.

3   System Architecture

The purpose of the APELS is to deliver recommended learning materials to learners who may have different backgrounds, learning styles and learning needs. The architecture is based on four main modules that will form the basis of the system. These include: student profile, student requirement, knowledge extraction system, and content delivery module as shown in Fig. 1.

3.1   Student Profile

This module contains two components; Student details and Learning style:

– *Student details:* This component stores the student's personal information.
– *Learning Style:* This component identifies the learning style of the learner using the VARK learning style [9].

3.2   Student Requirements

This module contains two components; Learning requirements and Learning area:

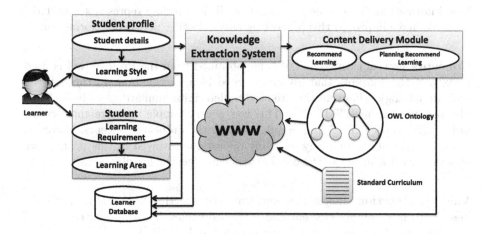

Fig. 1. System architecture

- *Learning Requirement:* This component gathers information related to the learners background and previous knowledge.
- *Learning Area:* In this component, the learner will choose the subject he/she wishes to study.

3.3   The Knowledge Extraction System

Once the details of the learner and his chosen area are known, the information extraction module is used to extract learning resources from the Web that are suitable for the learner. This Subsystem has the following 5 components: (i) Ontology, (ii) HTML2XML, (iii) Concepts Extraction, (iv) Value Extraction, (v) Matching Process.

The Ontology. The APELS system uses an ontology to help in extracting the required domain knowledge from the Web in order to improve the information retrieval process, organize and update learning resources specific to the user. To develop the ontology we used Protégé 2000 [3]. The ontology contains the standard is-a hierarchy of relevant domain concepts and relations between these concepts. Moreover, we exported the Web Ontology Language (OWL) which can represent domain knowledge using classes (concepts), properties, axioms and instances to support efficient reasoning and expressive power [12]. We will focus on concepts and relationships between them when building the Ontology. Avoiding the construction of a very large ontology that could be difficult to develop and use, we have adopted the development of smaller and specific ones that will support individual areas such as Computer Science, Law etc.

HTML2XML. Most Web content is written in HTML documents and it is difficult to extract the information from HTML. Therefore we have developed a tool, HTML2XML which will automatically create XML data sources from the collected HTML Web pages. Different approaches have been used for representing Web information [5,10]. The traditional XML file usually represents all HTML data. We have improved this process by only selecting pertinent information.

Concepts Extraction. The OWL file obtained from the Protégé tool is used to extract the concepts that are represented in a specific domain through the domain ontology. These entities are used to determine similarities with the XML files produced from HTML files. This was a very simple and straight forward process. A vector is used and populated with the ontology concepts. However, given the size of the ontology, in many operations involved in this project, we only deal with a subset of the ontology that models the learning area.

Values Extraction. In a similar way, when constructing XML files from HTML files, XML files contain element and attribute values. These are extracted and constitute the XML value vector and denoted as $V = [V_1, V_2, ..., V_m]$.

The Matching Process. The matching process involves the computation of a simple similarity measure between the subset of the ontology that models the learning domain D and the vector V for a website W. Given a set of relevant websites WS and their associated value vectors, the website with the highest similarity is selected as the best matching website for the learner satisfying his/her learning style. However, we do acknowledge the limitations of the current approach when running our experiments. Sometimes parts of the website only are relevant and appropriate for the learner and a combination of two or more websites contents will provide a better content. In addition, some concepts or terms may be given different names, although they have the same meaning. For instance, the equivalent terms for the concept "Calculus" include arithmetic, mathematics etc. This issue was solved by defining corresponding relations such as synonyms in the domain ontology. The similarity is calculated as follows: Given two vectors C and V defined as $C = [C_1, C_2, ..., C_m]$ where C_i represent an ontology Concept and $V = [V_1, V_2, ..., V_m]$ where V_j represent XML values. The similarity measure between vectors C and V is calculated as: $SC(C, V) = \frac{sc}{n} * 100$, where SC is the number concepts in C that are also present in V and n is the number of concepts in the subset of the ontology concepts dealing with the specific learning area of the learner.

4 Case Study and Evaluation

4.1 Description of the Case Study

To illustrate the functionality of the APELS system, we used some elements of the ACM/IEEE Computer Society Computer Science Curriculum, an internationally recognised and adopted standard in designing computer science related programs [11]. The IEEE/ACM Body of Knowledge (BoK) is organized into a set of 18 Knowledge Areas (KAs) corresponding to typical areas of study in computing such as Algorithms and Complexity, Operating Systems and Software Engineering. Each Knowledge Area (KA) is broken down into Knowledge Units (KUs). Each KU is divided into set of topics which are then classified into a tiered set of core topics (compulsory topics that must be taught) and elective topics (significant depth in many of the Elective topics should be covered). Core topics are further divided into Core-Tier1 topics and Core-Tier2 topics (Should almost be covered). The software development fundamentals area for example is divided into 4 KUs. The Algorithm and design KU is divided into 11 Core-Tier1 topics. Learning outcomes are then defined for each class of topics. We will specifically look at designing fundamental Programming Concepts in C++ module using the APELS system.

4.2 Use of the APELS System to Develop SE Learning Material

The learner will go through the various stages as described in Sects. 3.1 and 3.2 and will provide the required information. He will then choose the topic to study, here computer Science/Programming. Once Programming is chosen, the list of the KU will be displayed for the learner to choose from. Here for clarity we attempt to design a C++ programming module for beginners. Once a KU is chosen, APELS will start looking for material on the Web that will satisfy this specific learner and starts extracting the right material and organising it into smaller learning topics. The topics will then be mapped to hourly learning schedules.

4.3   Results and Evaluation

The launch of the APELS system will in a first instance return a list of websites for learning C++ language with the highest accuracy rating as shown in the Table 1 that identifies that the 4th website is the one that is most similar to the computer science Ontology represented as an OWL file. Together with the first website, they have better performances than the other websites as they have the highest similarities. The other websites show poor performance as the similarities with the required topics is low.

Table 1. Similarity matching with the OWL file

WWW	Concepts extraction	No of similarities	Accuracy
www.cplusplus.com/doc/tutorial/	9	7	77.78 %
www.penguinprogrammer.co.uk/	9	5	55.56 %
www.tutorialspoint.com/cplusplus/	9	4	44.44 %
www.doc.ic.ac.uk/wjk/C++Intro/	9	8	88.89 %
www.macs.hw.ac.uk/pjbk/pathways/cpp1	9	3	33.33 %
www.cprogramming.com/tutorial.html	9	2	22.22 %

5   Conclusion and Future Work

The work presented in this paper is the first stages in developing the APELS system that attempts to develop course contents based on freely available resources on the Web. The approach, functionality and architecture are improvements on existing E-learning systems. The system was evaluated using the ACM/IEEE Computing curriculum. The next steps in developing this system will include the user characteristics and background and also the learning outcomes of each Knowledge unit or modules that are not yet implemented in the current version of the system. The full version of the system will then be evaluated using domain experts and students.

References

1. Spero: Tele-informatics system for continuous collection, processing, diffusion of material for teacher training in special education. http://www.image.ntua.gr/spero
2. Baylari, A., Montazer, G.: Design a personalized e-learning system based on item response theory and artificial neural network approach. Expert Syst. Appl. **36**(4), 8013–8021 (2009)
3. Noy, N.F., Mcguinness, D.L.: Ontology development 101: a guide to creating your first ontology. Techinal report, KSL-01-05, Stanford Knowledge Systems Laboratory (2000)
4. Zhuge, H., Li, Y.: KGTutor: a knowledge grid based intelligent tutoring system. In: Yu, J.X., Lin, X., Lu, H., Zhang, Y. (eds.) APWeb 2004. LNCS, vol. 3007, pp. 473–478. Springer, Heidelberg (2004)
5. Breuel, T.M.: Information extraction from html documents by structural matching. In: Proceedings of the 2nd International Workshop on Web Document Analysis (2002)
6. El-Zayat, M.: An assessment of e-learning in egypt through the perceptions of egyptian university students: a field work survey. In: AUC 5th International Conference on E-Learning Applications (2007)
7. Yarandi, M., Tawil, A., Jahankhani, H.: Adaptive e-learning system using ontology. In: 22nd International Workshop on Database and Expert Systems Applications, pp. 511–516 (2011)
8. Cassin, P., Eliot, C., Lesser, V., Rawlins, K., Woolf, B.: Ontology extraction for educational knowledge bases. In: van Elst, L., Dignum, V., Abecker, A. (eds.) AMKM 2003. LNCS (LNAI), vol. 2926, pp. 297–309. Springer, Heidelberg (2004)
9. Honey, P., Mumford, A.: The Manual of Learning Styles, 3rd edn. Peter Honey, Maidenhead (1992)
10. Song, R., Liu, H., Wen, J.R., Ma, W.Y.: Learning block importance models for web pages. In: Proceedings of International WWW Conference (WWW-2004) (2004)
11. ACM/IEEE societies. Computer science curricula (2013). http://www.acm.org/education/CS2013-final-report.pdf
12. W3C. Owl 2 web ontology language primer, (2nd edn.) w3c recommendation, December 2012. http://www.w3.org/TR/owl2-primer/

A Controlled Natural Language for Business Intelligence Monitoring

Christian Colombo[⊠], Jean-Paul Grech, and Gordon J. Pace[⊠]

University of Malta, Msida, Malta
{christian.colombo,jean-paul.grech.11,gordon.pace}@um.edu.mt

Abstract. With ever increasing information available in social networks, the number of businesses attempting to exploit it is on the rise, particularly by keeping track of their customers' posts and likes on social media sites like Facebook. Whilst APIs can be used to automate the tracking process, writing scripts to extract information and process it requires considerable technical skill and is thus not an option for non technical business analysts. On the other hand, off-the-shelf business intelligence solutions do not provide the desired flexibility for the specific needs of particular businesses. In this paper, we present a controlled natural language enabling non-technical users to express their queries in a language they can easily understand but which can be directly compiled into executable code.

Keywords: Controlled natural languages · Social networks · Runtime verification

1 Introduction

Social media has provided the business community with a unique and unprecedented opportunity to engage with their customers, critics, competitors, etc. Yet, this comes at the cost of continuously monitoring various fora on which brand names may be mentioned, queries may be posted and products may be compared. Dealing effectively with social media in a context where even an hour can be considered as far too long a response time, is a challenging task.

Focusing in particular on Facebook, a typical business would have its own page as well as a strong interest in pages on which their products may be discussed or advertised. Typical events which are relevant for a business might include any mention of the brand or product, an advertising post by a competitor, a comment by a customer (particularly if negative or a question), and so on. To make the task of checking for these events manageable, dashboards [1–3] are available allowing users to specify events of interest so that a notification is received when such an event is detected (e.g., a notification when more than five comments are awaiting a response).

The problem with existing tools is that while they allow the specification of a number of events of interest, they do not offer the flexibility which might be required for the business' specific needs. For example, one might want to prioritise the notifications in order of urgency (e.g., a comment from a new customer

C. Biemann et al. (Eds.): NLDB 2015, LNCS 9103, pp. 300–306, 2015.
DOI: 10.1007/978-3-319-19581-0_27

might be given precedence over that of an existing customer); alternatively one might want to group them into batches (e.g., a notification per five comments unless a comment has been posted for more than three hours). Such flexibility usually comes at the price of a tailor-made solution which is generally expensive both if developed in house or by a third party.

One way of allowing a high degree of flexibility while providing an off-the-shelf solution would be to present a simple interface which would allow a business intelligence analyst the flexibility to express the desired events for notification. These would in turn be automatically compiled into Facebook monitors without any further human intervention. Whilst an automated compiler would struggle to handle natural language descriptions and a non-technical business analyst would struggle with a programming language, a domain-specific language presented to the user as a controlled natural language (CNL) [7] may act as an intermediary: it provides the feel of a natural language but constraints the writer to particular keywords and patterns.

In this paper, we present a CNL (Sect. 2) we have developed based on the results of interviews with business analysts, supporting the expression of requests such as *'Create an alert when the service page has a post and the post contains the keywords fridge, heater, and freezer'* and *'Create an alert when my page has a post and the post is negative and the post has 10 likes'*.

The language is given an operational semantics (Sect. 3), which in turn enables the compilation of specifications in the language into executable monitors which can analyse traces of Facebook events. Although runtime verification [4,8] is typically used for bug detection by matching the execution flow of a program to patterns encoded in terms of formally specified properties, runtime verification tools essentially provide specification of monitors independent of the main system. We have thus translated CNL specifications to be used by the runtime verification tool Larva [5] and then used an adapter to present relevant Facebook events as method calls in the control flow of a program.

Putting everything together, Fig. 1 depicts the proposed architecture. The user — depicted on the left — writes a specification in CNL and feeds it to the CNL-to-Monitor tool. Subsequently, this tool generates two elements: a monitoring specification for the Larva monitoring tool and an event bridge which harvests Facebook events and communicates them in an appropriate form to Larva. Finally, based on the output from Larva, a dashboard is updated to give feedback to the user. The architecture we have developed can be easily adapted to new data sources (e.g. Twitter) and to alternative monitoring tools, although some work would be required to modify the Facebook-specific parts of the CNL.

The CNL has been evaluated (see Sect. 4) from the point-of-view of non-technical users through a hands-on session and questionnaire involving users from an insurance company. The results, reported herein, indicate that the users managed to understand well ready-written rules and were able to express rules using the language without any syntactic checking and user-interface support.

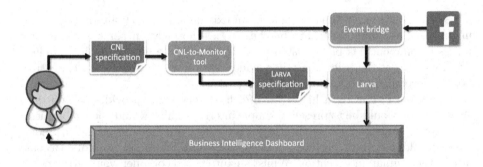

Fig. 1. The architecture of the proposed approach

2 A Controlled Natural Language for Business Intelligence

The business intelligence language being proposed, based on interviews with end-users, is at core a controlled natural language [7], to enable non-technical users to experiment with different rules and modify them without the need of going through a further cycle with a developer or technical expert.

To reduce the risk of syntax errors, at the top level, the language is largely a template-based one [6] allowing for the definition of alerts as in, for example, the declaration:

Create an alert when the service page p has a multiple of 5 check-ins, with priority 3.

This alert notifies the user whenever a particular page hits 5, 10, 15, etc. check-ins[1]. The priority identifies the severity of the alert, allowing us to have a tiered approach to alert handling.

Alerts can be triggered on three main types of events: (*i*) page-centric events; (*ii*) post-centric events; and (*iii*) message-centric events. All these events fire when a particular change of state happens. For instance, in the case of page-centric events, three types of event firings can be identified as shown in the following grammar fragment:

$$\langle PageEvent \rangle :: = \langle Page \rangle \text{ has a post } \langle Filter \rangle$$
$$| \quad \langle Page \rangle \text{ has } \langle Count \rangle \text{ check-ins}$$
$$| \quad \text{poster } \langle UserPageAction \rangle \langle Page \rangle$$

Thus, page-centric events can fire when (*i*) a new post appears on a page; (*ii*) a number of check-ins occur on a page; or (*iii*) a user (poster) performs an action on a page (e.g., likes or shares a page). As can be seen, these events can be easily extended by allowing for alternative templates at this level.

---

[1] Check-ins are events when users register their presence at a particular location or business premises.

However, at a finer level, our alert specification language loses its template-based flavour and allows for a slightly freer form of specification. Most events can be filtered by relevant features — for example, posts on a page can be filtered by placing a conjunction of constraints on the message or the poster, as can be seen in the following grammar fragment:

$$\langle MessageFilter \rangle ::= \text{has } \langle Count \rangle \langle UserPageRelation \rangle$$
$$| \quad \text{has keywords } \langle KeywordList \rangle$$
$$| \quad \text{is } \langle TypeOfComment \rangle$$
$$| \quad \ldots$$
$$\langle PosterFilter \rangle \quad ::= \text{poster } \langle UserPageAction \rangle \langle Page \rangle$$
$$| \quad \text{has left } \langle Count \rangle \text{ posts on } \langle Page \rangle$$
$$| \quad \text{is from } \langle Location \rangle$$
$$| \quad \ldots$$

Note that the filters take different forms, depending on the elements available for the event in question. While users posting comments can be filtered by their location, messages can be filtered by keywords. Of particular interest are: (i) filters based on linguistic analysis of the customer's post — the third option for a message filter in the grammar above is such a filter, allowing an alert to depend on whether a question was posted, or based on whether the comment was a positive or negative one by using an external sentiment analysis library; and (ii) filters based on temporal constraints, such as by day of the week, date ranges or time elapsed between two events.

Apart from alert definitions, the language also supports setting, and modifying parameters for the alert rules to trigger. For instance, a user may set the weight assigned to negative or positive posts on a page, to allow them firing when they exceed a particular threshold.

3 Monitoring Semantics

The CNL is given an operational semantics of the form $\sigma \xrightarrow{a} \sigma'$, where a is a timestamped Facebook event and σ is the configuration of the Facebook monitor, storing the relevant information, such as timers, the state of counters, etc. The rules for all the patterns supported by the CNL are represented by the rule CNL (see Fig. 2). These semantics can be readily translated into a specification for a runtime verification tool such as LARVA.

Note that rules of the form $\sigma \xrightarrow{a} \sigma'$ assume that monitoring can take place as soon as an event happens, i.e. in synchrony with events happening on the Facebook platform. Unfortunately, this is not always feasible in the case of Facebook, since one is not allowed to subscribe to notifications of pages one does not own. To address this constraint, we adopt a polling-based approach, in which the system regularly queries Facebook (through its API) and fetches relevant events which have gathered since the last query. For this reason, the monitor consumes events from a buffer rather than directly from Facebook as represented by rule MON.

$$\text{CNL} \ \frac{\cdots}{\sigma \xrightarrow{a} \sigma'} \qquad\qquad \text{MON} \ \frac{\sigma \xrightarrow{a} \sigma'}{(\text{fb}, \sigma)_{a:\text{buf}} \xrightarrow{a} (\text{fb}, \sigma')_{\text{buf}}}$$

$$\text{FB} \ \frac{\text{fb} \xrightarrow{W} \text{fb}'}{(\text{fb}, \sigma)_{\text{buf}} \xrightarrow{\text{poll}} (\text{fb}', \sigma)_{\text{buf}++\text{sort}(W)}}$$

Fig. 2. The semantic rules for monitoring Facebook events

Representing the progression of the state of Facebook pages is not as straightforward as it might appear, since Facebook can only be queried for events on a page-by-page basis, thus returning a set of unordered events in each case. Thus, mathematically, the trace fragment returned by a Facebook poll has to be chronologically sorted before being processed, as is shown in rule FB.

Connecting back to the architecture presented in the first section, rules MON and FB are embodied in the *Event bridge* component which replays the events to the LARVAmonitor (which in turn embodies the CNL rule) in the correct order. This logic has been successfully implemented and tested on two case studies. However, in this paper we focus on the language design aspect of this work and thus the next section describes how we evaluated the CNL in terms of its understandability and usability by non-technical users.

4 Evaluation

From an expressivity point of view, we ensured that the language supports the necessary aspects by interviewing business analysts. However, there were a number of interesting elements which were not easy to incorporate within the CNL without running into considerable complexities:

Social Awareness. It might be useful to distinguish between the people interacting with the business' online presence in terms of how much closely related they are to the business. For example, a like from a business employee or a close relative might be considered less important than that of a person with no links. Although a valid suggestion, we considered these social aspects to be outside the scope of our time-limited project.

Semantic Analysis of Posts. Another proposal emerging from the interviews was to enable semantic analysis of posts, identifying adverts by third parties, distinguishing between positive and negative posts, questions from non-questions, etc. Whilst we have successfully managed to integrate two third party projects[2] for these purposes, there are many other natural language techniques which can extract further useful information.

---

[2] http://sourceforge.net/projects/chatscript for question detection and http://sentiment.vivekn.com for sentiment analysis.

Apart from expressivity issues, our main concern for the proposed CNL was how easily a lay person would be able to understand expressions written in the CNL and subsequently how long it would take for the person to be able to write useful expressions in CNL. To this end, a questionnaire was used to interview thirteen non-technical persons from a local insurance company. The questionnaire was split into two main parts as follows:

Understanding the CNL. The participants were presented with a number of statements expressed in the proposed CNL (e.g., *'Create an alert when the competitor page* (www.facebook.com/competitor) *has a positive post and the poster has left posts on my page* (www.facebook.com/mybusiness) *before'*) and they were expected to explain the meaning of the statements in natural language without any supporting documentation. In each of the four cases, more than two-thirds of the respondents explained the statement correctly although some left out minor details in their explanations.

Expressing Statements in the Language. The second exercise involved the opposite: given a textual description (e.g., *'You want to know when someone leaves a question on your page'*), respondents were expected to write it in terms of the CNL without any support (except the language samples in the previous exercise). This proved to be harder but around 60 % of the respondents managed to produce an answer sufficiently close to be easily auto-correctable with the help of an appropriate user interface.

With these encouraging results, we feel that the presented work is a step in the right direction albeit requiring further development in order to make the approach more usable in practice.

5   Conclusions

In this paper, we have presented some initial experiments with the design of a controlled natural language to enable business analysts to customise a Business Intelligence dashboard. Although the language is rather contrived, it proved to be usable by non-technical experts, and can be used to effectively customise analysis of social media activity. Furthermore, the architecture can be easily adaptable to work of other streams of information e.g. Twitter, or to refer to other currently unhandled events from Facebook.

One way of increasing the usability of the CNL approach is by providing users with a richer interface (rather than simply a text editor) to write their business rules. By using a pull-down menu approach or a 'fridge magnet' interface [9] (in which the interface emulates fridge-magnet words, allowing the users to move the words to compose valid sentences on the fridge), would make the process less error prone and would avoid error-handling and syntax debugging to which non-technical persons can be averse.

The interviews we held with business analysts also suggest that there are areas of great interest which the CNL barely touches. We hope to extend this in the future by integrating with more NLP tools on the one hand, and extracting more information from the social network to provide further data points. Finally, by looking into other industrial case studies and by having more non-technical users we hope to obtain more feedback to fine-tune the CNL to improve its comprehensibility and utility.

References

1. Facebook real-time updates. March 2014. https://developers.facebook.com/docs/graph-api/real-time-updates
2. Geckoboard. December 2013. http://www.geckoboard.com
3. Tableau software. December 2013. http://www.tableausoftware.com
4. Colin, S., Mariani, L.: 18 run-time verification. In: Broy, M., Jonsson, B., Katoen, J.-P., Leucker, M., Pretschner, A. (eds.) Model-Based Testing of Reactive Systems. LNCS, vol. 3472, pp. 525–555. Springer, Heidelberg (2005)
5. Colombo, C., Pace, G.J., Schneider, G.: Larva – safer monitoring of real-time java programs (tool paper). In: Software Engineering and Formal Methods, pp. 33–37. IEEE (2009)
6. Esser, M.W., Struss, P.: Obtaining models for test generation from natural-language-like functional specifications. In: Workshop on Principles of Diagnosis, pp. 75–82 (2007)
7. Kuhn, T.: A survey and classification of controlled natural languages. Comput. Linguist. **40**(1), 121–170 (2014)
8. Leucker, M., Schallhart, C.: A brief account of runtime verification. J. Log. Algebr. Program. **78**, 293–303 (2009)
9. Ranta, A.: Grammatical framework. J. Funct. Program. **14**(2), 145–189 (2004)

Text Summarization and Speech Synthesis for the Automated Generation of Personalized Audio Presentations

Séamus Lawless[✉], Peter Lavin, Mostafa Bayomi, João P. Cabral, and M. Rami Ghorab

CNGL Centre for Global Intelligent Content, Knowledge and Data Engineering Group, School of Computer Science and Statistics, Trinity College Dublin, Dublin, Ireland
{seamus.lawless,peter.lavin,bayomim,cabralj, rami.ghorab}@scss.tcd.ie

Abstract. In today's fast-paced world, users face the challenge of having to consume a lot of content in a short time. This situation is exacerbated by the fact that content is scattered in a range of different languages and locations. This research addresses these challenges using a number of natural language processing techniques: adapting content using automatic text summarization; enhancing content accessibility through machine translation; and altering the delivery modality through speech synthesis. This paper introduces Lean-back Learning (LbL), an information system that delivers automatically generated audio presentations for consumption in a "lean-back" fashion, i.e. hands-busy, eyes-busy situations. These presentations are personalized and are generated using multilingual multi-document text summarization. The paper discusses the system's components and algorithms, in addition to initial system evaluations.

Keywords: Lean-back learning · Text summarization · Speech synthesis · Multilingual content adaptation, personalization

1 Introduction

The constantly connected nature of today's world places increasing demands on people's time. When coupled with the tsunami of information that is competing for individuals' attention on a daily basis, users typically face the challenge of having to consume large volumes of content in short periods of time. Moreover, the naturally distributed and inherently multilingual nature of the web creates a further challenge.

The terms "lean forward" and "lean back" can be used to characterize a user's engagement with a computing device. In *lean-forward* engagement a user is focused on the device and constantly interacts with the system. An example of this form of engagement is a user reading a report online or using spreadsheet software. In contrast, in *lean-back* engagement the user's focus can be elsewhere. Interaction with the device is minimal, yet they are still consuming information. Earlier forms of this type of engagement are radio and television. However, over recent decades, miniaturization coupled with widespread Internet connectivity has meant that media consumption with this low level of engagement can now also take place using mobile technology.

© Springer International Publishing Switzerland 2015
C. Biemann et al. (Eds.): NLDB 2015, LNCS 9103, pp. 307–320, 2015.
DOI: 10.1007/978-3-319-19581-0_28

Allowing users to consume traditional web content in a lean-back fashion begins to tackle the challenges outlined above. This paper introduces *Lean-back Learning (LbL)*, an intelligent and responsive information system built upon a set of web services. The system delivers automatically generated audio presentations for consumption in a lean-back fashion, i.e. hands-busy, eyes-busy contexts. Audio presentations are generated using automatic text summarization and speech synthesis. The audio output is provided in multiple languages and is compiled, translated and summarized using multilingual sources. The user can tailor presentations by selecting the preferred level of detail and output language, and specifying their available listening time.

Lean-back Learning allows a user who wishes to learn about a topic, to input one or more search terms. The system then presents the user with a summarized audio presentation, which they can listen to at their leisure. This application can be used in a variety of scenarios, for instance: a tourist who is about to visit a cultural site and who wants some last-minute background information; a student who wishes to get an overview of a new topic that they are about to learn; or a person who wishes to listen to a summary of a piece of news. The users can listen to this information, tailored to their needs, all while being able to carry out other tasks.

This paper discusses the components and algorithms of the proposed LbL system, and presents a usage scenario. It also presents an evaluation of the text summarization component of the framework. This is the first in a series of planned experiments which will evaluate the framework from quantitative and qualitative perspectives. The paper also addresses general questions on the feasibility of dynamically synthesizing summarized text. Areas such as practical and acceptable response times and management of users' expectations around the amount of content available are discussed.

The remainder of this paper is organized as follows. Section 2 presents state-of-the-art in the areas of Summarization and Speech Synthesis. Section 3 describes the proposed LbL framework (architecture, data flow, and user interface). Sections 4 and 5 discuss the two main underlying services of the framework: *SSC (Search, Summarize, and Combine)* and *SSyn (Speech Synthesis)*. Following that, as part of the research underpinning this framework, an evaluation of the framework components is presented in Sect. 6. Finally, insights and future work are explored Sect. 7.

2 Background and State-of-the-Art

As Text Summarization and Speech Synthesis are the core services used by the LbL framework, this section provides a state-of-the-art review of research in those areas.

2.1 Text Summarization

Automatic text summarization is a prominent research area in the field of Natural Language Processing (NLP). Research in this area led to the development of various summarization techniques as well as the application of those techniques in diverse domains and on a variety of content bases, such as scientific repositories and meeting recordings [1]. The two main approaches to text summarization are Abstraction [2] and

Extraction [3]. Abstraction is where the summarization system has to "understand" what the text means in order to build an internal semantic representation of the content. Natural language generation techniques are then used to create a summary that is deemed close to what a human being might create. A significant challenge that faces the Abstraction approach is that it requires accurate semantics and training data to be able to automatically interpret the meaning of the content. Because of this challenge, summarization techniques that are based on Abstraction have thus far only exhibited limited success [4]. On the other hand, the Extraction approach, such as LexRank [5] does not require that the summarization system understand the meaning of the text; rather, it attempts to identify the most important sentences in order to extract them into a summary. The system proposed in this paper is based on the Extraction approach as, to-date, it has been shown to be more effective.

Various methods have been discussed in the literature to improve extractive summarization systems. Early systems depended on simple methods, such as: Sentence Location [7], Cue Phrase [8], Most Frequent Words [9], and Sentence Length [8].

The LbL framework introduced in this paper uses *TextRank* [6] as its foundation. TextRank, one of the most prominent Extraction-based methods, is an unsupervised graph-based ranking model that summarizes text by extracting and ranking the most important sentences according to the number of overlapping words between them.

2.2 Speech Synthesis

The most popular speech synthesis approaches in the literature are Unit-Selection and Hidden Markov Model (HMM) based speech synthesis. A comparison of the two approaches is given in [10]. In summary, the Unit-Selection approach is based on the concatenation of pre-recorded speech units. This approach offers limited flexibility for controlling voice characteristics. Moreover, the type of voice is restricted to that of the recorded speech. On the other hand, the HMM-based approach is fully parametric and the acoustic modelling of speech is learned automatically by training the statistical models from a recorded speech corpus. Its advantage is that it permits the transformation of the synthesizer's voice type using a small amount of data from the target voice. HMM-based speech synthesis can produce high-quality speech but generally does not sound as natural as that of unit-selection systems [10].

There have been several recent attempts to combine the advantages of statistical approaches with the "naturalness" obtained using unit-selection. Current state-of-the-art systems, such as the system reported in [11], are hybrid systems that are based on the unit-selection approach but use HMMs for selection of the concatenation units. In spite of the relatively high quality voice output produced by these systems, there is a growing demand for increased expressiveness of synthetic speech that is beyond what can be currently produced [12], e.g. audiobooks, spoken dialogue systems, etc. For example, synthesizing voices with different speaking styles using a HMM-based system is one way to improve the speech expressiveness in audiobooks [13]. The prosodic aspects of the synthetic speech, such as pause duration and intonation along the phrase are also important for engaging the listener in the communication process and need to be better modelled by the synthesizers.

The LbL framework uses popular open source speech synthesis systems (HMM-based and unit-selection) for investigating the limitations of this technology in the generation of audio presentations. We aim to identify ways to improve the quality of the standard speech synthesis systems, particularly for this type of application. The HMM-based method is typically preferred in platforms such as LbL, because it is more suitable for mobile devices (faster speech synthesis and lower memory footprint) and it offers higher flexibility for personalizing the voice. For example, using HMM adaptation algorithms it is possible to build a new speaker's voice using a small amount of speech data from that speaker [14].

3   Lean-Back Learning Portal

LbL aims to provide an end-to-end platform for retrieving, adapting, and narrating concise knowledge. The system therefore orchestrates all the elements of the process: interfacing with the user, retrieving information, translation and summarization, and synthesizing voice narrations. This section discusses how the system controls the underlying components, the workflow and user interface. Details of the underlying algorithms of each component are given in the later sections.

The framework currently uses Wikipedia as its content source. The following reasons were behind the choice of Wikipedia: (1) it is an open and multi-domain content source; (2) it features content in multiple languages; and (3) Wikipedia articles are well and consistently structured, which facilitated producing tailored summaries that vary in the amount of content and level of detail (depending on the user's needs). The framework is, however, not functionally tied to Wikipedia as a content source. LbL could function just as effectively on any body of structured data.

3.1   Framework Architecture

The LbL framework is made up of three components, each of which can stand alone as an independent web service: *Search, Summarize & Combine (SSC)*; *Speech Synthesis (SSyn)*; and the *Sequence Controller (SC)*, which also incorporates the user interface (UI) component. The overview is shown in Fig. 1.

The framework's 'modular' design is intentional and uses standard Internet protocols (RESTful and HTTP posts) for communication between components. This allows components to be altered, added or replaced without impacting on other parts of the framework. The SC service also uses a database to store details of each presentation prepared. For design and performance monitoring purposes, it also records times taken for the SSyn and SSC components to complete their roles.

3.2   Workflow and User Interface

The LbL UI is developed using responsive design to ensure a consistent experience across all devices. The user has the ability to sign in using their Google + account. Once logged in, s/he can specify the topic using free-text keyword search, then select

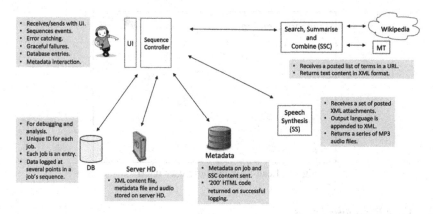

Fig. 1. Overview of lean-back learning architecture

the level of detail required: High-level Overview, Detailed Introduction, or All Details. The 'level of detail' selected governs the number of sections and the depth of sub-sections from the Wikipedia article to be included in the automatically generated summary. Next, the user selects their required output language (three languages are currently supported by the system: English, French, and German). The user's input is submitted to the SC component which passes it on to the SSC service. SSC carries out an initial evaluation for this topic and estimates the amount of content available. Three word-count values are returned, each one corresponding to one of the three levels of detail available in the UI.

Back at the UI, estimated minimum, intermediate and maximum durations for the audio presentation are calculated and presented to the user for selection; these are based on the aforementioned word-counts and use an empirically obtained 'word-count to audio seconds (duration)' ratio. Figure 2 shows the UI and a typical range of durations. The user can then select their desired duration. The presentation length selected is used within the SSC service to determine what level of summarization is applied to the sections used in the relevant article. For this work, summarization varies between 5 % and 80 % of the source text.

This sequence is interactive. If the user's initial expectations on presentation duration are not met or are exceeded, the user has two options as follows. They can select another level of detail as appropriate and thereby obtain a greater or lesser number of sections and sub-sections from the source article. Alternatively, they can select a different level of summarization by choosing different time options. The initial feedback provided, along with the user interaction around the amount of available content serves to ameliorate the unpredictability caused by variation in articles, both in terms of their length and depth of detail. When the user is satisfied and specifies their final desired presentation length, the details are submitted for processing. From this point onwards, the sequence of events is controlled by the SC. The topic strings, level of detail and selected presentation duration are sent to the SSC service as a tuple. The SSC then returns the summarized content in XML format.

Fig. 2. Lean-back learning user interface

3.3   User Response Times

For usability, it is desirable that the user has the shortest possible wait-time before audio playback begins. This time must include the time taken for content retrieval, summarization, and speech synthesis. A long wait-time here would have a significant impact on the usability of the system. To address this, rather than synthesizing the entire presentation at once, the XML output of the SSC component is parsed by the SC and divided into a number of small chunks of text. As soon as the first segment is created, it is immediately sent to the SSyn service to allow the user to start listening.

In LbL, Speech Synthesis takes place faster than audio playback. The first audio files are created relatively shorter than subsequent ones. This ensures that, once listening has commenced, there is always more synthesized audio available than the user has consumed. After the first segment is synthesized, listening and synthesis take place simultaneously and items on the playlist become available.

4   Search, Summarize, and Combine

This section describes the underlying process for retrieving, summarizing, translating, and combining content. The section also describes how the content that is output by the summarization process is prepared for speech synthesis.

The SSC service has a number of modules which interact in order to deliver the required summarized output. The service pipeline is shown in Fig. 3. An Information Retrieval (IR) module is used to search across Wikipedia and retrieve links to pages in multiple languages (English, French and German) which are relevant to the topic in question. Wikipedia APIs are used to retrieve the text of these pages. The information in Wikipedia articles of different languages is not necessarily the same (i.e. the articles are typically natively authored in each language and are not translations of each other). So the advantages of initially sourcing content in different languages are: (1) to extend the search to a wider content-base; and (2) obtain multiple perspectives on the same topics (e.g. political events). If this approach is extended to minority languages, detailed presentations could be generated where very limited source content exists. A Slicer module is used to slice each page into sections using the embedded Wikipedia structure. The Multilingual Summarizer component automatically generates summaries for each content slice in the source language of the text. The Machine Translation (MT) component[1] then translates the summaries from their source language into the output language that was specified by the user. The Merger combines the summaries from all the documents and merges those which contain information about the same aspect of the topic. Finally, the Multi-document Summarizer produces a final summary by eliminating redundant information that may exist after merging.

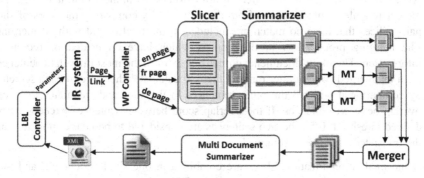

Fig. 3. The SSC service pipeline

The summarizer module uses a number of processing steps in order to identify and extract the most important sentences from an article. This extractive task is unsupervised and uses the TextRank algorithm. TextRank measures the similarity between sentences by calculating the number of overlapping words between them. Before the comparison of sentences takes place, stopwords are removed and then the remaining words are stemmed using the Lancaster stemmer (following on the successful approach presented in [23]). The sentence location approach [7] is applied to extract the first sentence from the original article to be used as the first sentence in the summary. The SSC service works as follows:

1. The SC sends four parameters to the SSC service: (a) the user's query; (b) desired level of detail; (c) desired presentation duration; and (d) output language.
2. The service starts by calling language detector API to identify the query language.
3. The IR module is then called to search across Wikipedia. The result is a Wikipedia link to a page that contains the article that the user searched for.
4. This link is used to retrieve links to the same article in the remaining two languages. The articles (one for each language) are retrieved as HTML.
5. The slicer segments each article into slices according to the page structure in Wikipedia (i.e. main sections, sub-sections and sub-sub sections). This is used to personalize the final presentation. For example, if a user requests a *"High-level*

---

[1] The MT service used in the experiments is Bing Translation API. The following are the reasons for choosing Bing: (a) it is generally known to perform relatively well in terms of translation quality and speed; (b) it supports a range of languages; and (c) it provides a well-defined RESTful Web service to communicate with it.

Overview", only the top-level hierarchical sections are used. The final output from the slicer is three groups of slices: all the slices of the article for each of the languages.

6. Each slice is summarized using the TextRank algorithm. The length of the summary generated is dictated by the desired duration that the user has specified.
7. After the slices are summarized, a machine translation module starts to translate slices from their source language to the output (user-specified) language.
8. After translation, the SSC has three groups of slices, all in the output language. The merger module combines slices from these groups by comparing the titles of slice pairs. Slices that have no match are included in the final output without merging. The merging process can produce slices which have duplicated or redundant information. The multi-document summarizer extracts the most important sentences from the merged content after ranking them. It calculates the score of each sentence by comparing it with all the other sentences in the slice and calculates the word overlap between sentences. If the overlap score between sentences exceeds a predefined threshold (0.5), the two sentences are considered to be near duplicates, and the translated sentence is removed.

After the final summarization, slices are combined in one single document and then converted into an XML file for delivery to the SC component.

5 The Speech Synthesis Process

The LbL framework is designed so that users can consume information in a lean-back fashion, i.e. hands-busy, eyes-busy contexts. Therefore, one of the main features of the implemented system is the generation of audio presentations based on the summarized content received from the previous step.

The speech synthesis process comprises a text analysis component which extracts the linguistic features that are used for building the synthetic voice and for processing input text at the speech synthesis stage. This module includes tools for text normalization, and Natural Language Processing (e.g. part-of-speech tagging, grapheme-to-phoneme conversion and intonation prediction). Typically, this part of the synthesizer is strongly language-dependent and it is necessary for the system to function effectively, regardless of the approach used (e.g. unit-selection or HMM-based).

The English HMM-based synthetic voice used in this framework was built using the HTS-2.2 toolkit [15]. The speech data used to build the English voice is the female US SLT subset of the CMU ARCTIC speech database [16]. The text analysis part of voice building was performed using Festival Multisyn [17]. The system uses the STRAIGHT vocoder [18] to extract spectrum and aperiodicity parameters from the signal during analysis. F0 is the other speech parameter which is estimated using the Entropic Speech Tools implementation of the RAPT algorithm [19].

For acoustic modelling, the system uses a five-state Hidden Semi-Markov Model (HSMM) structure. The F0 parameter vector (including its delta and delta-delta features) is modelled by multi-space probability distribution HMM (MSD-HSMM), whereas the spectrum and aperiodicity streams (including dynamic features) are modelled by HSMM using continuous distributions respectively. The F0, spectrum and aperiodicity parameters are clustered using different decision trees, because these parameters have their own contextual factors.

During synthesis, the speech parameters are generated from the input text and trained HSMMs, using a parameter generation algorithm based on the maximum likelihood criterion. Finally, the speech waveform is produced from the speech parameters using the STRAIGHT vocoder. For the provision of French and German languages in the service, the LbL framework currently uses the publically available Unit-Selection MARY TTS system [20].

6 Evaluation

Due to the multi-layered nature of the LbL system, a single method of evaluation would be insufficient to effectively evaluate the performance of each component. Therefore, the plan for evaluation consists of several phases that involve both quantitative and qualitative evaluation. The quantitative evaluation is concerned with the performance and effectiveness of the various components and algorithms that make up the framework. The qualitative evaluation is concerned with the user's perception of the service in a lean-back situation, the informational value of the generated summaries, and preference for synthetic audio presentations as compared to text form. The aim is to evaluate both the combination and the compartmentalization of all the elements that make up the LbL system. In this paper, we focus on the evaluation of the SSC component, which is a core component to the framework. We also provide an outline of the planned evaluation for the SSyn component.

6.1 SSC Component Evaluation

Throughout the evolution of approaches to automatic text summarization, many evaluation studies [3, 21] have been based on automated evaluation systems such as ROUGE [22]. ROUGE is a tool which includes several automatic evaluation methods that measure the similarity between summaries. However, automatic evaluation lacks the fine-grained, nuanced judgments of quality which can only be achieved through human evaluation. This includes quantified assessments of qualities such as *read-ability/understand-ability*, *informativeness*, *conciseness*, and the *overall quality* of the summary. To this effect, in this paper a human evaluation is carried out to assess the quality of the summaries produced by the SSC service. This is achieved via human judgments of the overall quality of the summaries produced as well as the quality of the summaries in specific subject domains. As different subject domains (e.g. Politics, Sports) use different terminology, language-styles, and structure, these factors would affect the performance of automatic summarization algorithms; hence, we evaluate the knock-on impact of this on users.

While one of the elements in the evaluation is the quality of the generated summary with respect to the source text, the quality of the source text itself is out of scope of this study. Furthermore, LbL functions independent of the structure of the underlying content source and is designed to allow different content sources to be utilised.

Experimental Setup. The aim of the experiment is to evaluate the quality of summaries in general and for specific subject domains. A document-set of 25 abstracts of Wikipedia articles was selected from six subject domains. Abstracts have different lengths (ranging from approx. 180 words to more than 560 words.) The subject domains were selected at random and the articles were randomly chosen from these domains. The selected domains were: Accidents, Natural Disasters, Politics, Famous People, Sports, and Animals. After selecting the articles from Wikipedia, a summary was generated for each abstract using the summarization module discussed earlier.

To conduct the evaluation of the summaries, a web application was developed. The articles were divided into groups; each group had five articles, each article from a different domain. When the first user (experiment participant) logged in, s/he was randomly assigned a group. The next user was then randomly assigned a group from the remaining unassigned groups. This continued until all the groups were assigned to users. The process was then repeated for the next set of users who logged in to the system. This ensured an even spread of assessment. Each article in the group that is presented to the user was followed by the generated summary. The users were asked to evaluate each summary according to the following characteristics:

1. Readability & Understand-ability: the user was asked to assess whether the grammar and the spelling of the summary is convenient or not.
2. Informativeness: assess how much information from the source text is preserved in the summary.
3. Conciseness: assess if this summary does not contain any unnecessary or redundant information.
4. Overall Quality: evaluate the overall quality of the summary.

The users were asked to evaluate each characteristic on a mean Likert scale (ranging from 1 to 6, where 1 is the lowest quality and 6 is the highest). An open call for participation in the experiment was made through mailing lists and social media. The participants came from different countries and were from academia and industry. They had different educational backgrounds and disciplines. Their ages ranged from 27 to 40.

Results and Discussion. Thirty eight users ultimately participated in the experiment. Each user evaluated at least four articles in different domains. The final result was analyzed regarding: (a) general summary quality and (b) domain-specific summary quality. Table 1 reports the mean and the standard deviation scores of the user evaluations. The scores show that, in general ("All Domains Combined"), users were satisfied with the summaries produced by our summarization system, as the summaries received an average mean score of over four in all criteria measured. The results also show the mean and the standard deviation scores of the user evaluations for each domain separately. Some domains exhibited higher mean scores than others (e.g. Accidents and Natural Disasters vs. the other domains).

Table 1. Quality of summaries (by domain and combined)

Domain	Mean & Std. Dev	Readability	Informativeness	Conciseness	Overall
Accidents	Mean	5.21	4.50	4.71	4.50
	SD	0.8926	0.7596	0.9945	0.6504
Natural disasters	Mean	5.00	4.60	4.33	4.56
	SD	0.8528	0.9145	1.044	0.8675
Politics	Mean	4.61	4.33	3.56	4.11
	SD	1.243	1.138	1.042	1.023
Famous people	Mean	4.79	4.21	4.34	4.31
	SD	1.048	0.9403	1.010	0.8906
Sports	Mean	4.63	4.43	4.27	4.30
	SD	0.9994	0.8172	1.172	0.9523
Animals	Mean	4.96	4.13	4.13	4.21
	SD	0.9991	0.7974	0.9470	0.8330
All domains combined	Mean	4.86	4.39	4.24	4.36
	SD	0.9936	0.9041	1.067	0.8856

6.2 SSyn Component Evaluation

The HTS system used in this work for synthesizing the English voice is a very popular HMM-based speech synthesizer, which performed very well against other speech synthesizers in the Blizzard Challenge 2005 [24].

The Blizzard Challenge is an annual event in which participants are provided with a speech corpus and have to synthesize a set of test utterances. Then, an overall evaluation of the synthesizers is conducted and the results can be examined in the Blizzard Challenge Workshop. The HTS system has been used as the benchmark HMM-based speech synthesizer in the Blizzard Challenge since 2006. For example, in the recent Blizzard Challenge 2013 [25], only two systems out of ten obtained better results in terms of naturalness and similarity to the target speaker than HTS and this benchmark system was in the group of the four equally-intelligible systems which obtained the best intelligibility results (among a total of 11 systems).

The MARY TTS system [27] which was also used in this work for synthesis of German and French has also taken part in several editions of the Blizzard Challenge evaluation, e.g. [26]. However, while the system itself has been evaluated, these languages have never been evaluated in the Blizzard Challenge. Therefore the system's speech quality for these languages will need to be evaluated against the benchmark systems in order to gain a complete picture of the performance of the SSyn service in LbL.

The current speech synthesizers used in the LbL framework acts as a baseline synthesizers in ongoing experiments which are being conducted to evaluate our own improved versions of the speech synthesis service in multiple languages. There are two hypotheses that we plan to test through the ongoing experimentation. One is to test the hypothesis that an audio presentation is better than the written one. This is being evaluated both in terms of user preference and listener comprehension, especially in the dual tasking situation where the user is listening and trying to understand a spoken summary while doing something else. Another evaluation is to test if the synthetic voice does not significantly affect the user satisfaction with the LbL service compared with a recorded voice from a person reading the summary. We also plan to conduct experiments to evaluate the quality of synthetic voice in the context of spoken summaries. This will be used to identify important factors which affect this type of audio presentation, such as pause position and length which are predicted from the text, speech naturalness and variation of intonation along and between sentences.

6.3   System-Level Evaluation

While this paper focuses on the evaluation of the two main components of the LbL system, this is not intended to downplay the importance of whole-system user-focused evaluation. A qualitative user study is planned which will evaluate the user perception of LbL with respect to the function of the user interface and the interaction mechanisms, the informational value of the summaries which are generated, and the quality of the synthesized voice and audio presentations that are delivered by the service.

7   Conclusion and Future Work

This paper introduced Lean-back Learning: an information system that delivers automatically generated audio presentations for consumption in hands-busy, eyes-busy contexts. The system is based on two core NLP components: an automatic text summarization service and a speech synthesis service. The paper presented the LbL architecture, its component services and the algorithms which are used by those services. In addition, a series of initial system evaluations were discussed.

There are a number of future areas of research which are planned, with the aim of improving the LbL framework. Features extracted from sentence parsing will be used to extend the feature set currently used in tree-based clustering for speech synthesis. A syntax parser can generate a large number of features from which the most appropriate can be selected to model prosody. For example, features could be derived from the syntax trees which are correlated with pause position and duration.

Further work is planned in the area of personalization. In the current version of the system, the audio presentation is adapted based upon explicit user input. This can be extended to utilize user history and inferred preferences to personalize both the content and synthesized voice used in generating the audio presentations. Lean-back Learning may also be used as a tool for evaluating the differences between the effects of learning from audio and written media.

Acknowledgements. This research is supported by the Science Foundation Ireland (grant 07/ CE/I1142) as part of the Centre for Next Generation Localisation (www.cngl.ie) at Trinity College, Dublin.

References

1. Murray, G., Renals, S., Carletta, J.: Extractive summarization of meeting recordings. In: Proceedings, Interspeech' 2005 - Eurospeech, 9th European Conference on Speech Communication and Technology. Lisbon, Portugal (2005)
2. Fiszman, M., Rindflesch, T.C.: Abstraction summarization for managing the biomedical research literature. In: Proceedings of the North American Chapter of the Association for Computational Linguistics: Human Language Technologies (HLT-NAACL) Workshop on Computational Lexical Semantics (CLS), pp. 76–83, Boston, Massachusetts (2004)
3. Vodolazova, T., Lloret, E., Muñoz, R., Palomar, M.: A comparative study of the impact of statistical and semantic features in the framework of extractive text summarization. In: 15th International Conference on Text, Speech Dialogue, (TSD), pp. 306–313, (2012)
4. Nenkova, A., Mckeown, K.R.: Automatic summarization. Found. Trends Inf. Retrieval **5**, 103–233 (2011)
5. Radev, D.R.: LexRank: graph-based lexical centrality as salience in text summarization. J. Artif. Intell. Res. **22**, 457–479 (2004)
6. Mihalcea, R., Tarau, P.: TextRank: Bringing order into texts. In: Lin D., Wu D. (eds) Proceedings of the 2004 Conference on Empirical Methods in Natural Language Processing (EMNLP). Association for Computational Linguistics, Barcelona, Spain, pp. 404–411 (2004)
7. Edmundson, H.P.: New methods in automatic extracting. J. ACM **16**(2), 264–285 (1969)
8. Teufel, S., Moens, M.: Sentence extraction as a classification task. In: ACL/EACL workshop on Intelligent and scalable Text summarization, pp. 58–65, Madrid, Spain (1997)
9. Luhn, H.P.: The automatic creation of literature abstracts. IBM J. Res. Dev. **2**(2), 159–165 (1958)
10. Black, A., Zen, H., Tokuda, K.: Statistical parametric speech synthesis. In: Proceedings of the International Conference on Acoustics, Speech, and Signal Processing (ICASSP 2007), pp. 1229–1232 (2007)
11. Ling, Z., Wang, R.: HMM-based hierarchical unit selection combining kullback-leibler divergence with likelihood criterion. In: Proceedings of the International Conference on Acoustics, Speech, and Signal Processing (ICASSP 2007), pp. 1245–1248 (2007)
12. Türk, O., Schröder, M.: Evaluation of expressive speech synthesis with voice conversion and copy resynthesis techniques. IEEE Trans. Audio, Speech, Lang. Proc. **18**(5), 965–973 (2010)
13. Székely, E., Cabral, J.P., Cahill, P., Carson-Berndsen, J.: Clustering expressive speech styles in audiobooks using glottal source parameters. In: Proceedings of Interspeech, Florence, Italy (2011)
14. Yamagishi, J., Kobayashi, T.: Adaptive training for hidden semi-Markov model. In: Proceedings of the International Conference on Acoustics, Speech, and Signal Processing (ICASSP 2005), Philadelphia, USA (2005)
15. Tokuda, K., Zen, H., Yamagishi, J., Black, A., Masuko, T., Sako, S.: The HMM-based speech synthesis system (HTS), version 2.1 (2009). http://hts.sp.nitech.ac.jp/
16. Kominek, J., Black, A.: The CMU arctic speech databases. In: Proceedings of 5th ISCA Speech Synthesis Workshop (SSW5), Pittsburgh, USA (2004)
17. Clark, R., Richmond, K., King, S.: Multisyn: Open-domain unit selection for the Festival speech synthesis system. Speech Commun. **49**, 317–330 (2007)
18. Kawahara, H., Masuda-Katsuse, I., Cheveigné, A.: Restructuring speech representations using a pitch-adaptive time-frequency smoothing and an instantaneous-frequency-based F0 extraction: possible role of a repetitive structure in sounds. Speech Commun. **27**, 187–207 (1999)

19. Talkin, D.: A robust algorithm for pitch tracking (RAPT). In: Kleijn W.B., Paliwal K.K. (eds.) Speech Coding and Synthesis, pp. 495–518. Elsevier Science, New York (1995)
20. Schröder, M., Trouvain, J.: The German text-to-speech synthesis system Mary: a tool for research, development and teaching. Int. J. Speech Technol. 6, 365–377 (2003)
21. Steinberger, J., Ježek, K.: Evaluation measures for text summarization. Comput. Inf. 28(2), 1001–1026 (2012)
22. Lin, C., Rey, M.: ROUGE: a package for automatic evaluation of summaries. In: Proceedings of Workshop on Text Summarization Branches Out, Post-Conference Workshop of ACL 2004, Barcelona, Spain (2004)
23. Augat, M., Ladlow, M.: An NLTK Package for Lexical-Chain Based Word Sense Disambiguation (2009)
24. Tofiloski, M., Julian, B., Maite, T.: A syntactic and lexical-based discourse segmenter. In: Proceedings of the Joint Conference of the 47th Annual Meeting of the Association for Computational Linguistics and the 4th International Joint Conference on Natural Language Processing of the Asian Federation of Natural Language Processing (ACL-IJCNLP 2009) - Short Papers. Association for Computational Linguistics, (2009)
25. Zen, H., Toda, T.: An overview of Nitech HMM-based speech synthesis system for blizzard challenge 2005. In: Blizzard Challenge Workshop, Lisbon, Portugal (2005)
26. King, S., Karaiskos, V.: The Blizzard Challenge 2013. In: Blizzard Challenge Workshop. Barcelona, Spain (2013)
27. Schröder, M., Pammi, S., Türk, O.: Multilingual MARY TTS participation in the Blizzard Challenge 2009. In: Blizzard Challenge Workshop, Edinburgh, UK (2009)

Text Classification and Extraction

Unsupervised Classification of Translated Texts

Sergiu Nisioi[✉]

Center for Computational Linguistics, University of Bucharest, Bucharest, Romania
sergiu.nisioi@gmail.com

Abstract. In our paper we investigate the possibility to use an unsupervised classifier to automatically distinguish between the translated and original novels of a multilingual writer (Vladimir Nabokov) and to determine whether the authorship of a translated document can be achieved. We employ a rank-based document vector representation using only function words as features. To extract the results, we propose a generalization of Ward's hierarchical clustering method that is compatible with any similarity metric.

1 Introduction

The research of automatic methods to measure stylistic similarities between texts has a long history, but one of the first successful studies in this direction is that of Mosteller and Wallace [22]. Their approach combined statistical models with linguistic information to infer the authorship of disputed Federalist papers. The linguistic information comprised of certain word categories extracted from the documents, concluding that the class of function words can act as an author's *fingerprint* for a text.

In our work we are interested to observe whether the style of an author can be preserved by translation, given that the style is defined by an author's use of function words. In this sense we compare the translated texts of a multilingual author (T) with the works originally written by the same author (O).

During translation, grammatical structures of the source language (O) can get printed unintentionally to the target/translation (T). In translation theory, this phenomenon is termed interference. *Language transfer* is a similar phenomenon in language acquisition theory which describes the influence carried from the mother tongue to the utterances in other languages spoken by an individual. Since we are investigating the works of a multilingual writer, both of these phenomena are likely to appear in the texts. Previous machine classification studies investigating interference [14,27,33] or language transfer [26,32] indicate that function words can be reliable, topic independent features that evidence these phenomena.

The author in our discussion is Vladimir Nabokov, a multilingual Russian-American novelist who wrote most of his Russian novels living in exile in Europe and switched to English after his departure in USA. We have constructed two significant Russian-English corpora containing both the original and the translated novels on which we attempt to apply a generalization of Ward's clustering method.

© Springer International Publishing Switzerland 2015
C. Biemann et al. (Eds.): NLDB 2015, LNCS 9103, pp. 323–334, 2015.
DOI: 10.1007/978-3-319-19581-0_29

2 The Nabokov Corpus - Interference and Language Transfer

The corpus is compiled out of ten Russian (O) novels and eight English (O) novels together with the corresponding translations (T) of each. The English translations have a better chance to preserve the original *fingerprint* of the author since he supervised and contributed to almost every work, while the Russian translations are more homogeneous, being translated by Sergey Ilyin. If the author is "more present" in the English translations, then we should expect the classifications to contain a larger degree of confusion between T and O in the English corpus. On the Russian side, *Lolita* is the only work translated into Russian by the author. Table 1 contains the details with respect to each novel included.

Both interference and language transfer could be present in Nabokov's translations. It is difficult to asses the amount of language transfer for a trilingual (Russian, English, French) author whose first reading language was probably [24] English. On one hand, Gorski [10] analyzing Nabokov's autobiographical works concludes that our author had near-native skills in English. On the other hand, from a second language acquisition perspective, Selinker and Rutherford [29] claim that a so-called fossilization intervenes for language learners. *Fossilization* designates the permanent cessation of target language (TL) learning before the learner has attained the TL norms at all levels of linguistic structure.

If such would be the case, then any of Nabokov's English novels as well as his translations into English would be, in fact, utterances of a fossilized interlanguage [29] - an independent linguistic system different from the mother tongue of an individual and from the languages acquired. Given the series of audio recordings of his English interviews, we can trace the presence of the open-mid front rounded vowel and other French specific phonological patterns [11] in a mix of British and Russian pronunciation of the voiced alveolar trill [r]. In this sense, we can observe an obvious effect of fossilization of the interlanguage at the phonological level. Nabokov himself claimed at the end of the English version of *Lolita* that he abandoned *my natural idiom, my untrammelled, rich, and infinitely docile Russian tongue for a second-rate brand of English* [23].

We are inclined to believe the translations in our corpus are *literal*, as the author puts it: *rendering, as closely as the associative and syntactical capacities of another language allow, the exact contextual meaning of the original. Only this is true translation* [25]. Under this assumption, interference should be visible in every translation that he approved or collaborated in English or Russian.

Although the works are written many years apart, there is no literary hypothesis to suggest that Nabokov went through a change of style after starting to write in English. Furthermore, the corpus is semi-aligned and the translators are varied, if similar results are extracted from English and Russian, we can be confident that the differences emerge due to a clear distinction between translator and author, including a possible connection with the language transfer phenomenon.

Table 1. The Russian-English corpora are represented in this table, on the left column the titles of original (O) and translations (T) are provided in Russian. The right column contains the English title and the translators who collaborated for that work. The year of writing/translating a certain novel is marked between parentheses. The size is measured as the number of tokens for each work.

Russian	Size	English	Size
Mashenka (1926) (O)	25,131	*Mary* (1970) (T: Michael Glenny and V. Nabokov)	34,359
Korol' Dama Valet (1928) (O)	55,149	*King, Queen, Knave* (1968) (T: Dmitri Nabokov)	83,975
Zashchita Luzhina (1930) (O)	52,173	*The (Luzhin) Defence* (1964) (T: Michael Glenny and V. Nabokov)	· 75,417
Sogliadatai (1930) (O)	16,007	*The Eye* (1965) (T: Dmitri Nabokov)	22,715
Podvig (1932) (O)	54,372	*Glory* (1971) (T: Dmitri Nabokov)	67,314
Camera Obskura (1933) (O)	43,566	*Laughter in the Dark* (1938) (T: V. Nabokov)	56,937
Otchayanie (1934) (O)	42,811	*Despair* (1965) (T: Vladimir Nabokov)	65,412
Priglasheniye na kazn (1936) (O)	40,434	*Invitation to a Beheading* (1959) (T: D. Nabokov and V. Nabokov)	56,081
Dar (1938) (O)	105,528	*The Gift* (1963) (T: Dmitri Nabokov)	115,265
Volshebnik (1939) (O)	12,106	*The Enchanter* (1986) (T: Dmitri Nabokov)	25,821
Podlinnaya zhizn Sebastyana Nayta (T: S. Ilyin)	49,435	*The Real Life of Sebastian Knight* (1941) (O)	62,390
Pod znakom nezakonnorozhdënnykh (T: S. Ilyin)	56,959	*Bend Sinister* (1947) (O)	73,075
Lolita (T: V. Nabokov)	107,271	*Lolita* (1955) (O)	117,185
Pnin (T: S. Ilyin)	46,584	*Pnin* (1957) (O)	52,628
Blednoye plamya (T: S. Ilyin)	76,924	*Pale Fire* (1962) (O)	85,164
Ada (T: S. Ilyin)	153,621	*Ada or Ardor: A Family Chronicle* (1969) (O)	181,346
Prozrachnyye veshchi (T: S. Ilyin)	23,852	*Transparent Things* (1972) (O)	29,073
Smotri na arlekinov! (T: S. Ilyin)	58,037	*Look at the Harlequins!* (1974) (O)	71,327
Russian Total	1,014,905	*English Total*	1,243,033

3   Unsupervised Classifier

An unsupervised classifier determines patterns in the data without making use of assigned labels, hence it can be considered a *more objective* method, the differences (if) discovered are more pronounced and generally, if labels are provided, a clustering result can be easily reproduced by a supervised classifier.

Nabokov's works can be regarded from multiple perspectives of linguistic phenomena which might go beyond the two languages that we consider here - Russian and English - possibly including French and other languages that the author might have had contact with. Therefore, we choose not to use a label-based supervised classifier to avoid having any prior expectation of the results. Our method is based on distance similarities between vector representations of documents, so the results are determined by the features and the similarity measure considered.

The classifier is based on a generalization of Ward's method [34] developed initially by Szekely and Rizzo [31] with a restriction for Euclidean distances. Our preliminary study [27] on a smaller corpus of Nabokov's novels already indicates a compatibility point with Burrows' Delta [2] similarity measure. However, in our previous study we do not provide the theoretical background behind the clustering algorithm in connection with any similarity metric.

The process starts with N clusters for each document and it consecutively merges two clusters at each step based on the minimum e distance. Given two classes $\mathscr{A} = \{A_1, \cdots, A_p\}$ and $\mathscr{B} = \{B_1, \cdots, B_q\}$ containing vector representations of documents, and $D \colon \mathbb{R}^m \times \mathbb{R}^m \to \mathbb{R}$ any similarity metric, the linkage criterion has the following mathematical formulation:

$$e^D(\mathscr{A}, \mathscr{B}) = \frac{pq}{p+q} \Big(\frac{2}{pq} \sum_{i=1}^{p} \sum_{j=1}^{q} D(A_i, B_j)$$

$$- \frac{1}{p^2} \sum_{i=1}^{p} \sum_{j=1}^{p} D(A_i, A_j) - \frac{1}{q^2} \sum_{i=1}^{q} \sum_{j=1}^{q} D(B_i, B_j)\Big) \tag{1}$$

The Lance-Williams parameters [19,31] for this linkage function are identical with the ones for Ward's method for any positive similarity measure D. A clustering result is usually rendered as a dendrogram - a binary tree in which the documents represent the leaves and the sub-clusters are defined by different subtrees starting from the root. In this paper we consider a cluster to be any part of a dendrogram tree, including the entire dendrogram. Our approach combines the single linkage criterion (by which two classes are merged given the smallest distance or nearest neighbor) with a custom objective function which in our case is the general e^D.

The sequential process of joining two clusters at the minimum e^D distance, induces an ultrametric over the space of documents, for which the triangle inequality has a stronger form: $e^D(\mathscr{A}, \mathscr{B}) < \max\{e^D(\mathscr{A}, \mathscr{C}), e^D(\mathscr{C}, \mathscr{B})\}$. Our approach is consistent with previous studies [3,21] which discuss the fact that single linkage and Ward's method always produce monotonic dendrograms, unlike other linkage criteria like UPGMC or WPGMC [7].

To evaluate the results, we make use of the maximum F_1 measure for each class [30]. For a cluster $\mathscr{C}$ and a class K, the precision (P) and recall (R) are defined as:

$$P(\mathscr{C}, K) = \frac{\# \text{ of elements of class } K \text{ in cluster } \mathscr{C}}{|K|}$$

$$R(\mathscr{C}, K) = \frac{\# \text{ of elements of class } K \text{ in cluster } \mathscr{C}}{|\mathscr{C}|}$$

where $|.|$ denotes the cardinal of a set.

The F_β measure is defined by the following formula:

$$F_\beta(\mathscr{C}, K) = (1 + \beta^2) \cdot \frac{P(\mathscr{C}, K) \cdot R(\mathscr{C}, K)}{(\beta^2 \cdot P(\mathscr{C}, K)) + R(\mathscr{C}, K)}$$

The parameter β is used to adjust the importance of precision and recall. For a hierarchical clustering algorithm the maximum precision is attained for any leaf-cluster while the maximum recall is obtained for the entire dendrogram-cluster. To equally weight precision and recall for each class, we select the maximum corresponding F_1 score.

The value of the F_1 score evaluates the degree of compactness of each class. If a class has elements dispersed in different clusters, the corresponding F measure will have a small value.

4   Ranked Lexical Features

Function words or the closed class words (conjunctions, prepositions, pronouns, determiners and particles) have long been studied for authorship attribution [16,17], proving to be a strong indicator of an author's fingerprint. Dinu et al. [4] used these words to uncover the pastiche of a Romanian writer who convinced the literary critics into believing he had discovered the lacking part of an unfinished novel. In such cases any additional authorship results may change the way an author is perceived, as Foucault [9] points out, the concept of *author* is a social construct which reaches beyond the limits of written texts.

The list of English function words, which we also employ in our study, was used to detect translation vs. original texts by Volansky et al. [33]. For Russian, we have constructed the list of function words with all their declensions by crawling Wiktionary [1] a collaborative on-line resource.

The documents are represented as a vector of ranks corresponding to each feature. Our previous approach on a smaller version of this corpus [27] offered good results as well as other previous studies on pastiche detection [5] or text similarity [28]. The idea is to translate the bag-of-words representation of the documents into a rank-vector representation by replacing the frequencies with their corresponding ranks in the document, such that the most frequent word is assigned rank one, the second most frequent rank two, and so on. We state that the ranks are tied when two or more frequencies are equal, in which case we assign the average between the competing, tied ranks. This type of weighting has its roots in Spearman's rank correlation coefficient which indicates the direction of association between two random variables. Forsyth and Sharoff [8] tested the quality of Spearman's correlation for text similarity demonstrating that the approach outperforms a multitude of standard methods.

Using ranks instead of frequencies on text similarity measurements is a good practice for two main reasons: (1) it reduces the bias arising from documents of different size and (2) all the obtained ranked vectors have the same L^1 norm: $\|X\|_1 := \sum_{i=1}^{n} |x_i|$. Where X is any vector of ranks obtained from the bag-of-words and x_i is the rank value corresponding to feature i. Geometrically,

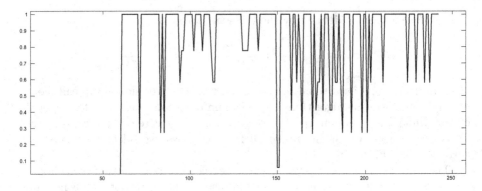

Fig. 1. Plot of the adjusted Rand index between consecutive clusters generated by adding one more word from the list of the first, most frequent function words in the entire English corpus.

the ranked vectors induce an n-dimensional grid, therefore a natural metric to use is the L^1 distance derived from the norm (also called taxicab distance or Manhattan distance):

$$D(X,Y) = \|X - Y\|_1 = \sum_{i=1}^{n} |x_i - y_i| \qquad (2)$$

4.1 Feature Selection

A common problem when carrying text classifications is related to the words that have good discriminative power [18]. In our work, the unsupervised classifier makes use of the pair-wise distances between documents to compute the final dendrogram, therefore, the distances are directly influenced by the features selected [12,20].

First we sort the entire list of function words by their frequency in the entire corpus. Starting from the first 60 function words, we investigate whether changes are produced in the clustering results by using additional features with lower frequencies. To observe the clustering variation, we make use of the adjusted Rand index [13] computed between the "current" and the "previous" cluster.

Given a set of n elements $|S| = n$, and two clustering results $\Psi = \{\mathscr{A}_1, \mathscr{A}_2, \ldots, \mathscr{A}_r\}$ and $\Xi = \{\mathscr{B}_1, \mathscr{B}_2, \ldots, \mathscr{B}_p\}$, construct a contingency table C of r rows and p columns with each value $c_{ij} = |\mathscr{A}_i \cap \mathscr{B}_j|$ being the number of common objects between cluster $\mathscr{A}_i$ and $\mathscr{B}_j$. Let $a_i = \sum_{j=1}^{s} c_{ij}$ be the sum of all the values from the row i and $b_j = \sum_{i=1}^{r} c_{ij}$ all the values from the column j. Then the adjusted Rand index as it is defined by Hubert and Arabie [13] is

$$ARI(\Psi, \Xi) = \frac{\sum_{ij} \binom{c_{ij}}{2} - [\sum_i \binom{a_i}{2} \sum_j \binom{b_j}{2}]/\binom{n}{2}}{\frac{1}{2}[\sum_i \binom{a_i}{2} + \sum_j \binom{b_j}{2}] - [\sum_i \binom{a_i}{2} \sum_j \binom{b_j}{2}]/\binom{n}{2}} \qquad (3)$$

If for example we add k consecutive function words for which we obtain identical clusters, then the k features are considered stable and we store them in a unique hash corresponding to the cluster produced. The hash key is obtained from the parenthesis representation of the binary tree, with all the subtrees being sorted lexicographically, for each dendrogram generated sequentially. This way we can have an exploratory technique to account which features contribute to which results. The final feature selection is not necessarily based on a label assignment, but rather it is decided based on the result with the maximal number of features [27]. In Fig. 1 we plot the sequential index values computed for the English corpus. We note that up to the first 150 function words, the clusters are more stable since sequences of different features produce similar results.

5   Similarity Measures and Results

5.1   Manhattan Distance

The most natural measure to be applied in an L^1 space is Manhattan distance. In combination with vectors of ranks it can also be encountered under the name of Spearman's foot-rule or Rank distance [6].

One important property of this metric is its *rank type invariance*: the distance remains unchanged if our tied ranked vectors are obtained by an ascending ordering relation (e.g. assign rank one to the most frequent function word, rank two to the second most frequent and so on) or by a descending ordering relation when rank one is assigned to the most infrequent word and so on. To prove this, we have to observe that for some frequencies $\{f_1 > f_2 > \cdots > f_n\}$, that generated an ascending tied rank $X_> = \{x_1, \cdots, x_n\}$, its descending tied rank can be obtained by the next equation from $X_>$:

$$X_< = (n - X_>) + 1 \tag{4}$$

We observe now that a reverse ranking is produced only with a geometric translation obtained by a subtraction and an addition. Manhattan distance remains unchanged if we translate all the points by the same constant.

This suggests that the use of ranks does not imply just a simple change of the weights, but rather a change of space in which distances between documents become more measurable and more stable (Fig. 2).

5.2   Delta Measure

Delta is a method of measuring stylistic similarities proposed by Burrows [2]. The standard equation for Delta has the following form:

$$\Delta(X, Y) = \frac{1}{n} \sum_{i=1}^{n} \left| \frac{x_i - y_i}{\sigma_i} \right| \tag{5}$$

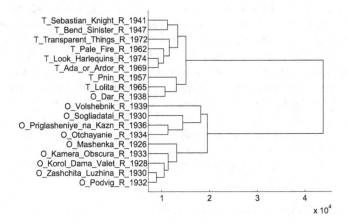

Fig. 2. Result obtained with e^{L^1} linkage criterion using the rankings extracted from the Russian corpus

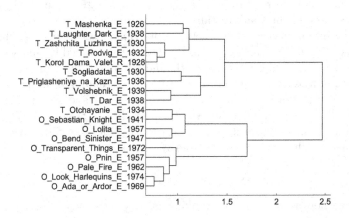

Fig. 3. Result obtained with e^{Δ} linkage criterion using the rankings extracted from the English corpus

where $X = \{x_1, \cdots, x_n\}$ and $Y = \{y_1, \cdots, y_n\}$ are vectors of ranks corresponding to words i and σ_i is the standard deviation of (the rank of) the word i in the given corpus.

Delta is incompatible with the strategy of selecting the entire list of function words due to possible zero standard deviation and other factors discussed by Jockers and Witten [15], in which case a feature selection method becomes almost mandatory. Furthermore, a significant improvement was observed with the use of ranks instead of frequencies. This measure is also invariant to ranking types, the final value of Delta depending on both the ranks of the words within one document and the standard deviation of ranks given all the other documents.

Table 2. Comparison of the F_1 evaluation scores for the entire documents and the texts split into 2000 tokens per chunk and the entire un-split documents.

		L^1		Delta	
		Rank	Freq	Rank	Freq
Russian	O	**0.92**	0.67	0.86	0.88
2000	T	**0.91**	0.65	0.82	0.88
Russian	O	0.94	0.72	**0.94**	0.75
	T	0.94	0.66	**0.94**	0.77
English	O	0.64	0.64	0.64	**0.77**
2000	T	0.69	0.69	0.69	0.69
English	O	0.94	0.73	**0.94**	0.63
	T	0.94	0.64	**0.94**	0.72

A cluster obtained from Delta applied on the English corpus is illustrated in Fig. 3.

Equation 5 is derived from Manhattan distance applied on z-scores of words. For a word i in a given corpus its z-score has the value $z(x_i) = \frac{x_i - \mu_i}{\sigma_i}$ where μ_i is the mean of frequencies x_i of word i. In this case we have the following L^1-like expression for delta measure between two documents X and Y:

$$\Delta(X, Y) = \frac{1}{n} \sum_{i=1}^{n} |z(x_i) - z(y_i)| \qquad (6)$$

5.3 F_1 Score

In order to evaluate the F_1 measure [30], we split the each document into smaller chunks of 2000 tokens. For every novel, we randomly extract the same number of chunks so that O and T are not biased by the presence of the larger novels. Moreover, since *Sogliadatai* and *Volshebnik* are considerably smaller in size, we decided to discard them from this analysis.

The F_1 scores obtained in Table 2 indicate that both L^1 and Delta are comparable in terms of results when the full documents are used since, in this scenario, the standard deviation does not have a large impact over the measured similarities. What is more, the use of ranks seems to greatly influence the compactness of the clusters (0.94) while standard frequencies barely score an F-measure above 0.7.

However, when the documents are split into chunks, we observe a significant drop in F-measure regardless of the ranking process. If for Russian the clusters are still quite compact - 0.92 for L^1 with ranks and 0.88 for Delta with frequencies, for English the best score (0.77) is obtained by using standard Delta in combination with frequencies.

We believe there are two causes for this behavior: (1) the chunks are smaller and the features that differentiate the author from translator are less frequent, making the distance-based similarities less prominent and (2) Nabokov's personal involvement in the translations from Russian may have determined his authorial *fingerprint* to be actively present in the English translations, thus creating a stronger resemblance between translation and original. A close inspection of the large dendrogram resulted shows that the translated English chunks are not homogeneous, but rather spread across different clusters of O. To conclude, interference seems to be more present in the Russian translations over which the author had a minimal contribution.

6   Conclusions

We propose an extension of Ward's hierarchical clustering method that is able to operate with custom user-defined objective functions that are not required to be metrics. Given the consistent results on two different languages, our combination of exploratory methods can be considered reliable for measuring distances between different text documents. Furthermore, our results indicate that ranks do improve the evaluation F-scores when the number of training examples is small. Both the L^1 metric and Delta are rank type invariant, which means the results are identical if we assign rank one to the most frequent feature and so on, or rank one to the most infrequent feature and so on.

Our adapted clustering algorithm was able to successfully distinguish between Nabokov's original novels and translations on two different languages with multiple translators involved. Compared to previous work investigating translation [14,33], our results further bring into discussion the influence of the author over the translation and a possible link between interference and language transfer. Hence, we show that it is difficult to correctly classify between author and author-as-translator, especially when the size of the documents is small and when a possible imprint of language transfer could influence the overall results. Translations highly depend on the choices a translator makes to reproduce the initial style of the text, but these decisions further depend on the O vs. T linguistic and cultural differences.

Last but not least, we further add a proof to the fact that the *fingerprint* of an author can be revealed by his use of function words, *fingerprint* which can get masked under the effect of translation.

References

1. Wiktionary. ru.wiktionary.org. Accessed in June 2013
2. Burrows, J.F.: Delta: a measure of stylistic difference and a guide to likely authorship. Literary Linguist. Comput. **17**(1), 267–287 (2002)
3. Carlsson, G.E., Mémoli, F.: Characterization, stability and convergence of hierarchical clustering methods. J. Mach. Learn. Res. **11**, 1425–1470 (2010)
4. Dinu, L.P., Niculae, V., Şulea, O.M.: Pastiche detection based on stopword rankings: exposing impersonators of a romanian writer. In: Proceedings of the Workshop on Computational Approaches to Deception Detection, EACL 2012, pp. 72–77. Association for Computational Linguistics, Stroudsburg (2012)

5. Dinu, L.P., Niculae, V., Şulea, O.M.: Pastiche detection based on stopword rankings: exposing impersonators of a romanian writer. In: Proceedings of the Workshop on Computational Approaches to Deception Detection, EACL 2012, pp. 72–77 (2012)
6. Dinu, L.P., Popescu, M.: Comparing statistical similarity measures for stylistic multivariate analysis. In: RANLP, pp. 349–354. Association for Computational Linguistics, Borovets (2009)
7. Everit, B., Landau, S., Leese, M.: Cluster Analysis. Hodder, London (2001)
8. Forsyth, R., Sharoff, S.: Document dissimilarity within and across languages: a benchmarking study. Literary Linguist. Comput. **29**, 6–22 (2014)
9. Foucault, M.: What Is an Author?. State University Press of New York, Albany (1987)
10. Gorski, B.: Nabokov vs. Набоков: A literary investigation of linguistic relativity. Vestnik, J. Russ. Asian Stud. (8) 56–78 (2010) http://www.sras.org/nabokov_vs_nabokov_linguistic_relativity
11. Hallé, P.A., Best, C.T., Levitt, A.: Phonetic vs. phonological influences on french listeners' perception of american english approximants. J. Phonetics **27**(3), 281–306 (1999)
12. Hoover, D.L.: Testing burrows's delta. Literary Linguist. Comput. **19**(4), 453–475 (2004)
13. Hubert, L., Arabie, P.: Comparing partitions. J. Classif. **2**, 193–218 (1985)
14. Ilisei, I., Inkpen, D., Corpas Pastor, G., Mitkov, R.: Identification of translationese: a machine learning approach. In: Gelbukh, A. (ed.) CICLing 2010. LNCS, vol. 6008, pp. 503–511. Springer, Heidelberg (2010)
15. Jockers, M.L., Witten, D.M.: A comparative study of machine learning methods for authorship attribution. Literary Linguist. Comput. **25**, 215–223 (2012)
16. Juola, P.: Authorship attribution. Found. Trends Inf. Retrieval **1**(3), 233–334 (2006)
17. Koppel, M., Schler, J., Argamon, S.: Computational methods in authorship attribution. J. Am. Soc. Inf. Sci. Technol. **60**(1), 9–26 (2009)
18. Koppel, M., Schler, J., Bonchek-Dokow, E.: Measuring differentiability: unmasking pseudonymous authors. J. Mach. Learn. Res. **8**, 1261–1276 (2007)
19. Lance, G.N., Williams, W.T.: A general theory of classificatory sorting strategies: 1. hierarchical systems. Comput. J. **9**(4), 373–380 (1967)
20. Madigan, D., Genkin, A., Lewis, D.D., Lewis, E.G.D.D., Argamon, S., Fradkin, D., Ye, L., Consulting, D.D.L.: Author identification on the large scale. In: Proceedings of the Meeting of the Classification Society of North America (2005)
21. Milligan, G.W.: Ultrametric hierarchical clustering algorithms. PSYCHOMETRIKA **44**(3), 343–346 (1979)
22. Mosteller, F., Wallace, L.D.: Inference in an authorship problem. J. Am. Stat. Assoc. **58**(302), 275–309 (1963)
23. Nabokov, V.: Lolita. Penguin Books Limited, UK (2012)
24. Nabokov, V.: Speak, Memory: An Autobiography Revisited. Vintage International, New York (1989)
25. Nabokov, V.: Eugene Onegin. A Translation from the Russian of Aleksandr Pushkin's (1833) Yevgeniy Onegin (1990)
26. Nisioi, S.: Feature analysis for native language identification. In: Gelbukh, A. (ed.) CICLing 2015, Part I. LNCS, vol. 9041, pp. 644–657. Springer, Heidelberg (2015)
27. Nisioi, S., Dinu, L.P.: A clustering approach for translationese identification. In: Proceedings of the International Conference Recent Advances in Natural Language Processing RANLP 2013, Hissar, Bulgaria, pp. 532–538, September 2013

28. Popescu, M., Dinu, L.P.: Comparing statistical similarity measures for stylistic multivariate analysis. In: RANLP 2009 Organising Committee/ACL RANLP, pp. 349–354 (2009)
29. Selinker, L., Rutherford, W.: Rediscovering Interlanguage. Applied Linguistics and Language Study. Routledge, New York (2014)
30. Steinbach, M., Karypis, G., Kumar, V.: A comparison of document clustering techniques (2000)
31. Szekely, G.J., Rizzo, M.L.: Hierarchical clustering via joint between-within distances: extending ward's minimum variance method. J. Classif. **22**, 151–183 (2005)
32. Tsvetkov, Y., Twitto, N., Schneider, N., Ordan, N., Faruqui, M., Chahuneau, V., Wintner, S., Dyer, C.: Identifying the l1 of non-native writers: the cmu-haifa system. In: Proceedings of the Eighth Workshop on Innovative Use of NLP for Building Educational Applications, pp. 279–287. Association for Computational Linguistics, Atlanta, June 2013
33. Volansky, V., Ordan, N., Wintner, S.: On the features of translationese. Digit. Scholars. Humanit. **30**(1) 98–118 (2015) doi:10.1093/llc/fqt031
34. Ward, J.H.: Hierarchical grouping to optimize an objective function. J. Am. Stat. Assoc. **301**(58), 236–244 (1963)

A Language-Independent Method for Detection and Correction of Alignment Errors in Parallel Corpora

Katarzyna Niżałowska[✉] and Urszula Markowska-Kaczmar

Department of Computational Intelligence, Wrocław University of Technology,
Wyb. Wyspiańskiego 27, 50-370 Wrocław, Poland
{katarzyna.nizalowska,urszula.markowska-kaczmar}@pwr.edu.pl

Abstract. We present a method for detection of the alignment errors in parallel corpora. The method is meant to be language-independent and was tested for pairs of English, Polish and Spanish languages. It utilizes automatically obtained dictionaries to perform the detection. A discussion about the origin of errors is included. An approach to correcting one of classes of errors is also described and tested. The proposed method has proven itself to be effective in improving the quality of Parallel Corpora. Conclusions of this study may be useful while dealing with errors in existing parallel data sources, as well as at the stage of aligning new parallel corpora.

Keywords: Machine Translation · Corpus analysis · Sentence alignment · Data cleaning

1 Introduction

A vast majority of modern information systems has a global range and thus is required to be multilingual. It results in dealing with plenty of problems related to the diversity of languages. Machine Translation (MT) systems provide solutions to cope with some of them. Most of recent MT systems, however, are data-driven and need huge datasets to learn how to translate texts across languages. The most vital resources for this task are Parallel Corpora.

Parallel Corpora are large sets of corresponding texts written in different languages (at least two). For such a corpus to be useful, it needs to be aligned. Analogous text fragments ought to be tagged as containing the same information. Aligning is the process of matching equivalent texts and can be performed on various levels. The absolutely mandatory alignment is the document level, however, most corpora are aligned at least at the level of sentences.

The alignment can be performed manually or automatically. Manual alignment is usually obtained when a document is translated sentence by sentence by a human translator. A pair of texts can be also aligned by a human, but it is not a common practice.

© Springer International Publishing Switzerland 2015
C. Biemann et al. (Eds.): NLDB 2015, LNCS 9103, pp. 335–346, 2015.
DOI: 10.1007/978-3-319-19581-0_30

Most of publicly available Parallel Corpora are aligned automatically. The reason why automatic alignment prevails is a huge amount of data to align, which exceeds capabilities of human translators. Due to the complexity of acquisition and processing of parallel data, the alignment errors are the common problem.

Another important aspect of creating Parallel Corpora is the source of data. It might have its origin in either corresponding texts, which were translated for some reason, Translation Memories, or databases used by various Multilingual Information Systems.

Altogether the diverse level of source data consistency and imperfection of automatic alignment methods have significant impact on Parallel Corpora quality. As the quality of publicly available Parallel Corpora varies a lot, the need of detection of those errors emerges.

This paper introduces a method for detecting sentence alignment errors in Parallel Corpora and correcting some of them. The method is designed to be language-independent, however it utilizes dictionaries automatically obtained from publicly available sources. The independence is preserved as long, as there are lexicons and dictionaries available for a given language.

The novelty of our method is using dictionaries as trustworthy language resources for error detection in Parallel Corpora of unknown quality while preserving language independence and moving away from pure statistical approaches, which do not perform well in all cases. The method focuses on detecting the most common errors present in existing Parallel Corpora, which were observed during preliminary studies.

The rest of the paper is organised in the following order. In Sect. 2, the previous research related to this topic is outlined. A discussion about types of errors present in Parallel Corpora is carried out in Sect. 3. In Sect. 4, the problem is defined and Sect. 5 is dedicated to the explanation of our method. The evaluation of its performance is presented in Sect. 6. The last section contains conclusions, that emerged during our study, as well as suggestions for further development.

2   Previous Works

Up to this day little effort has been put into development of universal methods for detecting errors in sentence alignment. A lot of researchers tolerate errors produced by an alignment algorithm basing on an assumption that error impact on translation quality is not significant. The majority of existing Machine Translation systems use Gale and Church method [9] or one of its modifications.

It should be stressed that this method is based on length similarity between sentences. Although in [12], it was shown that the selection of an alignment method does not influence the translation quality substantially, the test was conducted only for English-French language pair. However, not every language pair has such a similar sentence structure.

A research carried out by Ma [13] proved that for an English-Chinese pair, the number of errors produced by Gale and Church algorithm highly decreases the translation quality. While pure statistical Machine Translation systems may be robust to small amounts of alignment errors, in [10], it was shown that for rule extraction techniques used in MT, even a small amount of noise in training data results in high complexity of translation rules and decreases their accuracy.

The high number of alignment errors leads to deterioration of translation quality even in pure statistical systems. A study that tested a relation between these errors and the output performance of an existing statistical system was conducted by Khadivi and Ney [12]. The difference in translation for noisy and clean corpora was proven to be statistically significant.

A need to assess the alignment quality often emerges while creating a bilingual parallel corpora dedicated for a specific language pair. In such cases an effort is often put into eliminating the misalignments. A set of methods dedicated to solving observed problems can be developed such as those described in [8]. However, these methods are often tailored to suit a specific case and are not sufficiently general.

Sometimes a preferred cleaning strategy is performed on the document level instead of the sentence level. A strategy described in [14] rejects document pairs for which the file size differs significantly. However, in most scenarios it is not desired to reject whole documents since documents are usually believed to be aligned properly. This technique can be useful during mining huge resources of potentially parallel data.

Existing language-independent methods are based mainly on trivial length-based tests as in [8,12,16] or introspective Translation Likelihood-Based Filtering as in [12]. The main problem which the introspective methods face is the initially unknown corpora quality, which may happen to be very poor.

The problem may not be observable for a manually created corpora, but is apparent in case of publicly available, automatically generated resources such as in [15]. When the alignment quality is generally low, methods that base on outlier detection are not applicable. As shown in [12], the Translation Likelihood can be a very effective method, when the translation model is already trained on a trustworthy data. However, this is usually not the case during the corpus development.

High accuracy of alignment error detection can be achieved when sentences are additionally annotated with tags that can be used to match sentences. An example of a corpora annotated this way is OpenSubtitles2011. It consists of movie subtitles translated into a number of different languages. Since each line holds information about its time of display, the time matching can be performed accurately as specified in [15].

Narrow domain or language-pair specific error detection techniques prevail amongst published methods. The lack of accurate and language-independent solutions make the detection problematic during the development of parallel corpora for niche languages. There is a lot of parallel training data for many languages assembled already. One such example is OPUS [15]. However the quality of each document pair varies and is unknown. Misalignments are a common issue for many of them. Because the existing methods for error detection are not aggregated in a form of any publicly available toolset, the assessment of alignment quality of publicly available Parallel Corpora is a non-trivial task.

3 Errors in Parallel Corpora

Alignment errors emerge as a consequence of noisy data used to build a Parallel Corpora and flaws of alignment methods. There might be different types of errors. After analysing publicly available Parallel Corpora, we distinguished the following types of errors:

- *corrupted data structure,*
- *empty sentences,*
- *wrong language,*
- *wrong translation.*

Since detection of *empty sentences* is trivial, and *corrupted data structure* in most cases can be corrected easily, there is no need for a special method to deal with these errors.

In our work, we mostly focused on finding *wrong language*, and *wrong translation* errors and correcting some of the last one.

Wrong language errors frequently arise when a corpus is build for a number of languages simultaneously. Despite the existence of many reliable language detectors, errors of this type are still common in Parallel Corpora.

The second and the most problematic type of errors is *wrong translation*, which may emerge as a result of various conditions. Considering the origin of an error, we can distinguish:

- *poor translation,*
- *unrelated sentences,*
- *shifted alignment.*

Poor translation errors refer to the pairs of sentences with alike meanings, but not similar enough to consider them as translations. This type of errors is very common when the texts were not translated sentence by sentence, but as a whole and the translation strongly depends on a context. Examples of such translations can be found in books, lyrics, subtitles and many other sources.

Unrelated sentences are another example of errors in Parallel Corpora. They are usually results of errors during data preprocessing rather then sentence alignment itself. The sentences in a pair are completely unrelated in terms of meaning as well as their position in source and target texts.

Errors described as *shifted alignment* emerge as a result of low quality sentence alignment. They may sometimes be mistaken with *unrelated sentences*, but in contrary to them, the correct translations of source and target sentences from the erroneous pair are located in the texts near given wrong translations. It makes them possible to be automatically corrected.

4 Problem Description

The problem of detecting errors in Parallel Corpora can be defined as finding pairs of sentences, that do not correspond to each other semantically in order to remove them from the corpus. The semantic correspondence can be assessed accurately only by a human, however, automatic methods may try to imitate human judgement.

Although most corpora are aligned bidirectionally, we name sentences written in one language as source sentences, and sentences written in the other language as target sentences in order to avoid confusions.

The nature of some alignment errors and the alignment process itself, make it possible to automatically correct them instead of removing them from a corpus. As a result, not only the corpus quality increases, but also its size is not reduced unnecessarily.

We define the problem of correcting alignment errors as proposing an alternative alignment for a given sentence pair. It may be done by replacing either source or a target sentence with their neighbours.

5   Proposed Method

The general idea of our approach is to find sentence pairs that are not likely to be each other's translations in given languages. The proposed method uses monolingual lexicons and bilingual dictionaries in order to exclude pairs of sentences, that are considered errors in terms specified in Sect. 3, from a corpus.

The method consists of three parts. Each part focuses on eliminating other type of errors and can be performed independently. However, the best performance is achieved by applying all of them. The proposed way of combining these parts is shown in the Fig. 1, however it can be adjusted to fit individual needs.

In the first part, monolingual lexicons are utilized to find sentences, that are written in *wrong language*. Pairs containing at least one sentence in wrong language are excluded from a corpus.

The second part focuses on checking whether a target sentence can be a translation of a source sentence. The assessment is done using bilingual dictionaries. It is based on a number of words in both source and target sentences, that can be each others translation. This part is meant to detect all kinds of *wrong translation* errors.

The third part is designed to detect *shifted alignment* errors by comparing a number of translated words in source and target sentences with the number obtained by replacing any of them with its neighbour. Basing on the obtained scores, we also propose a method for correcting shift errors.

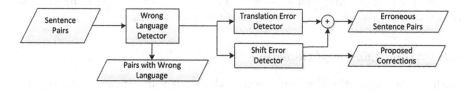

Fig. 1. The error detection pipeline

5.1 Dictionaries and Lexicons

Our bilingual dictionaries were obtained from Wiktionary [6], which is a free multilingual dictionary. Wiktionary is divided into separate sections for different languages. However each section contains information not only about words in its own language but also words from other languages.

Although the data in Wiktionary is structured, this structure varies across different languages. As a result, it is not possible to build a universal tool for creating dictionaries for all languages. There is a need for designing a specialised parser for each language. A similar approach to creating a bilingual dictionary from Wiktionary was presented in [7].

Process of building a bilingual dictionary for a source and target languages consists of two steps. In the first one, data for each language is extracted separately. The next step involves matching words in both languages, based on translations obtained for them.

For every word in a source language found in a source language's section of Wiktionary, all of its translations and conjugations are stored. The same procedure is executed for a target language. The translations from source language are matched with words in a target language, and links between them are created.

This operation is executed bidirectionally. Finally a dictionary is created by using all forms from source language as keys and lists of all possible forms of their translations as values. Thus, for a language pair two separate dictionaries are obtained.

Although Wiktionary is sufficient for assessment of a translation, for deciding whether a sentence is written in a given language a more complete lexicon is needed. We used Mozilla dictionaries [3]. They are available on open licences and were designed as spell checking lexicons so the completeness was a main goal during their compilation. They support 35 languages which is enough to use them in a language-independent method.

To improve our method's performance for Polish language, dictionary from RL-Button project [11] and conjugation list from Morfeusz analyzer [2] were used to extend a dictionary obtained from Wiktionary. A similar modifications could be applied for any language for which Wiktionary data is not complete enough.

Since our method was tested for pairs of languages including Polish, English and Spanish, we built dictionaries for these three languages. The distribution of a number of base forms for each language and a number of words in lexicons is presented in Table 1.

The number of translations (from each conjugated word to all of its possible translations) for each language pair is presented in Table 2.

5.2 Wrong Language Detection

During this phase, the decision whether the source and target sentences can come from given languages is made. Any language detector could be used for this task, but since there is no need for detecting the language, but only for

Table 1. The number of base forms and words in lexicons

Language	Number of base forms	Lexicon size
pl	21138	3638432
en	95842	136424
es	30307	670235

Table 2. The number of translations in dictionaries

Src Language	Tgt Language	Dictionary size (src to tgt)	Dictionary size (tgt to src)
en	pl	175585	857835
en	es	89892	573838
es	pl	164903	185823

deciding whether the sentence can be from a given language, we used lexicons instead of language detectors.

The language detection is performed for each sentence independently. Firstly, the list of words in a sentence is obtained. In the next step each word is checked whether it appears in a lexicon for a given language or not. If at least half of words from the sentence were found in the lexicon, then the sentence is considered as coming from the given language. Otherwise, the whole pair is marked as a *wrong language* error.

The optimal number of words required to mark a sentence as correct can differ depending on the completeness of a lexicon available for given language. Our preliminary studies have shown, that the half of words from a sentence is a good threshold in most cases.

The pairs marked as *wrong language* are eliminated from a corpus and are not taken into account in the next phases.

5.3 Translation Error Detection

In order to decide whether a sentence in the target language can be a translation of a sentence in the source language, we look for every word from source and target sentences in a bilingual dictionary counting the following statistics:

- w_{sd} - the number of words found in a source dictionary,
- w_{td} - the number of words found in a target dictionary,
- t_{st} - the number of words found in a source dictionary which translations were found in a target sentence,
- t_{ts} - the number of words found in a target dictionary which translations were found in a source sentence.

Subsequently, the match score can be calculated as expressed by the Eq. 1

$$score = \frac{t_{st} + t_{ts}}{w_{sd} + w_{td}}. \tag{1}$$

The score is not meant to be used as a measure of translation quality. It should be considered as a measure designed to separate pairs of sentences which can be each other translations from erroneous pairs.

The value of a threshold separating correct pairs of sentences from erroneous ones strictly depends on the size and the quality of the bilingual dictionary and should be adjusted experimentally. Our preliminary studies for English-Polish pair concluded that the value 0.25 is a good threshold. This observation was confirmed during tests for other language pairs.

5.4   Shift Error Detection

The detection of *shifted alignment* errors is performed using the score from Eq. 1. The score is calculated for the sentence pair, as well as for all its possible modifications.

Since in some corpora a single sentence pair can consist of more then one source and target sentences, through this paper the terms *source sentence* and *target sentence* referred to short text fragments, which not necessarily were single sentences. In this section, however, the term *sentence* will refer to a single sentence.

We proposed the following modifications to apply either to a source or a target text fragments:

- cB - delete first sentence,
- cE - delete last sentence,
- aB - add one sentence at the beginning,
- aE - add one sentence at the end,
- mF - move the text fragment forward (apply cB and aE),
- mB - move the text fragment backwards (apply aB and cE).

Not all of the above modifications are possible in every case. For example, if a text fragment consists of a single sentence, neither cB nor cE are possible to perform.

To detect *shifted alignment* errors, the score for the original sentence pair, as well as scores obtained for modifications are compared. We decided to detect an error if two conditions are satisfied. The first condition is for the highest score obtained for modifications to exceed a given threshold tr_{mod}. The second condition states, that the difference between the highest modification score and the original score must be higher than the other threshold tr_{df}.

If both conditions are satisfied, then the original sentence pair is considered erroneous. The technique for correction of such errors is described in the subsequent subsection.

Analogously to the previous phase, the thresholds depend on the size and the quality of used dictionaries. Our studies have shown, that the values $tr_{mod} = 0.3$ and $tr_{df} = 0.1$ result in very good performance of error detection.

5.5 Error Correction

The correction of errors detected in the previous subsection should be performed with caution. It simply involves applying to the sentence pair the modification with the highest score. However, if the quality of Parallel Corpora is much more important than its size, it is advisable to apply only modifications satisfying more rigorous conditions.

Tests performed during our studies have shown that $tr_{mod} = 0.5$ and $tr_{df} = 0.3$ result in correction accuracy close to 95 %.

6 Method Evaluation

The most trustworthy form of evaluation in case of translation-related tasks is always an assessment performed by at least one human translator. However, such evaluations are laborious and time-consuming and cannot be conducted for huge datasets. We designed an automatic assessment procedure and carried it out on a test corpus. Our method was also used to estimate the number of errors in other corpora.

6.1 Parallel Corpora Used

Three different publicly available corpora were selected for the evaluation. The most reliable corpus was Tatoeba [5]. It is composed of separate sentences translated into different languages. As a result it is accurately aligned at the sentence level.

Another used corpus was OpenSubtitles2011 [4], which is obtained from a movie subtitle database. Due to its large size, only a subset of documents was used for testing.

The last corpus was JRC-Acquis [1], which is composed of documents of European Union law applicable in EU Member States. All corpora were obtained from OPUS [15]. Their sizes expressed in numbers of sentence pairs are presented in Table 3.

Different types of errors were present in these corpora. From each corpus, we randomly selected 200 pairs of sentences. Each pair was analysed by a human.

Tatoeba was free from alignment errors, but exceptionally incorrect translations occurred. However, since the number of this errors did not exceed 0.5 % of the sample, we considered it as free of errors.

A subset taken from OpenSubtitles2011 suffers from common shift errors and non-literal translations.

In JRCAcquis, numerous shift errors were found as well as non-text noise occasionally found in documents. Additionally it is a domain specific corpora which leads to uncommon translations of phrases and makes error detection a more challenging task.

Table 3. The number of entries in all the Parallel Corpora used during the study

Language pair	Tatoeba	OpenSubtitles	JRC-Acquis
en-pl	21552	15297	1650478
en-es	150639	27907	814236
es-pl	3667	15763	1650406

Table 4. The evaluation results for modified Tatoeba

Lang pair	Accuracy	Precision	Recall	Number of corrections	Correction precision
en-pl	91 %	84 %	88 %	129	84 %
en-es	93 %	86 %	93 %	6026	94 %
es-pl	90 %	80 %	90 %	89	96 %

6.2 Results for Tatoeba

Since the valid evaluation of our method's performance can be performed only by a human, we decided to perform tests on a corpus containing known errors. As specified in the previous section, Tatoeba is an example of such a reliable source. In order to check the performance of our method, we injected errors of different types into the corpus and calculated the statistics of errors detected by our method.

For each language pair we automatically converted 10 % of the corpus to contain *wrong language* errors, another 10 % to contain *unrelated sentences*, and the other 10 % to contain *shifted alignment* errors of different type. The rest 70 % of a corpus remained correct.

We tested overall Accuracy, Precision and Recall of detecting errors by our method.

The second test, that we did was related to error correction. We tested the Precision of error correction performed by our method.

The experimental results are presented in Table 4. It is worth mentioning, that calculated Accuracy, Precision and Recall are very high, even though the parameters of the method were tuned for the English-Polish language pair only. We consider the results as very good, and visibly improving the quality of corpora.

The number of corrections applied for each corpus varied from 6 % of injected errors (for English-Polish pair) to 40 % (for English-Spanish). It is due to the fact, that the acceptance threshold for a correction to be applied was intentionally overstated in order to avoid applying false corrections. It is a reasonable approach as long as the quality of a corpus is the main criterion.

If a high number of errors is expected to be corrected, the semi-automatic approach is worth considering. In this case, the method should be used for suggesting corrections, which could be accepted or rejected by a human.

Table 5. Statistics obtained for JRC-Acquis and OpenSubtitles

Languages	Errors found in JRC-Acquis	Errors found in OpenSubtitles	Accuracy
en-pl	241740	4960	83 %
en-es	148131	6476	81 %
es-pl	433115	5745	75 %

6.3 Statistics for Other Corpora

In order to assess the performance of our method on JRC-Acquis and OpenSubtitles2011 we decided to manually annotate a small sample of sentence pairs. However, since JRC-Acquis is composed of highly domain-specific texts, a professional in-domain translator would be required to perform this task correctly.

For each language pair in OpenSubtitles2011 corpus, 200 sentence pairs were randomly chosen and annotated as either correct or not. The outcome of our method was then compared with these annotations to measure the detection accuracy.

The results of this experiment are presented in Table 5. The second and the third columns contain numbers of errors found by our method in JRC-Acquis and OpenSubtitles respectively. Since we were not able to verify all of them, they are not fully reliable. In the last column, there are listed Accuracies calculated for random samples from OpenSubtitles corpus.

7 Conclusions and Future Works

A dictionary-based method for detection and correction of sentence alignment errors in Parallel Corpora was presented. The outcome of the evaluation suggests that this approach can be successfully used for highly accurate error detection.

An important conclusion is, that sentence alignment errors are present in publicly available Parallel Corpora and the development of methods for detection and correction of these errors is reasonable. Existing corpora should be analysed to eliminate this sort of errors. Additionally, during the development of new Parallel Corpora, care should be taken to filter them out.

Since our method is dictionary-based, its performance depends on the quality of utilized dictionaries. However, our study has shown that dictionaries automatically obtained from Wiktionary are sufficient for this task.

Future development of our method should be focused on improving the correction part by taking global features into account, such as shifts of larger blocks of texts or whole paragraphs. A wider-perspective approach is expected to eliminate noise caused by incomplete dictionaries.

Conclusions that emerged during our study could be successfully applied in order to develop a new dictionary-based sentence alignment method. Promising results achieved by our error detection method suggest that this approach is likely to perform well.

References

1. JRC-Acquis (2014). https://ec.europa.eu/jrc/en/language-technologies/jrc-acquis
2. Morfeusz SGJP (2014). http://sgjp.pl/morfeusz/
3. Mozilla dictionaries (2014). https://wiki.mozilla.org/L10n:Dictionaries
4. OpenSubtitles2011 (2014). http://datahub.io/dataset/opus/resource/e5a441a7-73 d5-4f8c-a4b5-4bab42a739f2
5. Tatoeba (2014). http://tatoeba.org/pol/downloads
6. Wiktionary, the free dictionary (2014). http://en.wiktionary.org
7. Ács, J.: Pivot-based multilingual dictionary building using wiktionary. In: Proceedings of the Ninth International Conference on Language Resources and Evaluation (LREC 2014). European Language Resources Association (ELRA), Reykjavik, Iceland, May 2014
8. Bojar, O., Žabokrtský, Z.: CzEng 0.9: large parallel treebank with rich annotation. Prague Bull. Math. Linguist. **92**(1), 63–84 (2009)
9. Gale, W.A., Church, K.W.: A program for aligning sentences in bilingual corpora. Comput. Linguist. **19**(1), 75–102 (1993)
10. Galley, M., Hopkins, M., Knight, K., Marcu, D.: What's in a translation rule? In: HLT-NAACL, pp. 273–280 (2004). http://acl.ldc.upenn.edu/hlt-naacl2004/main/pdf/130_Paper.pdf
11. Kędzia, P.: Rl-button. http://nlp.pwr.wroc.pl/pl/narzedzia-i-zasoby/rl-button
12. Khadivi, S., Ney, H.: Automatic filtering of bilingual corpora for statistical machine translation. In: Montoyo, A., Muñoz, R., Métais, E. (eds.) NLDB 2005. LNCS, vol. 3513, pp. 263–274. Springer, Heidelberg (2005). http://dx.doi.org/10.1007/11428817_24
13. Ma, X.: Champollion: a robust parallel text sentence aligner. In: LREC 2006: Fifth International Conference on Language Resources and Evaluation, pp. 489–492 (2006)
14. Nie, J.Y., Cai, J.: Filtering noisy parallel corpora of web pages. In: 2001 IEEE International Conference on Systems, Man, and Cybernetics, vol. 1, pp. 453–458 (2001)
15. Tiedemann, J.: Parallel data, tools and interfaces in OPUS. In: Chair), N.C.C., Choukri, K., Declerck, T., Doğan, M.U., Maegaard, B., Mariani, J., Moreno, A., Odijk, J., Piperidis, S. (eds.) Proceedings of the Eight International Conference on Language Resources and Evaluation (LREC 2012). European Language Resources Association (ELRA), Istanbul, Turkey, May 2012
16. Vogel, S.: Using noisy bilingual data for statistical machine translation. In: Proceedings of the Tenth Conference on European Chapter of the Association for Computational Linguistics, vol. 2, pp. 175–178. Association for Computational Linguistics (2003)

High-Precision Person Name Extraction from Turkish Texts Using Wikipedia

Dilek Küçük[1]([⊠]) and Doğan Küçük[2]

[1] TÜBİTAK Energy Institute, Ankara, Turkey
dilek.kucuk@tubitak.gov.tr
[2] Gazi University, Ankara, Turkey
dogan.kucuk@gazi.edu.tr

Abstract. In this paper, we focus on person name extraction from diverse text types in Turkish and have compiled a large set of person names from Turkish Wikipedia. After automated post-processing to clean and extend it, we have performed extraction experiments using this resource on data sets of considerable sizes and achieved high precision rates. Next, we have shown that the use of non-local dependencies together with this Wikipedia resource improves recall, and hence F-Measure, considerably. Finally, we have tested the contribution of the resource and the scheme based on non-local dependencies to the person name extraction performance of a full-fledged named entity recognizer.

Keywords: Person name extraction · Turkish · Wikipedia · Named entity

1 Introduction

Extraction of proper person names from natural language texts is a subtask of named entity recognition (NER) which aims at identifying person, location, and organization names with some temporal and numeric expressions. For several languages, excluding the well-studied ones like English, person name extraction is still an important research problem as existing solutions are usually domain-specific and/or their performance rates are low.

In this paper, we present a person name extraction system for Turkish, based on a large person name list obtained from Wikipedia. The system is tested on text sets of diverse types, which were used for evaluating several previous NER proposals for Turkish, and the evaluation results are quite high in terms of precision. Additionally, we have considered the use of the person name extraction history that is obtained with the use of this list in order to introduce non-local features into the extraction procedure. This latter extension further improves the recall (and thereby the F-Measure) of the extractor. Finally, through experimentation with a full-fledged NER system, we have shown that the proposed approach can help improve the person name extraction performance of NER systems. As Wikipedia articles keep on increasing and are continuously updated, the

© Springer International Publishing Switzerland 2015
C. Biemann et al. (Eds.): NLDB 2015, LNCS 9103, pp. 347–354, 2015.
DOI: 10.1007/978-3-319-19581-0_31

compiled resource can also be automatically extended with new person names without manual intervention. With its use of the Wikipedia article titles, which can be linked to the original articles, the proposed approach not only extracts person names, but it also disambiguates them by linking them to the articles about people in the real-world. Furthermore, similar language resources can be compiled and extractors can be implemented for other resource-scarce languages.

2   Related Work

The earliest study regarding NER on Turkish texts is a language-independent system which is tested on Turkish along with four other languages [2]. In [15], a statistical name tagger based on HMMs is proposed while in [1], a person name extractor based on local patterns is presented. A rule-based NER system is proposed in [9] and a hybrid NER system built upon this rule-based system is described in [11]. The latter two systems are employed to extract named entities from video texts for semantic video indexing [10]. NER approaches based on CRFs are presented in [14,16] where the performance of the former system is reported to drop dramatically when evaluated on a tweet set [3]. A tweet data set in Turkish, annotated with named entities, is presented in [7] together with the evaluation results of a multilingual NER system. In [8], the NER system [9] is extended in several directions to be more applicable to tweets.

There are two previous studies using Wikipedia for NER in Turkish: in [5], the possible contribution of a classified subset of Wikipedia article titles to the NER task is investigated and in [6], this subset is used as the training data set of a kNN classifier to automatically compile person, location, and organization name lists from Wikipedia titles. In the current study, we target at high-precision person name extraction using an automatically obtained list of Wikipedia titles corresponding to proper person names using a heuristic-based approach. We have also considered the use of non-local dependencies to increase the coverage of the proposed extractor.

3   Person Name Extraction in Turkish Using Wikipedia

Person name extraction from Turkish texts can be considered a harder task compared to the extraction of other named entity types, reflected with low performance results. One of the reasons for the low precision is the homonymy of some common names with prevalent person forenames and surnames, and therefore employing lists of single person forenames/surnames and marking consecutively appearing entries in these lists as person names usually results in poor precision. The recall rates are again usually low, especially on informal texts, which is due to several reasons: (i) missing apostrophes that should separate the names from the attached suffixes, (ii) lack of case information, (iii) employment of the corresponding characters (*c, g, i, o, s,* and *u*) instead of Turkish characters with diacritics (*ç, ğ, ı, ö, ş,* and *ü*), and (iv) existence of foreign names. A plausible approach to address these problems is to maintain lists of real full person names and check for these names within the text. For this purpose, we have considered Wikipedia as an up-to-date and extensive information source, and used JWPL library [17] to access Turkish Wikipedia. Hence, our heuristic-based approach to automatically compile a person name list from Wikipedia is outlined below.

1. Based on the observation that Wikipedia articles on people usually include their dates of birth and that of death, with the JWLP library, we download the titles of person signifying pages under all Wikipedia categories of type "X doğumlular" ("those who were born in the year X") and "X yılında ölenler" ("those who died in the year X") where X is an integer ranging between 0 and 2014. After eliminating duplicate names listed under the considered 4,030 category pages, we obtain a list of person names of about 42,500 entries.
2. We remove single-token entries as they might be homonymous to common names, which in turn might lead to low precision during extraction.
3. For each entry having at least one Turkish character with a diacritic, we also include within the list a modified version of the entry in which all such characters are replaced with their ASCII counterparts. After this expansion scheme, we end up with a name list of about 55,100 entries.

We have implemented a person name extractor which marks person names within an input text by checking each possible candidate against the entries within this person name list; where all 7-grams, 6-grams, 5-grams, 4-grams, trigrams, and bigrams in the text are considered as candidates. The extractor checks these n-grams against the entries of the list in this sequential order and no overlapping extraction is allowed during the procedure.

4   Evaluation on Diverse Text Types

We have evaluated the person name extractor on Turkish text sets of diverse types and statistical information on these data sets is given in Table 1. The data sets employed correspond to almost all available text data sets in Turkish, proposed and used for NER evaluations so far, where the studies in which experiments on these data sets have been previously reported are given in parentheses in the first column. The last column displays the number of person names which span more than one token (hence, *multi-token person names*), which is provided since such names constitute the initial target domain of our extractor.

4.1   Extraction with the Compiled Wikipedia Resource

The performance evaluation of the proposed extractor on the data sets is performed using the metrics of precision (P), recall (R), and balanced F-Measure (F), as percentages. The evaluation results of the extractor are given in Table 2. The precision is quite high with values over 92 % for all data sets and reaching up to 98.11 % for *News Set–1*. The main reason leading to the rare false positives is the extraction of person name parts from location or organization names (hence, annotated as location or organization names in the gold standard) in

Table 1. Statistical information on the evaluation data sets.

Data set	# of tokens	# of person names	# of multi-token person names
News Set–1 [9]	~20K	398	169
News Set–2 [14,15]	~48K	1,596	743
News Set–3 [11]	~100K	3,288	1,488
Financial News Set [11]	~84K	1,115	468
Historical Text Set [9]	~20K	387	217
Tweet Set–1 [7]	~21K	457	149
Tweet Set–2 [3]	~50K	774	190
Overall	**~343K**	**8,015**	**3,424**

Table 2. Evaluation results of the person name extraction scheme.

Data set	P	R	F	R (Over multi-token names)	F (Over multi-token names)
News Set–1	98.11	26.13	41.27	61.54	75.64
News Set–2	95.54	20.11	33.23	43.2	59.5
News Set–3	97.25	21.5	35.22	47.51	63.84
Financial News Set	95.16	10.58	19.05	25.21	39.86
Historical Text Set	92.11	9.04	16.47	16.13	27.45
Tweet Set–1	96.67	12.69	22.44	38.93	55.50
Tweet Set–2	95.45	5.43	10.27	20.69	34.01

which the names of real people are given to the locations/organizations. For instance, "*Ali Sami Yen Stadı*" ("*Ali Sami Yen Stadium*") is a location name made of a person's name and annotated so within the gold standard while the extractor marks its first three tokens ("*Ali Sami Yen*") as a person name.

The recall and F-Measure rates are quite low, but since the extractor does not consider single-token person names, it is more plausible to consider the recall and F-Measure rates over multi-token names. For the three news sets, all from the METU Turkish corpus [13], the recall rates are promising as the articles usually report on well-known people who often have Wikipedia pages, and hence are included in our person name list. However, almost half of the person names within the news sets are not covered by the name list, most of them being foreign names. Another source of recall drop, albeit less frequent, is the appearance of some multi-token names in contracted forms while Wikipedia titles usually correspond to full person names. For example, the contracted person name "*M. Kemal Atatürk*" is missed by the extractor although the list includes the corresponding full name "*Mustafa Kemal Atatürk*".

Table 3. Evaluation results of the scheme also utilizing non-local dependencies.

Data set	P	R	F	ΔP	ΔR	ΔF
News Set–1	96.90	31.41	47.44	↓ 1.21	↑ 5.28	↑ 6.17
News Set–2	93.27	35.59	51.52	↓ 2.27	↑ 15.48	↑ 18.29
News Set–3	93.96	39.75	55.87	↓ 3.29	↑ 18.25	↑ 20.65
Financial News Set	97.23	28.34	43.89	↓ 2.07	↑ 17.76	↑ 24.84
Historical Text Set	87.50	9.04	16.39	↓ 4.61	*No change*	↓ 0.08
Tweet Set–1	93.94	13.57	23.71	↓ 2.73	↑ 0.88	↑ 1.27
Tweet Set–2	93.33	5.43	10.25	↓ 2.12	*No change*	↓ 0.02

The coverage of the extractor on financial news and historical texts is somewhat lower when compared to the coverage on news articles. Most of the missing cases are due to the fact that the owners of the missed names have not yet made their way to have individual Wikipedia pages and hence are not included in our name list. The recall rates on tweet sets are similarly low which is again due to the fact that some names are missed as they do not have Wikipedia pages. Additionally, spelling errors are quite common in this text type: some person names are misspelled while others are intentionally modified to show affection or emphasis, leading to missing of such names by the extractor.

To summarize, the precision of the presented extractor is quite high ranging from 92 % to 98 % on diverse text types. The precision rate of the initial form of the NER system for person name extraction on *News Set–1* was reported as 52.9 % [9] and the precision of its recent version is found to be 57.40 % as will be covered in Sect. 4.3. The system presented in [1] is reported to achieve a precision of 78.13 % on a news set of about 43 K tokens.

4.2 Extraction with the Resource and Non-local Dependencies

It is emphasized in the NER literature that multiple occurrences of the same entities within a discourse should be treated similarly and hence have identical labels [4,12]. The utilization of such non-local dependencies is known to improve the recognition performance. For instance, several approaches that use non-local dependencies are tested in [12]. Among them, *extended prediction history*, in which the label assignment distribution for all instances of the current token in the previous 1,000 words is recorded, outperforms the others on two data sets [12]. Following this approach, our extractor is modified as follows: when the extractor (with its original name list) detects a person name within the text, then the name list is extended to include the last token (surname part) of the extracted name and henceforth unigrams are also considered as candidates.

The evaluation results of the extended person name extractor are provided in Table 3. Within the columns 5–7 of Table 3, the performance differences between the columns 2–4 of this table and the columns 2–4 of Table 2 are given. The recall rates for the first four data sets are considerably improved. This result

Table 4. Initial and joint (with the Wikipedia resource and non-local dependencies) evaluation results of the person name extraction performance of the NER system [9].

Data set	Initial			Joint with the Wikipedia resource				Joint with the resource and non-local dependencies			
	P	R	F	P	R	F	ΔF	P	R	F	ΔF
News Set–1	57.40	71.11	63.52	59.57	70.35	64.52	↑1.00	60.58	73.37	66.36	↑1.84
News Set–2	53.85	45.99	49.61	62.88	54.76	58.54	↑8.93	64.53	63.60	64.06	↑5.52
News Set–3	61.77	64.57	63.14	66.48	66.06	66.27	↑3.13	66.08	70.86	68.39	↑2.12
Financial News Set	39.34	52.47	44.97	40.95	54.80	46.87	↑1.9	43.31	60.63	50.52	↑3.65
Historical Text Set	18.89	39.53	25.56	22.24	46.77	30.14	↑4.58	22.18	46.77	30.09	↓0.05
Tweet Set–1	36.73	43.33	39.76	38.67	50.77	43.90	↑4.14	38.70	50.98	44.00	↑0.10
Tweet Set–2	35.61	38.37	36.94	36.16	40.83	38.35	↑1.41	36.11	40.83	38.33	↓0.02

confirms that utilizing non-local dependencies improves the coverage of person name extraction in news articles. In all data sets, a slight decrease in precision is observed, which is an expected outcome, since it is likely that the newly-considered single-token person names may lead to a limited number of false positives. In the historical text and tweet sets, either a slight or no improvement in recall is observed. This is an expected result for the tweet sets, because we consider each tweet set as a coherent discourse although it is not the case. The recall decrease on the historical texts is due to that in few cases the recorded surnames coincide with historical dynasty names which are annotated as organization names in the gold standard.

4.3 Extraction Through Joint Utilization of the Resource and Non-local Dependencies with a Named Entity Recognizer

To observe the contribution of the Wikipedia resource and the use of non-local dependencies on the person name extraction performance of a full-fledged NER system, we execute the system in [9] first on the bare forms of the data sets, next on the pre-annotated versions of the sets with person names extracted using the Wikipedia resource only, and finally on the pre-annotated versions of the sets using both the resource and the scheme based on non-local dependencies. In these experiments, the system is configured not to use the capitalization clue and the evaluation results are provided in Table 4. Within the 8th and 12th columns of the table, the differences between the F-measure rates of the current settings and that of the immediately preceding settings are given.

In line with the results given in the previous sections, the recognizer consistently achieves better results with the joint utilization of the resource compiled from Wikipedia and the non-local dependencies, compared to its initial performance rates on the bare forms of the data sets (columns 2–4). Hence, it can be concluded that the automatic compilation of named entity resources from Wikipedia helps improve named entity recognition performance.

5 Conclusion

In this paper, we present a high-precision person name extraction scheme for Turkish texts, based on a list of names automatically compiled from Wikipedia. The scheme is tested on data sets of diverse text types, resulting in high precision rates ranging from 92 % to 98 % and the best performance rates are obtained on news texts. We have also shown that the use of non-local dependencies together with this Wikipedia resource improves recall rates considerably. Following similar procedures, name lists can be created for other resource-scarce languages, hence future work includes compiling name lists for other languages and using them together with the existing list for Turkish during extraction.

References

1. Bayraktar, Ö., Taşkaya-Temizel, T.: Person name extraction from Turkish financial news text using local grammar based approach. In: Proceedings of the International Symposium on Computer and Information Sciences, pp. 1–4 (2008)
2. Cucerzan, S., Yarowsky, D.: Language independent named entity recognition combining morphological and contextual evidence. In: Proceedings of the Joint SIG-DAT Conference on Empirical Methods in Natural Language Processing and Very Large Corpora, pp. 90–99 (1999)
3. Çelikkaya, G., Torunoğlu, D., Eryiğit, G.: Named entity recognition on real data: a preliminary investigation for Turkish. In: Proceedings of the 7th International Conference on Application of Information and Communication Technologies (2013)
4. Krishnan, V., Manning, C.D.: An effective two-stage model for exploiting non-local dependencies in named entity recognition. In: Proceedings of the International Conference on Computational Linguistics, pp. 1121–1128 (2006)
5. Küçük, D.: Utilizing annotated Wikipedia article titles to improve a rule-based named entity recognizer for Turkish. In: Larsen, H.L., Martin-Bautista, M.J., Vila, M.A., Andreasen, T., Christiansen, H. (eds.) FQAS 2013. LNCS, vol. 8132, pp. 683–691. Springer, Heidelberg (2013)
6. Küçük, D.: Automatic compilation of language resources for named entity recognition in Turkish by utilizing Wikipedia article titles. Comput. Stand. Interfaces **41**, 1–9 (2015)
7. Küçük, D., Jacquet, G., Steinberger, R.: Named entity recognition on Turkish tweets. In: Proceedings of the Language Resources and Evaluation Conference, pp. 450–454 (2014)
8. Küçük, D., Steinberger, R.: Experiments to improve named entity recognition on Turkish tweets. In: Proceedings of the EACL Workshop on Language Analysis for Social Media, pp. 71–78 (2014)
9. Küçük, D., Yazıcı, A.: Named entity recognition experiments on Turkish texts. In: Andreasen, T., Yager, R.R., Bulskov, H., Christiansen, H., Larsen, H.L. (eds.) FQAS 2009. LNCS, vol. 5822, pp. 524–535. Springer, Heidelberg (2009)
10. Küçük, D., Yazıcı, A.: Exploiting information extraction techniques for automatic semantic video indexing with an application to Turkish news videos. Knowl.-Based Syst. **24**(6), 844–857 (2011)
11. Küçük, D., Yazıcı, A.: A hybrid named entity recognizer for Turkish. Expert Syst. Appl. **39**(3), 2733–2742 (2012)

12. Ratinov, L., Roth, D.: Design challenges and misconceptions in named entity recognition. In: Proceedings of the 13th Conference on Computational Natural Language Learning, pp. 147–155 (2009)
13. Say, B., Zeyrek, D., Oflazer, K., Özge, U.: Development of a corpus and a treebank for present-day written Turkish. In: Proceedings of the 11th International Conference of Turkish Linguistics (2002)
14. Şeker, G.A., Eryiğit, G.: Initial explorations on using CRFs for Turkish named entity recognition. In: Proceedings of the International Conference on Computational Linguistics, pp. 2459–2474 (2012)
15. Tür, G., Hakkani-Tür, D., Oflazer, K.: A statistical information extraction system for Turkish. Nat. Lang. Eng. **9**(2), 181–210 (2003)
16. Yeniterzi, R.: Exploiting morphology in Turkish named entity recognition system. In: Proceedings of the ACL Student Session, pp. 105–110 (2011)
17. Zesch, T., Müller, C., Gurevych, I.: Extracting lexical semantic knowledge from Wikipedia and Wiktionary. In: Proceedings of the Language Resources and Evaluation Conference, pp. 1646–1652 (2008)

A Hybrid Approach for Extracting Arabic Persons' Names and Resolving Their Ambiguity from Twitter

Omnia H. Zayed[(✉)] and Samhaa R. El-Beltagy

Center of Informatics Science, Nile University, Giza, Egypt
omnia.zayed@gmail.com, samhaa@computer.org

Abstract. Tweets offer a novel way of communication that enables users all over the world to share real-time news and ideas. The massive amount of tweets, generated regularly by Arabic speakers, has resulted in a growing interest in building Arabic named entity recognition (NER) systems that deal with the informal colloquial Arabic. The unique characteristics of the Arabic language make Arabic NER a challenging task, which, the informal nature of tweets further complicates. The majority of previous works addressing Arabic NER were concerned with formal modern standard Arabic (MSA). Moreover, taggers and parsers were often utilized to solve the ambiguity problem of Arabic persons' names. Although, previously developed approaches perform well on MSA text, they are not suited for colloquial Arabic. This paper introduces a hybrid approach to extract Arabic persons' names from tweets in addition to a way to resolve their ambiguity using context bigram patterns. The introduced approach attempts not to use any language-dependent resources. Evaluation of the presented approach shows a 7 % improvement in the F-score over the best reported result in the literature.

1 Introduction

Named entity recognition (NER) is an Information Extraction (IE) task that aims to discern entities such as persons' names from unstructured text. NER has become a crucial constituent of different kinds of natural language processing (NLP) and text mining applications such as sentiment analysis, machine translation and text summarization.

Twitter, which is the most popular micro-blogging application in the world, presents a new type of social media content which is produced at unprecedented volumes. Specialized and up-to-date news can be drawn from tweets. The sheer amount of regularly generated tweets has encouraged researchers in many fields to analyse their content automatically for event detection and opinion mining. The informal nature of messages exchanged within this platform, poses new challenges for natural language processing (NLP) applications, as their content tends to be noisy and to deviate from known grammatical rules.

Studies have shown that Arabic was the fastest growing language on Twitter in 2011, and was the 6th most used language on Twitter in 2012 [1]. This rapid increase in online social media usage by Arabic speakers resulted in a growing interest in building Arabic NLP applications that deal with informal colloquial Arabic. Since NER is an

© Springer International Publishing Switzerland 2015
C. Biemann et al. (Eds.): NLDB 2015, LNCS 9103, pp. 355–368, 2015.
DOI: 10.1007/978-3-319-19581-0_32

important component of many text analysis applications, any application aimed at dealing with Twitter content, needs to employ an NER system capable of addressing Twitter specific challenges.

Research in the area of Arabic NER is still in its early phases compared to that of English NER [2], with the focus of most of that research being aimed at MSA Arabic since to date, only a few attempts have addressed the extraction of named entities from informal colloquial Arabic text.

Arabic is challenging when it comes to automatic text analysis not only due to its inflective nature but also due to its complex linguistic structure and rich morphology [3] as well as its inherent ambiguity. Ambiguity is in fact, one of the major challenges in detecting Arabic Persons' names as detailed in Sect. 2. Previous approaches that have tackled the problem of Arabic NER, have heavily depended on Arabic parsers, taggers and morphological analysers combined with a huge set of gazetteers and sometimes large training sets. But while these approaches are applicable to formal MSA text, they cannot handle informal colloquial Arabic with acceptable precision [4]. The unstructured nature of the colloquial language used in tweets degrades the performance of NER systems which are trained on formal text as shown by experimental results presented in [5].

The aim of this work is to extract Arabic persons' names from tweets in addition to resolving their ambiguity. The proposed hybrid approach combines a set of simple rules to extract single or full persons' names, with a supervised machine learning (ML) model that could classify a given ambiguous entity as either a name of a person or not. The classification technique determines whether an ambiguous entity is an actual person's name or not based on surrounding patterns. We introduce a training dataset of tweets to build the training model. No similar datasets are currently available for NER research; we plan to make the developed dataset available to boost the research in the area of Arabic NER. It worth mentioning that, the proposed approach does not use any language-dependent resources such as parsers or taggers. Thus, it can be migrated to other domains, text genres, and languages. Moreover, the proposed system does not depend on extensive lexical resources. The system depends, initially, on persons' names dictionaries obtained from publicly available resources such as Wikipedia [6].

The rest of the paper is structured as follows: Sect. 2 discusses Arabic specific challenges that affect extracting persons' names from social media content; Sect. 3 overviews the proposed approach in depth. In Sect. 4, experiments and results are described. Section 5 gives a brief overview on previous work pertained on Arabic NER with focus on Arabic NER from micro-blogs. Finally, the conclusion and future work are presented in Sect. 6.

2 The Effect of Arabic Specific Challenges on Person's Names Extraction Within the Social Media Context

The unique characteristics of the Arabic language makes the extraction of Arabic named entities a challenging task, to which, the nature of tweets adds new dimensions. Among the challenging characteristics, are the rich morphology, complex orthography, and the different levels of ambiguity. Tweets are usually written in dialectical Arabic, with dialects from all over the Arab World being represented, adding to the complexity of an already difficult problem. These characteristics were reviewed in detail in [2–4], and are revisited here with examples as follows.

- The Arabic language can be categorized into MSA and colloquial Arabic. MSA is the official language used throughout the Arab world and is used in official documents, newspapers and scientific books as well as in formal spoken occasions. The majority of Arabic NLP applications, including NER target MSA. Colloquial Arabic on the other hand, is very commonly used within all social media platforms. Colloquial Arabic is comprised of multiple spoken Arabic dialects used for daily communication in different Arab countries. Colloquial Arabic varies regionally from one Arabic speaking country to another. Common dialects amongst geographical regions have been grouped together into: Egyptian Arabic, Levantine Arabic, Gulf Arabic, North African (Maghrebi) Arabic, Iraqi Arabic, and Yemenite Arabic [2]. There are significant differences between Arabic dialects with respect to phonological, morphological, lexical and syntactic features. These differences also exist between these dialects and MSA [7]. Colloquial Arabic adds challenges to NER due to its ungrammatical and error prone nature.
- The Arabic Language has a complex morphology due to its agglutinative and inflective nature in which suffixes, infixes, and prefixes can be attached to the root of a word. This aspect creates semantic ambiguity in which one word could imply a variety of meanings [4]. A lot of examples can be found frequently in tweets such as, the word "هنادي" which may imply the colloquial phrase (I will call out), or the female name (Hanadi). Another example is the word "وفيه" which could mean the phrase (and in it), the adjective (loyal), or the female name (Wafia). This example also illustrates another problem which is the attaching of particles; here clitics, such as conjunctions and prepositions, could directly attach to any given word. This problem is not confined to the example above, in which the conjunction "and" attached to the proposition "in" creating a term that could be mistaken for a name or a particle, but extends to cases where conjunctions, invocation particles or other connection letters, attach to Arabic named entities [2]. For example, the invocation particle "يا" (O) can be found frequently in tweets attached to a name such as "يانبيل" (O Nabil), similarly the attachment of the conjunction letter "و" (and) such as "ونبيل" (and Nabil). This characteristic only serves to complicate the task of extracting persons' names from Arabic text.
- Arabic has no capital letters. Capitalization is a distinctive feature when it comes to NER [2, 4]. Additionally, Arabic has no letters dedicated to short vowels; instead diacritics[1] are used. However, these diacritics are rarely used in contemporary writings; yet, it is possible for a native speaker to infer the missing diacritics [3]. The absence of diacritics presents a challenge when it comes to NER, as it causes structural and lexical ambiguity in which a word could belong to more than one part of speech with different meanings [3, 4]. For example, the word "امنية" without diacritics could imply the female name (Omneya), or the noun (wish) or the adjective (security).

---

[1] Diacritics are special marks placed above or below the letters.

- Many persons' names are either derived from adjectives or can be confused with other nouns sharing the same surface form. Examples of ambiguous Arabic male names include [Nabil, Shreif, Kareem, and Radi] their different polysemies are [Noble, Virtuous, Generous, and Satisfied]. Examples of some ambiguous female names include [Ibtesam, Ilham, Alia, and Rahma] which could be interpreted as [Smile, Important or Inspiration, High, and Mercy]. Examples of some ambiguous family names are [Adeeb, Salama, Sorour, and Al-Arabi] which may be translated to [Writer, Safety, Happiness and The Arab] [4].
- Moreover, some Arabic persons' names match with verbs or prepositions such as [Ali, Yasser, Emam, and Waked] their different polysemies are [On, Imprisons, in front of, and Emphasized]. In addition, some foreign persons' names transliterated to Arabic could be interpreted as prepositions or pronouns such as [Ho, Anna, Ann, and Lee] which could be interpreted as [He, I, That, and Mine] [4]. Interestingly, some colloquial dialectal words may match with foreign persons' names such as [Wein, Mo, and Abby] which are polysemies of [Where, Not, and I want] in the Algerian/Tunisian, Saudi and Kuwaiti dialect, respectively.

The combination of the above-mentioned factors makes Arabic persons' names the most challenging Arabic named entity to extract without any morphological processing. A system that depends on straightforward matching using dictionaries will perform poorly if the various levels of ambiguity of the Arabic language are neglected. The properties of colloquial Arabic will degrade the performance of parsers and morphological analysers as a traditional solution for such challenges [4]. In this paper, the ambiguity problem is addressed using a supervised ML model trained on the bigram patterns surrounding ambiguous names.

3 Overview of the Proposed Approach

In this work, we introduce an approach to extract Arabic Persons' names which helps in resolving their ambiguity without employing any morphological analysis or language-dependant features. The proposed hybrid approach integrates a supervised ML model to resolve the ambiguity of a given entity with a set of simple rules.

The system incorporates two phases, namely a "Training Phase" to train the supervised model using patterns that surround an ambiguous entity and an "Extraction Phase" to extract persons' names and use the pre-trained model to resolve their ambiguity. A traditional Naïve Bayes classifier is trained to build the model using in-domain data and then the output from this classifier is integrated within a set of rules to extract persons' names. The dictionaries utilized by the proposed system are introduced in the next sub-section. After that, both of these phases will be described in depth.

3.1 Persons' Names Dictionaries

This system depends on a persons' names gazetteer that was developed in our previous work [4]. The creation of this gazetteer consisted of two stages. In the first stage, persons' names were collected from public resources, namely Wikipedia [8] and Kooora [9]. In the second stage, lists of first, male/middle and family persons' names were built automatically from the collected resources. The technique followed in gathering the names and building these lists is described extensively in [4]. These lists are utilized to create the rules for extracting persons' names as will be detailed later.

In addition to the previously developed lists, which are available on-line,[2] two more lists were manually created. The first list consists of ambiguous first names or names that are polysemies of other nouns or other Arabic entities/words while the second list consists of ambiguous family names. An example of a word that appears in the first list is "آية" (Aya) which could mean the first female name (Aya) or the noun (miracle) or the Egyptian interrogative pronoun (what) which is used frequently on Twitter. An example of a word that appears in the second list is "الأسد" which could mean the noun (the lion) or the family name (Al-Asad).

3.2 Training Phase: Building a Supervised Model

The first phase in the proposed system aims at classifying a given Arabic ambiguous person's name into one of two classes: "Name" or "NotName". A word should be classified as "Name" if it is an actual person's name, and "NotName", if it not.

A traditional Naïve Bayes classifier has been used and bigram patterns, around an ambiguous entity under inspection, were selected as features. Since our primary target is to extract persons' names from colloquial Arabic which is used nowadays in social media communication, the training dataset is built using in-domain data from Twitter.

The Training Dataset. To build the training model, we had to create our own training dataset of tweets, since no similar dataset is currently available.

The training dataset was collected by querying the Twitter Search API [10] using a subset of names[3] from the ambiguous names lists descried in Sect. 3.1. Each name was used as a query to get Arabic tweets using the language parameter "lang:ar" and the geo-code parameter "30.0500,31.2333,350 km". This geo-code specifies the location of the retrieved tweets to be Cairo with a radius of 350 km. Using this geo-code allows us to get the majority of tweets from Egypt and a small amount, from Saudi Arabia, Jordan and Palestine.

After getting the tweets, pre-processing steps are carried out to omit unwanted features. Essentially, the pre-processing steps involved removing the re-tweet symbol, English hash-tags and mentions, in addition to omitting hyper-links and English words. Moreover, diacritics were removed and text normalization was applied. Finally, duplicate tweets, due to re-tweets, were eliminated to guarantee the uniqueness of tweets. To carry out this step, a similarity check was performed by employing the cosine similarity technique [11] with a threshold value of 0.72.

---

[2] http://tmrg.nileu.edu.eg/downloads.html (under the title: Persons' Names Dictionaries).

[3] The rest is used to collect a test dataset.

Each tweet, now containing either a name, or a word that can be confused for a name was manually labelled as either "Name" or "NotName" according to its context in the tweet. Table 1 shows examples of the tweets in the training dataset with their corresponding annotations.

Table 1. Examples from the labeled tweets in the training dataset

Name used as a Query	Label	Tweet after Pre-processing	English Translation
احلام Ahlam (polysemy of Dreams)	Name	- #وائل كفوري يسخر من احلام علي الهواء	- #Wael Kafoury is making fun of **Ahlam**[a] on air
	NotName	- الليبراليين كلهم مساكين عايشين احلام اليقظه	- All liberals are poor they are living day **dreams**
كريم Kareem (polysemy of Cream and Generous)	Name	- بس يا كريم بقي مش كلكوا عليا :(	- That is enough **Kareem** do not be all against me :(
	NotName	- خدت حقنه . وكتبلي علي كريم مضاد حيوي - صباح راقي طيب كريم شريف لحضرتكوا جميعا نينا اختي يارب يحفظك ويسعدك واصدقائك	- I took an injection and he prescribed an anti-biotic **cream** for me - Elegant, kind, **generous,** and virtuous morning to all of you. Nina my sister, may God protect and bless you and your friends

[a]Bold words in the English translation correspond to either a name, or the English translation of the word that can be confused for a name

The final dataset consisted of nearly 9500 unique tweets in which 4550 are annotated as "Name" and 4920 are annotated as "NotName".

Building the Model. To build the model, bigram patterns around ambiguous entities were extracted from the annotated dataset. These patterns were used as features for classification. A pattern consists of $< word_1$ before ambiguous name $>$, $< word_2$ before ambiguous name $>$, $< word_1$ after ambiguous name $>$, $< word_2$ after ambiguous name $>$. Table 2 highlights a pattern around an example ambiguous entry.

Other examples of training patterns are shown in Table 3, along with their English translations. B1 and B2 are the bigram pattern before the ambiguous entity and A1 and A2 are the bigram pattern after it.

Sometimes, the inspected entity may be either at the beginning or at the end of the tweet; in that case a pattern does not exist before or after that entity. Therefore, a "null" value is added instead. Moreover, specific words, such as honorifics, unambiguous first names, and unambiguous family names, are replaced with their corresponding label (Honorific, UnambigFamilyName, UnambigFirstName, etc.) as shown by italicized entries in Table 3.

Table 2. Patterns around the given ambiguous name كريم "Kareem" (a polysemy of cream)

Label	Tweet after Pre-processing	Patterns
Name	بس يا كريم بقى مش كلكوا عليا): That is enough **Kareem**, do not be all against me :(	Bigram pattern before: يا بس (That is enough) Bigram pattern after: مش بقى (do not be) Represented as: مش, بقى, يا بس
NotName	خدت حقنه . وكتبلى على كريم مضاد حيوى I took an injection and he prescribed an anti-biotic **cream** for me	Bigram pattern before: على وكتبلى (pre-scribed for me) Bigram pattern after: مضاد حيوى (an anti-biotic[a]) Represented as: حيوى مضاد , على وكتبلى

[a]Words are re-arranged due to the English translation

Table 3. Examples of the training bigram patterns used as features with english translation

Class	B1	B2	A1	A2
Name	اوي very much	يا invocation particle	وربنا And God	يوفقك help you
Name	null	Honorific	بنطلب Ask	منك you
NotName	هجوم attack	الفولو follow	.	لم not
NotName	هو he	انسان human	.	null

Around 9800 patterns were extracted from the annotated dataset divided into patterns that surrounds a name and patterns that surrounds a polysemy of that name (i.e. NotName). These patterns were used as the training features to the classifier. Training was carried out using the Naïve Bayes classifier in WEKA[4] which is an open source suit for ML techniques [12] in order to build a classification model. The Evaluation of this classification model is detailed in the next section.

Ambiguity Model Evaluation. The purpose of this evaluation was to determine how well the classification model is capable of determining whether an ambiguous entry, is a name or not. Table 4 demonstrates the Naïve Bayes[5] classifier performance using two

---

[4] Waikato Environment for Knowledge Analysis.

[5] We experimented with SVM and BayesNET. Although the results in the classification phase were better but they were significantly lower in the extraction phase.

Table 4. Evaluation of the Naïve Bayes classifier through WEKA

Test Mode	Precision	Recall	F-score
10-folds CV	77.9 %	76.8 %	76.7 %
Unseen tweets test set	77.1 %	76 %	76.1 %

different test modes in WEKA [12]. The obtained results are in terms of weighted averaged precision, recall and F-score measures. The two different test modes are 10-folds cross validation and a supplied unseen test set respectively. We created a test dataset of around 1200 tweets in the same way the training dataset was built (described in Sub-Sect. 3.2). The set consists of 550 tweets classified as "Name" and 664 tweets classified as "NotName". Less than 6 % of the names that were used to create this dataset were seen before in the training set. Moreover, we made sure that the tweets in the test dataset were completely different from the tweets in the training dataset.

3.3 Extraction Phase: Rule Based Names Extraction with Resolving Ambiguity

In this phase, a person's name is extracted by plugging the classification model into a rule based approach.

A naive approach based on straightforward matching of persons' names using dictionaries can be used to identify previously unseen names by stating that a full name is composed of a first name followed by zero or more male names followed by (a male name or a family name). However, the inherent ambiguity of Arabic names, does not lend itself to such a simplistic solution [4].

One of the problems of simple matching is the possibility of incorrectly extracting a full name which consists of an ambiguous first name, which is not actually a proper name, followed by a family name. For example, given the tweet "لقاء بسيوني على قناة البحرين" (meeting with Basiouny on Bahrain channel) using straightforward matching rule would result in the extraction of the full name "لقاء بسيوني" (Lekaa Basiouny) even though (Lekaa) here means (meeting).

In addition, mentioning single names (only the first or family name) of a person is used frequently in social media communication due to the limited number of characters allowed in Twitter messages. Therefore, another problem of simple matching would result in the extraction of single first or family name which is not a name. For example, given the tweet "هي ايه الاستراتيجيه؟" (**What** is the strategy?), the word "ايه" (what) could be mistakenly extracted as the first female name (Aya).

In order to avoid such cases, we used the trained classification model in conjunction to rules to decide whether to consider a term as part of a name or not. Honorifics and other indicators are used as part of the employed rules. An example of the rules used to extract persons' names and resolve their ambiguity is shown in the pseudo code below.

```
full_name = "";
For each word w_i in tweet_words_list:
  if w_i in (ambiguous_first_names_list || ambigu-
ous_family_names_list)
      class = resolve_ambiguity(w_i); //Invokes the classifier
      if class == 'Name`
        extract(w_i);
        full_name = get_middel_names(i,tweet_words_list);
      else
        i++;
  else if w_i in (unambiguous_family_names_list || unambi-
guous_male_names_list)
      extract(w_i);
      full_name = get_middel_names(i,tweet_words_list);
```

It should be noted that, the utilized set of rules is used to match known names and not to learn new names. Because the used dictionaries have a good coverage, this does not pose a serious limitation.

4 Experiments and Results

4.1 Extraction Phase Evaluation

Evaluation Metrics. In order to evaluate the proposed approach, CoNLL's NER standard evaluation script [13] was used. As discussed extensively in [2], CoNLL's evaluation methods are one of the most aggressive methods as no partial credit is given for a partially extracted named entity. The results after running CoNLL's evaluation script are given in terms of precision, recall and f-score for each NER class [14]; in our case there is only one class: "persons' names". Each of these measures is defined as follows:

- Precision is the ratio of correctly retrieved persons' names to all the retrieved persons' names by the system.
- Recall is the ratio of correctly retrieved persons' names to the relevant persons' names in the gold standard dataset.
- Finally, the F-score is the micro-averaged harmonic mean of the precision and recall.

The Evaluation Dataset. In order to compare our proposed approach to [15], the test dataset of tweets developed by the authors of [15] was used. This test set is referred to here as Darwish's test set. Darwish's test set [15] consists of 1,423 tweets with nearly 26k tokens. Arabic and English named entities are both tagged in this dataset. The persons' names class comprises around 3 % of the total tokens among which 6 % of are English names. The annotation of the dataset followed the Linguistic Data Consortium (LDC) [16] guidelines for tagging; Since the only constraint on the retrieval of tweets in this dataset was that they are written in Arabic ("lang:ar"), the tweets were written in different Arabic dialects. Around 72 % out of the total tweets were written in Gulf

Table 5. Evaluation results for the proposed approach

System		Precision	Recall	F-score
Supervised ML approach presented in [15]		67.1	47.8	55.8
Our Proposed Approach	E1: (mistakes + English Entities)	67.20	53.53	59.59
	E2: (**no** mistakes + English Entities)	71.24	57.24	63.47
	E3: (mistakes + **no** English Entities)	66.49	58.74	62.38
	E4: (**no** mistakes + **no** English Entities)	**69.92**	**64.15**	**66.91**

Arabic dialects such as Saudi, Yemeni, Kuwaiti, Palestinian, Syrian...etc., and nearly 19.4 % were written in Egyptian colloquial Arabic; and the rest of the tweets were written in MSA.

Results. We carried out four different experiments to evaluate our system using Darwish's test set. The first experiment was done using the dataset as it is. The next experiment was done after fixing some annotation mistakes discovered in the dataset. The mistakes could be due to either not tagging a person's name or mistakenly tagging a word as a person's name. Two final experiments were conducted to test the effect of removing the entities written in English, which comprise around 6 % of the persons' names entities as mentioned in the previous sub-section, with and without the correction of the annotation mistakes.

Table 5 shows the results of the four experiments carried out using our system. It also shows the results obtained from the supervised ML system used to extract named entities from tweets presented in [15]. This system serves as a frame of reference for the results.

From this comparison, it can be seen that the presented approach achieves a 7.1 % improvement in F-score over the approach presented [15]. Fixing the annotation mistakes improved the results by around 4 %. Excluding the English entities improved the recall by 5.21 %. Since the system does not address the extraction of English entities, it is not entirely fair to include those when evaluating it.

5   Related Work

Arabic NER systems are broadly classified into ones that address formal MSA or those that address informal colloquial Arabic text. MSA is considered the predominant text class targeted by previous work addressing Arabic NER. On the other hand, extracting named entities from informal colloquial Arabic text has not received as much attention despite its extensive use in social media communication [4]. This section highlights some of the previous work done on Arabic NER focusing on recent attempts of NER from micro-blogs. Additionally, the survey in [2] presents an extensive review for numerous works done previously to recognize Arabic named entities. A brief review, is also provided in [15] to highlight the work done on English NER in the social media context.

Generally, "Rule Based NER" combines hand-crafted grammar rules with gazetteers to extract the named entities [4]. Among the earlier prominent works that use the rule based approach to extract persons' names from MSA text is PERA [17].

The system employs grammar rules to identify persons' names by utilizing trigger words to form a window around a person's name. Those rules are also supported by gazetteers including persons' names "Whitelist". A filtration mechanism is used as a final stage to get rid of any invalid persons' names. The evaluation of PERA was done on a reference dataset built by selecting random texts from within the ACE and Treebank news corpora. The system obtained a precision of 85.5 %, a recall of 89 %, and an F-score of 87.5 %. The system has been generalized as NERA [18, 19] to extract ten types of named entities.

According to [20], the machine learning (ML) approach is the most frequently used approach for NER from MSA text. In this approach, text features are used to classify input text depending on an annotated dataset. Different ML techniques and configurations [21–26] have been explored. The best performing of these (the ANER system) makes use of optimized feature sets [22]. N-grams and a maximum entropy classifier were employed to develop ANERsys [25] which represents the initial system developed in this series of research. A training and test corpora (ANERcorp) and gazetteers (ANERgazet) were developed to train, evaluate and boost the performance of the implemented technique. ANERcorp is currently considered the benchmark dataset for testing and evaluating NER systems applied on MSA text [4].

A number of hybrid approaches, that combine ML techniques, statistical methods and predefined rules, have been proposed. The best performing hybrid NER system, targets MSA text and uses a rule based NER component integrated with a ML classifier [27]. The system employs morphological and contextual features to extract eleven different types of named entities including persons, locations and organizations. The reported results of the system are significantly better than pure rule-based systems and pure ML classifiers. In addition this system's reported results are better than the state of the art Arabic NER system based on conditional random fields (CRF) [26].

A recent attempt to extract Arabic named entities from micro-blogs (namely, tweets) was introduced in [15]. The authors developed different "language-independent" approaches to extract person, location and organization named entities from Arabic tweets. The baseline system employed a CRF classifier which makes use of various features such as leading and trailing characters in a word, in addition to the current, previous, and next words in their raw forms. The system does not employ any linguistic processing. Two training corpora were used to train the baseline system: an out-of-domain corpus (ANERcorp [21]) and an in-domain one(a tweets corpus of 3,646 tweets). The baseline system was improved in three ways: first by utilizing large gazetteers from Wikipedia (Wikigaz), then by using domain adaptation (Adapt), and finally using a two-pass semi-supervised approach (2Pass). The effectiveness of each approach was tested individually on an annotated test set of 1,423 tweets. The semi-supervised method improved the baseline system's overall performance by 15 %. Finally, the overall results, for all named entities addressed by the system which are: person, location and organization, obtained by combining all three approaches were 76.8 %, 56.6 %, and 65.2 % for precision, recall, and F-score respectively. The results for detecting persons' names, as shown in Table 5, were 67.1 %, 47.8 %, and 55.8 %, for precision, recall, and F-score respectively.

Earlier work [5], applied a news trained NER system on tweets. The system utilized cross-lingual features from English to improve Arabic NER. The reported performance on the same tweets test dataset presented in [15], was very poor with an overall f-score of 39.9 % for all targeted entities. The system's poor performance can be attributed to the fact that the system was originally trained using a news dataset. The performance of the system for persons' names was 40.5 %, 39.2 %, and 39.8 % for precision, recall, and F-score respectively. The authors attributed the poor performance on persons' names to the various levels of ambiguity of Arabic persons' names. Our proposed approach attempted to overcome this problem by employing a supervised ML model to resolve the ambiguity of a given persons' name.

A second recent attempt to extract named entities from dialectal Arabic contents focusing on the Egyptian Dialect is presented [28]. The system adopts the supervised ML approach by using a CRF sequence labeller to extract persons', locations' and organizations' names. Two baseline systems were introduced. The first one, called (BAS1), uses the features reported in [29] which are: previous and next word, in addition to the leading and trailing 2, 3, and 4 letters in a word. Such features do not utilize any morphological analysers or taggers, thus, they are "language-independent" and are applicable to colloquial Arabic. The second baseline system, called (BAS2), used similar morphological features as [27] but employed the MADAMIRA [30] morphological analyser. In order to improve the baseline systems, various lexical, contextual, and morphological features have been used. Furthermore, ANERgazet [21] and Wikigaz [15] gazetteers have been utilized in addition to a window of 1 or 2 words as the distance from specific keywords such as honorifics, invocation particle and nicknames. Finally, the Brown Cluster [31] has been used to group together the semantically similar words. The effect of different permutations of features has been tested. The best performance was achieved when combining all features except Brown clustering. The system obtained overall results, for all named entities addressed by the system which are: locations and persons,[6] of 86.533 %, 62.302 %, and 70.305 % for precision, recall, and F-score respectively, in which persons' names achieved 78.947 %, 35.714 %, and 49.18 % for precision, recall, and F-score respectively. The used test set consists of a manually annotated portion of an Egyptian dialect corpus collected and provided by the LDC from web blogs[7] and contains nearly 40 k tokens. A performance comparison between our system and this system will be imperfect because of using different test datasets.

6   Conclusion and Future Work

This paper introduced a hybrid approach to extract persons' names from the informal colloquial Arabic text. The paper also presented a solution for the persons' names ambiguity problem without depending on extensive morphological or linguistic analysis. Thus, the proposed system can be ported easily to other text genres, languages and domains. Moreover, it does not depend on extensive lexical resources. The system relies on dictionaries of persons' names obtained from open-source resources to build a

---

[6] The authors omitted the organizations' names class because it has small frequency in the annotated data.

[7] LDC2012T09: GALE Arabic-Dialect/English Parallel Text.

set of rules. These rules were integrated with a supervised ML model to resolve the ambiguity of a given persons' name. A traditional Naïve Bayes classifier was utilized to build the training model by selecting bigram patterns surrounding a given ambiguous entity, as features. System evaluation showed that the performance of the presented technique outperformed a recent ML approach to extract named entities from tweets on their test dataset of tweets. Although the majority of tweets used to train our system were written in Egyptian colloquial Arabic, the system was able to discern the ambiguity of persons' names in the test set of tweets which were written in different Arabic dialects from various regions.

In the future, we plan to do further experimentation on other colloquial/MSA datasets for further testing and improving the system's behaviour. Also, we intend to extract other named entities such as locations and organizations.

References

1. Semiocast: Geolocation analysis of Twitter accounts and tweets by Semiocast. http://semiocast.com/en/publications/2012_07_30_Twitter_reaches_half_a_billion_accounts_140m_in_the_US
2. Shaalan, K.: A survey of arabic named entity recognition and classification. Comput. Linguist. **40**, 469–510 (2014)
3. Farghaly, A., Shaalan, K.: Arabic natural language processing: challenges and solutions. ACM Trans. Asian Lang. Inf. Process. **8**, 1–22 (2009)
4. Zayed, O., El-Beltagy, S., Haggag, O.: An approach for extracting and disambiguating arabic persons' names using clustered dictionaries and scored patterns. In: Métais, E., Meziane, F., Saraee, M., Sugumaran, V., Vadera, S. (eds.) NLDB 2013. LNCS, vol. 7934, pp. 201–212. Springer, Heidelberg (2013)
5. Darwish, K.: Named entity recognition using cross-lingual resources: arabic as an example. In: Proceedings of the 51st Annual Meeting of the Association for Computational Linguistics, pp. 1558–1567. Association for Computational Linguistics, Sofia (2013)
6. Wikipedia. https://www.wikipedia.org/
7. Habash, N.Y.: Introduction to Arabic Natural Language Processing. Mogran & Claypool, San Rafael (2010)
8. Wikipedia People Category. http://ar.wikipedia.org/wiki/تصنيف:تراجم
9. Kooora: Arabic Sports Web Site. http://www.kooora.com/default.aspx?showplayers=true
10. Twitter Search API. https://dev.twitter.com/rest/public/search
11. Singhal, A.: Modern information retrieval: a brief overview. Bull. IEEE Comput. Soc. Tech. Comm. DATA Eng. **24**, 35–43 (2001)
12. Witten, I.H., Frank, E., Trigg, L., Hall, M., Holmes, G., Cunningham, S.J.: Weka: practical machine learning tools and techniques with Java implementations. In: Kasabov, N., Ko, K. (eds.) Proceedings of the ICONIP/ANZIIS/ANNES 1999 Workshop on Emerging Knowledge Engineering and Connectionist-Based Information Systems, Dunedin, New Zealand, pp 192–196 (1999)
13. CoNLL's Standard NER Evaluation Script. http://www.cnts.ua.ac.be/conll2000/chunking/conlleval.txt
14. De Sitter, A., Calders, T., Daelemans, W.: A formal framework for evaluation of information extraction, Antwerp (2004)

15. Darwish, K., Gao, W.: Simple effective microblog named entity recognition: arabic as an example. In: Proceedings of the Ninth International Conference on Language Resources and Evaluation (LREC 2014), pp. 2513–2517. European Language Resources Association (ELRA), Reykjavik (2014)

16. Linguistic Data Consortium (LDC). https://www.ldc.upenn.edu

17. Shaalan, K., Raza, H.: Person name entity recognition for Arabic. In: Proceedings of the 5th Workshop on Important Unresolved Matters, pp. 17–24. Association for Computational Linguistics, Prague (2007)

18. Shaalan, K., Raza, H.: Arabic named entity recognition from diverse text types. In: Nordström, B., Ranta, A. (eds.) GoTAL 2008. LNCS (LNAI), vol. 5221, pp. 440–451. Springer, Heidelberg (2008)

19. Shaalan, K., Raza, H.: NERA: named entity recognition for Arabic. J. Amer. Soc. for. Inf. Sci. Technol. **60**, 1652–1663 (2009)

20. Abdallah, S., Shaalan, K., Shoaib, M.: Integrating rule-based system with classification for Arabic named entity recognition. In: Gelbukh, A. (ed.) CICLing 2012, Part I. LNCS, vol. 7181, pp. 311–322. Springer, Heidelberg (2012)

21. Benajiba, Y., Rosso, P., BenedíRuiz, J.M.: ANERsys: an arabic named entity recognition system based on maximum entropy. In: Gelbukh, A. (ed.) CICLing 2007. LNCS, vol. 4394, pp. 143–153. Springer, Heidelberg (2007)

22. Benajiba, Y., Diab, M., Rosso, P., Valencia, D.: Arabic named entity recognition using optimized feature sets. In: Proceedings of the Conference on Empirical Methods in Natural Language Processing, EMNLP 2008, pp. 284–293. Association for Computational Linguistics, Morristown (2008)

23. Benajiba, Y., Diab, M., Rosso, P.: Arabic named entity recognition: a feature-driven study. IEEE Trans. Audio Speech Lang. Process. **17**, 926–934 (2009)

24. Benajiba, Y., Diab, M., Rosso, P.: Arabic named entity recognition: an SVM based approach. In: Proceeding of the 2008 Arab International Conference on Information Technology (ACIT) (2008)

25. Benajiba, Y., Rosso, P.: ANERsys 2 . 0 : conquering the NER task for the arabic language by combining the maximum entropy with POS-tag information. In: IICAI, pp. 1814–1823 (2007)

26. Benajiba, Y., Rosso, P.: Arabic named entity recognition using conditional random fields. In: Proceedings of Workshop on HLT & NLP within the Arabic World, LREC, vol. 8 (2008)

27. Shaalan, K., Oudah, M.: A hybrid approach to Arabic named entity recognition. J. Inf. Sci. **40**, 67–87 (2014)

28. Zirikly, A., Diab, M.: Named entity recognition system for dialectal Arabic. In: Proceedings of the EMNLP 2014 Workshop on Arabic Natural Language Processing (ANLP), pp. 78–86. Association for Computational Linguistics, Doha (2014)

29. Abdul-hamid, A., Darwish, K.: Simplified feature set for Arabic named entity recognition. In: Proceedings of the 2010 Named Entities Workshop (NEWS 2010), pp. 110–115. Association for Computational Linguistics, Stroudsburg (2010)

30. Pasha, A., Al-Badrashiny, M., Diab, M., Kholy, A.El, Eskander, R., Habash, N., Pooleery, M., Rambow, O., Roth, R.: MADAMIRA: a fast, comprehensive tool for morphological analysis and disambiguation of Arabic. In: Calzolari, N., Choukri, K., Declerck, T., Loftsson, H., Maegaard, B., Mariani, J., Moreno, A., Odijk, J., Piperidis, S. (eds.) Proceedings of the Ninth International Conference on Language Resources and Evaluation (LREC 2014), pp. 26–31. European Language Resources Association (ELRA), Reykjavik (2014)

31. Brown, P.F., DeSouza, P.V., Mercer, R.L., Dellapietra, V.J., Lai, J.C.: Class-based n-gram models of natural language. Comput. Linguist. **18**, 467–479 (1992)

Extracting Relations from Unstructured Text Sources for Music Recommendation

Mohamed Sordo[1]([⊠]), Sergio Oramas[2], and Luis Espinosa-Anke[2]

[1] Center for Computational Science, University of Miami, Coral Gables, FL, USA
`msordo@miami.edu`
[2] Department of Information and Communication Technologies,
Universitat Pompeu Fabra, Barcelona, Spain
{`sergio.oramas,luis.espinosa`}`@upf.edu`

Abstract. This paper presents a method for the generation of structured data sources for music recommendation using information extracted from unstructured text sources. The proposed method identifies entities in text that are relevant to the music domain, and then extracts semantically meaningful relations between them. The extracted entities and relations are represented as a graph, from which the recommendations are computed. A major advantage of this approach is that the recommendations can be conveyed to the user using natural language, thus providing an enhanced user experience. We test our method on texts from *songfacts.com*, a website that provides facts and stories about songs. The extracted relations are evaluated intrinsically by assessing their linguistic quality, as well as extrinsically by assessing the extent to which they map an existing music knowledge base. Finally, an experiment with real users is performed to assess the suitability of the extracted knowledge for music recommendation. Our method is able to extract relations between pair of musical entities with high precision, and the explanation of those relations to the user improves user satisfaction considerably.

1 Introduction

Music consumption has changed dramatically in the last few years. The rise of digital audio and streaming services means users are now one click away from accessing millions of songs by more than a million artists [8]. Yet this vast availability has posed a serious problem: how can a user explore or discover preferred music from all the available content? Traditionally, users have relied on their friends, their favorite music radio host, a music expert in their local retail store, etc. to obtain recommendations on artists or albums they might like. Although this traditional approach is still valid and used by many people, its ability to cover the vast amount of available music nowadays is seriously hindered. Automatic approaches to music recommendation have become necessary. According to [8], there is no clear formula for providing *good* recommendations to a user. There are however some key elements that should be taken into account: *novelty, familiarity* and *relevance*. The authors also emphasize the importance of

© Springer International Publishing Switzerland 2015
C. Biemann et al. (Eds.): NLDB 2015, LNCS 9103, pp. 369–382, 2015.
DOI: 10.1007/978-3-319-19581-0_33

providing an *explanation* to the user, as to why a music item has been recommended. The latter can provide an enhanced user experience, helping the user gain confidence in the recommendation system.

In this paper we propose a method that exploits unstructured text sources from the web to provide music recommendations. It does so by identifying music-related entities in the text (such as *song, band, person, album* and *music genre*) and extracting relations between these entities. The resulting knowledge graph can be used to not only provide recommendations but also to give an explanation of the recommendations to the user by using natural language.

2   Related Work

2.1   Music Recommendation

Music Recommendation is a relatively young but continuously growing research topic, in both MIR and RecSys communities [7]. There are many methods for recommending music. In this paper we only concentrate on Context-based filtering methods. Context-based filtering methods use information extracted from text sources to obtain similarity between artists or songs. Approaches based on this technique typically compute some sort of term weighting, like TF-IDF [30], or exploit co-occurrences between musical entities [16,26].

Most research in Music Recommendation has been dedicated to developing algorithms that provide *good* and *useful* recommendations [7], yet very few approaches (at least to our knowledge) provide explanations of the recommendation to the users [24,25]. According to [8], giving explanations of the recommendations provides transparency to the recommendation process and increases the confidence of the user in the system. In [24] Passant proposes a Music Recommendation system that uses the dataset of structured information DBpedia as a backbone for finding similar artists. Explanations are given to the user based on the shared sub-classes of the DBpedia ontology between two artists (e.g.: voice type, instrument, death place, etc.). In this paper we propose a method to extract explanations from unstructured text sources.

2.2   Relation Extraction

Relation Extraction (RE) approaches are often classified according to the level of supervision involved. Supervised learning is a core-component of a vast number of RE systems, as they offer high precision and recall. However, the need of hand labeled training sets makes these methods not scalable to the thousands of relations found on the Web [18]. More promising approaches, called semi-supervised approaches, bootstrapping approaches, or distant supervision approaches do not need big hand labeled corpus, and often rely on existent knowledge base to heuristically label a text corpus (e.g., [6,18]) Open Information Extraction methods do not require a pre-specified vocabulary, as they aim to discover all possible relations in the text [2]. However, these methods have

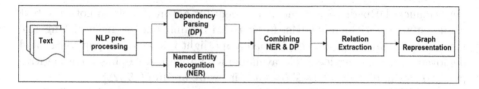

Fig. 1. Workflow of the proposed method.

to deal with uninformative and incoherent extractions. In ReVerb [13] part-of-speech based regular expressions are introduced to reduce the number of these incoherent extractions. Less restrictive pattern templates based on dependency paths are learned in OLLIE [21] to increase the number of possible extracted relations. Unsupervised approaches do not need any annotated corpus. In [12] verb relations involving a subject and an object are extracted, using simplified dependency trees in sentences with at least two named entities. These approaches can process very large amounts of data, however, the resulting relations are hard to map to ontologies [19].

In this paper we use a technique called Dependency Parsing to extract relations from text. Dependency Parsing provides a tree-like syntactic structure of a sentence based on the linguistic theory of Dependency Grammar [29]. One of the outstanding features of Dependency Grammar is that it represents binary relations between words [1], where there is a unique edge joining a node and its parent node (see Fig. 2 for the full parsing of an example sentence). Dependency relations have been successfully incorporated to RE systems. For example, [5] describe and evaluate a RE system based on shortest paths among named entities. Reference [10] focus on the smallest dependency subtree in the sentence that captures the entities involved in a relation, and [14] propose a rule-based dependency-parsing Open IE system.

3   Methodology

3.1   NLP Pre-processing

Figure 1 depicts the work-flow of the proposed method. Given a text input (e.g., a collection of web documents) the pre-processing module segments it into single sentences. Each sentence is subsequently divided into a sequence of words or tokens. In this paper we use the Stanford NLP tokenizer[1].

3.2   Named Entity Recognition (NER)

Although NER is not a solved problem [20], there are many available tools with good enough performance ratios [15]. Among those tools, we tried AIDA [31] and DBpedia Spotlight [22]. Although AIDA has a higher recall [15], in terms of CPU and memory consumption DBpedia Spotlight provided a better

---

[1] http://nlp.stanford.edu/software/tokenizer.shtml.

performance. DBpedia Spotlight is a system for automatically annotating text documents with DBpedia URIs, finding and disambiguating natural language mentions of DBpedia resources. DBpedia Spotlight is shared as open source and deployed as a Web service freely available for public use[2]. It has a competitive performance and evaluations show an F-measure around 0.5 [22].

Our NER module receives a list of sentences as input and uses DBpedia Spotlight to find DBpedia entities in the sentences. The entities are then annotated with their corresponding URI and type. In the current approach we only consider 5 different types that are relevant to the music domain: *song*, *band*, *person*, *album* and *music genre*. The rest of the recognized entities are ignored.

3.3 Dependency Parsing (DP)

Our DP module uses the implementation by [3] and produces a tree for each sentence. Each node in the tree represents a single word of the sentence, together with additional linguistic information like Part-of-Speech[3] and syntactic function. For instance, in Fig. 2 the word *Freedom* is the subject (SBJ) of the root word *was*. The definition of all these syntactic functions is given in [28]. In our case, however, we want to find relations between music-related entities, which can consist of more than one word. The next module takes care of this.

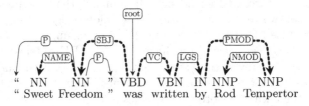

Fig. 2. Example sentence with dependency parsing tree.

3.4 Combining NER and DP

The aim of this module is to combine the output of the two previous modules. For each recognized music-related entity in the NER module (Sect. 3.2), the combination module merges all the nodes in the dependency tree of the sentence that correspond to that entity into a single node. Figure 3 shows how the example sentence from Fig. 2 is modified by merging the nodes that correspond to the recognized entities (in this case, the album *"Sweet Freedom"* and the person *Rod Temperton*) into single nodes.

3.5 Relation Extraction (RE)

The Relation Extraction module analyzes the modified dependency trees from the combination module (Sect. 3.4), and extracts relations between pairs of recognized music-related entities. Two entities (nodes) in a tree are considered to be

---

[2] https://github.com/dbpedia-spotlight/dbpedia-spotlight/wiki/Web-service.
[3] http://ling.upenn.edu/courses/Fall_2003/ling001/penn_treebank_pos.html.

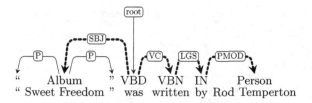

Fig. 3. Example sentence with its modified dependency tree, after merging nodes that correspond to an entity.

related if there is a path between them that does not contain any other entity. Since dependency trees are directed trees, there is no guarantee in finding a path. Therefore we use an undirected version of the tree to obtain the path between the aforementioned pair of entities. Interestingly, the nodes that are part of the path between the two entities are considered by our method to represent the actual relation between the entities. In the example of Fig. 3, the resulting path between *"Sweet Freedom"* and *Rod Temperton* contains the words *was, written, by*. These words are used to define the relation between *"Sweet Freedom"* and *Rod Temperton*.

By analyzing the output of a subset of the Songfacts dataset (explained in Sect. 4.1), we observed that not all relations between pairs of music-related entities made sense linguistically. Thus, we empirically introduced a set of rules to filter out irrelevant relations. A rule in our case is defined as a sequence (or a regular expression) of word types that can appear between a pair of music-related entities. The word types are represented by their part of speech tags. For instance, the previous example (Fig. 3) has a relation between an entity of type *Album* (*"Sweet Freedom"*) and an entity of type *Person* (*Rod Temperton*). The relation *Album-Person* in this case contains the terms *VBD*, *VBN* and *IN*, which are part of speech labels meaning verb past tense, verb past participle and preposition, respectively. The complete list of rules is shown in Fig. 4.

3.6 Graph Representation

After the list of relations between entities is filtered, the method creates a graph representation of it, where the nodes are the music-related entities and the edges represent the relations (i.e., the path) between pairs of entities. The graph contains five chosen types of nodes corresponding to the 5 music-related types: *song, band, person, album* and *music genre*.

4 Evaluation

We tested our method against a dataset gathered from *songfacts.com* (Sect. 4.1). The output of the method has been evaluated from two different standpoints, namely: (1) a linguistically motivated evaluation of the extracted relations and (2) a data-driven evaluation of the extracted knowledge. The linguistic evaluation

```
('Album', 'Band'): ['(VBD)(NN)', '(VBZ)(NN)', '(VBD)(NN)(IN)(NN?)', '(NN?)(IN)(NN?)', '(VBD)(VBN)(IN)'],
('Album', 'Person'): ['(NN)(NN?)', '(VBD?)(NN)', '(NN?)(IN)(NN?)', '(VBD)(VBN?)(IN)(NN?)'],
('Album', 'Song'): ['(NN?)(VBD)', '(VBD)(NN)(IN)'],
('Band', 'Album'): ['(VBD?)(NN)', '(VBD)(IN)', '(NN)(VBD)', '(NN)(IN)(NN)', '(NN?)(VBD)(IN)(NN)'],
('Band', 'MusicGenre'): ['(VBZ)(NN)', '(VBD)(NN)', '(VBP)(NN)'],
('Band', 'Person'): ['(NNS)', '(NN)(NN?)', '(NN)(VBD)', '(NN)(CC)(NN)', '(VBD)(IN)(NN?)', '(VBD)(NN)'],
('Band', 'Song'): ['(NN?)(VBD)', '(VBD?)(NN)(IN)', '(NN)(VBD)(NN)(IN)'],
('MusicGenre', 'Band'): ['(NN)(NN?)'],
('MusicGenre', 'Person'): ['(NN)(NN?)', '(NN)(CC)(NN)'],
('Person', 'Album'): ['(NN)(NN?)', '(NN)(IN)(NN?)', '(NN?)(VBD)(IN?)(NN?)'],
('Person', 'Band'): ['(NN?)(IN)', '(VBD)(IN?)(NN?)', '(VBD)(NN)(IN)', '(NN)(IN)(NN)'],
('Person', 'Person'): ['(NN)', '(NN)(CC)(NN)', '(VBD)(IN)(NN?)', '(VBD)(CC)(VBD)'],
('Person', 'Song'): ['(VBD)(NN)', '(VBD)(IN?)', '(NN)(IN)', '(VBD)(NN)(VBG)', '(NN)(VBZ)(IN)', '(VBZ)(NN)(IN)',
                      '(NN)(VBD)(NN?)(IN?)', '(VBD)(IN?)(NN)(IN?)'],
('Song', 'Album'): ['(VBD)(VBN)', '(VBZ)(NN)(NN)', '(VBD)(VBN?)(IN)(NN?)'],
('Song', 'Band'): ['(VBZ)(NN)', '(IN)(NN)', '(VBD)(VBN?)(IN)(NN?)', '(VBZ)(VBN?)(IN)(NN?)'],
('Song', 'MusicGenre'): ['(VBZ)(NN?)', '(VBZ)(NN)(IN)(NN?)', '(VBD)(VBN)(IN)(NN?)'],
('Song', 'Person'): ['(VBZ)(NN)', '(IN)(NN?)', '(VBD?)(NN)', '(VBZ)(IN)(NN?)', '(VBD)(VBN?)(IN)(NN?)', '(VBZ)(NNS)(IN)',
                     '(VBZ)(NN)(IN)(NN?)', '(NN)(VBD)(VBN)(IN)'],
```

Fig. 4. Part-of-speech rules that represent the relation between music entity types

quantifies the correctness of a relation by comparing it to a reference annotation manually crafted by a Computational Linguistics expert. Data-driven evaluation compares the extracted knowledge with a reference knowledge-base. A final experiment involving real users was also performed to assess the suitability of the extracted knowledge for music recommendations, and to test the effect produced by textual explanations in the recommendations given to the user. The following subsections provide a detailed description of the dataset and the experimental results.

4.1 Dataset

Songfacts[4] is an online database that collects, stores and provides facts and stories about songs. These stories are collected in a crowd-sourcing way by registered users and they are reviewed by the website staff. The web site contains information about more than 30.000 songs belonging to nearly 6.000 artists. Songfacts tidbits are little pieces of information telling stories about a song, such as what the song is about or who wrote it, who produced it, who collaborated with whom or who directed the video clip, etc. Therefore, a huge amount of information about the actors involved in the creative process of a song is present in the aforementioned tidbits.

We crawled the whole song dataset from Songfacts in mid-January 2014. Before applying our method, we did some adjustments in the NER module due to the specificity of the dataset. Identification of song titles in text is a challenge, since titles are often short and ambiguous [17]. Fortunately, due to the nature of the Songfacts website — which provides a separated web page for each song, with its corresponding metadata and facts — the complexity of the identification process of song titles is reduced considerably, under the assumption that ambiguity is less probable in this scenario. Thus, apart from using the DBpedia Spotlight NER, we searched and matched each song title in the facts. Moreover, further analysis of the facts showed that usually the facts refer to the song in question using expressions such as "the song" or "this song". Therefore, we looked for these structures and treated them as detected song entities. We chose

---

[4] http://www.songfacts.com.

only the songs whose title had been recognized by our system as a song entity and only those among them who were involved in at least one relation with another recognized entity. Finally, we also used the artist metadata provided by Songfacts to add a relation between entities of type artist and entities of type song with the label "by". After applying all the steps of our method, we obtained 12838 entities and 16341 relations between them. Among the detected entities, 6116 were songs, and those songs were related to 1483 different artists.

4.2   Linguistic Evaluation

We base our evaluation on previous work in Relation Extraction. For example, in [13,21], the general approach is to assess the automatically extracted relations in terms of correctneess according to human judgement. Additionally, [2] describes a finer-grained analysis, adding a prior step in which relations are judged as being *concrete* or *abstract*. Our evaluation sample amounts to 205 relations extracted from 155 randomly selected sentences. Two human judges marked a relation as "correct" if the information contained in the sentence implied or connoted that the relation was true. An "incorrect" label was assigned otherwise.

The results obtained were very high with regard to the observed agreement. Our results indicate that out of 205 relations, both evaluators agreed in judging 146 relations as correct and 23 as incorrect. This means that the overall observed agreement reaches 82.43 %, while the agreement only on correct relations is 71.21 %. We also computed the Cohen's Kappa [9] agreement measure in order to have an additional viewpoint of the reliability of this evaluation. Our computation of Cohen's Kappa was 40.68, which is a reasonably high value considering that the evaluation only consists of two classes ("correct" and "incorrect"), and this metric strongly punishes the chances of two evaluators to agree by chance.

We illustrate these results with an example that showcases a case of agreement (the first case) and a case of disagreement (the second) in the same sentence.

Sentence: [*Weezer*] frontman [***Rivers Cuomo***] wrote this song for and about Jamie Young , the band's first lawyer.
Entities: Band ↔ Person.
Extracted Relation: Weezer (frontman) Rivers Cuomo.

Sentence: Weezer frontman [***Rivers Cuomo***] wrote [*this song*] for and about Jamie Young, the band's first lawyer.
Entities: Person ↔ Song
Extracted Relation: Rivers Cuomo (frontman wrote) Weezer_-_Jamie

In the first example both evaluators agreed in assigning a "correct" label to the relation. In the second example one evaluator found it to be incorrect. We argue that this can be due to the distracting presence of "frontman", which can be considered to be a property of the first entity, rather than an element of the

Table 1. Results per pairs for the token-wise evaluation. The analyzed entities are: P (person), S (song), B (band), A (album), M (Music Genre)

	P - P	P - S	B - P	B - A	P - A	B - S	S - A	M - B	M - P	S - M
Precision	59.09	95.52	88.88	94.87	81.13	**95.74**	95.55	85.18	62.5	96.55
Recall	48.14	60.95	49.23	31.62	45.74	55.55	47.25	**71.87**	66.66	57.14
F-Score	53.55	74.41	63.36	47.43	58.49	70.36	63.23	**77.96**	64.51	71.79

relation. While this dichotomy has been addressed in previous work ([13] evaluated a relation to be correct where critical information was dropped from the relation but included in the second argument), we propose a lexically motivated approach, which compares the relations extracted by the system with those that would be extracted by a human expert. The idea was to be able to compare the wording of a relation between system and human. Precision and Recall are computed by looking at word-overlap. For instance, in the above case, the relation for the pair Person↔Song would get a score of P = 0.5 and R = 1 because the human evaluator extracted the relation *Rivers Cuomo (wrote) Weezer_-_Jamie*. "Frontman" would be a false positive. Table 1 provides results for the full evaluation dataset and for each pair of entity types.

It is worth noting how our approach has performed very well in certain pairs, especially in the MusicGenre ↔ Band, Person ↔ Band, Band ↔ Song and Song ↔ MusicGenre pairs. This might be due to the many straightforward one-word relations among these entities, as shown in the following examples:

Sentence: The [*Christian Metal*] band [*Stryper*] recorded this song for their 1990 album Against the Law and made a video for it.
Entities: MusicGenre ↔ Band.

Sentence: Jessie Lacey of Brand New's girlfriend cheated on him with [*John Nolan*] of [*Taking Back Sunday*].
Entities: Person ↔ Band.

Lower scores were obtained in relations like Person ↔ Album or Band ↔ Album. A closer look at these relations shows that there are many cases where an album is preceded by a number of adjectives and other noun modifiers. These modifiers are often described as sibling nodes of the relation in the dependency tree, and thus do not appear in the path between the two related entities.

4.3   Data-Driven Evaluation

The output of our system can be regarded as a knowledge base of music related information. This knowledge base consists of entities and relations, two building blocks of a simple, non-taxonomic ontology. According to [11], a learned ontology can be evaluated in three different ways: in the context of an application, by domain experts or by comparing it with a predefined reference ontology (i.e., a

Gold Standard). In this section we use the latter approach as it allows a certain amount of automation of the evaluation process [27]. The Gold Standard with which to evaluate our learned knowledge base was obtained from MusicBrainz[5], the most complete and accurate open knowledge base of music information. Instances of musical entities such as Recording, Artist, Release, etc. are identified by a universally unique identifier, a MusicBrainz ID. We extracted a subset of the MusicBrainz database containing all the entities that could be mapped to entities in our knowledge base, along with their corresponding relations. This mapping was accomplished as follows: for those entities in our knowledge base with a DBpedia URI (such as entities of type *person*, *band* and *album*) we obtain their MusicBrainz ID by first mapping the DBpedia URIs to Freebase (another open knowledge-base system) and then mapping Freebase IDs to MusicBrainz. This is because currently MusicBrainz IDs cannot be resolved directly from DBpedia. Regarding entities of type *song*, since we do not have a URI, we query the MusicBrainz API[6] by using song and artist name strings. Entities of type *musicgenre* were not considered for this evaluation as there is no corresponding concept in MusicBrainz. Finally, relations between the mapped entities were obtained using the aforementioned MusicBrainz API.

Of the 12838 entities in our knowledge base we could map 11740 entities in MusicBrainz, which represent a 91.4 %. In order to evaluate both knowledge bases we removed those entities that could not be mapped to MusicBrainz. To facilitate the evaluation process we represented both our knowledge base and the Gold Standard as graphs, where nodes correspond to musical entities and edges represent relations between those entities. Some pairs of entities could have more than one relationship. For example, *artist* "Bob Ezrim" is related to *album* "The Wall" as orchestration and producer in MusicBrainz. In our case we simplified this by merging all these relation terms into a single edge with multiple labels. As a result of this process, our knowledge base and the MusicBrainz Gold Standard contained 13165 and 10595 edges, respectively.

As a first evaluation we calculated the overlap of edges between the two graphs, regardless of the labels (i.e., the relation concepts) of those edges. We obtained an overlap of 5236 edges, which represented a 49.4 % of the Gold Standard relations and a 39.8 % of our extracted knowledge base. Once this overlap is obtained, the next step is to assess how our knowledge base "fits" the MusicBrainz Gold Standard [4]. Evaluating two ontologies, or in this case two knowledge bases, is an arduous task. Traditional Information Retrieval evaluation measures such as precision and recall cannot be easily used in their strict sense, as there is no clear definition of what knowledge is acquired [4]. The main problem in our case is that the vocabularies used in the two knowledge bases are different. Nevertheless, even though the vocabularies are different, many of their terms refer to similar music-related concepts. Hence, finding a conceptual equivalence between relation terms in our knowledge base and the MusicBrainz Gold Standard is fundamental in order to evaluate our approach in terms of precision and recall. Of the overlapping 5236 edges, we selected all the distinct

---

[5] http://musicbrainz.org/.

[6] http://musicbrainz.org/doc/Development/XML_Web_Service/Version_2.

combinations of the MusicBrainz relation terms (i.e., labels) and our knowledge base relation terms that co-occur in the same edges and grouped them by relation type. A relation type in this case is defined as a relation between a pair of types of entities. For example, *person* ↔ member of band ↔ *band* is a relation type, where member of band is the relation between an entity of type *person* and an entity of type *band*. A closer look at the Gold Standard graph shows that many relations in this graph do not have labels. For example, many artists are related to recordings in MusicBrainz without any explicit relation concept. Also, as mentioned previously in Sect. 4.1, our knowledge graph had some artificially added relations (the "by" relation between songs and artists). We thus decided to ignore these relations from our evaluation.

The grouping of the relation terms in relation types resulted in 727 different combinations, for which an equivalence had to be computed. A MusicBrainz relation type is considered to have an equivalent relation type in our knowledge base if their relation terms are conceptually similar. For example, the relation term married in MusicBrainz is conceptually implicit in the relation term husband in our knowledge base. Futhermore, MusicBrainz also organizes its relation terms in tree-like taxonomies, where conceptually similar terms are grouped in the same tree branch[7]. This can be used to decide whether a term in our knowledge base can be mapped to a term in the MusicBrainz relation taxonomies. In order to compute the equivalence of the 727 combinations we asked three human annotators to vote whether the two relation terms are conceptually similar. Due to lack of space, we made the votings available at http://goo.gl/uOGjlo.

Once this equivalence is obtained, precision and recall can be computed at an edge level. For this evaluation we only use a subset of the graphs. The subset is defined by all the overlapping edges in both graphs with at least one relation term. For each edge in the graphs, precision refers to how many relation terms in our knowledge base edge have an equivalence in the Gold Standard edge, whilst recall refers to how many relation terms in the Gold Standard edge have an equivalence in our knowledge base edge. Lets use the previous example of *artist* "Bob Ezrim" related to *album* "The Wall" as orchestration and producer in MusicBrainz. The relation between "Bob Ezrim" and "The Wall" in our knowledge base is defined by the single term producer. In this case, precision will be 1, but recall will be 0.5. We computed the average precision and recall over the 1143 total overlapping edges and obtained a score of 0.74 and 0.72, respectively[8]. These scores show a high correlation between MusicBrainz and our approach, which can confirm the veracity of many relations in the *songfacts.com* website. This could suggest that a combination of both knowledge bases might increase the completeness of metadata in MusicBrainz. The assessment of such assumption is though left for future work.

4.4 Recommendation Experiment

The aim of this experiment is to check the suitability of the extracted knowledge for music recommendation, and test the utility of explaining relations between

---

[7] https://musicbrainz.org/relationships.

[8] The individual precision and recall scores are available at http://goo.gl/C4Coj3.

songs. Although there are several approaches to compute recommendations using knowledge graphs with proven good performance [23], our approach was reduced to finding shortest paths between entities of type *song* in the graph. This baseline approach was selected for simplicity reasons, as the aim of the experiment was not to measure the performance of the recommendation system.

The experiment involved 30 participants, 24 males and 6 females, from 24 to 51 years old and with different musical background and listening habits. Most of the participants affirmed that they had previous experience with recommendation systems. In this experiment a set of recommended songs with textual explanations is presented to the participants. First, the participant is asked to choose a list of 10 songs from different artists she likes among all the songs in our dataset. Then, a new page with a list of the 10 seed songs is displayed, along with one recommended song per seed song and a textual explanation showing the relation between the two songs. The following is an example of how the explanations are given to the users:

Seed song: *Cloud Of Unknowing* by Gorillaz
Explanation:
Bobby Womack (performance on) Cloud Of Unknowing
Bobby Womack (legend played on) Shake
Recommended song: *Shake* by Sam Cooke.

Participants could listen to a 30 s preview of the songs. They were asked to rate the recommendations and the explanations (with a 1–5 rating scale), and to select whether the explanations influenced their ratings. Finally, participants were asked whether they were familiar with the recommended songs.

A total of 279 answers (corresponding to individual song recommendations) were collected[9], from which the participants knew only 81 recommended songs. The experiment yielded an average recommendation rating score of 3.13 ± 1.12 and an explanation rating score of 3.18 ± 1.21. Recommendation scores around 3 are typical average ratings for unknown recommendations [8]. Interestingly, the authors of [8] emphasize the need for adding context to the recommendations. In this experiment we provide our users with explanations of the recommendations. From the total of 279 answers, 41.22 % of the explanations were marked by participants to positively influence their recommendation ratings, while 13.98 % were marked as negative influence and 44.48 % as to not having influenced the ratings at all. Indeed, Table 2 shows that when the explanations are positively influencing the ratings, the average recommendation rating score increases by 0.82 (from 3.13 to 3.95). Furthermore, we also calculated the correlation between the influence of the explanations in the ratings and the familiarity of the user with the recommended songs, as shown in Table 3. It is interesting to note that the number of ratings with a positive influence is about 10 % higher when the recommended song is unknown to the user. This might suggest that explanations are indeed helping users to appreciate the recommendations more.

---

[9] Some participants did not rate all the 10 recommended songs.

Table 2. Mean and standard deviation of ratings

	Total	Influence of the explanations		
		Positive	Negative	No influence
Rec.	3.13 ± 1.32	3.95 ± 1.02	2 ± 0.91	2.74 ± 1.23
Exp.	3.18 ± 1.21	3.99 ± 0.81	2.18 ± 1.3	2.74 ± 1.02

Table 3. Percentage of influence of the explanations

	Positive	Negative	No influence
Known song	34.12 %	15.29 %	50.59 %
Unknown song	**45.00 %**	12.50 %	42.50 %

5   Conclusions

In this paper we presented a method for the creation of datasets for Music Recommendation that exploits information extracted from unstructured text sources. The method identifies music-related entities in the text (such as *songs, bands, persons, albums* and *music genres*) and extracts relations between these entities using an unsupervised rule based approach. The entities and relations are then represented as a graph from where song recommendations can be computed. A good characteristic of our approach is that a recommender system may provide explanations of the recommendations using natural language. We tested our method with a dataset gathered from *songfacts.com*, an online database of facts and stories about songs. We evaluated the extracted relations from a linguistic perspective and the extracted knowledge by comparing it with an existing knowledge base. We also performed a music recommendation experiment based on the extracted knowledge with real users. Evaluation results showed that our method is able to extract relations with a high linguistic and conceptual precision. It also shows that provide explanations with recommendations influence user satisfaction positively, especially when the recommendations are unknown to the user.

Still, there are many avenues for future work. Although the evaluation of our relation extraction system shows good values in terms of precision, recall is low between several pairs. One possible improvement of our approach is to introduce a prior step consisting in syntactic simplification. This would enable capturing potentially noisy relations (which are frequent in text featuring high variability such as the one displayed in Songfacts). Exploring alternative techniques to extract and represent relations between two or more entities is also crucial. In addition, new extracted knowledge could be used to enhance existing ontologies (such as MusicBrainz). All in all, our method provides a first attempt towards exploiting Knowledge Acquisition for Music Recommendation.

Acknowledgments. The authors would like to thank Miguel Ballesteros for his valuable advice and the subjects of the online experiment for their feedback.

References

1. Ballesteros, M., Nivre, J.: Going to the roots of dependency parsing. Comput. Linguist. **39**(1), 5–13 (2013)
2. Banko, M., Cafarella, M.J., Soderland, S., Broadhead, M., Etzioni, O.: Open information extraction from the web. In: International Joint Conferences on Artificial Intelligence, pp. 2670–2676 (2007)
3. Bohnet, B.: Very high accuracy and fast dependency parsing is not a contradiction. In: Proceedings of the 23rd International Conference on Computational Linguistics, pp. 89–97 (2010)
4. Brewster, C., Alani, H., Dasmahapatra, S., Street, P., Wilks, C.B.Y.: Data driven ontology evaluation. In: International Conference on Language Resources and Evaluation (2004)
5. Bunescu, R.C., Mooney, R.J.: A shortest path dependency kernel for relation extraction. In: Proceedings of the Conference on HLT/EMNLP, pp. 724–731 (2005)
6. Carlson, A., Betteridge, J., Wang, R.C., Hruschka Jr, E., Mitchell, T.M.: Coupled semi-supervised learning for information extraction. In: Proceedings of the Third ACM WSDM, pp. 101–110 (2010)
7. Celma, Ò.: Music Recommendation and Discovery - The Long Tail, Long Fail, and Long Play in the Digital Music Space. Springer, Heidelberg (2010)
8. Celma, Ò., Herrera, P.: A new approach to evaluating novel recommendations. In: Proceedings of the 2008 ACM Conference on Recommender Systems, pp. 179–186. ACM (2008)
9. Cohen, J.: Weighted kappa: nominal scale agreement provision for scaled disagreement or partial credit. Psychol. Bull. **70**(4), 213 (1968)
10. Culotta, A., Sorensen, J.: Dependency tree kernels for relation extraction. In: Proceedings of the 42nd Annual Meeting on Association for Computational Linguistics. Association for Computational Linguistics (2004). http://dx.doi.org/10.3115/1218955.1219009
11. Dellschaft, K., Staab, S.: On how to perform a gold standard based evaluation of ontology learning. In: Cruz, I., Decker, S., Allemang, D., Preist, C., Schwabe, D., Mika, P., Uschold, M., Aroyo, L.M. (eds.) ISWC 2006. LNCS, vol. 4273, pp. 228–241. Springer, Heidelberg (2006)
12. Eichler, K., Hemsen, H., Neumann, G.: Unsupervised relation extraction from web documents. In: Proceedings of the 6th International Conference on Language Resources and Evaluation. ELRA (2008)
13. Fader, A., Soderland, S., Etzioni, O.: Identifying relations for open information extraction. In: Empirical Methods in Natural Language Processing (2011)
14. Gamallo, P., Garcia, M., Fernández-Lanza, S.: Dependency-based open information extraction. In: Proceedings of the Joint Workshop on Unsupervised and Semi-Supervised Learning in NLP, ROBUS-UNSUP 2012, pp. 10–18. Association for Computational Linguistics, Stroudsburg (2012)
15. Gangemi, A.: A comparison of knowledge extraction tools for the semantic web. In: Cimiano, P., Corcho, O., Presutti, V., Hollink, L., Rudolph, S. (eds.) ESWC 2013. LNCS, vol. 7882, pp. 351–366. Springer, Heidelberg (2013)
16. Geleijnse, G., Korst, J.H.: Web-based artist categorization. In: Proceedings of ISMIR, pp. 266–271 (2006)

17. Gruhl, D., Nagarajan, M., Pieper, J., Robson, C., Sheth, A.: Context and domain knowledge enhanced entity spotting in informal text. In: Bernstein, A., Karger, D.R., Heath, T., Feigenbaum, L., Maynard, D., Motta, E., Thirunarayan, K. (eds.) ISWC 2009. LNCS, vol. 5823, pp. 260–276. Springer, Heidelberg (2009)

18. Hoffmann, R., Zhang, C., Ling, X., Zettlemoyer, L., Weld, D.S.: Knowledge-based weak supervision for information extraction of overlapping relations. In: Proceedings of the 49th Annual Meeting of the Association for Computational Linguistics: Human Language Technologies - Volume 1, HLT 2011, pp. 541–550. Association for Computational Linguistics, Stroudsburg (2011). http://dl.acm.org/citation.cfm?id=2002472.2002541

19. Augenstein, I., Maynard, D., Ciravegna, F.: Relation extraction from the web using distant supervision. In: Janowicz, K., Schlobach, S., Lambrix, P., Hyvönen, E. (eds.) EKAW 2014. LNCS, vol. 8876, pp. 26–41. Springer, Heidelberg (2014)

20. Marrero, M., Urbano, J., Sánchez-Cuadrado, S., Morato, J., Gómez-Berbís, J.M.: Named entity recognition: fallacies, challenges and opportunities. Comput. Stand. Interfaces 35(5), 482–489 (2013). http://linkinghub.elsevier.com/retrieve/pii/S0920548912001080

21. Mausam, Schmitz, M., Bart, R., Soderland, S., Etzioni, O.: Open language learning for information extraction. In: Conference on Empirical Methods in Natural Language Processing and Computational Natural Language Learning (2012)

22. Mendes, P.N., Jakob, M., García-silva, A., Bizer, C.: DBpedia spotlight: shedding light on the web of documents. In: Proceedings of the 7th International Conference on Semantic Systems (2011)

23. Ostuni, V.C., Di Noia, T., Mirizzi, R., Di Sciascio, E.: A linked data recommender system using a neighborhood-based graph kernel. In: Hepp, M., Hoffner, Y. (eds.) EC-Web 2014. LNBIP, vol. 188, pp. 89–100. Springer, Heidelberg (2014)

24. Passant, A.: dbrec — music recommendations using DBpedia. In: Patel-Schneider, P.F., Pan, Y., Hitzler, P., Mika, P., Zhang, L., Pan, J.Z., Horrocks, I., Glimm, B. (eds.) ISWC 2010, Part II. LNCS, vol. 6497, pp. 209–224. Springer, Heidelberg (2010)

25. Passant, A., Raimond, Y.: Combining social music and semantic web for music-related recommender systems. In: Social Data on the Web Workshop (2008)

26. Schedl, M., Widmer, G., Knees, P., Pohle, T.: A music information system automatically generated via web content mining techniques. Inf. Process. Manag. 47, 426–439 (2011)

27. Serra, I., Girardi, R., Novais, P.: Evaluating techniques for learning non-taxonomic relationships of ontologies from text. Expert Syst. Appl. 41(11), 5201–5211 (2014)

28. Surdeanu, M., Johansson, R., Meyers, A., Màrquez, L., Nivre, J.: The CoNLL-2008 shared task on joint parsing of syntactic and semantic dependencies. In: Proceedings of the Twelfth Conference on Computational Natural Language Learning, pp. 159–177. Association for Computational Linguistics (2008)

29. Tesnière, L.: Elements de syntaxe structurale. Editions Klincksieck (1959)

30. Whitman, B., Lawrence, S.: Inferring descriptions and similarity for music from community metadata. In: Proceedings of the 2002 International Computer Music Conference, pp. 591–598 (2002)

31. Yosef, M.A., Hoffart, J., Bordino, I., Spaniol, M., Weikum, G.: AIDA: an online tool for accurate disambiguation of named entities in text and tables. In: Proceedings of the 37th International Conference on Very Large Databases, pp. 1450–1453 (2011)

Posters and Demonstrations

Simulating Misreading

Armin Hoenen[✉]

Texttechnology Lab, University of Frankfurt, Frankfurt, Germany
hoenen@em.uni-frankfurt.de

Abstract. Physical misreading (as opposed to interpretational misreading) is an unnoticed substitution in silent reading. Especially for legally important documents or instruction manuals, this can lead to serious consequences. We present a prototype of an automatic highlighter targeting words which can most easily be misread in a given text using a dynamic orthographic neighbour concept. We propose measures of fit of a misread token based on Natural Language Processing and detect a list of short most easily misread tokens in the English language. We design a highlighting scheme for avoidance of misreading.

Keywords: Misreading · Lectio difficilior · Reading simulation

1 Introduction and Literature

Misreading is the process of mistaking one visual element of a text for another with the word as a primary target unit. It results in the unconscious integration of the wrong item into the semantic thread. We assume, this item is most often a (more frequent) *orthographic neighbour*. Reference [4] has defined an orthographic neighbour as a word with one letter difference to the target word, roughly, a Levenshtein distance of 1, [11]. Reference [23] mentions, that the density of neighbours and their frequency affect target word processing. We develop the concept of dynamic orthographic neighbour, relating to the results of psycholinguistic research by [24]. We embed our model in an application (the **Mis**reading **H**ighlighter (MisH)) which produces a reading aid in highlighting the most easily misread words. Although modelling should ideally take account of each reader in each reading event individually, we believe that with the right choice of representative resources we account for the vast majority of readers in the vast majority of reading events.

Publications centering on silent misreading proper are largely missing. In philology however, miscopyings are analyzed to be mainly caused by confusing similar looking letters, [19], and letter confusability research in turn received wide attention in psychology, compare [14].

A famous philological principle is called *lectio difficilior potior* (LD). It postulates that scribes had a tendency to replace obscure typically low frequency (LF) items by more comprehensible typically high frequency (HF) ones leading to a higher probability of LF items to be authorial wordings. A possible explanation by [19, p.222] is: "careless copying or a desire to simplify a diffcult passage"

© Springer International Publishing Switzerland 2015
C. Biemann et al. (Eds.): NLDB 2015, LNCS 9103, pp. 385–389, 2015.
DOI: 10.1007/978-3-319-19581-0_34

by the scribes "sometimes consciously, sometimes inadvertently". LD implicitly postulates an optimal scibal strategy in always writing the more frequent variant comitting only errors of the type $LF \rightarrow HF$. Errors $HF \rightarrow LF$ a posteriori appear to be dispreferred.

LD has a striking parallel in reading research where HF tokens are a priori processed quicker than LF items (for a discussion see [17, p.54/55] on lexical decision times). As for cognitive letter processing, the two most well known models of letter identification are *template matching* and *feature detection*. Recent findings of [6,12,15] and [8] agree in feature detection being the more probable model. References [15] and [12] find through ERP[1] analyses that the letter identification process most probably has *three* time steps. Furthermore, according to [17, p.64/65]: "The [herein previous] experiments [...] appear to rule out the hypothesis that letters in words (or even in short nonwords) are processed serially". Other findings are that letters in context of a word are recognized quicker than when presented alone, the so-called *word superiority effect* [3,18] and that contextual priming facilitates word access (consider [17, p.154 ff.]). In this paper, we first develop construct features. Then we develop a simulation, where up to 12 letters are processed at once (letter identification span motivated by [21]) in three subsequent processing steps. Finally, we present original-misread candidate pairs and their distance, acompanied by discussion and outlook.

2   Feature Construction

No feature list has ultimately become the dominant model of letter decomposition and none models time steps. We construct a three step letter recognition model from a cluster analyses given in the only paper known to us dealing with mixed case: [1]. For instance, one group consists of <m>,<n>,<h> and <r>, where we take the feature *upperbow* to be their first basic feature. The second and third features are constructed top-down from the node where the groups converge. We built our list, an excerpt seen in Fig. 1, as a temporary solution until better groupings and architectures are available.

3   Implementation

We implement the process of reading as an alternating *jump and zoom-in/extract* process: As is known from eye-tracking research, a skip (or saccade) is followed by a rest of the eye (or fixation) and then by another subsequent skip and so forth. In our model, during fixation the eye zooms in and extracts ever more features of all letters fixated (a bonus feature from parafoveal preview applies to the first and last letters). At the same time all lexemes, which match the feature profile of the extracted letters of the vocabulary represented by a frequency list (published by [2] extracted from the Wikipedia) are collected. The process as a whole allows for a successive jump to be conducted at a moment when not yet all visual

---

[1] Event Related Potentials.

Letter	Feature 1	Feature 2	Feature 3	Letter	Feature 1	Feature 2	Feature 3
Z	zigzag	sharp	largeVertical	U	spade	top	smoothlarge
z	zigzag	sharp	smallVertical	u	spade	top	smoothsmall
N	zigzag	sharp	largeHorizontal	y	spade	top	descender
S	zigzag	soft	large	Y	spade	top	middle
s	zigzag	soft	small	A	spade	invert	bar

Fig. 1. Letter features (examples) inspired by [1,16] and [8].

Extraction	Features	letters	lexeme
1. (2,1,2)	(egg,rnd);(up);(egg,asc)	(a,e);(m,n,h,r);(d,b,t)	and, end, ant, art
2. (3,2,3)	(egg,rnd, op);(up,sym);(egg,asc,ri)	(a);(m,n);(d)	and
3. interrupt	/	/	only one winner item

Fig. 2. Example of three time-step feature extraction of "and".

information has been extracted by the eye. From feature decomposition it follows that at each zoom step, different letter sets are confusable. In consequence, the distance between two letters and between two words is not fixed, as distance models operationally presuppose but dynamic or shrinking and depends on the time of exposure. Whilst in the first step, a <C> can be confused with an <O>, in the second step, this is no longer possible.

Note, that in our example in Fig. 2, the third zoom-in step is omitted/skipped, although the third feature for the middle letter has not yet been extracted by vision. The algorithm collects a list of possible candidate words present in the lexicon simulating their simultaneous activation. If one of those words in comparison to all other such candidates has an exceedingly high frequency or is the only survivor it will be taken as the winner token and semantic integration will start.

Hereafter, we attemprt to test if the item is integrateable, allowing the misread to happen, assuming the properties of the original item are required. We apply four preliminary and potentially bias-loaden measures. We collect and compare the **part-of-speech** of the candidate tokens from the integrated tagger in the python nltk library.[2]

A generic term can usually replace a more specialized one without a loss of semantic and syntactic integrity. We use WordNet[3] to calculate the **WordNet path length** between original and candidate token, as described in [20].

We use the python nltk library collocations to extract all significant collocations of the previous token from the Brown corpus, [7]. If the misread is included, we assume **co-occurrence fit**.

We substitute the original token by the candidate token and paste the so-obtained current sentence into a search engine. The number of results compared to the original sentence is stored as **search engine sentence fit**.

The aforementioned measures are all indicative but not definitive. We use their votes for the coloring scheme of the MisH aiming at highlighting in a non disruptive way the most probable misreads. There are some dimensions of a visualization, which can be used, such as text color, text size, font, underlining

[2] http://www.nltk.org/.

[3] http://wordnet.princeton.edu/.

Original(LF)	Replacement(HF)	DL	Original(LF)	Replacement(HF)	DL
signer	singer	1	Whistler	Minister	4
Shaping	shading	2	evicted	allowed	5
murder	number	3	celibacy	capacity	6
trees	teams	3	ordinances	characters	8

Fig. 3. Computed misreading candidate pairs, elicited examples.

etc. Considering [9,10,13] and results from search term highlighting, we use boldfacing, italicization and red text and come up with the following schema:

- if a possible misread has been detected but none of the measures indicated its fit, the item is italicized
- if the misread has the same pos, the original word will be written in bold face
- if the misread matches any two of the aforementioned criteria, the original word will be displayed in red
- if the misread matches two of the above criteria including pos, the original word will be displayed in red and bold

4   Results, Discussion and Outlook

Apart from programming the highlighter, we calculated all possible misreads of words of length 4–5 letters in our frequency list. We added the Damerau-Levenshtein distance (DL), [5], of the detected tokens using the R *stringdist* library by [22]. We ended up with a list of pairs, examples in Fig. 3. The unevaluated measure produces reasonable candidates. An interesting observation is that morphological patterns repeat (singular and plural). With pairs such as morality/majority or declined/directed, we find candidates, which would receive a large DL and not pop up in contemporary applications although they appear to be probable misreads. The average DL for all token pairs was 2.99 ranging from 1 to 10. The next step is the production of a Gold Standard for a thorough evaluation and a publicly accessible user-interface. Unusual letter shapes, and strong contextual priming should be remodelled along.

References

1. Boles, D.B., Clifford, J.E.: An upper- and lowercase alphabetic similarity matrix, with derived generation similarity values. Behav. Res. Meth. Instrum. Comput. **21**, 579–586 (1989)
2. vor der Brück, T., Mehler, A., Islam, Z.: ColLex.en: automatically generating and evaluating a full-form lexicon for english. In: Calzolari, N., Choukri, K., Declerck, T., Loftsson, H., Maegaard, B., Mariani, J., Moreno, A., Odijk, J., Piperidis, S. (eds.) Proceedings of the Ninth International Conference on Language Resources and Evaluation (LREC 2014), European Language Resources Association (ELRA), Reykjavik, Iceland, May 2014, pp. 3756–3760 (2014). http://www.lrec-conf.org/proceedings/lrec2014/pdf/1099_Paper.pdf, aCL Anthology Identifier: L14-1075
3. Cattell, J.: The time it takes to see and name objects. Mind **11**, 63–65 (1886)

4. Coltheart, M., Davelaar, E., Jonasson, T., Besner, D.: Access to the internal lexicon. In: Dornic, S. (ed.) Attention and Performance VI. Lawrence Erlbaum Associates, Hillsdale, NJ (1977)
5. Damerau, F.J.: A technique for computer detection and correction of spelling errors. Commun. ACM **7**, 171–176 (1964)
6. Fiset, D., Blais, C., Ethier-Majcher, C., Arguin, M., Bub, D., Gosselin, F.: Features for identification of uppercase and lowercase letters. Psychol. Sci. **19**(11), 1161–1168 (2008)
7. Francis, W.N., Kucera, H.: Brown corpus manual. Technical report, Department of Linguistics, Brown University, Providence, Rhode Island, US (1979). http://icame.uib.no/brown/bcm.html
8. Grainger, J., Rey, A., Dufau, S.: Letter perception: from pixels to pandemonium. Trends Cogn. Sci. **12**, 381–387 (2008)
9. Hearst, M.A.: Search User Interfaces, 1st edn. Cambridge University Press, New York (2009)
10. Landauer, T., Egan, D., Remde, J., Lesk, M., Lochbaum, C., Ketchum, D.: Enhancing the usability of text through computer delivery and formative evaluation: the superbook project. In: McKnight, C., Dillon, A., Richardson, J. (eds.) Hypertext: A Psychological Perspective, pp. 71–136. Ellis Horwood (1993)
11. Levenshtein, V.I.: Binary codes capable of correcting deletions, insertions, and reversals. Doklady Akademii Nauk SSSR **163**(4), 845–848 (1965). English in: Sov. Phys. Dokl. **10**(8), 707–710 (1966)
12. Madec, S., Rey, A., Dufau, S., Klein, M., Grainger, J.: The time course of visual letter perception. J. Cogn. Neurosci. **24**(7), 1645–1655 (2012). http://dblp.uni-rier.de/db/journals/jocn/jocn24.html#MadecRDKG12
13. Mazza, R.: Introduction to Information Visualization, 1st edn. Springer, London (2009)
14. Müller, S., Weidemann, C.: Alphabetic letter identification: effects of perceivability, similarity, and bias. Acta Psychol. **139**(1), 19–37 (2011)
15. Petit, J.P., Midgley, K.J., Holcomb, P.J., Grainger, J.: On the time course of letter perception: a masked priming ERP investigation. Psychonom. Bull. Rev. **13**(4), 674–681 (2006)
16. Podgorny, P., Garner, W.: Reaction time as a measure of inter- and intraobject visual similarity: letters of the alphabet. Percept. Psychophys. **26**, 37–52 (1979)
17. Rayner, K., Pollatsek, A., Ashby, J., Clifton Jr, C.: Psychology of Reading. Psychology Press, New York/Hove (2012)
18. Reicher, G.: Perceptual recognition as a function of meaningfulness of stimulus material. J. Exp. Psychol. **81**, 275–280 (1969)
19. Reynolds, L., Wilson, N.: Scribes and Scholars: A Guide to the Transmission of Greek & Roman Literatures. Oxford University Press, Oxford (2013)
20. Bird, S., Klein, E., Loper, E.: Natural Language Processing with Python. O'Reilly, USA (2009)
21. Underwood, N.R., McConkie, G.W.: Perceptual span for letter distinctions during reading. Read. Res. Q. **20**, 153–162 (1985)
22. van der Loo, M.: The stringdist package for approximate string matching. R J. **6**, 111–122 (2014). http://CRAN.R-project.org/package=stringdist
23. Van Heuven, W.J.: Bilingual interactive activation models of word recognition in a second language. In: Bassetti, B., Cook, V. (eds.) Second Language Writing Systems, pp. 260–288. Multilingual Matters (2005)
24. Wheeler, D.: Processes in word recognition. Cogn. Psychol. **1**, 59–85 (1970)

(German) Language Processing for Lucene

Bastian Entrup[✉]

Applied and Computational Linguistics,
Justus-Liebig-Universität Gießen, Giessen, Germany
bastian.entrup@germanistik.uni-giessen.de

Abstract. This paper introduces an open-source Java-package called *German Language Processing for Lucene* (glp4lucene). Although it was originally developed to work with German texts, it is to a large degree language independent. It aims at facilitating four language processing steps for working with non-English texts and Apache Lucene/Solr: lemmatizing words, weighting terms based on their part-of-speech, adding synonyms and decompounding nouns, without the necessity of a thorough understanding of natural language processing.

1 Introduction

In recent years, Apache Lucene and Apache Solr[1], the search platform based on Lucene, have gained a lot of popularity not only in industrial applications, but also in searching websites, and for academic purposes.

Lucene offers many interesting, expandable features and a number of resources and literature on how to use and apply Lucene to different purposes (e.g. [12]) are available. Although Lucene offers quite some possibilities to incorporate natural language processing (NLP) or methods from computational linguistics, these are usually only available in English by default. For example, German language support is basically limited to stemming or to language-independent methods, such as providing a stop-word list or a dictionary for compound splitting.

The open-source package[2] *German Language Processing for Lucene* (glp4lucene) described here aims at facilitating four obstacles that one might encounter when working with non-English texts in Lucene: lemmatization, synonym expansion, decompounding, and part-of-speech (POS) weighting. Despite its name, it is not only applicable to German, but to other languages as well; it is basically language independent. It was developed within a Digital Humanities project, where a number of German texts where to be processed, and where synonym expansion and decompounding were of special importance for the performance of the search platform.

---

[1] http://lucene.apache.org/, http://lucene.apache.org/solr/.
[2] https://sourceforge.net/projects/glpforlucene.

© Springer International Publishing Switzerland 2015
C. Biemann et al. (Eds.): NLDB 2015, LNCS 9103, pp. 390–394, 2015.
DOI: 10.1007/978-3-319-19581-0_35

2   Motivation

While it is easy to do very basic indexing and searching with almost no experience in Lucene/Solr, applying NLP methods requires more in-depth knowledge of Lucene. This packages aims at facilitating the usage of these methods.

Lemmatization, i.e., the reduction of inflected word forms to a base form, is, from a linguistic point of view, more meaningful than stemming, the reduction of a word to its stem. Language is, per se, ambiguous (cf. cases of homonymy or polysemy). Using stemming, this ambiguity is further increased since different, not related words are reduced to the same word stem. Even words that are not homograph, like German *Bauer* (farmer) and *Bau* (a homonymous word with the meanings building, construction site, and jail), are reduced to the same stem: *Bau*. Lemmatization, on the other hand, leads to different lemmas of the words, while still reducing different inflected word forms (such as *Bauers* (of the farmer) and *Bauern* (the farmers)) to one base form.

Despite these assumptions, studies have shown that stemming increases the precision in German information retrieval (IR) compared to using neither stemming nor lemmatization. Reference [3] found an increase between 11 and 23 %, while [5] found that stemming improved the precision for German by 7.3 % and lemmatization only by 6 %.

Nonetheless, lemmatizing words is a necessity to look up synonymous words from GermanNet [4], a German WordNet [13] counterpart. Adding synonyms is commonly expected to increase recall [11]. To be able to find all relevant documents the search-engine has to identify the concept in question even if other words than the query term where used to refer to it.

While in the beginning of IR stop word lists were commonly applied, todays state-of-the-art web search engines do not use stops words, since they are problematic, e.g., when it comes to finding song titles or proverbs[3]. Weighting terms instead seems to be a plausible approach: as [7] shows, nouns are the most commonly searched for terms. Weighting terms by their POS is inspired by Jespersen's Rank Theory [6], which states that the open POS, nouns, adjectives, adverbs, and verbs, are more content bearing, while the class of closed POS are, roughly speaking, more or less empty and fulfill mainly grammatical or deictic purposes. This idea has been applied to IR before [9,10]. Unlike these proposals, only ranking of single words is supported here: one can increase the weight of a term depending on its POS. Given a query such as *Café in Paris*, documents containing *Café* and *Paris* are more likely hits than documents containing *in*. One could thus decide to weight nouns more heavily than prepositions. This is, of course, totally up to the user. If no weights are given, or if the method is not used at all, all terms are weighted equally.

3   Implementation

The following three modules of `glp4lucene` can be used to analyze and filter the input during indexing or search time. While GermaNet requires the input to

---

[3] Think of Shakespeare's *To Be or not to Be*, where almost every token is a stop word, and one cannot just ignore them altogether.

be lemmatized, all parts of the package can freely be combined, left out, or used together with Lucene's standard analyzers or stemming

Lemmatization: The implementation of the lemmatization presented here is language independent and can be used in the same way and out of the box for other languages as well. It only requires a MATE-tool [2] model appropriate for the given language[4]. This model, for German the model described in [14] can be used, is used to assign an appropriate lemma to each token of the field it is applied to.

Synonym Expansion: The `glp4lucene` packages extends the existing synonym interface[5] of Lucene and comes with three pre-defined implementations of this interface.

To add appropriate synonyms for words in the search index or the query term, GermaNet can be used. GermaNet is a manually compiled resource that establishes different semantic and lexical relations between words, or senses. The building block of GermaNet is the synset, a set of synonymous word senses. The German word *Gefängnis* (prison) for example is found in a synset together with *Bau* (jail).

Since GermaNet is a proprietary software, the `glp4lucene` package includes two other methods to assign synonyms: The first is to take a list of semantic similarity of words of the form `focus-word synonym <similarity>`. The second is to use a list of synonyms, compiled from whatever resource one has at hand, of the form `focus-word synonym`. These two methods are, of course, also language independent. Furthermore, one can use the interface to implement further methods to add synonyms.

Instead of looking up each word that is encountered in the texts during indexing or searching, a map between each word form found and its synonyms is built and stored for later re-use. This list can directly be (re-)used in Lucene's decompounding class. Splitting up compounds has been found to be very helpful for many Germanic but also for other languages [5].

POS-based Term Weighting: The implementation uses the Stanford Maxent Tagger [15]. Again, if presented the correct model[6], this implementation is language independent. Besides the model, one also has to provide a list of POSs and the according weight. If no weight is given, the POS is treated as being neutral, no weight is stored in the index.

---

[4] Models for French, Spanish, Chinese, English, and German are available from https://code.google.com/p/mate-tools/.

[5] The interface follows the implementation found in [12], extends it by new methods, and is adapted to the newer Lucene versions 4.x. It has been tested using versions 4.6 to 4.8.1.

[6] Models for English, Arabic, Chinese, French, Spanish, and German are available at http://nlp.stanford.edu/software/tagger.shtml.

4 Conclusion

The package is meant to assist developers without a thorough knowledge of NLP tools, to build a Lucene index applying different techniques from NLP. The source of the texts can be a database or text files in any format. The resulting index can be used in web search servers such as Apache Solr or it can be used to speed up and facilitate database queries. Using the morphological information given in the index, as well as the lemmatization, building a linguistic corpus based on Lucene is an other possible application of the software.

Using the source code, provided along with the compiled JAR-file, the implementation can be adapted to one's needs. For example, it might be interesting to assign not only synonyms but also hyponyms or meronyms as described in [8]: Searching for *Hund* (dog) should perhaps also find texts about *Dackel* (dachshund) as well.

Other possibilities include using distributional thesauri[7] to find related, though maybe not strictly synonymous, words and using those instead of the information provided by GermaNet. This approach can also be used to create a second index that provides the end-user with alternative query terms. Also one can use other resources, e.g., Wikitionaries, to generate and use lists of synonymous words. If resources for other languages are available, the implementation given in this package can be used as is for not only German texts, but for other languages as well.

Acknowledgemets. This package was created for and within the GeoBib project to facilitate searching the project's data set and will be used in the planed website. GeoBib is funded by the German Federal Ministry of Education and Research (grant no. 01UG1238A-B).

References

1. Biemann, C., Riedl, M.: Text: now in 2D! a framework for lexical expansion with contextual similarity. J. Lang. Model. **1**(1), 55–95 (2013)
2. Bohnet, B.: Very high accuracy and fast dependency parsing is not a contradiction. In: Proceedings of the 23rd International Conference on Computational Linguistics, COLING 2010, pp. 89–97. Association for Computational Linguistics, Stroudsburg (2010)
3. Braschler, M., Ripplinger, B.: How effective is stemming and decompounding for german text retrieval? Inf. Retr. **7**(3–4), 291–316 (2004)
4. Hamp, B., Feldweg, H.: GermaNet - a lexical-semantic net for german. In: Proceedings of ACL Workshop Automatic Information Extraction and Building of Lexical Semantic Resources for NLP Applications, pp. 9–15 (1997)
5. Hollink, V., Kamps, J., Monz, C., de Rijke, M.: Monolingual document retrieval for european languages. Inf. Retr. **7**(1–2), 33–52 (2004)

---

[7] For example, for German http://sourceforge.net/projects/jobimtext/files/data/ models/de_news70M_pruned.zip/download; based on 70 million sentences from a news corpus extracted using the system described in [1].

6. Jespersen, O.: The Philosophy of Grammar. Chicago Studies in Ethnomusicology Series. University of Chicago Press, Chicago (1992)
7. Kraaij, W., Pohlmann, R.E.: Viewing stemming as recall enhancement. In: Proceedings of the 19th Annual International ACM SIGIR Conference on Research and Development in Information Retrieval, pp. 40–48 (1996)
8. Leveling, J.: University of hagen at CLEF 2003: natural language access to the GIRT4 data. In: Peters, C., Gonzalo, J., Braschler, M., Kluck, M. (eds.) CLEF 2003. LNCS, vol. 3237, pp. 412–424. Springer, Heidelberg (2004)
9. Lioma, C., Blanco, R.: Part of speech based term weighting for information retrieval. In: Boughanem, M., Berrut, C., Mothe, J., Soule-Dupuy, C. (eds.) ECIR 2009. LNCS, vol. 5478, pp. 412–423. Springer, Heidelberg (2009)
10. Lioma, C., van Rijsbergen, C.K.: Part of speech based term weighting for information retrieval. In: Revue Franaise de Linguistique Applique, vol. 1 (2008)
11. Manning, C.D., Raghavan, P., Schütze, H.: Introduction to Information Retrieval, vol. 2. Cambridge University Press, Cambridge (2008)
12. McCandless, M., Hatcher, E., Gospodnetic, O.: Lucene in Action, Second Edition: Covers Apache Lucene 3.0. Manning Publications Co., Greenwich (2010)
13. Miller, G.A.: WordNet: a lexical database for english. Commun. ACM **38**, 39–41 (1995)
14. Seeker, W., Kuhn, J.: Making ellipses explicit in dependency conversion for a german treebank. In: LREC, pp. 3132–3139 (2012)
15. Toutanova, K., Klein, D., Manning, C.D., Singer, Y.: Feature-rich part-of-speech tagging with a cyclic dependency network. In: Proceedings of the 2003 Conference of the NAACL on Human Language Technology, NAACL 2003, pp. 173–180. Association for Computational Linguistics, Stroudsburg (2003)

Optimized Uyghur Segmentation for Statistical Machine Translation

Chenggang Mi[1,2(✉)], Yating Yang[1], Rui Dong[1,2], Xi Zhou[1],
Lei Wang[1], Xiao Li[1], Tonghai Jiang[1], and Turghun Osman[1]

[1] Xinjiang Technical Institute of Physics and Chemistry of Chinese Academy
of Sciences Urumqi, Xinjiang 830011, China
michenggang@gmail.com, {yangyt,dongrui,zhouxi,
wanglei,xiaoli,jth}@ms.xjb.ac.cn, turghunjan@sina.com
[2] University of Chinese Academy of Sciences Beijing, Beijing 100049, China

Abstract. In this paper, we propose an optimized method to segment the
Uyghur word. We consider the optimization as a classification problem; the
features are extracted from Uyghur-Chinese bilingual corpus. Experimental
results show that with our method the performance of Uyghur-Chinese machine
translation improved significantly.

Keywords: Uyghur segmentation · Data sparsity · Character tagging · Segmentation optimization · Uyghur-Chinese machine translation

1 Introduction

To mitigate the effects of data sparsity in natural language processing, a frequently used
technique is word segmentation (stemming) [1]. However, most researches treat the
suffixes of words useless, and discard them during segmentation. This may work in
some tasks, but weakens the ability of Uyghur-Chinese machine translation model.
Nguyen et al. [2] proposed a generative Bayesian model to describe the generation of a
sentence pair for tokenization, which gave improvements on Arabic-English and
Chinese-English translation tasks.

In this paper, we propose an optimized Uyghur segmentation method for Uyghur-
Chinese machine translation. According to this method, the Uyghur words are first
segmented by a character tagging based model, which is trained based on a conditional
random field (CRF) model [3], then, we extract several features from Uyghur-Chinese
parallel corpus, and combine these features into a logistic regression model [4], the
output of this model is a label indicates that whether we should retain the suffixes of
current Uyghur word. Compare with previous works, our method based on several
features which can present the corresponding between Uyghur and Chinese sentence.

2 Optimized Word Segmentation Based on Bilingual Corpus

Our work relies on the initial segmentation of a character-tagging based model (CRF).
The features we used in our optimization are defined as follows:

© Springer International Publishing Switzerland 2015
C. Biemann et al. (Eds.): NLDB 2015, LNCS 9103, pp. 395–398, 2015.
DOI: 10.1007/978-3-319-19581-0_36

2.1   Features

Bilingual Sentence Length Feature. If the length of Uyghur sentence is much shorter than the Chinese sentence, the suffixes of Uyghur words can be retained. Here, we define a threshold according to models training.

$$f_{bl} = \frac{|l_{uyg} - l_{chn}|}{\max(l_{uyg}, l_{chn})} \tag{1}$$

The l_{uyg} and l_{chn} are length of Uyghur sentence and Chinese sentence, respectively.

Bilingual Dictionary Feature. In this section, we suggest a bilingual dictionary feature, which count to the translated word pairs in every Uyghur-Chinese sentences. And compute the ratio of number of translated word pairs to the length of two sentences. The feature can be described as follows:

$$f_{bd} = \frac{\#uyghur\_null + \#chinese\_null}{\max(l_{uyg}, l_{chn})} \tag{2}$$

$\#uyghur\_null$ indicates the number of times Uyghur word can't find corresponding Chinese word in current sentence pair, and $\#chinese\_null$ represents the number of times Chinese word can't find corresponding Uyghur word.

Word Alignment Feature. We collect the word alignment feature to calculate the ratio of number of asymmetrical word alignment (1-to-n or n-to-1) to the length of Uyghur sentence:

$$f_{wa} = \frac{|\#1\_to\_1 - (\#1\_to\_n + \#n\_to\_1)|}{2 * \max(l_{uyg}, l_{chn})} \tag{3}$$

$\#1\_to\_1$, $\#1\_to\_n$ and $\#n\_to\_1$ are the number of times 1-to-1, 1-to-n and n-to-1 word alignment in initial alignment results, respectively.

2.2   Optimized Word Segmentation Model

We consider the optimization as a classification problem, and the goal of logistic regression based classification is to fit the regression curve according to the training data collected. The features (f_{bl}, f_{bd}, f_{wa}) we suggested in Sect. 2.1 are inputs in logistic regression model, and whether to retain suffixes is the class label. We select 200 sentence pairs which are annotated by dealing with the suffixes of Uyghur words, to train the optimization model.

3   Evaluation

We use the data provided by the CWMT 2013. The **original** (Uyghur words are not segmented) BLEU of Uyghur-Chinese machine translation is: **32.27** (PB, Phrase-based model), **33.08** (HPB, Hierarchal Phrase-based model).

Table 1. Results on Uyghur-Chinese machine translation (BLEU: %)

	Retain		Delete		Optimized	
	PB	HPB	PB	HPB	PB	HPB
CRF	30.07	31.87	32.42	33.50	32.40	33.52
CRF + Dict	31.22	32.51	33.05	34.20	**33.25**	**34.62**

In Table 1, we show the results of Uyghur-Chinese machine translation experiments, the column "**retain**" indicates suffixes of Uyghur words are all retained as words, and the performance reduced significantly (compared with original). The column "**delete**" shows the performance of machine translation model which delete all suffixes during segmentation, all of these models are outperformed the "**retain**" column and "**original**". In "**optimized**" column, we present the results given by our optimized segmentation. Compare with others, our method take the features of Uyghur-Chinese bilingual corpus into account, which give us more information; the character tagging-based (the baseline is CRF) segmentation model with the bilingual dictionary achieved the best performance in all experiments. The hierarchal phrase-based machine translation model outperforms phrase-based models in all experiments; an important reason is that the hierarchal phrase-based models have the abilities of generalization and local reordering.

4 Conclusion

In this paper, we presented a novel approach to optimize the Uyghur words segmentation for machine translation. Our method based on a character-tagging segmentation model. We collect features from Uyghur-Chinese bilingual corpus, and combine them into a logistic regression model to determine whether we keep suffixes of the current Uyghur word. Remarkably, this method outperforms others. In our feature work, we will focus on co-segmentation of Uyghur and Chinese for statistical machine translation.

Acknowledgements. This work is supported by the National High Technology Research and Development Program of China (No. 2013AA01A607), Strategic Priority Research Program of the Chinese Academy of Sciences (No. XDA06030400), West Light Foundation of Chinese Academy of Sciences (No. XBBS201216), and Key Project of Knowledge Innovation Program of Chinese Academy of Sciences (No. KGZD-EW-501).

References

1. Larkey, L.S., Ballesteros, L., Connell, M.E.: Improving stemming for Arabic information retrieval: light stemming and co-occurrence analysis. In: Proceedings of the 25th ACM SIGIR, pp. 275–282 (2002)

2. Nguyen, T.L., Vogel, S., Smith, N.A.: Nonparametric word segmentation for machine translation. In: Proceedings of the 23rd COLING, pp. 815–823 (2010)
3. Lafferty, J., McCallum, A., Pereira, F.C.N.: Conditional random fields: Probabilistic models for segmenting and labeling sequence data. In: Proceedings of the 8th ICML, pp. 282–289 (2001)
4. McDonald, J.H.: Handbook of Biological Statistics. 2nd edn pp. 173–181 (2009)

Management and Publishing of Multimedia Dictionary of the Czech Sign Language

Adam Rambousek[✉] and Aleš Horák

Natural Language Processing Centre Faculty of Informatics,
Masaryk University, Botanická 68a, 60200 Brno, Czech Republic
{rambousek,hales}@fi.muni.cz

Abstract. This paper describes the development of a multimedia dictionary writing system for the Czech Sign Language dictionary, prepared in cooperation of several institutions dedicated to the sign language research and study. The presented dictionary system takes the advantage of electronic format and strongly relies on the use of multimedia evidences. The dictionary system has to deal with large amount of video recordings of both a human narrator and digital avatar. Since the dictionary is prepared by several remotely located groups, the dictionary system provides support for complex publishing processes. The dictionary writing system is used in active preparation of the Czech Sign Language dictionary and already serves as a publisher for the dictionary data to the general public.

Keywords: Dictionary writing system · Sign language · Multimedia dictionary · DEB platform

1 Introduction

Preparation of a new dictionary of the Czech Sign Language is one of the key activities of the *Network of Expert Centres Providing Inclusion in Tertiary Education*[1] project. The dictionary is being created in cooperation of five Czech universities and organizations dedicated to the sign language issues. The whole project is co-ordinated by the Teiresiás Centre, Masaryk University. At the current stage of the process, the aim is to create an extensive dictionary of the Czech sign language and the explanatory dictionary of the Czech language, both inter-connected to serve also as a bilingual dictionary. More languages (both sign and spoken) may be added later to form a multilingual dictionary.

2 DEB Platform

Utilizing the experience from several lexicographic projects, we have designed and implemented a universal dictionary writing system that can be exploited in various lexicographic applications to build large lexical databases. The system is

---

[1] http://www2.teiresias.muni.cz/expin/en.

© Springer International Publishing Switzerland 2015
C. Biemann et al. (Eds.): NLDB 2015, LNCS 9103, pp. 399–403, 2015.
DOI: 10.1007/978-3-319-19581-0_37

Fig. 1. Entry in the Czech Sign Language dictionary.

called Dictionary Editor and Browser, or the DEB platform [1]. Since 2005 the DEB platform was applied in more than 10 large international research projects, see for example [2–5]. The DEB platform is based on the client-server architecture, which brings along a lot of benefits. All the data are stored on a server and a considerable part of the functionality is also implemented on the server-side, while the client application can be very lightweight. This approach provides very good tools for editor team cooperation; data modifications are immediately seen by all the users. Server also provides authentication and authorization tools.

3   Entry Representation

Creating the dictionaries of sign languages has always been challenging in the printed form, this may change dramatically with electronic dictionaries. Kristoffersen and Troelsgard [6] have described their experience in building the Danish Sign Language Dictionary: *The overriding challenge in sign language lexicography is how to render signs in the absence of a written language. ... the use of video recordings has predominated. This approach is obviously limited to electronic dictionaries.*

The Czech Sign Language dictionary relies heavily on video recordings, they are used not just for the sign representation (front view and side view), but also for meaning explanations or usage examples. See Fig. 1 for the example of an entry in the Czech Sign Language dictionary. However, the video is not the only way to present the sign to the dictionary users. The dictionary also provides tools to enter and display two methods of formal sign transcription: *SignWriting* [7,8] (iconic script system which encodes hand shapes and movements with image symbols) and *Hamburg Notation System (HamNoSys)* [9] (transcription system similar to IPA [10]).

4   Access Control and Validations

The dictionary is produced by several teams at different institutions, each covering different parts of a dictionary entry or different domain. Users may get various roles (e.g. editor, reviewer, supervisor, ...) and the DEB platform access management module combines user role with the group, resulting in complex access control system. Furthermore, each part of the dictionary entry (e.g. grammatical information, definitions, translations, ...) have to be approved by the supervisor before it may be published for public presentation.

Validations are also needed to check the completeness and correctness of input information for each dictionary entry. First stage of validations is provided during the editing, for example the correctness of the references between the sign language and the Czech language. More validations are run after the batch of entries is handed over by a group of authors to the reviewers. For example, the reviewer may check for entries that contain video recording of the lemma and the definition, but do not contain a usage example or translations.

5   Data Import

Each institution or department involved in the project specializes on one task. For most of the entries in the initial phase, the following process was exploited. One department recorded the videos for a batch of signs. The batch was then handed over to the lexicographic team that added grammatical information, translations, etc. Meanwhile, another team provides SignWriting transcriptions. To speed up the creation of new entries and updates of existing entries, the dictionary system supports complex tools for batch import of video recordings.

The import system scans the batch of video recordings and detects the type of video (sign, definition, usage example) and checks whether the file is an update of an existing video, completely new entry, or an additional video to an existing entry. The system is able to detect whether to add a definition video to a certain entry meaning. Translations to the Czech language are proposed, linking with existing Czech entries or creating new Czech equivalents.

As for the Czech language dictionary in the project, the goal was not to create a completely new dictionary, but re-use and update existing resources. For that reason, the three large monolingual dictionaries were included into the system – see [11–13]. Each of the dictionaries follows different entry structure. During the import, all the information had to be normalised to the common entry structure while not losing the information. To enrich the dictionary entries for the Czech language with "real world" usage examples, the dictionary writing system provides examples extracted from the CzTenTen12 corpus [14], rated by the GDEX tool [15] to get the most suitable examples.

6   Searching

Searching in the sign language dictionary is a challenging issue. For the Czech Sign Language dictionary, several methods are provided and users may choose

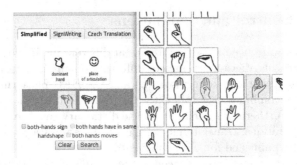

Fig. 2. Example of hand shape selection in "iconic search".

their preferred method. Thanks to the cross-references between sign language and spoken language dictionaries, users may search for a word in its written form and the sign language translations are displayed. Experienced users may enter the SignWriting or HamNoSys transcription.

The most convenient search method is the "iconic search". Users may select hand shapes or positions from the set of images. The selection is internally converted to the SignWriting representation. Users are able to enter even very complex queries just by clicking on a few icons. The graphical interface for this search method is inspired by the Dutch Sign Language Dictionary [6], but provides new options to exactly specify the signs (e.g. sign symmetricity, or hand positions). See the shape selection in Fig. 2.

7 Current Results and Future Work

The dictionary writing system was successfully implemented and tested. The system is used for the dictionary editing on a daily basis. Already approved parts of the dictionary are publicly accessible at the following link http://www.dictio.info. Currently, the dictionary contains 11,477 entries for the Czech Sign Language, with 26,040 video recordings, and 121,354 entries for Czech. After the final public release of the Czech dictionary, multiple languages will be included to the dictionary.

Acknowledgements. This work has been partly supported by the Ministry of Education of CR within the LINDAT-Clarin project LM2010013.

References

1. Horák, A., Rambousek, A.: DEB platform deployment - current applications. In: RASLAN 2007: Recent Advances in Slavonic Natural Language Processing, Brno, Czech Republic, Masaryk University, pp. 3–11 (2007)
2. Horák, A., Rambousek, A.: PRALED – a new kind of lexicographic workstation. In: Przepiórkowski, A., Piasecki, M., Jassem, K., Fuglewicz, P. (eds.) Computational Linguistics. SCI, vol. 458, pp. 131–141. Springer, Heidelberg (2013)

3. Horák, A., Vossen, P., Rambousek, A.: A distributed database system for developing ontological and lexical resources in harmony. In: Gelbukh, A. (ed.) CICLing 2008. LNCS, vol. 4919, pp. 1–15. Springer, Heidelberg (2008)
4. Maarouf, I.E., Bradbury, J., Baisa, V., Hanks, P.: Disambiguating verbs by collocation: corpus lexicography meets natural language processing. In: Calzolari, N (Chair)., Choukri, K., Declerck, T., Loftsson, H., Maegaard, B., Mariani, J., Moreno, A., Odijk, J., Piperidis, S. (eds.) Proceedings of the Ninth International Conference on Language Resources and Evaluation (LREC 2014). European Language Resources Association (ELRA), Reykjavik, Iceland (2014)
5. Hanks, P., Coates, R., McClure, P.: Methods for studying the origins and history of family names in britain. In: Facts and Findings on Personal Names: Some European Examples, Uppsala, Acta Academiae Regiae Scientiarum Upsaliensis, pp. 37–58 (2011)
6. Kristoffersen, J.H., Troelsgård, T.: The electronic lexicographical treatment of sign languages: the Danish sign language dictionary. In: Granger, S., Paquot, M. (eds.) Electronic Lexicography, pp. 293–315. Oxford University Press, Oxford (2012)
7. Sutton, V.: SignWriting basics. In: Center for Sutton Movement Writing, Incorporated (2009)
8. Kato, M.: A study of notation and sign writing systems for the deaf. Intercult. Commun. Stud. **17**(4), 97–114 (2008)
9. Hanke, T.: HamNoSys-representing sign language data in language resources and language processing contexts. In: Streiter, O., Vettori, C. (eds.) LREC 2004 Workshop proceedings: Representation and Processing of Sign Languages, pp. 1–6. ELRA, Paris (2004)
10. International Phonetic Association: Handbook of the International Phonetic Association. Cambridge University Press, Cambridge (1999)
11. Filipec, J., et al.: Slovník spisovné češtiny (SSČ), 1st edn. Academia, Praha (1995). elektronická verze, LEDA, Praha
12. Martincová, O., et al.: Nová slova v češtině 1. Academia, Praha (2004)
13. Kraus, J., Petráčková, V., et al.: Akademický slovník cizích slov (SCS). Academia, Praha (1999). elektronická verze. LEDA, Praha
14. Suchomel, V.: Recent czech web corpora. In: Horák, A., Rychlý, P. (eds.) 6th Workshop on Recent Advances in Slavonic Natural Language Processing, pp. 77–83. Brno, Tribun EU (2012)
15. Rychlý, P., Husák, M., Kilgarriff, A., Rundell, M., McAdam, K.: GDEX: automatically finding good dictionary examples in a corpus. In: Proceedings of the XIII EURALEX International Congress, Barcelona, Institut Universitari de Lingüística Aplicada, pp. 425–432 (2008)

Automatic Detection of Modality
with ITGETARUNS

Rodolfo Delmonte[(✉)]

Department of Language Studies, Department of Computer Science,
Ca' Foscari University, 30123 Venice, Italy
delmont@unive.it

Abstract. In this paper we present a system for modality detection which is then used for Subjectivity and Factuality evaluation. The system has been tested lately on a task for Subjectivity and Irony detection in Italian tweets (http://www.di.unito.it/~tutreeb/sentipolc-evalita14/index.html), where the performance was 10th and 4th, respectively, over 27 participants overall. We will focus our paper on an internal evaluation where we considered three national newspapers *Il Corriere, Repubblica, Libero*. This task was prompted by a project on the evaluation of press stylistic features in political discourse. The project used newspaper articles from the same sources over a period of three months, thus including latest political 2013 governmental crisis. We intended to produce a similar experiment and evaluate results in comparison with previous 2011 crisis. In this evaluation, we focused on Subjectivity, Polarity and Factuality which include Modality evaluation. Final graphs at the end of the paper will show results confirming our previous findings about differences in style, with *Il Corriere* emerging as the most atypical.

1 Introduction

In this paper we present a system for modality detection which uses the output of a deep dependency parser of Italian. The system focuses on the semantics associated to the Verbal Complex in order to tell whether the event described and the participants associated to the event are related to a fact that took place in the world or not. Modality in our system can also refer to the degree of certainty a speaker associated to the current proposition or sentence. This is turned into a set of features which are then used for Subjectivity (see, [11, 12, 16–18]) and Factuality evaluation.

Modality in the Italian Verb Complex can be expressed by Modal Verbs and by fully inflected verbs, thus referrable by Mood and Tense morphological features. Classical subdivision of modality is into four main classes: POSSIBILITY, NECESSITY, OBLIGATION and PERMISSION. In addition to these classes Mood and Tense may express modality content when associated to Aspectual information. We tap this information directly from our lexicon of Italian, MIDUV. Additional information is gathered from special periphrastic constructions for PROGRESSIVE made of STARE PER, STARE Verb + gerund. Typical modality triggering Moods are CONDITIONAL and SUBJUNCTIVE; as to modal Tense, we consider all those related to the FUTURE or Irrealis mode. In addition we also assign modality markers to QUESTIONS and

C. Biemann et al. (Eds.): NLDB 2015, LNCS 9103, pp. 404–411, 2015.
DOI: 10.1007/978-3-319-19581-0_38

IMPERATIVES. Other typical modality markers are assigned to ADVERBIALS appropriately classified in our lexicon of Italian for modality. Modality is an attribute of the event expressed by a proposition that is crucial for the assessment of FACTU-ALITY. However, we also consider other attributes related to the assessment of SUBJECTIVITY of paramount importance for Modality detection. We assume that Modality is always related to an attitude by the speaker to express some degree of uncertainty in the statement he/she is formulating. In particular we are referring to specific classes of verbs like Mental Activity verbs, Presuppositional verbs, Opacity inducing or Intensional verbs: Hope, Want, Wish, Seem, Appear, Desire, Believe, Think etc. which are also coincident with the class of Attitudinal verbs. Modality related grammatical structures include Hypotheticals or Conditionals clauses triggered by the presence of a discourse marker (see [8, 15]).

The paper is organized as follows: in Sect. 2, we briefly present the system; in Sect. 3 we discuss and propose annotation examples from the database of sentences we evaluated; in Sect. 4 we discuss previous work and the experiment we did with 6000 sentences. Then the evaluation and some conclusion.

2 The System ITGETARUNS

In this section we present a detailed description of the system for Italian that we used in this experiment. The system is derived from GETARUNS, a multilingual system for deep text understanding with limited domain dependent vocabulary and semantics, that works for English, German and Italian and has been developed in the past 20 years or so and documented in several publications and conference presentations [3–7]. The current version used for Italian has been made possible by the creation of the needed semantic resources, in particular a version of SentiWordNet [8] adapted to Italian and heavily corrected and modified. SentiWordNet derives from the English WordNet and the mapping of sentiment weights has been done automatically starting from the linguistic content of WordNet glosses. However, this process has introduced a lot of noise in the final results, and many entries have a totally wrong opinion evaluation. So we started to correct and optimize the resource for our domain, i.e. newswire politically oriented text.

We modified the classification in order to characterize uniquely all those entries that have a "generic" or "commonplace" positive, or negative meaning associated to them in the specific domain. This was deemed the only possible solution to the problem of semantic ambiguity, which could only be solved by introducing a phase of Word Sense Disambiguation which was not part of the system. However this was not possible for all entries. So, we decided to erase all entries that had multiple concepts associated to the same lemma, and had conflicting sentiment values. We also created and added an ad hoc lexicon for the majority of concepts (some 3000) contained in the texts we analysed, in order to increase the coverage of the lexicon. This was done again with the same approach, i.e. labelling only those concepts which were uniquely intended as one or the other sentiment, restricting reference to the domain of political discourse.

The system has been lately documented by our participation in the EVALITA (*Evaluation of NLP and Speech Tools for Italian*) challenge.[1] It works in a usual NLP pipeline: the system tokenizes the raw text and then searches for Multiwords. The creation of multiwords is paramount to understanding specific domain-related meanings associated to sequences of words. This procedure is then extended to NER (*Named Entity Recognition*), which is performed on the basis of a big database of entities, lately released by JRC (*Joint Research Centre*) research centre.[2] We also use our own list of entities and multiwords some 100 K entries. Words that are not recognized by simple matching procedures in the big word form dictionary (500 K entries), are then passed to the morphological analyser. In case also this may fail, the guesser is activated, which will at first strip the word of its affixes. It will start by stripping possible prefixes and then analysing the remaining portion; then it will continue by stripping possible suffixes. If none of these succeeds, the word will be labelled by a back off procedure, as foreign word if the final character is not a vowel; a noun otherwise. We then perform tagging and chunking. In order to proceed to the semantic level, each nominal expression is classified at first on the basis of the assigned tag: proper nouns are classified in the NER task. The remaining nominal expressions are classified using classes derived from ItalWordNet (*Italian WordNet*).[3] In addition to that, we have compiled specialized terminology databases for a number of specific domains including: medical, juridical, political, economic, and military. These lexica are used to add a specific class label to the general ones derived from ItalWordNet. And in case the word or multiword is not present there, to uniquely classify them. The output of this semantic classification phase is a vector of features associated to the word and lemma, together with sentence index and sentence position. Semantic mapping is then produced by a linguistically based dependency parser. In particular, we use a subcategorized lexicon of Italian verbs of some 17 K entries to choose between argument labels like SUBJ, OBJ2, OBL which are used for core arguments, and ADJ which is used for all adjuncts requires some additional information related to the type of governing verb. The first element for Modality annotation is the Verbal Complex(hence VC), which contains all linguistic items that may contribute to its semantic interpretation, including auxiliaries, modals, adverbials, negation, clitics. We then distinguish passive from active diathesis and we use the remaining information available in the feature vector to produce a full-fledged semantic classification at propositional level. Semantic mapping includes, beside diathesis:

– Change in the World; Subjectivity and Point of View; Speech Act; Factuality; Polarity.

At first we compute Mood and Tense from the VC which, as said before, may contain auxiliaries, modals, clitics, negation and possibly adverbials in between. From Mood_Tense we derive a label that is the compound tense and this is then used together with Aspectual lexical properties of the main verb to compute Change_in_the_World.

---

[1] http://www.evalita.it/.

[2] http://irmm.jrc.ec.europa.eu/.

[3] http://www.ilc.cnr.it/iwndb/iwndb_php/.

This results in a subclassification of events into three subclasses: Static, Gradual, Culminating. From Change_in_the_World we compute (Point_of_)View, which can be either Internal (Extensional/Intensional) or External. Internal View then allows a labeling of the VC as Subjective for Subjectivity and otherwise, Objective (more details below). Eventually, we look for negation which can be produced by presence of a negative particle or be directly in the verb meaning as lexicalised negation. Negation, View and Semantic Class, together with presence of absence of Adverbial factual markers are then used to produce a Factuality labeling.

One important secondary effect that carries over from this local labeling, is a higher level propositional level ability to determine inferential links intervening between propositions. Whenever we detect possible dependencies between adjacent VCs we check to see whether the preceding verb belongs to the class of implicatives. We are here referring to verbs such as "refuse, reject, hamper, prevent, hinder, etc." on the one side, and "manage, oblige, cause, provoke, etc." on the other. In the first case, the implication is that the action described in the complement clause is not factual. In the second case, the opposite will apply.

3 Modality Annotation: Some Examples

Modality annotation hinges mainly upon lexical information, but has also the need of morphological and syntactic processing of the input sentence [1, 2, 10, 13]. It is attached to a clause or a proposition, thus including both tensed and untensed VP structures. The main annotation items are three VIEW, FACTUALITY and CHANGE, and they are organized as shown here below:

VIEW: Function/Semantic_Role, Speech_Act, Semantic_Category, Diathesis, Verb_Predicate, Support_Verb
FACTUALITY: Verb_Predicate, Tense, Mood, Function/Semantic_Role, Support_Verb
CHANGE: Speech_Act, Polarity, Support_Verb, View, Mood, Tense, Aspect

In the examples below, will use the term Factivity and factive rather than Factuality and factual. Annotations may contain more than one Modality attribute, so MODAL1 may include additional features presented above: "forse"/maybe, will be annotated modal1 = "probable"; "mica"/at_all as modal1 = negated; deontic DOVERE/Must or Has_to, as modal1 = "deontic". All degree modifiers are also included under MODAL1 with a feature derived from [13] and taken from the following list: Intensifier, Maximizer, Booster, Approximator, Compromiser, Diminisher, Minimizer, Exclusivizer. They all contribute to Subjective interpretation. Notice that Subjective interpretation will also follow from the union of two components: View = internal and Factivity = nonfactive. Here below some Italian examples and the related classification: examples are shown after they have been tokenized and multiwords have been automatically created by the system.

<testo_frase id = "repubblica_1911822" > , "Ancora ricordo l'incontro che feci in Transatlantico con Giorgio_Amendola." </testo_frase >/I still remember the meeting I had in "Transatlantico" with Giorgio_Amendola.

Lemma	View	Word	Modal	Change	Factivity	Mood tense
ricordare	internal	ricordo		gradual	factive	present
fare	internal	feci		culminated	factive	past_tense

Below an example of deontinc modality and one case of Modal1: a deontic double modality example. Notice that even though dependency from a nonfactive matrix clause induces nonfactivity in a dependent clause, when a relative clause appears, this is not inherited.

<testo_frase id = "corriere_318263" > , "Sui Trasporti, può essere comodo scaricare l'aumento delle tariffe sulla Regione, ma una città come Roma deve poter decidere per conto proprio." </testo_frase >/About Transportation, it could be convenient to charge the rise in tarifs on the Region, but a city like Rome has to be able to decide by herself.

Lemma	View	Word	Modal	Modal1	Change	Factivity	Mood tense
essere	external	essere	potere		culminated	nonfactive	perfect
scaricare	external	scaricare			null	nonfactive	infinitive
decidere	internal	decidere	potere	deontic	null	nonfactive	infinitive

4   Previous Analysis and the Experiment

We will focus now on the evaluation of newswire articles where we classified Subjectivity and Factuality which include Modality evaluation. Final graphs at the end of the paper show results and evaluation, confirming our previous findings in differences in style, with *Il Corriere* emerging as the most atypical. We decided to evaluate manually the data produced by our system and this was the topic of a Master thesis which was also checked personally by myself. The experimental setup required a smaller amount of data to be checked manually and a clear indication of choices to be made when annotating different types of modality. Instructions to the annotator were as follows:

- differentiate tensed propositions which can be computed as factives from untensed ones
- differentiate tensed propositions were modality was present in one or double feature and compute them as nonfactive
- differentiate gerundives and participles which must be computed as factives from infinitivals
- differentiate simple infinitivals from past or complex infinitivals which can be computed as factives
- differentiate propositions which are dependent from a nonfactive matrix clause from the rest
- check for lexically triggered subjectivity – semantically marked verb classes.

The general quantitative data presented here below in fact show a similar situation to the previous 2011 evaluation. In fact, even though the database created was much smaller, only 6000 sentences compared to 20000 of the previous experiment, we can clearly see that the number of nonfactive and subjective propositions in *Il Corriere* is much higher in absolute numbers than the ones of the other two newspapers. It constitutes the 37 % against the 29 % of *Libero* and the 34 % of *Repubblica*. Similar proportions can be found for Subjectivity, where *Corriere* has again 36 % against 30 % of *Libero* and 34 % of *Repubblica*.

Table 1. Quantitative overall data of the experiment for subjectivity and nonfactivity evaluation

Newspapers	Tot. Subject	Tot. Nonfact	Errs. Nonfact	Errs. Subject	No. Sents	No. Propos. Structs.
Corriere	1377	2504	236	196	1804	5514
Libero	1142	1971	159	47	1965	4424
Repubblica	1290	2264	152	36	2042	5048
TOTALS	3809	6739	547	279	5811	14986

Results in the form of weighted data of the evaluation are shown in the graphs in Figs. 1 and 2. In Table 1 we also counted mistakes under Errs. since mistakes in tagging and in dependency parsing may affect the final outcome. Mistakes in automatic annotation of semantic features are strongly related to error propagation in the pipeline that constitutes the system. Additional errors are caused by problems in the semantic predicate-argument structure building process where in some cases verbs have been wrongly collapsed into one single Verb Complex even though they constituted sepa-rated items. However, error percentages for nonfactivity is overall 8.2 %; while errors percentages for subjectivity is slightly lower, at 7.35 %. As can be noticed from Table 1 and graphs below, Corriere is by far the more difficult newspaper to analyse in terms of semantic features. The great majority of errors are present in Corriere which also has the highest number of propositions but the lowest number of sentences. This amounts to saying that sentences in Corriere are much longer and more complex to read. When compared to number of propositions we see a different distribution of data, with *Il Corriere* having the highest number of nonfactive proposition but *Libero* having the highest number of Subjective propositions.

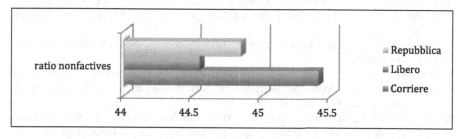

Fig. 1. Proportion of nonfactive propositions for the three newspapers

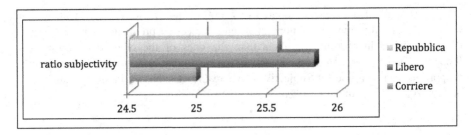

Fig. 2. Proportion of subjective propositions for the three newspapers

5   Conclusion and Future Work

We have presented work carried out to organize experiments intended to evaluate the ability of a system for deep dependency parsing of Italian to detect two semantic features, Factuality and Subjectivity, which are particularly sensitive to presence of modality at propositional level. We have presented the system and previous work done. We have shown in detail how the automatic annotation works and the types of different modality operators that can be represented by the system. Finally we have presented the experiment and the results obtained. In the future we intend to improve building of the Verbal Complex and the parser and tagger output.

References

1. Baker, K., Bloodgood, M., Dorr, B.J., Filardo, N.W., Levin, L., Piatko, C.: A modality lexicon and its use in automatic tagging. arXiv preprint arXiv:1410.4868 (2014)
2. Bracewell, D., Hinote, D., Monahan, S.: The author perspective model for classifying deontic modality in events. In: The Twenty-Seventh International Flairs Conference, March 2014
3. Delmonte, R.: Extracting opinion and factivity from Italian political discourse. In: Sharp, B., Zock, M. (eds.) Proceedings 10th International Workshop NLPCS, Natural Language Processing and Cognitive Science, Marseille, pp. 162–176 (2013)
4. Delmonte, R., Gifu, D., Tripodi, R.: Opinion and factivity analysis of Italian political discourse. In: Basili, R., Sebastiani, F., Semeraro, G. (eds.) Proceedings of 4th Italian Information Retrieval Workshop, IIR2013, Pisa. CEUR Workshop Proceedings, vol. 964, pp. 88–99. CEUR-WS.org (2013). ISSN 1613-0073, http://ceur-ws.org
5. Delmonte, R., Pallotta, V.: Opinion mining and sentiment analysis need text understanding. In: Pallotta, V., Soro, A., Vargiu, E. (eds.) DART 2011. SCI, vol. 361, pp. 81–95. Springer, Heidelberg (2011)
6. Delmonte, R.: Computational Linguistic Text Processing. Nova Science Publishers, New York (2007)
7. Delmonte, R.: Text understanding with GETARUNS for Q/A and summarization. In: Proceedings of ACL 2004 - 2nd Workshop on Text Meaning & Interpretation, Barcelona, pp. 97–104. Columbia University (2004)
8. Doddington, G.R., Mitchell, A., Przybocki, M.A., Ramshaw, L.A., Strassel, S., Weischedel, R.M.: The automatic content extraction (ace) program-tasks, data, and evaluation. In: LREC (2004)

9. Frawley, W. (ed.): The Expression of Modality. Mouton de Gruyter, Berlin (2006)
10. Kim, S., Hovy, E.: Automatic identification of pro and con reasons in online reviews. In: COLING/ACL (2006)
11. Kobayashi, N., Iida, R., Inui, K., Matsumoto, Y.: Opinion mining on the web by extracting subject-attribute-value relations. In: Proceedings of AAAI-CAAW 2006 (2006)
12. Nuyts, J.: Modality: overview and linguistic issues. In: Frawley, W. (ed.) The expression of modality, pp. 1–26. Mouton De Gruyter, Berlin (2006)
13. Walker, C., Strassel, S., Medero, J., Maeda, K.: Ace 2005 Multilingual Training Corpus. Linguistic Data Consortium, Philadelphia (2006)
14. Wilson, T., Wiebe, J., Hwa, R.: Just how mad are you? Finding strong and weak opinion clauses. In: AAAI 2004 (2004)
15. Wiebe, J., Mihalcea, R.: Word sense and subjectivity. In: ACL 2006 (2006)
16. Wiebe, J., Riloff, E.: Creating subjective and objective sentence classifiers from unannotated texts. In: Gelbukh, A. (ed.) CICLing 2005. LNCS, vol. 3406, pp. 486–497. Springer, Heidelberg (2005)

Gathering Knowledge for Question Answering Beyond Named Entities

Piotr Przybyła[✉]

Institute of Computer Science, Polish Academy of Sciences, Warsaw, Poland
P.Przybyla@phd.ipipan.waw.pl

Abstract. This paper presents an entity recognition (ER) module for a question answering system for Polish called RAFAEL. Two techniques of ER are compared: traditional, based on named entity categories (e.g. person), and novel Deep Entity Recognition, using WordNet synsets (e.g. impressionist). The latter is possible thanks to a previously assembled entity library, gathered by analysing encyclopaedia definitions. Evaluation based on over 500 questions answered on the grounds of Wikipedia suggests that the strength of DeepER approach lies in its ability to tackle questions that demand answers beyond the categories of named entities.

1 Introduction

A Question Answering (QA) system is a computer program capable of understanding questions in a natural language, finding answers to them in a knowledge base and providing answers in the same language. All factoid QA solutions yield information about simple facts, but differ with respect to the form of a returned answer: a document, paragraph, sentence or a single entity name. A QA system for Polish, called RAFAEL (*RApid Factoid Answer Extraction aLgorithm*) [1,2], returns answers as short strings, which requires an entity recognition (ER) module. Usually this step is realised by Named Entity Recognition (NER) tools, capable of finding names of several types, such as persons or places.

Herein, a generalization of the NER approach, called Deep Entity Recognition (DeepER), is proposed. This solution, instead of assigning each entity to one of several predefined NE categories, assigns it to a WordNet synset. For example, let us consider a question: *Which exiled European monarch returned to his country as a prime minister of a republic?* In the classical approach, we would recognise the question as concerning a person and treat all persons found in texts as potential answers. Using DeepER, it is possible to limit the analysis to persons being monarchs, which results in more accurate answers. In particular, we could utilise information that *Simeon II* (our answer) is a tsar; thanks to WordNet relations we know that it implies being a monarch. When using synsets instead

The study was supported by research fellowship within "Information technologies: research and their interdisciplinary applications" agreement number POKL. 04.01.01-00-051/10-00.

© Springer International Publishing Switzerland 2015
C. Biemann et al. (Eds.): NLDB 2015, LNCS 9103, pp. 412–417, 2015.
DOI: 10.1007/978-3-319-19581-0_39

of NE categories, answering new questions becomes possible, e.g. *Which bird migrates from the Arctic to the Antarctic and back every year? Arctic tern* is not recognized as named entity by NER systems, but DeepER tool labels it as a seabird and includes among possible answers.

The entity recognition process requires an *entity library*, containing known entities, their text representations (different names) and WordNet synsets, to which they belong. Similar libraries already exist: *Freebase* [3], *BabelNet* [4], *DBpedia* [5] or *YAGO* [6], but they mostly contain English names. Adapting them to another language is far from trivial, especially in case of Slavonic languages with complex NE inflection [7]. Therefore, a new approach is proposed here, which allows to automatically create such resource by analysing definitions of entries found in encyclopaedia (in this case the Polish Wikipedia), also taking into account redirect and disambiguation pages.

2 Related Work

Elements of the approach outlined above have already been used in tasks of natural language processing. Firstly, comparing synsets assigned to a question and a possible answer has been proposed by Mann [8]. An entity library is generated by extracting certain expressions from newswire texts. Analysing encyclopaedic definitions to get this type of information has been applied to other tasks, such as enriching ontologies [9,10], building a gazetteer [11] or a NE recognizer [12].

Other researchers dealt with classifying Wikipedia entries to NER categories using features typical for this resource, such as contexts of entity mentions [13] or article categories [14]. Categories assigned to entity definitions were also used as features in NER [15]. This approach could also lead to transforming Wikipedia into a high-quality NER training corpus [16].

3 Entity Library

An entity library contains information about entities that is necessary for deep entity recognition. For example description of entity #9751 (the Polish president, Bronisław Komorowski) contains the following:

- Main name: *Bronisław Komorowski,*
- Other names (aliases): *Bronisław Maria Komorowski, Komorowski,*
- WordNet synsets:
  - <polityk.1> (politician),
  - <wicemarszałek.1> (vice-speaker of the Sejm, the Polish parliament),
  - <prezydent.1, prezydent miasta.1> (president of a city, mayor),
  - and 5 more.

The process of entity library extraction, i.e. converting the first paragraph of a Polish Wikipedia entry into a list of WordNet synsets, is performed as follows. First, all unessential parts of the paragraph are omitted. This includes text in brackets or quotes, but also introductory expressions like *jeden z* (*one of*) or *typ* (*type of*). Then, an entity name is detached from the text by matching one of definition patterns, such as dash. Next, separators (full stops, commas and semicolons) are used to divide the text into chunks. The following step employs shallow parsing annotation – only chunks that begin with nominal groups are passed on. Finally, we split the coordination groups and examine each group to check whether their lemmas correspond to any lexemes in WordNet. If not, the process repeats with the group replaced by its semantic head. In case of polysemous words, only the first word sense (usually the most common) is taken into account. A new entity descriptor is created, containing synsets extracted from definition and names associated with the article (from redirect pages). Disambiguation pages are divided into paragraphs corresponding to different meanings, and each of them is treated in the way explained above.

A library built for deep entity recognition in RAFAEL, based on Polish Wikipedia, contains 809,786 entities with 1,169,452 names (972,592 unique) and 1,264,918 synsets (31,545 unique). In order to assess quality of the entity library, its content has been compared with synsets manually extracted from randomly selected 100 Wikipedia articles. 95 of them contain a description of an entity in the first paragraph. Among those, DeepER entity library includes 88. 135 synsets have been manually assigned to those entities, while the corresponding set in library contains 133 items. 106 of them are equal, while 13 differ only by word sense. 16 of manually extracted synsets hove no counterpart in the entity library, which instead includes 14 irrelevant synsets.

4   Entity Recognition

The entity recognition step aims at selecting all entity mentions that match a question type synset in a given annotated document. The document is searched for strings that resemble any of the entity names using PATRICIA trie. This process takes into account lemmata of words and syntactic groups, but also allows imperfect matching because of complicated inflection of proper names in Polish.

Given a list of entity mentions, RAFAEL checks their compatibility with the question. The question focus synset needs to be a (direct or indirect) hypernym of one of synsets assigned to an entity. For example, list of synsets assigned to entity *Symeon II* contains <car.1> (tsar), so it matches a question focus <monarcha.1, koronowana głowa.1> (monarch).

5   Evaluation

To evaluate the new approach *in vivo*, the performance of the whole QA system is assessed using different entity recognition techniques: traditional NER tools for Polish, using CRFs (*NERF* 0.1 [17] and *Liner2* 2.3 [18]), deep entity recognition, or hybrid approach, where entity mentions were gathered from all the above sources.

Table 1. Question answering accuracy of RAFAEL with different entity recognition strategies: traditional NER (*Nerf*, *Liner2*), deep entity recognition (*DeepER*) and their combination (*Hybrid*).

	Recall	Precision	F1 measure	MRR
Nerf	56.25 % ± 2.12 %	34.88 % ± 2.73 %	0.4306 ± 0.0213	33.66 % ± 2.29 %
Liner2	45.31 % ± 2.05 %	**39.08 %** ± 2.90 %	0.4197 ± 0.0188	41.36 % ± 2.70 %
DeepER	72.92 % ± 1.88 %	35.24 % ± 2.23 %	0.4751 ± 0.0214	32.80 % ± 1.99 %
Hybrid	**89.58 %** ± 1.24 %	33.14 % ± 2.01 %	0.4838 ± 0.0221	35.57 % ± 1.88 %

The Polish Wikipedia from 03.03.2013, converted into plaintext, serves as a knowledge base. The questions that are to be answered come from an open dataset for Polish QA systems, gathered from *Did you know...* column of Polish Wikipedia [19]. From this dataset, 576 entity questions have been chosen at random, and expected answer strings have been assigned manually to them.

Table 1 shows results of the final evaluation, expressed as recall (percentage of questions, to which RAFAEL gives any answer), precision (percentage of questions answered correctly), F1 measure and Mean Reciprocal Rank (MRR)[1]. Standard deviations of these values have been obtained by bootstrap resampling of the test set. As we can see, NER-based solutions answer slightly more (*Nerf*) or less (*Liner2*) than a half of the questions. When using DeepER, the recall ratio rises to 73 % while the precision does not change significantly. That is because questions beyond NE categories account for a substantial part of the test set. The maximum recall is obtained by the hybrid solution (90 %) but it comes at a cost of lower precision (33 %).

6   Conclusion

The main strength of DeepER compared to NER, according to the results presented in Table 1, is much higher recall. Thanks to the new technique the system may answer questions concerning animals, devices, chemical substances, ideas, and other concepts beyond reach of named entity recognition. However, this approach does not seem to improve precision. A part of wrong answers was inspected and most of the errors of entity recognition result from lack of word sense disambiguation, which impedes WordNet-based inference. Even having a correct entity library, it may be hard to decide which entity is referenced in a particular context. For example, consider a word *kot*, which means *a cat*. However, it is also a name of a journal, a lake, a village, a badge (*KOT*), a surname of 10 persons in Polish Wikipedia and more. Dealing with such ambiguities seems to be the most promising direction for future research.

---

[1] Evaluated automatically due to the length of ranking lists.

References

1. Przybyła, P.: Question analysis for polish question answering. In: 51st Annual Meeting of the Association for Computational Linguistics, Proceedings of the Student Research Workshop, Sofia, Bulgaria, pp. 96–102. Association for Computational Linguistics (2013)
2. Przybyła, P.: Odpowiadanie na pytania w języku polskim z użyciem głębokiego rozpoznawania nazw. Ph.D. thesis, Instytut Podstaw Informatyki PAN (to appear) (2015)
3. Bollacker, K., Evans, C., Paritosh, P., Sturge, T., Taylor, J.: Freebase: a collaboratively created graph database for structuring human knowledge. In: Proceedings of the 2008 ACM SIGMOD International Conference on Management of Data, SIGMOD 2008, pp. 1247–1250. ACM Press (2008)
4. Navigli, R., Ponzetto, S.P.: BabelNet: building a very large multilingual semantic network. In: Proceedings of the 48th Annual Meeting of the Association for Computational Linguistics, pp. 216–225. Association for Computational Linguistics (2010)
5. Bizer, C., Lehmann, J., Kobilarov, G., Auer, S., Becker, C., Cyganiak, R., Hellmann, S.: DBpedia - a crystallization point for the web of data. J. Web Sem. **7**(3), 154–165 (2009)
6. Suchanek, F.M., Kasneci, G., Weikum, G.: Yago: a large ontology from wikipedia and WordNet. In: Proceedings of the 16th International Conference on World Wide Web - WWW 2007, pp. 697–706. ACM Press (2007)
7. Przepiórkowski, A.: Slavonic information extraction and partial parsing. In: Proceedings of the Workshop on Balto-Slavonic Natural Language Processing Information Extraction and Enabling Technologies - ACL 2007. Association for Computational Linguistics (2007)
8. Mann, G.S.: Fine-grained proper noun ontologies for question answering. In: Proceedings of the 2002 Workshop on Building and Using Semantic Networks (SEMANET 2002), vol. 11. Association for Computational Linguistics (2002)
9. Ruiz-Casado, M., Alfonseca, E., Castells, P.: Automatic assignment of wikipedia encyclopedic entries to WordNet synsets. In: Szczepaniak, P.S., Kacprzyk, J., Niewiadomski, A. (eds.) AWIC 2005. LNCS (LNAI), vol. 3528, pp. 380–386. Springer, Heidelberg (2005)
10. Toral, A., Muñoz, R., Monachini, M.: Named entity WordNet. In: Proceedings of the International Conference on Language Resources and Evaluation, LREC 2008 (2008)
11. Toral, A., Muñoz, R.: A proposal to automatically build and maintain gazetteers for named entity recognition by using wikipedia. In: Proceedings of the 11th Conference of the European Chapter of the Association for Computational Linguistics. Association for Computational Linguistics (2006)
12. Kazama, J., Torisawa, K.: Exploiting wikipedia as external knowledge for named entity recognition. In: Proceedings of the Joint Conference on Empirical Methods in Natural Language Processing and Computational Natural Language Learning, pp. 698–707. Association for Computational Linguistics (2007)
13. Dakka, W., Cucerzan, S.: Augmenting wikipedia with named entity tags. In: Proceedings of the Third International Joint Conference on Natural Language Processing (IJCNLP 2008). Association for Computational Linguistics (2008)
14. Ponzetto, S.P., Strube, M.: Deriving a large scale taxonomy from wikipedia. Artif. Intel. **22**, 1440–1445 (2007)

15. Richman, A.E., Schone, P.: Mining wiki resources for multilingual named entity recognition. In: Proceedings of the 46th Annual Meeting of the Association for Computational Linguistics (ACL 2008). Association for Computational Linguistics (2008)
16. Balasuriya, D., Ringland, N., Nothman, J., Murphy, T., Curran, J.R.: Named entity recognition in wikipedia. In: Proceedings of the 2009 Workshop on The People's Web Meets NLP: Collaboratively Constructed Semantic Resources, pp. 10–18. Association for Computational Linguistics (2009)
17. Savary, A., Waszczuk, J.: Narzędzia do anotacji jednostek nazewniczych. In: Narodowy Korpus Języka Polskiego [Eng.: National Corpus of Polish], pp. 225–252. Wydawnictwo Naukowe PWN (2012)
18. Marcińczuk, M., Janicki, M.: Optimizing CRF-based model for proper name recognition in polish texts. In: Gelbukh, A. (ed.) CICLing 2012, Part I. LNCS, vol. 7181, pp. 258–269. Springer, Heidelberg (2012)
19. Marcińczuk, M., Ptak, M., Radziszewski, A., Piasecki, M.: Open dataset for development of polish question answering systems. In: Proceedings of the 6th Language & Technology Conference: Human Language Technologies as a Challenge for Computer Science and Linguistics, Wydawnictwo Poznańskie, Fundacja Uniwersytetu im. Adama Mickiewicza (2013)

MaNER: A MedicAl Named Entity Recogniser

Isabel Moreno[1]($\boxtimes$), Paloma Moreda[1], and M.T. Romá-Ferri[2]

[1] Department of Software and Computing Systems, University of Alicante,
Alicante, Spain
{imoreno,moreda}@dlsi.ua.es
[2] Department of Nursing, University of Alicante, Alicante, Spain
mtr.ferri@ua.es

Abstract. This paper describes a medicinal products and active ingredients named entity recogniser (MaNER) for Spanish technical documents. This rule-based system uses high quality and low-maintenance lexicons. Our results (F-measure 90 %) proves that dictionary-based approaches, without any deep natural language processing (e.g. POS tagging), can achieve a high performance in this task. Our system obtains better results when compared to similar systems.

Keywords: Named Entity Recognition · Lexicon · Medicinal product · Active ingredient · Spanish

1 Introduction

There is a huge amount of information concerning health, stored in heterogeneous sources. Employing this information, most of which is in textual form, is critical for all healthcare aspects [8]. For example, pharmacological treatments prescription is mainly related with key concepts such as medicinal products (a preparation that treats or prevents a disease) and active substances (a substance giving a drug its effect). Nevertheless, its analysis is: (i) unmanageable to physicians [9]; and (ii) inaccessible for machines [8,11]. These issues could be improved applying Natural Language Processing (NLP) techniques, which automatically transform relevant information from texts into structured data to be used by computer processes [8]. For instance, the objective of the Named Entity Recognition (NER) task is to identify key elements appearing explicitly in a text and assign them into a predefined category [7]. The purpose of this work is to propose a NER system to identify mentions of medicinal products and active substances for Spanish to enhance available resources in this language. The next section reviews the state of the art. Then our approach is described and evaluated. Finally, the conclusions are drawn.

2 Background

Lately, medicinal products and active ingredients extraction (among other substances) has received considerable attention. Specifically for English, some

© Springer International Publishing Switzerland 2015
C. Biemann et al. (Eds.): NLDB 2015, LNCS 9103, pp. 418–423, 2015.
DOI: 10.1007/978-3-319-19581-0_40

shared-tasks have promoted it, such as the i2b2 Medication challenge [17] or the DDIExtraction 2013 [14], but there is only one system [15] for Spanish. Reference [15] and most systems from this challenges are rule-based and make extensive use of lexicons [5, 6, 10, 13, 15, 16, 19] with good results. These dictionaries were built combining several sources: (i) biomedical knowledge resources [5, 6, 10, 13, 15, 16], such as the ATC Classification System [18]; (ii) training data [10, 16, 19]; (iii) unlabelled data [10]; and (iv) the web [15, 16]. Given their lexicons, these systems employ pattern matching and regular expressions [5, 10, 16, 19], join an spell checker with an existing NER system (*MedEx*) [6], combine rules from experts with a concept recognition tool (*Mgrep*) [13] or integrate a dictionary-based NER (*Textalytics*) and GATE gazetteers [15]. Two main conclusions can be drawn: (i) there is a scarcity of NER tools for Spanish language; and (ii) these dictionaries, gathered from different sources, require more maintenance. To overcome them, our contribution focuses on creating a medical rule-based NER using high quality and low-maintenance lexicons, for Spanish technical documents.

3   MaNER Description

MaNER (MedicAl Named Entity Recogniser) performs medicinal products and active ingredients mentions extraction using lexicons and rules in two stages:

- An off-line processing step to build and update our dictionaries from only one trustworthy source: March 2011 Nomenclator DIGITALIS [1], a database with medicinal products and medical devices authorised in Spain plus complementary data (i.e. active ingredients). Since DIGITALIS is updated monthly, we developed an automatic process to maintain our dictionaries, following recommendations from [2]. Two independent lexicons were built:
  - Medicinal products lexicon (MePLex) excludes terms referring to medical devices, whose ATC [18] code starts with letters from V to Z by definition. These terms include the brand name followed by extra information such as strengths. Since we were only interested in the brand name, a method to extract it automatically was implemented following the patterns from [3]. Our lexicon contains 14572 unique brand names.
  - Active ingredients lexicon (ActILex) was created by querying: (i) ATC codes, which are unique; and (ii) the name of each active ingredient, which can be shared by several codes due to ATC internal organization. We stored names and codes separately, keeping 3245 unique names and 3583 codes.
- A real-time processing for NER: There is a rule to extract medicinal products and other for active ingredients. These rules are Java Regular Expressions (RE) generated automatically from our lexicons to perform string matching.
  - Medicinal products rule is a case insensitive RE, which allows us to match complete names. For instance, a fragment of this RE recognises two medicinal products from our lexicon: ".*(^|\\p{Punct}|\\ s)
    (|AMOXICILINA CINFA |AMOXICILINA DR FERRER)(\\p{Punct}|\\s|\$).*"

Table 1. MaNER Evaluation

NE	Recall	Precision	F-measure
Active ingredient	0.87	0.88	0.87
Medicinal product	0.85	1.00	0.92
MACRO	0.86	**0.94**	**0.90**
MICRO	**0.87**	0.88	0.88

- Since active ingredients mentions can be names or codes, its rule has two conditions combined with an OR logical operator: (i) a case insensitive RE that matches complete names; and (ii) a case sensitive RE to recognise complete codes which are upper-cased.

4   Evaluation

Our NER tool was evaluated against DrugSemantics corpus [12], a collection of Spanish Summaries of Product Characteristics (SPC) manually annotated. It contains 670 sentences and 19279 tokens with 34 medicinal products and 582 active ingredients.

We computed an inexact-match F-Measure [4] for active ingredients and medicinal products independently over the complete corpus, as well as micro and macro averaged them. Here is an example of inexact match: our gold standard has "*etinil estradiol*" marked, whereas our system only tags "*estradiol*".

MaNER obtains very acceptable results with a 90 % F-measure Macro-average (see Table 1). The best recognised entity was medicinal product. This indicates that our lexicons (MePLex and ActILex) are relatively complete and include most terms in our corpus.

Error Analysis. Missing medicinal products were due to the lack of certain brand names in the DIGITALIS version employed. For instance, "*Ácido acetil-salicílico Ratiopharm*" is not in March 2011 DIGITALIS and is not in our lexicon, but this brand name is included in newer versions.

Concerning missing and spurious active ingredients, our analysis revealed errors in:

- Inversion (33.3 %): in our lexicon acids and hydroxides are in a reverse order (e.g. "*ácido valproico*") when compared to SPCs ("*valproico ácido*");
- Not active ingredients (29.2 %): some substances that are not active ingredients (i.e. "*almidón de patata*") are included in our lexicon.
- Ambiguity: (i) in SPCs there are drug families that can also refer to substances produced by our body (1.4 %), such as "*litio*" (litium); (ii) in these texts there are active ingredients names that also represent substances outside the scope of this study (around 2.1 %), like "*etinil estradiol*" (Ethinyl estradiol) which is an estrogen; and (iii) some phrases contain an active ingredient name as part

Table 2. MaNER versus top four NER rule-based, ranked by F-measure

System	Lang	Recall	Precision	F-measure	Corpus	Entities
MaNER	ES	0.86	**0.94**	**0.90**	DrugSemantics	616
OpenU [19]	EN	-	-	*0.89*	i2b2	-
Vanderbilt [6]	EN	**0.87**	0.90	0.88	i2b2	-
BME-Humboldt [16]	EN	0.82	0.92	0.87	i2b2	-
SpanishADRTool [15]	ES	0.80	0.87	*0.83*	SpanishADR	188

Acronyms: (i) ES: Spanish; (ii) EN: English; (iii) Lang: Language; and (iv) -: Unknown

of a bigger phrase (17.4 %), i.e. active group names: MaNER recognises *"proteasa"* (protease) where *"inhibidores de la proteasa del VIH"* (HIV-protease inhibitor) should be extracted as an active ingredient group name (which is an entity outside this paper scope).

– Lack of synonyms and lexical variations (more than 16.6 %): ActILex contains a large amount of active ingredients, however their synonyms and certain variations are not included. For instance: (i) *"nicotínico ácido"* is an entry in our dictionary but *"niaciana"*, its synonym, is not; and (ii) our lexicon has *"ezetimibA"*, but *"ezetimibE"* is found in our corpus.

Comparison with the State of the Art. Comparing our system to existing ones is difficult because either the evaluation corpus or the target language are different. To prove that our results are in-line with the state of the art, we could have evaluated Spanish systems (MaNER and [15]) on the same corpus. However, results would not be comparable since their granularity differs. Therefore, Table 2 contains a brief comparison between our best result and the top four systems in Sect. 2. MaNER achieved slightly better F-measure when compared to these systems (1–5%). Our good results are explained by two facts: (i) compilation of high quality and low-maintenance lexicons; and (ii) our fine grained approach allows to define specific dictionaries for each pharmacological substance.

5   Conclusions and Future Work

In this paper, MaNER, a MedicAl Named Entity Recogniser, was presented. This rule-based approach, which is language independent, performs extraction of medicinal products and active substances using lexicons. It was evaluated with Spanish technical documents. Our results (F-measure 90 %) proves that dictionary-based systems without any deep natural language processing obtain high performance in this task. Also our system obtains better results when compared to the state of the art, up to 5 %. As future work, we would like to fix the issues discovered by the error analysis process, also we plan to enhance our system to identify other relevant entities (i.e. dosages forms) and their relations.

Acknowledgments. This paper has been partially supported by the Spanish Government (grant no. TIN2012-38536-C03-03 and TIN2012-31224)

References

1. AEMPS: NOMENCLATOR DE PRESCRIPCION. http://www.aemps.gob.es/cima/pestanias.do?metodo=nomenclator
2. Cimino, J.J.: Desiderata for controlled medical vocabularies in the twenty-first century. Meth. Inf. Med. - Author manuscript; available in PMC 2012 August 10 **37**(4—5), 394–403 (1998)
3. Cruanes Vilas, J.: Una aproximación léxico-semántica para el mapeado automático de medicamentos y su aplicación al enriquecimiento de ontologías farmacoterapéuticas. Ph.D. thesis, Universida de Alicante (2014). http://hdl.handle.net/10045/42146
4. Cunningham, H., Maynard, D., Bontcheva, K., Tablan, V., Aswani, N., Roberts, I., Gorrell, G., Funk, A., Roberts, A., Damljanovic, D., Heitz, T., Greenwood, M., Saggion, H., Petrak, J., Li, Y., Peters, W., Al, E.: Developing Language Processing Components with GATE Version 7 (A User Guide), vol. 8 (2012). http://gate.ac.uk
5. Deléger, L., Grouin, C., Zweigenbaum, P.: Extracting medical information from narrative patient records: the case of medication-related information. J. Am. Med. Inf. Assoc.: JAMIA **17**(5), 555–558 (2010). doi:10.1136/jamia.2010.003962
6. Doan, S., Bastarache, L., Klimkowski, S., Denny, J.C., Xu, H.: Integrating existing natural language processing tools for medication extraction from discharge summaries. J. Am. Med. Inf. Assoc. **17**(5), 528–531 (2010). doi:10.1136/jamia.2010.003855
7. Feldman, R., Sanger, J.: The Text Mining Handbook: Advanced Approaches in Analyzing Unstructured Data, 2009th edn. Cambridge University Press, New York (2009). doi:10.1017/CBO9780511546914
8. Friedman, C., Rindflesch, T.C., Corn, M.: Natural language processing: state of the art and prospects for significant progress, a workshop sponsored by the National Library of Medicine. J. Biomed. Inf. **46**(5), 765–773 (2013). doi:10.1016/j.jbi.2013.06.004
9. González-González, A.I., Sánchez Mateos, J., Sanz Cuesta, T., Riesgo Fuertes, R., Escortell Mayor, E., Hernández Fernández, T.: Estudio de las necesidadesde información generadas por los médicos de atención primaria (proyecto ENIGMA)*. Atención primaria **38**(4), 219–224 (2006). http://www.sciencedirect.com/science/article/pii/S0212656706704814
10. Hamon, T., Grabar, N.: Linguistic approach for identification of medication names and related information in clinical narratives. J. Am. Med. Inf. Assoc. **17**(5), 549–554 (2010). doi:10.1136/jamia.2010.004036
11. Meystre, S.M., Thibault, J., Shen, S., Hurdle, J.F., South, B.R.: Textractor: a hybrid system for medications and reason for their prescription extraction from clinical text documents. J. Am. Med. Inf. Assoc. **17**(5), 559–562 (2010). doi:10.1136/jamia.2010.004028
12. Moreno, I., Moreda, P., Romá-Ferri, M.: Reconocimiento de entidades nombradas en dominios restringidos. In: Actas del III Workshop en Tecnologías de la Informática, pp. 41–57. Alicante, Spain (2012)

13. Sanchez-Cisneros, D., Aparicio Gali, F.: UEM-UC3M: an ontology-based named entity recognition system for biomedical texts. In: Proceedings of the Seventh International Workshop on Semantic Evaluation (SemEval 2013), vol. 2, pp. 622–627 (2013). http://www.aclweb.org/anthology/S13-2104

14. Segura-Bedmar, I., Martnez, P., Herrero-Zazo, M.: SemEval-2013 task 9: extraction of drug-drug interactions from biomedical texts (DDIExtraction 2013). In: Proceedings of the Seventh International Workshop on Semantic Evaluation (SemEval 2013), vol. 2011 (2013). www.aclweb.org/anthology/S13-2056

15. Segura-Bedmar, I., Revert, R., Martínez, P.: Detecting drugs and adverse events from spanish health social media streams. In: Proceedings of the 5th International Workshop on Health Text Mining and Information Analysis (Louhi) @ EACL 2014, pp. 106–115 (2014). https://www.aclweb.org/anthology/W/W14/W14-1117.pdf

16. Tikk, D., Solt, I.: Improving textual medication extraction using combined conditional random fields and rule-based systems. J. Am. Med. Inf. Assoc. **17**(5), 540–544 (2010). doi:10.1136/jamia.2010.004119

17. Uzuner, O., Solti, I., Cadag, E.: Extracting medication information from clinical text. J. Am. Med. Inf. Assoc. **17**(5), 514–518 (2010). doi:10.1136/jamia.2010.003947

18. WHO collaborating center for drug statistics methodology: guidelines for ATC classification and DDD assignment (2015). http://www.whocc.no/atc_ddd_publications/guidelines/

19. Yang, H.: Automatic extraction of medication information from medical discharge summaries. J. Am. Med. Inf. Assoc. **17**(5), 545–548 (2010). doi:10.1136/jamia.2010.003863

Upper Bound for Cross-Lingual Concept Mapping with External Translation Resources

Mamoun Abu Helou$^{(\boxtimes)}$ and Matteo Palmonari

University of Milano Bicocca, Milan, Italy
{mamoun.abuhelou,matteo.palmonari}@disco.unimib.it

Abstract. One way to achieve semantic interoperability when data lexicalized in different languages is by means of cross-lingual linking. Translation resources are used as an intermediate step to reduce the language barriers. The key challenge is to select the correct mapping among candidate matches. We define an experiment to study upper bounds for the correctness of a cross-lingual ontology matching system. We highlight different lexical characteristics that can support the selection step. We believe that our findings can be useful in the design of cross-lingual mapping algorithms.

Keywords: Cross-lingual linking · WordNets · Machine translation · Web-based lexicons

1 Introduction

When data are expressed (lexicalized) in a certain language, they are not easily accessible by speakers of other languages. Accessing or integrating data lexicalized in different languages is a challenge [9]. To make the semantics of the data published on the web explicit, several knowledge organization systems are used. *Ontologies* are often used to model and classify data instances, e.g., [17,21,25]. When data sources that use different ontologies have to be integrated, mappings, i.e., correspondences, between the concepts described in the different ontologies have to be found. This task is also called *ontology mapping* and has two main sub tasks: in *candidate match retrieval*, a first set of potential matches are found; in *mapping selection*, a subset of the potential matches is included in a final alignment. While the methods used in mapping selection strongly depend on the specific matching problem, candidate match retrieval is often based on matching methods that use the concepts' lexicalizations [20].

Most of the approaches to map two ontologies lexicalized in different languages include a step in which the concepts' lexicalizations of one ontology are translated into the language of the other ontology. To translate concepts' lexicalizations external translation *resources* have to be used. Various translation-based techniques are used to reduce the language barriers. *Manual* approach was adopted in several works [8,12,16,19,24]. The mappings generated by such

© Springer International Publishing Switzerland 2015
C. Biemann et al. (Eds.): NLDB 2015, LNCS 9103, pp. 424–431, 2015.
DOI: 10.1007/978-3-319-19581-0_41

Table 1. Words by category: quantity and percentage

Words	English		Arabic		Italian		Slovene		Spanish	
$Monosemous(M)$	120433	(81.8)	10025	(72.3)	29816	(74.2)	28635	(71.6)	30106	(81.6)
$Polysemous(P)$	26873	(18.2)	3841	(27.7)	10362	(25.8)	11350	(28.4)	6774	(18.4)
$Simple(S)$	83118	(56.4)	8953	(64.6)	33133	(82.5)	29943	(74.9)	22630	(61.4)
$Collection(C)$	64188	(43.6)	4913	(35.4)	7045	(17.5)	10042	(25.1)	14250	(38.6)
$M\&S$	59021	(40.1)	5361	(38.5)	22987	(57.2)	19223	(48.1)	16212	(44.0)
$M\&C$	61412	(41.6)	4664	(33.6)	6827	(17.0)	9412	(23.5)	13894	(37.7)
$P\&S$	24097	(16.4)	3592	(26.0)	10146	(25.3)	10720	(26.8)	6418	(17.4)
$P\&C$	2776	(01.9)	249	(01.8)	218	(00.5)	630	(01.6)	356	(00.9)

Table 2. The synsets categories

Categoriy	Synset name	Definition "a synset that has..."
all_M	*All word mounsemous*	Only monosemous words
all_P	*All word polysemous*	Only polysemous words
MWS	*Many-word*	Two or more synonyms
OWS	*One-word*	Only one word
MIX	*MIXed*	Monosemous and polysemous synonyms
$M\&OWS$	*Monosemous and OWS*	Only one word, which is also a monosemous word
$M\&MWS$	*Monosemous and MWS*	Two or more synonyms, which are all monosemous words
$P\&OWS$	*Polysemous and OWS*	Only one word, which is also a polysemous word
$P\&MWS$	*Polysemous and MWS*	Two or more synonyms, which are all polysemous words

systems are likely to be accurate and reliable. However, this is often an effort-intensive and time-consuming task specially for maintaining large and complex ontologies. *Machine translation tools* are widely adopted for cross-lingual ontology matching [7,22,23]. Recently, the Web has been used as a corpus of background linguistic knowledge [11]. *Wiktionary* [30] and *Wikipedia* [29] *inter-lingual links* are used as an intermediate resource for cross-lingual linking tasks [4,10,13]. In *BabelNet* [14] a graph-based approach is used to establish a mapping between the English Wikipedia pages and the English WordNet synsets. Then they enriched the WordNet synsets with lexicons for various languages using Wikipedia inter-lingual links, and they automatically translate a set of English sense-tagged sentences, the most frequent translation is detected and included as a variant for the mapped senses in the given language. Later, more lexical resources were also integrated; Wiktionary, Wikidata [33], OmegaWiki [34], and Open Multilangauge Wordnet [3].

In this paper we present a study on the effectiveness and quality of translations returned by translation resources to match concepts lexicalized in different languages. We believe that our findings help to better understand the candidate match retrieval problem and the usefulness of integrating different translation resources so as to design better cross-lingual mapping systems. Next, in Sect. 2 the experimental designing, the translation resources, the evaluation, and the discussion are given. In Sect. 3 we conclude and outline potential future steps.

2   Experiment Design

In our experiment four non-English wordnets, which are *manually* assembled and mapped to the English WordNet, are analyzed to study the coverage and the correctness of *automatic* translation resources on mapping concepts lexicalized in different languages. First an analytical statistics for the used wordnets is given. Then we perform our experiment and discuss the obtained results.

Gold Standards: Mapped Concepts in Different Languages. We use wordnets for English [5], Arabic [18], Italian [16], Slovene [6] and Spanish [8], whose size, respectively, in terms of words is: 147306, 13866, 40178, 39985, and 36880. In terms of word senses is: 206941, 23481, 61588, 70947, 57989. In terms of synsets is: 117659, 10349, 33731, 42583, and 38702.

The wordnets are selected to cover different approaches that are used to build wordnets. The selected languages are considered to cover different families of languages; the *Germanic* languages (English), the *Ramoance* languages (Italian and Spanish), the *Balto-Slavic* languages (Slovene), and the *Semitic* languages (Arabic). Moreover, Spanish, English, and Arabic languages are considered because they are among the top five spoken languages in the world [26], and to the increasing interest in the research community on these languages. Italian and Slovene languages are selected to our interest in integrating datasets published in *COMSODE* open data publication platform.

In Table 1 we provide details on the words distribution for each wordnet from two dimensions: The ambiguity: *Monosemous* word (M), a word that has only one sense(meaning), or *Polysemous* word (P), a word that can have two or more senses. The complexity: *Single* word (S), a string that has no spaces or hyphens, or *Collection* word (C), a string that consists of two or more simple words, connected by spaces or hyphens. For example, *"tourism"* is a monosemous and simple word, $M\&S$, *"tabular array"* is a monosemous and collection word, $M\&C$, *"table$^+$"* is a polysmouse and simple word, $P\&S$, and *"break up$^+$"* is a polysemous and collection word, $P\&C$. The superscript *"+"* indicates a polysemous word. *Observe that the vast majority of the collection words are monosemous words, on average only 1.3 % of the collection words are polysemous words.* In Table 2 we classify the synsets based on the words that form the synsets, we consider the words ambiguity and the number of synonym words they are lexicalized with. The non-English wordnets, in this study, are mapped to the English WordNet, there exist mappings between synsets of any possible

Table 3. Percentage of mappings by synset category

English	M&OWS	M&MWS	MIX	P&OWS	P&MWS	M&OWS	M&MWS	MIX	P&OWS	P&MWS
	Arabic					Italian				
M&OWS	32.9	19.2	5.1	5.4	2.3	36.2	20.9	10.6	9.4	4.1
M&MWS	15.1	28.6	5.1	2.5	1.5	21.2	34.9	10.3	4.6	2.8
MIX	17.2	28.7	37.7	15.5	22.6	17.8	27.2	38.7	22.5	26.8
P&OWS	27.4	14.8	21.7	57.3	29.5	17.9	10.7	18.4	43.0	29.0
P&MWS	7.3	8.7	30.4	19.4	44.2	6.9	6.3	22.0	20.5	37.4
	Slovene					Spanish				
M&OWS	23.4	25.2	14.2	9.0	6.8	42.6	10.7	7.8	8.4	3.1
M&MWS	47.8	39.7	13.0	4.4	4.3	22.2	63.1	7.7	3.3	1.9
MIX	18.1	27.5	48.7	20.2	27.1	14.5	17.1	44.1	19.4	24.2
P&OWS	7.1	4.0	8.4	45.3	25.7	17.8	5.4	15.1	48.5	25.9
P&MWS	3.5	3.7	15.7	21.1	36.1	2.9	3.8	25.3	20.4	44.9

Table 4. Word sense and synset coverages by translation resources

Translation	Arabic		Italian		Slovene		Spanish	
	Words	Synsets	Words	Synsets	Words	Synsets	Words	Synsets
BNcxt	19.9	37.4	40.0	62.5	28.8	44.2	33.9	44.7
BN	30.8	51.3	51.7	72.8	35.9	52.0	39.8	49.0
MT_fromEn	51.3	69.9	60.2	81.9	40.2	60.0	56.1	67.8
MT_toEn	57.9	76.1	65.4	83.9	49.6	67.2	67.0	77.0
MT	59.2	77.7	68.1	87.6	53.8	72.4	69.4	79.7
MT&BNcxt	60.8	79.2	69.8	89.0	55.8	74.2	71.5	81.3
MT&BN	62.5	80.2	72.2	89.9	57.5	75.2	72.3	81.7

categories. The percentage of the mapped synsets categories between the non-English wordnets and the English WordNet is reported in Table 3. The results reveals that concepts shared in different languages might have different ways (i.e., different synsets categories) to express their meanings. *This explains why the monosemous heuristic translation strategy, which is adopted by several mapping systems, including BabelNet, BabelNet, holds for a large number of concepts but does not hold for a significant number of concepts.*

Translation Resources. We use two translation resources: *Google Translate* [32], which is an automatic translation service, and *BabelNet* [27], which is a multilingual ontology. We chose these resources because they cover a large number of languages and have been frequently used in several cross-lingual ontology matching approaches, e.g., OAEI campaigns [28]. In addition, previous work suggested that Google Translate performs better than other Web translation services [15], and BabelNet integrates a large number of lexical resources. Google translate is used to obtain bilingual dictionaries from English to non-English (*MT_fromEn*), and form non-English

Table 5. Word sense coverage and correctness by category

Words	Arabic						Italian					
	BN		MT		MT&BN		BN		MT		MT&BN	
M	20.2	(63.6)	45.1	(36.9)	48.0	(56.4)	49.0	(65.6)	65.8	(47.0)	69.8	(62.1)
P	53.1	(38.0)	83.3	(22.8)	85.2	(40.9)	71.5	(44.9)	89.8	(31.9)	91.3	(45.1)
S	38.1	(48.3)	67.0	(27.3)	70.0	(49.1)	54.4	(57.0)	73.0	(41.0)	75.9	(55.5)
C	13.3	(63.4)	35.1	(43.9)	37.0	(53.8)	56.9	(65.8)	67.3	(47.5)	72.6	(62.8)

Words	Slovene						Spanish					
	BN		MT		MT&BN		BN		MT		MT&BN	
M	45.2	(66.1)	63.6	(47.8)	66.6	(60.8)	28.1	(61.6)	68.4	(48.1)	70.8	(56.9)
P	42.6	(39.6)	73.0	(30.1)	75.4	(33.7)	74.8	(38.9)	92.1	(28.4)	93.7	(41.4)
S	43.7	(56.3)	67.0	(41.4)	69.7	(49.9)	48.9	(51.0)	78.8	(41.0)	81.0	(53.3)
C	46.7	(66.0)	64.3	(45.0)	67.3	(60.3)	17.4	(62.3)	63.1	(48.4)	65.7	(53.5)

to English (MT_toEn). A bilingual dictionary is also created by mering both directions (MT). Two settings are used to obtain bilingual dictionaries encoded in BabelNet. Translations that are obtained from the *context-translation* approach [14] including Wikipedia inter-lingual links ($BNcxt$), and the whole translations encoded in BabelNet (BN). We excluded translations obtained from the Open Multilingual WordNet [3], which are used as the gold standards.

Evaluation. The non-English wordnets synsets are mapped to their corresponding English synsets. Each non-English sense has a set of English translations, which is given by the set of synonym words in the corresponding English synset (and, vice versa). We call this a gold standard translation. A word in language L_2, w^{L_2}, is considered to be *correct translation* of word in language L_1, w^{L_1}, if w^{L_2} belongs to the corresponding set of synonym words in L_2 in the gold standard. We quantify the *translation correctness* of word w^{L_1} into a language L_2, using F_1-*measure*, which determines the harmonic mean of precision and recall. *Precision* is defined as the number of correct translations given by a translation resource over the total number of translations given by the translation resource. *Recall* is defined as the number of correct translations given by a translation resource over the total number of translations to be given in the gold standard. We quantify the *translation coverage* in terms of word senses and synsets, similarly as defined in [14]. For word senses, the coverage is defined as the ratio of translated words that occur in the corresponding synsets in the gold standard to the over all number of senses in the gold standard. For synsets, the coverage is defined as the ratio of translated synsets that share at least one correctly translated word of the corresponding synsets in the gold standard.

Results and Discussion. The word senses and synsets coverage with the different translation settings is reported in Table 4. Observe that *Machine translation systems performs asymmetrically*. The combination of both machine translation directions MT (i.e., MT_toEn and MT_fromEn) performs better than

Table 6. Synsets coverage by category

Synsets	Arabic			Italian			Slovene			Spanish		
	BN	MT	MT&BN	BN	MT	MT&BN	BN	MT	MT&BN	BN	MT	MT&BN
M&OWS	32.2	58.0	61.5	63.8	83.5	86.1	60.8	78.3	80.3	32.4	74.7	76.2
M&MWS	40.3	73.4	76.0	81.6	92.6	94.5	54.3	81.4	83.2	40.1	88.0	89.0
MIX	59.8	86.9	88.6	80.7	94.3	95.8	53.1	76.5	79.0	65.5	85.8	87.6
P&OWS	55.8	75.4	79.0	71.0	83.3	86.4	43.4	60.5	64.5	57.9	77.1	80.2
P&MWS	61.9	90.5	92.0	80.8	93.7	95.1	47.3	74.7	77.6	75.5	88.8	90.9

considering each direction alone. $BNcxt$ and BN translations have less coverage than the machine translation system for all wordnets. BabelNet translations is influenced by the errors of the mapping phase, and due to the limited coverage of the non-English wordnets words in Wikipedia, which is concerned mostly with named entities than concepts [14]. The best results are obtained when combining translations form BabelNet and the machine translation system, $MT\&BN$. The percentage of the word senses coverage and word translation correctness by category are given in Table 5. $MT\&BN$ has improved the quality of the translated words while preserving the high coverage *Observe that the correctness of the translation tasks for the monosemous and the collection words are higher than the polysemous and the simple words.* Table 6 reports the percentage of the synsets coverage by category. The results show that the MWS synsets has higher coverage than the OWS synsets in all translation settings. However, the mapping selection task is much easier for OWS as they have less number of candidate mappings. Of course, this might be not the case if we consider more contextual knowledge in more efficient way for selecting the correct senses. For instance, with majority voting we can better disambiguate the MWS [1], this is due to the fact that, synonym words when translated might give same candidate matches.

3   Conclusions and Future Works

In this study four large-scale mapped concepts datasets in different languages, which are mapped to their equivalent concepts in English, are analyzed to investigate the correctness and the coverage of automatic translation resources and their impacts on the matching retrieval step in the cross-lingual mapping tasks. The mapping selection (disambiguation) task is difficult without the involvement of *contextual knowledge.* An interesting direction is to study the impact of the automatic translation resources on the mapping selection tasks. A natural subsequent step is also to undertake the study observations in building a cross-lingual mapping system, and to examine its impacts on real-world scenarios (e.g., in [2] and [31]).

References

1. Abu Helou, M.: Towards constructing linguistic ontologies: mapping framework and preliminary experimental analysis. In: 13th AIIA-DC (2014)
2. Abu Helou, M., Palmonari, M., Jarrar, M., Fellbaum, C.: Towards building linguistic ontology via cross-language matching. In: 7th GWC (2014)
3. Bond, F., Foster, R.: Linking and extending an open multilingual wordnet. In: 51st Annual Meeting of the Association for Computational Linguistics: ACL-2013, Sofia, pp. 1352–1362 (2013)
4. Bouma, G.: Cross-lingual ontology alignment using EuroWordNet and wikipedia. In: LREC (2010)
5. Fellbaum, C.: Wordnet: An Electronic Lexical Database. MIT Press, Cambridge (1998)
6. Fiser, D.: Leveraging parallel corpora and existing wordnets for automatic construction of the slovene wordnet. In: Proceedings of the 3rd Language and Technology Conference, vol. 7, p. 35 (2007)
7. Fu, B., Brennan, R., O'Sullivan, D.: A configurable translation-based cross-lingual ontology mapping system to adjust mapping outcomes. J. Web Sem. **15**, 15–36 (2012)
8. Gonzalez-Agirre, A., Laparra, E., Rigau, G.: Multilingual central repository version 3.0. In: 8th LREC, Istambul, Turkey (2012)
9. Gracia, J., Montiel-Ponsoda, E., Cimiano, P., Gmez-Prez, A., Buitelaar, P., McCrae, J.: Challenges for the multilingual web of data. JWS **11**, 63–71 (2012)
10. Hertling, S., Paulheim, H.: WikiMatch - using wikipedia for ontology matching. In: OM. Proceedings (2012)
11. Hovy, E., Navigli, R., Ponzetto, S.: Collaboratively built semi-structured content and artificial intelligence: the story so far. AI **194**, 2–27 (2013)
12. Liang, A., Sini, M.: Mapping AGROVOC & the chinese agricultural the saurus: definitions, tools procedures. In: New Review of Hypermedia & Multimedia (2006)
13. Lin, F., Krizhanovsky, A.: Multilingual ontology matching based on wiktionary dataaccessible via SPARQL endpoint. In: Proceedings of the 13th All-Russian Conference Digital Libraries: Advanced Methods and Technologies, Digital Collections (2011)
14. Navigli, R., Ponzetto, S.: BabelNet: the automatic construction, evaluation and application of a wide-coverage multilingual semantic network. AI **193**, 217–250 (2012)
15. Oliver, A., Climent, S.: Building WordNets by machine translation of sense tagged corpora. In: Proceedings of the 6th International Conference on Global WordNet (2012)
16. Pianta, E., Bentivogli, L., Girardi, C.: MultiWordNet: developing an aligned multilingual database. In: Proceedings of the 1st International GWC (2002)
17. Po, L., Sorrentino, S.: Automatic generation of probabilistic relationships for improving schema matching. Inf. Syst. **36**(2), 192–208 (2011)
18. Rodrguez, H., Farwell, D., Farreres, J., Bertran, M., Alkhalifa, M., Antonia Mar, M., Black, W., Elkateb, S., Kirk, J., Pease, A., Vossen, P., Fellbaum, C.: Arabic WordNet: current state and future extensions. In: Proceedings of the GWC (2008)
19. Tufis, D., Cristea, D., Stamou, S.: BalkaNet: aims, methods, results and perspectives: a general overview. Sci. Technol. **7**(1–2), 943 (2004)
20. Shvaiko, P., Euzenat, J.: Ontology matching: state of the art and future challenges. IEEE Trans. Knowl. Data Eng. **25**(1), 158–176 (2013)

21. Sorrentino, S., Bergamaschi, S., Gawinecki, M., Po, L.: Schema label normalization for improving schema matching. Data Knowl. Eng. **69**(12), 1254–1273 (2010)
22. Spohr, D., Hollink, L., Cimiano, P.: A machine learning approach to multilingual and cross-lingual ontology matching. In: Aroyo, L., Welty, C., Alani, H., Taylor, J., Bernstein, A., Kagal, L., Noy, N., Blomqvist, E. (eds.) ISWC 2011, Part I. LNCS, vol. 7031, pp. 665–680. Springer, Heidelberg (2011)
23. Trojahn, C., Fu, B., Zamazal, O., Ritze, D.: State-of-the-art in multilingual and cross-lingual ontology matching. In: Buitelaar, P., Cimiano, P. (eds.) Towards the Multilingual Semantic Web, pp. 119–135. Springer, Heidelberg (2014)
24. Vossen, P.: EuroWordNet: a multilingual database of autonomous and language-specific wordnets connected via an Inter-Lingual-Index. IJL **17**(2), 161–173 (2004)
25. Zhang, Z.: Towards efficient and effective semantic table interpretation. In: Mika, P., Tudorache, T., Bernstein, A., Welty, C., Knoblock, C., Vrandečić, D., Groth, P., Noy, N., Janowicz, K., Goble, C. (eds.) ISWC 2014, Part I. LNCS, vol. 8796, pp. 487–502. Springer, Heidelberg (2014)
26. http://en.wikipedia.org/wiki/List_of_languages_by_number_of_native_speakers
27. BabelNet version 2.5. http://babelnet.org/download
28. http://oaei.ontologymatching.org/
29. https://www.wikipedia.org/
30. https://www.wiktionary.org/
31. http://www.comsode.eu/
32. https://translate.google.com/
33. http://www.wikidata.org/
34. http://www.omegawiki.org/

Generating Logical Representations for Natural Language Requirements Using Syntactic Dependencies and Norm Analysis Patterns

Richa Sharma[1](✉) and K.K. Biswas[2]

[1] School of Information Technology, IIT, Delhi, India
sricha@gmail.com
[2] Department of Computer Science, IIT, Delhi, India
kkb@cse.iitd.ernet.in

Abstract. Requirements expressed in Natural Language are often ambiguous, inconsistent and, not amenable to automated analysis and validation. Formal approaches like mathematical or logical formalism to requirements representation offer possible solution to these problems. However, formal specifications are not widespread in industry as analysts and business users find them difficult to comprehend. In this paper, we present an approach to translate Natural Language representation of requirements to Logical representations. We have used Courteous logic, a non-monotonic form of logic, for the purpose. Our approach is based on syntactic dependency analysis of requirements statements, Norm Analysis Patterns and Grammatical Knowledge Patterns. The analyzed information is stored in frame-based structured representation for the requirements. These structured representations are translated to courteous logic form. We also report the effectiveness of our approach through the case-studies conducted.

Keywords: Requirements specification · Norm analysis patterns · Syntactic analysis

1 Introduction

Requirements analysis phase serves as the basis for subsequent phases of software development and is, therefore, crucial to the success of the software. However, the requirements to be analyzed are often ambiguous, inconsistent and incomplete in nature. The end-users are often not able to articulate their expectations from the software. Secondly, requirements gathered are expressed in the form of Natural Language (NL), which is inherently ambiguous. Moreover, requirements expressed in NL cannot be put to automated reasoning and analysis. Therefore, requirement analysis becomes dependent on analysts' expertise.

Formal approaches like mathematical formalism and logical formalism to represent requirements offer possible solutions to these problems. Several languages based on formal approaches have been proposed like Z [1], VDL [2], RML [3], Telos [4],

© Springer International Publishing Switzerland 2015
C. Biemann et al. (Eds.): NLDB 2015, LNCS 9103, pp. 432–436, 2015.
DOI: 10.1007/978-3-319-19581-0_42

HCLIE [5] etc. Z and VDL are based on mathematical formalism. RML, Telos, HCLIE are based on First-order Logic. These formal languages, however, are less preferred in industry as industry practitioners find corresponding expressions difficult to express and understand. Nevertheless, it is a widely acknowledged fact that formal approaches to requirements representation have been quite successful in uncovering ambiguities, inconsistencies and incompleteness issues. Formal descriptions can be assigned well-defined semantics which in turn, prove useful in adjudicating among different inter-pretations of a given model as well as serving a basis for various ways of reasoning with models, either through consistency-checking or by supporting question-answering [6].

Motivated by the concerns of NL representation of requirements and the advantages of using formal representation of requirements, we propose an approach for bridging the gap between NL requirements and their corresponding formal logical representa-tions. The logical formalism adopted in our work is based on Courteous Logic [7], a non-monotonic form of logic. We have used Courteous logic in our work as the adequacy of courteous logic for requirements representation and consistency checking has been proved in our earlier works [11].

Our translation approach is based on syntactic dependency analysis of requirements statements using Stanford Parser [8], Norm Analysis Patterns [9] and Grammatical Knowledge Patterns (GKP) [10]. Our contribution lies in combining these patterns, dependency analysis and logic in an eloquent way for automated translation of NL requirements to courteous logic representations. Our focus in this paper is on functional requirements that represent business rules. We shall use the term 'requirements state-ment' in this paper to refer only to such functional requirements. It is to be noted that our objective in this paper is to present an automated translation process; and therefore, we shall not be proving consistency handling in this work (as presented earlier in [11]).

The rest of the paper is organized as follows. Section 2 describes in detail our approach to transform NL requirements to courteous logic expressions. We present the details of the case-studies conducted in Sect. 3 and comparison with related existing work. This is followed by discussion and conclusion in Sect. 4.

2   Our Approach

We now present our approach for automated translation of NL representation of requirements to corresponding courteous logic form. Our approach is based on the assumption that the NL requirements have been analyzed and processed for the pres-ence of ambiguity concern, if present. Our approach can be summarized in two major steps as presented below:

(1) Frame-Based Structured Representation Generation. The structured represen-tations of the requirements statements are inspired from Minsky's frames [12]. Min-sky's frames are slot-filler data structure and, therefore, are suitable for capturing the pattern information for the statement under study. The statement can possibly have more than one GKP pattern and also, more than one norm pattern (identified by precondition and marker GKP). GKPs are combinations of grammatical categories, i.e. part of-speech [10]. We have designed separate frame structure for each of the GKPs with *frame keys* to capture semantics of the statement. The statements having only active-voice or passive-voice patterns are simple in nature whereas, the statements having more than one GKP are considered to be complex in nature. Frames for complex statements are generated by taking union of frames for simple part of the statement (active/passive voice pattern) and the frames for other GKPs present in that complex statement. The process of identifying GKPs, classifying the statements into simple and complex category and then, generating frames structures for the statements has been presented in detail in our earlier work [13].

(2) Courteous Logic Translation. The frame-based structured representation of the requirements acts as an intermediary output that is transformed to courteous logic expressions using following rules: (i) *if simple sentences (i.e. only active/passive voice pattern)*: If 'ACTION' key is populated, generate binary clause with first argument as 'ACTOR' and second argument as 'OBJECT'. Add modifiers for actor and object to this binary clause in conjunction as unary clauses. Else if 'ACTION' key is not populated, then generate unary clauses for frame keys – 'ACTOR' and the 'ACTOR MODIFIER', if present. (ii) *if precondition or marker present*: process both the condition clause and the consequence clause separately in the similar manner as in step (i). Join the condition clause and the consequence clause using 'IF – THEN' construct of courteous logic. (iii) *if preposition present*: add preposition object and modifier, if present to the associated clause in conjunction as unary clauses. (iv) *if coordination is present between two words (noun or verb)*: then clauses are generated for each of these in the similar manner as described in step (i).

3   Case Study

We have considered the requirements statements for library system where the requirements for library are based on library norms mentioned in [14]. Liu and Dix [15] have proposed norm pattern to be of the form: *"If < condition > then < consequence>"*. However, a statement can be written in multiple ways. We have, therefore, rewritten the book-borrowing norms from [14] in various possible ways as presented below to check if our approach is able to process all such different ways of writing requirements statements describing a business rule. Let us consider the norm for borrowing a book from library whose frame-based structured representation (based on the identified GKP and norm pattern) is shown in Table 1:

R1: *If the book is available **then** the member can borrow the book.*
Applying the rules for courteous logic translation as presented in Sect. 2, the generated courteous logic representation for R1 is as below:

C1: *if book(?Book) and available(?Book) then borrow(?Member, ?Book)*
Our approach generates similar frame structure as illustrated in Table 1 for different forms of R1 like:

1. **In case** the book is available **then** the member can borrow the book.
2. **When** the book is available **then** the member can borrow the book.
3. **Once** the book is available **then** the member can borrow the book.
4. The member can borrow the book **provided** the book is available.

Table 1. Frame Structure for requirement statement, R1

Frame key	Values
Active Voice	
Actor	Member
Action	Borrow
Object	Book
Precondition	
Precondition	If
Precondition on action	Borrow
Subject of Precondition	Book
Modifier of subject	Available

5. **Only** available book can be borrowed by the member.
6. The book must be available **in order to** be borrowed by the member.
7. The book must be available **must** to be borrowed by the member.

Since the intermediary output (i.e. the frame structure) to courteous logic translation process remains same for different versions of R1, therefore our approach always generates C1 as final courteous logic representation for different forms of R1.

Automated translation of NL text to logical form has been of interest in context of several problems like learning from text [16], question-answering [17] and requirements analysis [6, 18]. Most of such approaches are based on syntactic analysis of NL text using parser and, often domain-dependent parsers. In context of RE where business rules are of major concern, authors in [6] have used FOL form derived from frame-like structures of requirements statements. However, FOL is monotonic in nature and cannot handle conflicts. The authors in [18] have used default logic formalism to handle conflicts but their translation approach from NL to default form is based on domain-dependent parser, CICO. Default logic fails to handle conflicts in the presence of multiple defaults. We have strived towards being domain-independent and ability to handle multiple conflicts in our approach as discussed above.

4 Discussion and Conclusion

In this paper, we have proposed an automated translation approach to convert NL representation of functional requirements to courteous logic. Our approach generates frame-based structured representations of the requirements statements using syntactic dependency analysis, norm patterns and GKP. These structured representations are, in turn, used for automated translation purpose. We have shown through library case-study how these structured representations can prove effective in capturing the essence of the requirements statement irrespective of the way the statement is written. We, therefore, believe that our approach will be able to bridge the gap between NL and formal specification of requirements and, that bridging this gap will substantially improve software requirements analysis and consequently, software development. Our approach is generic across different domains as Stanford parser is domain-independent. We intend to develop tool based on our approach in future.

References

1. Spivey, M.: The Z Notation: A Reference Manual, 2nd edition. Prentice Hall International Series in Computer Science, Upper Saddle River (1992)
2. Wegner, P.: The Vienna Definition Language. In: ACM Computing Surveys (CSUR) Surveys Homepage archive, vol. 4 Issue 1, pp. 5–63 ACM New York, NY, USA (1972)
3. Greenspan, S., Borgida, A., Mylopoulos, J.: A requirements modeling language and its logic. Inf. Sys. 11(1), 9–23 (1986)
4. Mylopoulos, J., Borgida, A., Koubarakis, M.: Telos: representing knowledge about information systems. ACM Trans. Inf. Sys. 8(4), 325–362 (1990)
5. Tsai, J.J.P., Weigert, T.: HCLIE: a logic based requirement language for new software engineering paradigms. Softw. Eng. 6(4), 137–151 (1991)
6. Mylopoulos, J., Borgida, A., Yu, E.: Representing software engineering knowledge. Autom. Softw. Eng. 4(3), 291–317 (1997)
7. Grosof, B.N.: Courteous Logic Programs: prioritized conflict handling for rules. IBM Research Report RC20836. IBM Research Division, T.J. Watson Research Centre, New York (1997)
8. Marneffe, M.C., de MacCartney, B., Manning, C.D.: Generating Typed Dependency Parses from Phrase Structure Parses. In: LREC (2006)
9. Liu, K., Sun, L., Dix, A., Narasipuram, M.: Norm-based agency for designing collaborative information systems. Inf. Sys. 11, 229–247 (2001)
10. Marshman, E., Morgan, T., Meyer, I.: French patterns for expressing concept relations. Terminology 8(1), 1–29 (2002)
11. Sharma, R., Biswas, K.K.: Using courteous logic based representations for requirements specification. In: 4th International Workshop on Managing Requirements Knowledge (MaRK 2011), in conjunction with 19th IEEE International Requirements Engineering Conference (RE'11), pp. 12–16 (2011)
12. Minsky, M.: A framework for representing knowledge. In: Haugeland, J. (ed.) Mind Design. MIT Press, Cambridge (1981)
13. Bhatia, J., Sharma, R., Biswas, K.K., Ghaisas, S.: Using grammatical knowledge patterns for structuring requirements specifications. In: 3rd IEEE International Workshop on Requirements Patterns (RePa'13), in conjunction with 21st IEEE International Requirements Engineering Conference (RE'13), pp. 31–34 (2013)
14. Salter, A., Liu, K.: Using semantic analysis and norm analysis to model organisations. In: 4th International Conference on Enterprise Information Systems (ICEIS), Spain (2002)
15. Liu, K., Dix, A.: Norm Governed Agents in CSCW. In: First International Workshop on Computational Semiotics, Paris (1997)
16. Mulkar, R., Hobbs, J.R., Hovy, E.: Learning from Reading Syntactically Complex Biology Texts. In: AAAI (2007)
17. Hirschman, L., Gaizauskas, R.: Natural language question answering: the view from here. Nat. Lang. Eng. 7(4), 275–300 (2001)
18. Gervasi, V., Zowghi, D.: Reasoning about inconsistencies in natural language requirements. ACM Trans. Softw. Eng. Method. 14(3), 277–330 (2005)

Random Indexing Revisited

Behrang QasemiZadeh[(✉)]

National University of Ireland, Galway and University of Passau, Passau, Germany
behrang.qasemizadeh@uni-passau.de

Abstract. Random indexing is a method for constructing vector spaces at a reduced dimensionality. Previously, the method has been proposed using Kanerva's sparse distributed memory model. Although intuitively plausible, this description fails to provide mathematical justification for setting the method's parameters. The random indexing method is revisited using the principles of sparse random projections in Euclidean spaces in order to complement its previous delineation.

Keywords: Random indexing · Dimensionality reduction techniques · Vector space models · Random projections

1 Introduction

In order to model any aspect of language, data-driven approaches to natural language processing exploit patterns of co-occurrences. For example, distributional semantic models collect patterns of co-occurrences and investigate similarities in these patterns to quantify meanings. Vector spaces are mathematically well-defined models that are often employed to serve this purpose [18].

In a vector space model (VSM), each element $\vec{s}_i$ of its standard basis—informally, each dimension of the VSM—represents a contextual element. Given n context elements, linguistic entities are expressed using vectors $\vec{v}$ as linear combinations of $\vec{s}_i$ and scalars $\alpha_i \in \mathbb{R}$ such that $\vec{v} = \alpha_1\vec{s}_1 + \cdots + \alpha_n\vec{s}_n$. The value of α_i is acquired from the frequency of the co-occurrences of the entity that $\vec{v}$ represents and the context element that $\vec{s}_i$ represents. Therefore, the values assigned to the coordinates of a vector—that is, α_i—exhibit the correlation of an entity and context elements in an n-dimensional real vector space $\mathbb{R}^n$. In this VSM, a distance function, therefore, is employed for the discovery of similarities. Amongst several choices of distance metrics, the Euclidean distance is an innate choice. A VSM is endowed with the ℓ_2 norm to estimate distances between vectors, which is accordingly called a Euclidean VSM (denoted by $\mathbb{E}^n$). A classic document-by-term model is, perhaps, the most familiar example of the models described above for constructing VSMs [17].

B. QasemiZadeh—This publication has emanated from research conducted with the financial support of Science Foundation Ireland under Grant Number SFI/12/RC/2289.

C. Biemann et al. (Eds.): NLDB 2015, LNCS 9103, pp. 437–442, 2015.
DOI: 10.1007/978-3-319-19581-0_43

In distributional approaches to text analysis, when the number of entities in a VSM increases, the number of context elements employed for capturing similarities between them surges. As a result, high-dimensional vectors, in which most elements are zero, represent entities. But, the proportional impact of context elements on similarities declines when their number increases. In a high-dimensional model, except vectors vary in most dimensions, it becomes difficult to distinguish similarities [2]. Moreover, the high-dimensionality of vectors hampers the computation of distances. These setbacks are known as the *curse of dimensionality*. A *dimensionality reduction* technique is often employed to solve these problems.

Dimensionality reduction can be achieved using a number of methods as an auxiliary process followed by the construction of a VSM. This process improves the computational performance by reducing the number of context elements employed for the construction of a VSM. In its simple form, dimension reduction can be performed by choosing a subset of context elements using a heuristic-based *selection process*. That is, a number of context elements that account for the most discriminative information in VSM are chosen using a heuristic such as a statistical weight threshold. Alternatively, a *transformation* method can be employed. This process maps $\mathbb{R}^n$ onto a $\mathbb{R}^m$, $m \ll n$, in which $\mathbb{R}^m$ is the best approximation of $\mathbb{R}^n$ in a *sense*. For example, the well-known latent semantic analysis method employs singular value decomposition (SVD) truncation, in which $\mathbb{R}^m$ gives the best approximation of the Euclidean distances in $\mathbb{R}^n$ [7].

The use of these dimension reduction methods is hindered by a number of factors. Firstly, a VSM at the original high dimension must be first constructed. The VSM's dimension is then reduced in an independent process. Hence, the VSM at a reduced dimensionality is available for processing only after the whole sequence of these processes. Construction of the VSM at its original dimension is computationally expensive and a delay in access to the VSM at the reduced dimension is not desirable.

Secondly, reducing the dimension of vectors using the methods listed above is resource intensive. For instance, SVD truncation demands a process of the time complexity $O(n^2m)$ and space complexity $O(n^2)$. Similarly, depending on the employed heuristic, a selection process can be resource intensive too. Last but not least, these methods are *data-sensitive*: if the structure of the data being analysed changes—that is, if either the entities or context elements are updated—the dimensionality reduction process is required to be repeated and reapplied to the whole VSM in order to reflect the updates. As a result, these methods may not be desirable in several applications, particularly when dealing with frequently-updated big text-data.

Random projections (RPs) are employed to implement alternative dimensionality reduction methods. In the remaining of this paper, I describe the use of RPs in Euclidean spaces, which consequently arrives to the well-known random indexing (RI) technique, which has been employed in a number of applications (e.g., [3,5,19]). I then suggest a guideline for setting the method's parameters.

2   Random Projections in Euclidean Spaces

In Euclidean spaces, RPs are elucidated using the Johnson and Lindenstrauss lemma (JL lemma) [9]. Given an ϵ, $0 < \epsilon < 1$, the JL lemma states that for any set of p vectors in an $\mathbb{E}^n$, there exists a mapping onto an $\mathbb{E}^m$, for $m \geq m_0 = O(\log p / \epsilon^2)$, that does not distort the distances between any pair of vectors, with high probability, by a factor more than $1 \pm \epsilon$. This mapping is given by

$$\mathbf{M}'_{p \times m} = \mathbf{M}_{p \times n} \mathbf{R}_{n \times m}, \; m \ll p, n, \tag{1}$$

where $\mathbf{R}_{n \times m}$ is called the RP matrix, and $\mathbf{M}_{p \times n}$ and $\mathbf{M}'_{p \times m}$ denote the p vectors in $\mathbb{E}^n$ and $\mathbb{E}^m$, respectively. According to the JL lemma, if the distance between any pair of vectors $\vec{v}$ and $\vec{u}$ in $\mathbf{M}$ is given by the $d_{\text{Euc}}(\vec{v}, \vec{u})$, and their distance in $\mathbf{M}'$ is given by $d'_{\text{Euc}}(\mathbf{v}, \mathbf{u})$, then there exists an $\mathbf{R}$ such that $(1 - \epsilon) d'_{\text{Euc}}(\mathbf{v}, \mathbf{u}) \leq d_{\text{Euc}}(\mathbf{v}, \mathbf{u}) \leq (1 + \epsilon) d'_{\text{Euc}}(\mathbf{v}, \mathbf{u})$. Accordingly, instead of the original high-dimensional $\mathbb{E}^n$ and at the expense of negligible amount of error ϵ, the distance between $\vec{v}$ and $\vec{u}$ can be calculated in $\mathbb{E}^m$ to reduce the computational cost of processes.

The JL lemma does not specify the projection matrix $\mathbf{R}$. Establishing a random matrix $\mathbf{R}$ is therefore the most important design decision when using RPs. In [9], the lemma was proved using an orthogonal projection. Subsequent studies simplified the original proof that resulted in projection techniques with enhanced computational efficiency (see [4] for references). Recently, it is shown that a sparse $\mathbf{R}$, whose elements r_{ij} are defined as

$$r_{ij} = \sqrt{s} \begin{cases} -1 & \text{with probability } \frac{1}{2s} \\ 0 & \text{with probability } 1 - \frac{1}{s} \,, \\ 1 & \text{with probability } \frac{1}{2s} \end{cases} \tag{2}$$

for $s \in \{1, 3\}$, results in a mapping that also satisfies the JL lemma [1]. Subsequent research showed that $\mathbf{R}$ can be constructed from even sparser vectors than what is suggested in [1]. In [12], it is proved that in a mapping of an n-dimensional real vector space by a sparse $\mathbf{R}$, the JL lemma holds as long as $s = O(n)$, such as $s = \sqrt{n}$ or even $s = n/\log(n)$. The sparseness of $\mathbf{R}$ consequently enhances the time and space complexity of the method by the factor $\frac{1}{s}$.

Another benefit when computing $\mathbf{M}'$ is obtained using the linearity of matrix multiplication. As stated earlier, each vector $\vec{v}_{e_i}$ in $\mathbb{E}^n$ (i.e., the ith row of $\mathbf{M}$) is given by a linear combination of the basis vectors $\vec{v}_{e_i} = w_{i1}\vec{s}_{c_1} + \cdots + w_{in}\vec{s}_{c_n}$ ($i \leq p$ and $j \leq n$). By the basic properties of the matrix multiplication, the projection of $\vec{v}_{e_i}$ in $\mathbf{M}'$ is given by $\vec{v}'_{e_i} = \vec{v}_{e_i}\mathbf{R} = w_{i1}\vec{s}_{c_1}\mathbf{R} + \cdots + w_{in}\vec{s}_{c_n}\mathbf{R}$. In turn, since by definition all the elements of $\vec{s}_{c_k}$ are zero except the kth element (i.e., 1), $\vec{v}'_{e_i}$ can be equally written as

$$\vec{v}'_{e_i} = w_{i1}\vec{r}_1 + \cdots + w_{in}\vec{r}_n, \tag{3}$$

where $\vec{r}_j$ is the jth row of $\mathbf{R}$. Equation 3 means that row vectors $\mathbf{v}'_{e_i}$, thus $\mathbf{M}'$, can be computed directly without necessarily constructing the whole matrix $\mathbf{M}$.

The jth row of $\mathbf{R}_{n \times m}$ represents a context element in the original VSM that is located at the jth column of $\mathbf{M}_{p \times n}$. Therefore, an entity at a reduced dimension can be computed directly by accumulating the row vectors of $\mathbf{R}$ that represent the context elements that co-occur with the entity.

The explanations above results in a two-step procedure similar to what is earlier suggested as the RI technique [10,16]: the construction of (a) *index vectors* and (b) *context vectors*. In the first step, each context element is assigned *exactly* to one *index vector*. [16] indicates that index vectors are high-dimensional randomly generated vectors, in which most of the elements are set to 0 and only *a few* to 1 and -1. In the second step, the construction of *context vectors*, each target entity is assigned to a vector of which all elements are zero and has the same dimension as the index vectors. For each occurrence of an entity (represented by $\vec{v}_{e_i}$) and a context element (represented by $\vec{r}_{c_k}$), the context vector is accumulated by the index vector (i.e., $\vec{v}_{e_i} = \vec{v}_{e_i} + \vec{r}_{c_k}$). The result is a vector space model constructed directly at reduced dimension. As can be understood, the first step of RI is equivalent to the construction of the random projection matrix $\mathbf{R}$, whose elements are given by Eq. 2. Each index vector is a row of the random projection matrix $\mathbf{R}$. The second step of RI deals with the computation of $\mathbf{M}'$. Each context vector is a row of $\mathbf{M}'$, which is computed by the iterative process justified in Eq. 3.

Compared to the justification of RI, which are based on Kanerva's sparse distributed memory (e.g., [10,16]), and whereas in previous research the method's parameters are left to be decided through experiments (e.g., [13,14]), we leverage the adopted mathematical framework to provide a guideline for setting these parameters. In an RI-constructed VSM at reduced dimension m (i.e., $\mathbb{E}^m$), the degree of preservation of distances in $\mathbb{E}^n$ and $\mathbb{E}^m$ is determined by the number of vectors in the model and the value of m. If the number of vectors is fixed, then the larger m is, the better the Euclidean distances are preserved at the reduced dimension m. In other words, the probability of preserving the pairwise distances increases as m increases. Hence, m can be seen as the capacity of an RI-constructed VSM for accommodating new entities. Compared to $m = 4000$ suggested in [10] or $m = 1800$ in [16], depending on the number of entities that are modelled in an experiment, m can be set to a smaller value such as 400.

Based on the proofs in [12], when embedding $\mathbb{E}^n$ onto $\mathbb{E}^m$, the JL lemma holds as long as s in Eq. 2 is $O(n)$. In text processing applications, the number of context elements (i.e., n) is often very large. When using RI, therefore, even a careful choice such as $s = \sqrt{n}$ in Eq. 2 results in highly-sparse index vectors. Hence, by setting only 2 or 4 non-zero elements in index vectors, distances in the RI-constructed $\mathbb{E}^m$ resembles distances in $\mathbb{E}^n$. If the dimension of index vectors (i.e., m) is fixed, then increasing the number of non-zero elements in index vectors causes additional distortions in pairwise distances. For index vectors of fixed dimensionality m, if the number of non-zero elements increases, then the probability of the orthogonality between index vectors decreases; hence, it stimulates distortions in pairwise distances (although in some applications, distortions in pairwise distances can be beneficial).

Lastly, it is important to note that RI-constructed VSMs can be only used for estimating similarity measures that are derived from the ℓ_2 norm. For instance, the use of RI-constructed VSMs for estimating city block distances (e.g., as suggested in [11]) is not justified, at least mathematically. Hence, techniques other than RI must be used (e.g., see [6,8,20,21]).[1]

References

1. Achlioptas, D.: Database-friendly random projections. In: Proceedings of the Twentieth ACM SIGMOD-SIGACT-SIGART Symposium on Principles of Database Systems, PODS 2001, pp. 274–281. ACM, New York (2001)
2. Beyer, K., Goldstein, J., Ramakrishnan, R., Shaft, U.: When is nearest neighbor meaningful? In: Beeri, C., Bruneman, P. (eds.) ICDT 1999. LNCS, vol. 1540, pp. 217–235. Springer, Heidelberg (1998)
3. Damljanovic, D., Petrak, J., Lupu, M., Cunningham, H., Carlsson, M., Engstrom, G., Andersson, B.: Random indexing for finding similar nodes within large RDF graphs. In: Proceedings of the 8th International Conference on the Semantic Web, ESWC 2011, pp. 156–171. Springer, Heidelberg (2012). http://dx.doi.org/10.1007/978-3-642-25953-1_13
4. Dasgupta, S., Gupta, A.: An elementary proof of a theorem of Johnson and Lindenstrauss. Random Struct. Algorithms **22**(1), 60–65 (2003)
5. De Vries, C.M., De Vine, L., Geva, S.: Random indexing k-tree (2010). CoRR abs/1001.0833, http://arxiv.org/abs/1001.0833
6. De Vries, C.M., Geva, S.: Pairwise similarity of TopSig document signatures. In: Proceedings of the Seventeenth Australasian Document Computing Symposium, ADCS 2012, pp. 128–134. ACM, New York (2012)
7. Deerwester, S.C., Dumais, S.T., Landauer, T.K., Furnas, G.W., Harshman, R.A.: Indexing by latent semantic analysis. J. Am. Soc. Inf. Sci. **41**(6), 391–407 (1990). http://citeseerx.ist.psu.edu/viewdoc/summary?doi=10.1.1.49.7546
8. Geva, S., De Vries, C.M.: TOPSIG: topology preserving document signatures. In: Proceedings of the 20th ACM International Conference on Information and Knowledge Management, CIKM 2011, pp. 333–338. ACM, New York (2011)
9. Johnson, W., Lindenstrauss, J.: Extensions of lipschitz mappings into a Hilbert space. In: Conference in Modern Analysis and Probability (New Haven, Connecticut, 1982), Contemporary Mathematics, vol. 26, pp. 189–206. American Mathematical Society (1984). http://www.ams.org/books/conm/026/
10. Kanerva, P., Kristoferson, J., Holst, A.: Random indexing of text samples for latent semantic analysis. In: Proceedings of the 22nd Annual Conference of the Cognitive Science Society, pp. 103–106. Erlbaum (2000). http://www.rni.org/kanerva/cogsci2k-poster.txt
11. Lapesa, G., Evert, S.: Evaluating neighbor rank and distance measures as predictors of semantic priming. In: Proceedings of the Fourth Annual Workshop on Cognitive Modeling and Computational Linguistics (CMCL), pp. 66–74. Association for Computational Linguistics, Sofia, Bulgaria, August 2013. http://www.aclweb.org/anthology/W13-2608
12. Li, P., Hastie, T.J., Church, K.W.: Very sparse random projections. In: Proceedings of the 12th ACM SIGKDD International Conference on Knowledge Discovery and Data Mining, KDD 2006, pp. 287–296. ACM, New York (2006)

---

[1] An extension to this discussion and some empirical experiments can be seen in [15].

13. Lupu, M.: On the usability of random indexing in patent retrieval. In: Hernandez, N., Jäschke, R., Croitoru, M. (eds.) ICCS 2014. LNCS, vol. 8577, pp. 202–216. Springer, Heidelberg (2014)

14. Polajnar, T., Clark, S.: Improving distributional semantic vectors through context selection and normalisation. In: Proceedings of the 14th Conference of the European Chapter of the Association for Computational Linguistics (EACL 2014). ACL, Gothenburg, Sweden (2014). http://www.cl.cam.ac.uk/%7Esc609/pubs/eacl14tam.pdf

15. QasemiZadeh, B.: Random indexing explained with high probability (2015)

16. Sahlgren, M.: An introduction to random indexing. In: Methods and Applications of Semantic Indexing Workshop at the 7th International Conference on Terminology and Knowledge Engineering, TKE 2005 (2005). http://soda.swedish-ict.se/221/1/RI_intro.pdf

17. Salton, G., Wong, A., Yang, C.S.: A vector space model for automatic indexing. Commun. ACM 18(11), 613–620 (1975)

18. Turney, P.D., Pantel, P.: From frequency to meaning: vector space models of semantics. J. Artif. Int. Res. 37(1), 141–188 (2010). http://dl.acm.org/citation.cfm?id=1861751.1861756

19. Zadeh, B.Q., Handschuh, S.: Evaluation of technology term recognition with random indexing. In: Calzolari, N., Choukri, K., Declerck, T., Loftsson, H., Maegaard, B., Mariani, J., Moreno, A., Odijk, J., Piperidis, S. (eds.) Proceedings of the Ninth International Conference on Language Resources and Evaluation (LREC 2014). European Language Resources Association (ELRA), Reykjavik, Iceland, May 2014. http://www.lrec-conf.org/proceedings/lrec2014/pdf/920_Paper.pdf, aCL Anthology Identifier: L14–1703

20. Zadeh, B.Q., Handschuh, S.: Random Manhattan indexing. In: 25th International Workshop on Database and Expert Systems Applications, DEXA 2014, pp. 203–208. IEEE (2014). http://dx.doi.org/10.1109/DEXA.2014.51

21. Zadeh, B.Q., Handschuh, S.: Random Manhattan integer indexing: incremental L1 normed vector space construction. In: Proceedings of the 2014 Conference on Empirical Methods in Natural Language Processing (EMNLP), pp. 1713–1723. Association for Computational Linguistics (2014). http://aclweb.org/anthology/D14-1178

On Developing Extraction Rules for Mining Informal Scientific References from Altmetric Data Sources

Waqas Khawaja[1]([✉]), Michael Taylor[2], and Brian Davis[1]

[1] Insight Centre for Data Analytics, National University of Ireland,
Galway, Republic of Ireland
{waqas.khawaja,brian.davis}@insight-centre.org
http://www.insight-centre.org
[2] Elsevier, Oxford, UK
Mi.Taylor@elsevier.com
http://www.elsevier.com

Abstract. Altmetrics measure scientific impact outside of traditional scientific literature. We identify mentions of scientific research or entities like researchers, academic or research organizations in a corpus containing blogs, articles, news items etc. We first manually analyse the corpus for patterns of such informal mentions and then apply text mining techniques by developing extraction rules for mining informal mentions. We apply them to our development corpus and present our results. This work takes us closer to developing concrete altmetrics for determining research impact on news and public discourse ultimately leading to measuring impact of scientific research on government policies.

Keywords: Text mining · Altmetrics · Informal scientific references

1 Introduction

Citation count had been the foundation of measuring research impact for a long time. The more recent measures like H-Index [7,10] still rely on citation count to measure impact. While more recent, they do not provide ways to determine the impact research created on media, public discourse or even government policies.

We look for mentions of research itself or research related entities for example, scientists, research organizations, or research and development departments of commercial entities in a heterogeneous corpus of sources such as news, articles, blogs and official government pages. We manually annotated these mentions and crafted JAPE grammar rules to extract the same in General Architecture for Text Engineering (GATE) [2]. JAPE(Java Annotations Pattern Engine) is a pattern matching language over features and annotations implemented as a finite state transducer [3]. Our task is somewhat similar to scenario template extraction in Information Extraction but our intention is to convert the problem to sentence/relation classification task. We first look at the existing work done in

© Springer International Publishing Switzerland 2015
C. Biemann et al. (Eds.): NLDB 2015, LNCS 9103, pp. 443–447, 2015.
DOI: 10.1007/978-3-319-19581-0_44

the domain and then explain the corpus we used. We then highlight the grammar development process and present our initial results.

2   Altmetrics and Related Work

Social media has attracted a lot of attention from scientists in search of altmetrics in the recent years. They looked at Linkedin, Facebook, Twitter, blogs and review websites etc. and adopted different methods to develop and evaluate altmetrics [1,11].

A Twitter study found that 6 % of studied tweets contained first or second order link to research articles [1]. Researchers concluded that Twitter is mostly using for making idea popular [5]. Unlike Twitter, blogs are thought to be an effective medium for initiating discourse [8].

While text mining has focused on traditional scientific databases and publications, methods from text mining/analytics [4] have seen adoption for analyzing opinions and sentiments as well as to determine the impact of scientific research on other research [6].

3   Implementation and Experiment

We collected a corpus of around 500 documents reaching to 130 MB. As a use case for scholarly discourse, we choose the anti viral drug *Tamiflu* and indexed the documents from the web against this keyword. The corpus includes news reports, articles, and reports among others. We indexed the corpus in Mimir [9] which is a semantic search platform. Our queries looked for mentions of entities using a variety of combinations. We manually annotated this corpus with 232 mentions of scientific research or entities. We also devised a list of 30 trigger phrases that frequently appear in these mentions.

We compiled all our triggers in finite state custom gazetteer. We included the default *VP chunker* (rule based) in our pipeline to account for different forms of verb phrases. We first process lookup triggers that overlap only with verbs for simple disambiguation. We put this all together in a corpus pipeline in GATE which consists of some default shallow linguistic processing resources from the standard ANNIE information extraction pipeline that includes a Sentence Splitter and POS Tagger. We replaced the ANNIE NER with Stanford NER [12] based on a simple comparison experiment[1] and added Verb Group (VG) chunker and finally two custom JAPE grammars to identify mentions in our corpus based on the trigger words and their lexico-syntactic/semantic context.

The following JAPE grammar rule looks for mentions in text where an entity appears first that may be followed by one or more organizations and finally a trigger phrase before the end of the sentence.

---

[1] http://goo.gl/bvpqTG.

```
Rule:Reference3
(
  ({Person} | {Organization})+ {Trigger} {Split}
):bind
-->
:bind.TempMention = {rule=Reference3, type=reference}
```

Our entities and trigger phrases can appear anywhere in a sentence. Consider for example the mention, *Tamiflu is an antiviral medication that blocks the actions of the influenza virus in the body, says Dr. Sterkel.* The trigger phrase and entity are the last two in the sentence highlighted with green and red respectively. In order to capture the complete sentence as a mention, we first create a *TempMention* and finally the JAPE grammar given below creates the annotates the complete sentence as a mention.

```
Rule: Reference1
(
{Sentence contains TempMention}
):bind
-->
:bind.Mention = {rule=Reference2, type=reference}
```

4 Results

Our manually annotated corpus was used for analysis of trigger phrases as well as for developing the JAPE grammars. Our results presented here are obtained from testing the JAPE grammars on our development corpus. The number of annotations and precision, recall, and F-measure information can be seen in Tables 1 and 2 respectively.

We expanded our basic custom gazetteer using synonyms from Wordnet lookup. While this enabled capture of more mentions, it also introduced some false positives which needed to be corrected.

Table 1. Annotation Results of Pipeline

	Key	System
Total	321	329
Match	162	
Only Key	76	
Only System	84	
Overlap	83	

Table 2. Precision, Recall & F-Measure

Recall	Strict	0.50
	Lenient	0.76
	Average	0.63
Precision	Strict	0.49
	Lenient	0.74
	Average	0.61
F measure	Strict	0.49
	Lenient	0.75
	Average	0.62

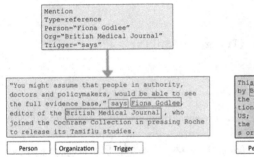

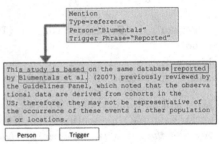

Fig. 1. Example of a mention with person, organization and trigger phrase (Color figure online)

Fig. 2. Another captured mention with person and trigger (Color figure online)

There are some erroneous mentions as well that are annotated by our JAPE grammars. Consider the following mention annotation because R&D has been annotated as an organization which is to be expected from an off the shelf NER and a trigger phrase *reported* is found although subject verb dependency is clearly incorrect as the rule is too relaxed. The trigger phrase is highlighted in green and the organization is highlighted in red:

However, these figures should be taken with caution as they are usually taken from pharmaceutical industry **reports** *reports which are known for the lack of transparency in relation to the cost of* **R&D** *and there are difficulties for verifying the figures reported.*

Further example of correct mentions annotated by our JAPE grammars can be seen in Figs. 1 and 2.

5 Future Work

Our goal is to measure the impact of scientific research in government literature and policies. We have already gathered a corpus of government documents indexed from government health related websites. A more fine grained annotation schema must be developed that would later be used in development of a gold standard model corpus with inter annotator agreement and involving at minimum three annotators. The rule based extraction offers high precision over recall that is more suited for boot strapping machine learning in the absence of a training corpus. We also plan to tune our rules and augment the rule based extraction patterns with machine learning to enhance our recall. Finally, we will look at linking the extracted mentions of entities to unique identifiers in to scientific databases such as Scopus[2].

Acknowledgments. This publication has emanated from research conducted with the financial support of Science Foundation Ireland (SFI) under Grant Number SFI/12/RC/2289 and by targeted project funding from Elsevier.

---

[2] http://www.scopus.com.

References

1. Bollen, J., Van de Sompel, H., Hagberg, A., Chute, R.: A principal component analysis of 39 scientific impact measures. PloS one **4**(6), e6022 (2009)
2. Cunningham, H.: Gate, a general architecture for text engineering. Comput. Humanit. **36**(2), 223–254 (2002)
3. Cunningham, H., Maynard, D., Tablan, V.: JAPE: a Java annotation patterns engine, 2nd edn. Research Memorandum CS-00-10, Department of Computer Science, University of Sheffield, November 2000
4. Elder IV, J., Hill, T.: Practical text mining and statistical analysis for non-structured text data applications. Academic Press, UK (2012)
5. Holmberg, K., Thelwall, M.: Disciplinary differences in twitter scholarly communication. Scientometrics **101**(2), 1027–1042 (2014)
6. Kostoff, R.N., Temixco, M., Humenik, M.M.J.A., Rockville, M., Ramírez, M.L.A.M.: Citation mining
7. Moed, H.F.: Citation Analysis in Research Evaluation, vol. 9. Springer Science and Business Media, New York (2006)
8. Shema, H., Bar-Ilan, J., Thelwall, M.: How is research blogged? a content analysis approach. J. Assoc. Inf. Sci. Technol. **65**(5), 1018–1027 (2014)
9. Tablan, V., Bontcheva, K., Roberts, I., Cunningham, H.: Mímir: An open-source semantic search framework for interactive information seeking and discovery. Web Semant. Sci. Serv. Agents World Wide Web (2014)
10. Thelwall, M.: Bibliometrics to webometrics. J. Inf. Sci. **34**(4), 605–621 (2008)
11. Zahedi, Z., Costas, R., Wouters, P.: How well developed are altmetrics? a cross-disciplinary analysis of the presence of alternative metrics in scientific publications. Scientometrics **101**(2), 1491–1513 (2014)
12. Finkel, J., Grenager, T., Manning, C.: Incorporating non-local information into information extraction systems by gibbs sampling. In: Proceedings of the 43rd Annual Meeting on Association for Computational Linguistics (2005)

Lemonade: A Web Assistant for Creating and Debugging Ontology Lexica

Mariano Rico[1](✉) and Christina Unger[2]

[1] Universidad Politécnica de Madrid, Madrid, Spain
mariano.rico@upm.es
[2] Bielefeld University, Bielefeld, Germany
cunger@cit-ec.uni-bielefeld.de

Abstract. The current state of the art in providing lexicalizations for ontologies is the *lemon* model. Based on experiences in creating a *lemon* lexicon for the DBpedia ontology in English and subsequently porting it to Spanish and German, we show that creating ontology lexica is a time consuming, often tedious and also error-prone process. As a remedy, this paper introduces *Lemonade*, an assistant that facilitates the creation of lexica and helps users in spotting errors and inconsistencies in the created lexical entries, thereby 'sweetening' the otherwise 'bitter' lemon.

Keywords: Ontology lexicon · Lemon · Grammatical Framework · DBpedia

1 Introduction

One of the major challenges in providing natural language access to Semantic Web data – be it through question answering, OWL and RDF summarization, or SPARQL query verbalization – is relating natural language expressions and corresponding vocabulary elements. One possibility to specify this relation are ontology lexica [4]. The current state of the art for specifying ontology lexica is the *lemon* model[1] [2], which provides a standard format for capturing linguistically rich information about how the vocabulary elements of a particular ontology or dataset are verbalized in natural language, in particular covering different verbalization variants, possibly in multiple languages. The resulting lexica are themselves expressed as RDF data, so that they can be shared in accordance with linked data principles and can be re-used across applications.

Although the process of creating ontology lexica can be automated to a certain extent [7], creating a wide coverage and high precision lexical resource still requires a significant manual effort. This presupposes familiarity with RDF, and even though there are tools like the *lemon* design patterns library[2] [3] to support

---

M. Rico—LIDER (EU FP7 project No. 610782) and MINECO's JdC Grant (JCI-2012-12719) and INFRA (UNPM13-4E-1814).

[1] http://lemon-model.net.
[2] http://github.com/jmccrae/lemon.patterns.

© Springer International Publishing Switzerland 2015
C. Biemann et al. (Eds.): NLDB 2015, LNCS 9103, pp. 448–452, 2015.
DOI: 10.1007/978-3-319-19581-0_45

and facilitate lexicon creation, still the process of creating lexica manually is time-consuming, and often also tedious and error-prone.

To illustrate this, we take as starting point the creation of an English lexicon for DBpedia [6] (available at http://github.com/cunger/lemon.dbpedia) and our efforts to port this lexicon to German and Spanish. In total, the DBpedia lexicon contains 1,217 lexicalizations for the most important classes and properties of the DBpedia ontology, specified using the *lemon* design pattern macros, which correspond to more than 50,000 RDF triples.

Although using the *lemon* design patterns almost completely frees the lexicon engineer from writing verbose RDF code, it also has several limitations. First, writing lexicalizations by hand is error-prone. Typos, for instance, give rise to errors when converting the macros into RDF, and you typically run the conversion at least as many times as you have errors. Therefore, the time required to remove all errors is very high. Second, and more importantly, validation of lexica is currently only possible with respect to the well-formedness of the RDF code (by means of RDF validators or editors) and its conforming to the *lemon* ontology (e.g. by means of RDFUnit [1]), but cannot be performed on the level of lexical consistency and correctness. For example, at one point you can specify the grammatical gender of a particular word to be feminine, and at another point you can specify it to be masculine, without noticing the inconsistency. Another example concerns mistakes in the argument mapping. For instance, one of the English lexicalizations of the DBpedia property `writer` is specified as follows:

```
StateVerb("write", dbpedia:writer,
          propSubj = DirectObject)
          propObj  = Subject)
```

This instantiation of the *lemon* design patterns `StateVerb` macro specifies the written representation of the verb to be "write" and its meaning with respect to the datatset to refer to the property `writer`. In addition, it establishes that the subject of the property corresponds to the direct object in syntactic structures, and that the object corresponds to the syntactic subject. That is, the triple (`Macbeth`, `writer`, `Shakespeare`) can be expressed as "Shakespeare wrote Macbeth". However, if we accidentally swapped `DirectObject` and `Subject`, then the same triple would be expressed as "Macbeth wrote Shakespeare". These kinds of errors are very hard to spot when all you get is a huge, automatically generated RDF file.

In this paper we therefore present a system, *Lemonade*, that assists users in creating lexica by means of an easy-to-use web interface, and furthermore provides support for spotting errors and inconsistencies in the created lexicon. In particular, we suggest that showing natural language sentences that would result from using the specified entries significantly helps in detecting erroneous and inappropriate entries.

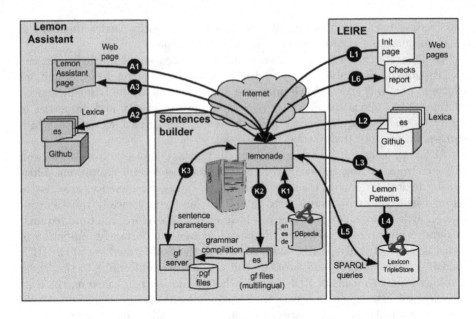

Fig. 1. Architecture of the *Lemonade* system.

2   Architecture of *Lemonade*

Figure 1 shows the architecture of *Lemonade*. At its very core, it is a library written in R that interfaces with *lemon* for the creation of lexica and with Grammatical Framework [5] for the construction of example sentences based on already created lexicalizations. On top of this library we have developed two applications intended to assist users in the creation of ontology lexica:

- The *lemon assistant*, shown in the left side of Fig. 1, is a web interface for creating lexicalizations of classes and properties. It covers the most common *lemon* design patterns, in particular common nouns (e.g. "mountain"), relational nouns (e.g. "capital of"), state verbs (e.g. "to write"), as well as intersective and relational adjectives (e.g. "Dutch" and "similar to").
- The *lemon lint remover* (LEIRE), shown in the right side of Fig. 1, then reads created lexicalizations and implements several consistency checks for each design pattern, such as checking for multiple plural forms for nouns. In addition it creates a natural language sentence that illustrates a possible use of the created entry. The result of the analysis is published as a web page on GitHub, which users can check in order to realize possible errors and inconsistencies.

We instantiated the system for DBpedia and three languages – English, Spanish, and German – but it can easily be ported to other datasets (given they provide instance data with human-readable labels) and a wide range of other languages.

The web interface is shown in Fig. 2. For example, if we choose to create a state verb for the DBpedia property `writer` in English, the assistant prompts

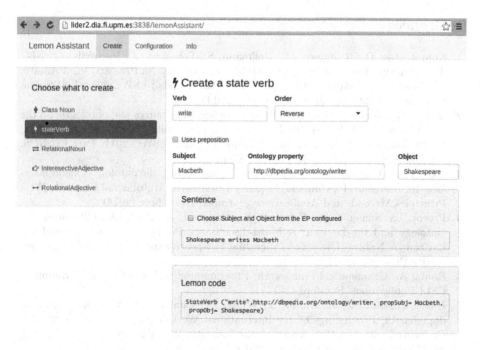

Fig. 2. Screenshot of the web interface to create an English state verb for the DBpedia property `writer`.

us to provide the necessary information, in this case the infinitive form of the verb ("write"), the order of the argument mapping ("Linear" if the subject of the property in the triplestore corresponds to the grammatical subject of the sentence and the object of the property corresponds to the grammatical object, or "Reverse" if the subject of the property corresponds to the grammatical object and the object of the property corresponds to the grammatical subject). Based on this information the assistant then creates a sentence, in this case "Shakespeare writes Macbeth".

The user can know the triples stored in the language-specified DBpedia endpoint with the property `writer`. In addition it shows the *lemon* design patterns macro, so that expert users can directly check the lexicon code that is created.

If the user validates the information, the lexical entry in stored in the GitHub repository underlying the project.

The process of creating lexical entries and storing them in the repository is described in Fig. 1 by the sequence A1-K1-K2-K3-A2-A3. The process of reading lexicalizations from the repository and creating example sentences is described in Fig. 1 by the sequence L1-L2-L3-L4-L5-K1-K2-K3-L6. Information about the tool and links to the web application can be found at https://github.com/cunger/lemon.dbpedia/blob/master/test/LemonadeTools.md.

References

1. Kontokostas, D., Brümmer, M., Hellmann, S., Lehmann, J., Ioannidis, L.: NLP data cleansing based on linguistic ontology constraints. In: Presutti, V., d'Amato, C., Gandon, F., d'Aquin, M., Staab, S., Tordai, A. (eds.) ESWC 2014. LNCS, vol. 8465, pp. 224–239. Springer, Heidelberg (2014)
2. McCrae, J., Spohr, D., Cimiano, P.: Linking lexical resources and ontologies on the semantic web with lemon. In: Antoniou, G., Grobelnik, M., Simperl, E., Parsia, B., Plexousakis, D., De Leenheer, P., Pan, J. (eds.) ESWC 2011, Part I. LNCS, vol. 6643, pp. 245–259. Springer, Heidelberg (2011)
3. McCrae, J., Unger, C.: Design patterns for engineering the ontology-lexicon interface. In: Buitelaar, P., Cimiano, P. (eds.) Towards the Multilingual Semantic Web: Principles Methods and Applications. Springer, Heidelberg (2014)
4. Prévot, L., Huang, C.R., Calzolari, N., Gangemi, A., Lenci, A., Oltramari, A.: Ontology and the Lexicon: A Multi-disciplinary Perspective. Ontology and the Lexicon: A Natural Language Processing Perspective, pp. 3–24. Cambridge University Press, Cambridge (2010)
5. Ranta, A.: Grammatical Framework: Programming with Multilingual Grammars. CSLI Publications, Stanford (2011)
6. Unger, C., McCrae, J., Walter, S., Winter, S., Cimiano, P.: A lemon lexicon for dbpedia. In: Proceedings of 1st International Workshop on NLP and DBpedia, Co-located with the 12th International Semantic Web Conference (ISWC 2013), 21–25 October, Sydney, Australia (2013)
7. Walter, S., Unger, C., Cimiano, P.: M-ATOLL: a framework for the lexicalization of ontologies in multiple languages. In: Mika, P., Tudorache, T., Bernstein, A., Welty, C., Knoblock, C., Vrandečić, D., Groth, P., Noy, N., Janowicz, K., Goble, C. (eds.) ISWC 2014, Part I. LNCS, vol. 8796, pp. 472–486. Springer, Heidelberg (2014)

A Comparative Study on Twitter Sentiment Analysis: Which Features are Good?

Fajri Koto[✉] and Mirna Adriani

Faculty of Computer Science, University of Indonesia,
Depok 16423, Jawa Barat, Indonesia
fajri91@ui.ac.id, mirna@cs.ui.ac.id
http://www.cs.ui.ac.id

Abstract. In this paper, investigations of Sentiment Analysis over a well-known Social Media Twitter were done. As literatures show that some works related to Twitter Sentiment Analysis have been done and delivered interesting idea of features, but there is no a comparative study that shows the best features in performing Sentiment Analysis. In total we used 9 feature sets (41 attributes) that comprise punctuation, lexical, part of speech, emoticon, *SentiWord* lexicon, *AFINN*-lexicon, *Opinion* lexicon, *Senti-Strength* method, and Emotion lexicon. Feature analysis was done by conducting supervised classification for each feature sets and continued with feature selection in subjectivity and polarity domain. By using four different datasets, the results reveal that AFINN lexicon and *Senti-Strength* method are the best current approaches to perform Twitter Sentiment Analysis.

Keywords: Twitter · Sentiment Analysis · Comparative study · Polarity · Subjectivity

1 Introduction

In general the goal of Sentiment Analysis is to determine the polarity of natural language text by performing supervised and/or unsupervised classification. This sentiment classification can be roughly divided into two categories: Subjectivity and Polarity [1]. The difference between subjectivity and polarity classification is the class involved in conducting training and testing stage. Sentiment of subjectivity comprises of *subjective* and *objective* class [2]. Whereas polarity classification involves classes of *positive, negative* and *neutral* [3].

Many approaches [4–12] have been addressed to classify sentiment over Twitter[1]. However, based on the previous study there is no a comparative study that shows the good feature in performing Sentiment Analysis. Whereas, this information will be necessary especially for today's business that concern with social media analysis in running their work. Driven by this fact, we first derive all possible features and then investigate the cases by performing supervised classification for each feature set.

---

[1] http://www.twitter.com.

© Springer International Publishing Switzerland 2015
C. Biemann et al. (Eds.): NLDB 2015, LNCS 9103, pp. 453–457, 2015.
DOI: 10.1007/978-3-319-19581-0_46

Table 1. List of all feature sets for Twitter Sentiment Analysis

Set	#Attr	List of attribute	Description
Punctuation [3], range = $\{0,1,..,n\}$	5	Number of "!", "?", ".", ",", and special character	Number of corresponding punctuation in a tweet
Lexical, range1 = $\{0,1,..,n\}$, range2 = $\{false,true\}$	9	(1) tweetLength, #lowercase, #uppercase, Aggregate$\{min, max, avg\}$ of #letterInWord, #hashtag	The corresponding number of attributes
		(2) haveRT	True if the tweet contains "RT" phrase, False otherwise
Part of Speech, extracted by NLTK Python [16] range1 = $\{0,1,..,n\}$, range2 = $\{false,true\}$	8	(1) #noun, #verb, #adjective, #adverb, #pronoun	Number of corresponding POS tag in a tweet
		(2) hasComparative, hasSuperlative, hasPastPartciple	True if the tweet contains a comparative/superlative adjective or adverb; or a past participle, False otherwise
Emoticon, obtained from [3,5] and Wikipedia range = $\{-n,.0,1.,n\}$	1	emoticonScore	Increasing the score by +1 and −1 for positive and negative emoticon respectively, initiated by 0
SentiWord Lex. [8], range = $\{0,1,..,n\}$	2	sumpos, sumneg	Sum of the scores for the positive or negative words that matches the lexicon
AFINN Lex. [9,10], range1 = $\{0,1,..,n\}$, range2 = $\{-n,..,-1,0\}$	2	(1) APO	Sum of the scores for the positive words that matches the lexicon
		(2) ANE	Sum of the scores for the negative words that matches the lexicon
Opinion Lex. (OL), range = $\{0,1,..,n\}$	4	(1) Wilson (positive words, negative words) [6]	Sum of the scores for the positive or negative words that matches the lexicon
		(2) Bingliu (positive words, negative words) [7]	
Senti-Strength (SS) [12], range1 = $\{-5,-4,..-1\}$ range2 = $\{1,2,..,5\}$	2	(1) ssn	Method score for negative category
		(2) ssp	Method score for positive category
NRC Emotion Lex. [11,13,14], range = $\{0,1,..,n\}$	8	joy, trust, sadness, anger, surprise, fear, disgust, anticipation	Number of words that matches with corresponding emotion class word list

2   Experiment with Feature of Sentiment Analysis

The experiment was conducted in two sentiment domains: polarity and subjectivity. There are 4 different datasets: (1) *Sanders* [1], (2) *Health Care Reform* (HCR) [15], (3) *Obama-McCain Debate* (OMD)[6] [15], and (4) *International*

Table 2. Balanced dataset

Subjectivity	Sanders	HCR	OMD	SemEval	Polarity	Sanders	HCR	OMD	SemEval
#neutral	1190	280	800	2256	#negative	555	368	800	896
#objective	1190	280	800	2256	#positive	555	368	800	896
#total	2380	560	1600	4512	#total	1110	736	1600	1792

Table 3. Classification result for each feature set

Feature	Subjectivity															
	SemEval				Sanders				HCR				OMD			
	A	B	C	D	A	B	C	D	A	B	C	D	A	B	C	D
Punct.	56.4	56.7	55.4	57.4	57.6	57.3	57.3	59.3	56.1	59.6	58.6	62.7	56.1	62.1	56.4	60.5
Lexical	51.3	51.8	51.7	54.9	56.7	55.9	55.9	55.4	59.1	59.5	59.3	58.8	52.8	71.8	58.4	68.3
POS	52.4	56.3	55.1	57.4	60.6	60.2	61.4	61.9	59.6	57.0	59.6	59.3	51.1	49.9	50.7	50.8
Emoticon	54.7	53.0	53.4	53.4	53.4	51.3	50.0	48.2	50.7	51.6	51.3	50.5	50.6	49.8	49.8	50.4
SentiWord	58.6	60.6	60.2	60.4	60.4	62.6	60.7	61.1	56.1	57.4	56.3	54.3	50.8	50.2	50.6	47.6
AFINN	64.3	68.8	68.7	68.8	61.2	65.1	64.0	64.8	60.7	63.2	62.3	61.9	51.1	52.1	50.8	51.3
OL	62.0	62.4	62.3	62.7	60.5	63.8	58.9	63.1	61.6	60.5	59.6	61.6	66.4	66.3	63.9	66.1
SS	63.6	66.8	65.9	65.9	62.7	64.5	63.8	64.6	60.9	58.9	60.2	60.7	55.9	55.1	56.4	55.9
Emotion	57.0	58.1	58.6	59.1	58.2	56.3	57.2	57.1	56.1	59.8	52.5	55.9	51.2	50.1	51.0	50.6
Feature	Polarity															
	SemEval				Sanders				HCR				OMD			
	A	B	C	D	A	B	C	D	A	B	C	D	A	B	C	D
Punct.	61.8	59.8	61.5	62.1	57.7	57.5	56.2	57.8	63.6	62.2	60.2	64.7	58.8	59.1	59.9	59.2
Lexical	54.3	55.9	54.5	56.4	59.6	59.4	58.7	62.7	54.2	59.6	52.8	60.5	50.5	53.6	55.2	55.3
POS	55.7	55.7	56.1	55.8	57.6	62.6	61.0	60.6	49.9	49.5	50.8	49.1	57.6	56.9	57.0	56.5
Emoticon	55.0	55.3	55.1	55.1	52.8	52.1	52.2	53.3	51.4	49.6	48.8	49.6	49.6	50.3	50.6	50.6
SentiWord	60.9	60.7	60.6	60.4	58.7	56.1	59.3	56.2	56.3	54.3	54.5	55.0	52.7	52.1	52.7	53.9
AFINN	74.3	75.2	75.2	75.2	69.8	70.9	70.6	71.1	60.5	58.7	60.3	60.1	62.7	62.8	62.5	62.8
OL	68.5	70.2	70.1	69.8	70.0	68.2	69.2	70.2	59.6	59.3	61.0	61.7	60.3	62.9	61.1	58.9
SS	72.9	75.2	73.0	74.9	72.3	71.8	72.2	72.3	59.7	60.5	59.7	58.7	62.5	62.6	61.7	62.5
Emotion	66.4	66.2	66.4	68.5	65.7	65.1	63.9	66.5	55.8	52.6	55.3	55.9	59.1	57.9	57.6	59.2

Workshop Sem-Eval 2013 (SemEval)[2] data (see Table 2) that were used in this work. In total we used 9 feature sets (41 attributes) that comprise punctuation, lexical, part of speech, emoticon, *SentiWord* lexicon, *AFINN*-lexicon, *Opinion* lexicon, *Senti-Strength* method, and Emotion lexicon (see Table 1). For pre-processing stage, it was adjusted based on the type of feature. It comprises: removing username, *url*, *RT* phrase, special character, stopwords; converting to lowercase; stemming and lemmatization. For the first experiment, we conducted binary classification for each feature set on each dataset. We then also performed feature selections of all feature sets (by merging the features into a set of 41 attributes) to all row of datasets.

---

[2] http://www.cs.york.ac.uk/semeval-2013/.

Table 4. Feature selection result

Feature	#Attr	Subjectivity				Polarity			
		A	B	C	D	A	B	C	D
Punct.	5	1	3	1	1	2	2	2	2
Lexical	9	4	1	2	-	1	2	3	1
POS	8	2	-	1	-	3	-	1	-
Emoticon	1	-	1	-	-	-	1	1	1
SentiWord	2	1	1	1	-	-	1	1	-
AFINN	2	2	1	2	2	1	2	2	2
OL	4	1	2	1	1	-	-	2	1
SS	2	2	2	2	2	1	1	1	1
Emotion	8	-	3	2	-	-	-	2	1
Accuracy		65.5	67.4	63.4	66.0	71.5	73.9	73.5	75.0

Results of these experiments are summarized in Tables 3 and 4. Letter A, B, C, and D in both tables represent the classifiers of *Naive Bayes*, *Neural Network*, SVM, and *Linear Regression* consecutively. In the first experiment (see Table 3), the colored cells are the top-5 of features according to its accuracy. For both classifications, it reveals that AFINN, Senti-Strength, and Opinion lexicon are the feature sets that are often found as the top-5 on each dataset. Whereas, the well-known lexicon, SentiWord, is not able to beat these lexicons. It affirms that SentiWord is not compatible for Twitter Sentiment Analysis. Our result also shows that emotion and punctuation are good features for Twitter Sentiment Analysis, especially in polarity classification.

In Table 4 we show the result of our second experiment, feature selection. The column of each classifier (A, B, C and D) is filled by number of corresponding attributes of each feature sets that arise based on feature selection. The table reveals that punctuation, AFINN and Senti-Strength are the most selected features either in subjectivity and polarity classification. It is quite similar with the previous experiment and affirms that AFINN and Senti-Strength are the current best feature to conduct Twitter Sentiment Analysis. Thus they are very good to use as current baseline for Twitter Sentiment Analysis.

3   Conclusion and Future Work

In this work, a comparative study between various features of Twitter Sentiment Analysis was done by using four different datasets and nine feature sets. Our experiment reveals that AFINN and Senti-Strength are the current best features for Twitter Sentiment Analysis. According to the results, the other features such as punctuation, Opinion lexicon and emotion are also important to consider. Future research may be conducted along with new idea of features released and investigated.

References

1. Bravo-Marquez, F., Mendoza, M., Poblete, B.: Combining strengths, emotions and polarities for boosting Twitter sentiment analysis. In: Proceedings of the Second International Workshop on Issues of Sentiment Discovery and Opinion Mining, vol. 2 (2013)
2. Raaijmakers, S., Kraaij, W.: A shallow approach to subjectivity classification. In: ICWSM (2008)
3. Aisopos, F., Papadakis, G., Tserpes, K., Varvarigou, T.: Content vs. context for sentiment analysis: a comparative analysis over microblogs. In: Proceedings of the 23rd ACM Conference on Hypertext and Social Media, pp. 187–196 (2012)
4. Go, A., Bhayani, R., Huang, L.: Twitter sentiment classification using distant supervision. In: CS224N Project Report, Stanford (2009)
5. Agarwal, A., Xie, B., Vovsha, I., Rambow, O., Passonneau, R.: Sentiment analysis of twitter data. In: Proceedings of the Workshop on Languages in Social Media, pp. 30–38 (2011)
6. Wilson, T., Wiebe, J., Hoffmann, P.: Recognizing contextual polarity in phrase-level sentiment analysis. In: Proceedings of the Conference on Human Language Technology and Empirical Methods in Natural Language Processing, pp. 347–354 (2005)
7. Liu, B., Hu, M., Cheng, J.: Opinion observer: analyzing and comparing opinions on the web. In: Proceedings of the 14th International Conference on World Wide Web, pp. 342–351 (2005)
8. Baccianella, S., Esuli, A., Sebastiani, F.: SentiWordNet 3.0: an enhanced lexical resource for sentiment analysis and opinion mining. In: LREC, vol. 10, pp. 2200–2204 (2010)
9. Bradley, M.M., Lang, P.J.: Affective norms for English words (ANEW): instruction manual and affective ratings. In: Technical report C-1, The Center for Research in Psychophysiology, University of Florida, pp. 1–45 (1999)
10. Nielsen, F.A.: A new ANEW: evaluation of a word list for sentiment analysis in microblogs. in: arXiv preprint arXiv: 1103.2903 (2011)
11. Mohammad, S.M., Turney, P.D.: Crowdsourcing a wordemotion association lexicon. Comput. Intell. **29**(3), 436–465 (2013)
12. Thelwall, M., Buckley, K., Paltoglou, G.: Sentiment strength detection for the social web. J. Am. Soc. Inf. Sci. Technol. **63**(1), 163–173 (2012)
13. Ekman, P.: An argument for basic emotions. Cogn. Emot. **6**(3–4), 169–200 (1992)
14. Plutchik, R.: The Psychology and Biology of Emotion. HarperCollins College Publishers, New York (1994)
15. Speriosu, M., Sudan, N., Upadhyay, S., Baldridge, J.: Twitter polarity classification with label propagation over lexical links and the follower graph. In: Proceedings of the First workshop on Unsupervised Learning in NLP, pp. 53–63 (2011)
16. Bird, S.: NLTK: the natural language toolkit. In: Proceedings of the COLING/ACL on Interactive Presentation Sessions, pp. 69–72 (2006)

Author Index

Printed in the United States
by Bookmasters

Printed in the United States
By Bookmasters